Edited by
Miguel Vaz Júnior,
Eduardo A. de Souza Neto, and
Pablo A. Muñoz-Rojas

Advanced Computational Materials Modeling

Related Titles

Arendt, Wolfgang, Schleich, Wolfgang, P. (eds.)

Mathematical Analysis of Evolution, Information, and Complexity

2009

ISBN: 978-3-527-40830-6

Velten, Kai

Mathematical Modeling and Simulation

Introduction for Scientists and Engineers

2008

ISBN: 978-3-527-40758-3

Kalos, Malvin, H., Whitlock, Paula A.

Monte Carlo Methods

2nd revised and enlarged edition

2008

ISBN: 978-3-527-40760-6

Öchsner, A., Murch, G. E., de Lemos, M. J. S. (eds.)

Cellular and Porous Materials

Thermal Properties Simulation and Prediction

2008

ISBN: 978-3-527-31938-1

Gottstein, G. (ed.)

Integral Materials Modeling

Towards Physics-Based Through-Process Models

2007

ISBN: 978-3-527-31711-0

Dronskowski, R.

Computational Chemistry of Solid State Materials

A Guide for Materials Scientists, Chemists, Physicists and others

2005

ISBN: 978-3-527-31410-2

Raabe, D., Roters, F., Barlat, F., Chen, L.-Q. (eds.)

Continuum Scale Simulation of Engineering Materials

Fundamentals - Microstructures - Process Applications

2004

ISBN: 978-3-527-30760-9

Boresi, A. P., Chong, K. P., Saigal, S.

Approximate Solution Methods in Engineering Mechanics

2003

ISBN: 978-0-471-40242-8

Huebner, K. H., Dewhirst, D. L., Smith, D. E., Byrom, T. G.

The Finite Element Method for Engineers

2001

ISBN: 978-0-471-37078-9

Springborg, M. (ed.)

Density-Functional Methods in Chemistry and Materials Science

Hardcover

ISBN: 978-0-471-96759-0

Edited by Miguel Vaz Júnior, Eduardo A. de Souza Neto, and Pablo A. Muñoz-Rojas

Advanced Computational Materials Modeling

From Classical to Multi-Scale Techniques

WILEY-VCH

WILEY-VCH Verlag GmbH & Co. KGaA

The Editors

Prof. Miguel Vaz Júnior
State Univ. of Santa Catarina
Dept. of Mech. Engineering
Univ. Campus Avelino Marcante
89223-100 Joinville SC
Brasil

Prof. Eduardo A. de Souza Neto
Swansea University
Civil & Comp. Eng. Centre
Singleton Park
Swansea SA2 8PP
United Kingdom

Prof. Dr. Pablo A. Muñoz-Rojas
State Univ. of Santa Catarina
Ctr. Techn. Sciences
Univ. Campus Avelino Marcante
89223-100 Joinville SC
Brasil

Library of Congress Card No.: applied for

British Library Cataloguing-in-Publication Data
A catalogue record for this book is available from the British Library.

Bibliographic information published by the Deutsche Nationalbibliothek
The Deutsche Nationalbibliothek lists this publication in the Deutsche Nationalbibliografie; detailed bibliographic data are available on the Internet at <http://dnb.d-nb.de>.

Cover Formgeber, Eppelheim
Typesetting Laserwords Private Ltd., Chennai, India
Printing and Binding Fabulous Printers Pte Ltd, Singapore

Printed in Singapore
Printed on acid-free paper

ISBN: 978-3-527-32479-8

Contents

Advanced Computational Materials Modeling: From Classical to Multi-Scale Techniques.
Edited by Miguel Vaz Júnior, Eduardo A. de Souza Neto, and Pablo A. Munoz-Rojas

ISBN: 978-3-527-32479-8

Preface

The systematic analysis of solid mechanics problems using numerical techniques can be traced back to the 1960s and 1970s following the development of the finite element method. The early approaches to elastic materials and, to a certain extent, inelastic problems, paved the way to an all-encompassing discipline known today as *computational materials modelling.*

As computer technologies have evolved, placing portable computers on the desk of virtually every university staff and graduate student, numerical techniques and algorithms have experienced extraordinary advances in a wide range of engineering fields. The development of new computational modelling strategies, especially those based on the finite element method, has prompted new applications such as crystal plasticity, damage and multi-scale formulations, semi-solid, particulate, porous and functionally graded materials amongst others.

This book was conceived in an attempt to congregate innovative modelling approaches so that graduate students and researchers, both from academia and industry, can use it as a springboard to further advancements. It is also important to say that this book is by no means exhaustive on the subject of materials modelling and some advanced readers would probably have appreciated the inclusion of further details on the underlying mathematical formulations. For the sake of objectivity, we have focussed on topics which show not only new and innovative modelling strategies, but also on sound physical foundations and both promising and direct application to engineering problems. Emphasis is placed on computational modelling rather than materials processing, although illustrative examples featuring some process applications are also included. A review of the state-of-the-art modelling approaches as well as a discussion on future trends and advancements is also presented by the contributors.

Finally we would like to sincerely thank all the authors for their time and commitment to produce such high quality chapters. We really appreciate their contribution.

July 2010

Miguel Vaz Jr.
Eduardo A. de Souza Neto
Pablo A. Muñoz-Rojas

Advanced Computational Materials Modeling: From Classical to Multi-Scale Techniques.
Edited by Miguel Vaz Júnior, Eduardo A. de Souza Neto, and Pablo A. Munoz-Rojas

ISBN: 978-3-527-32479-8

List of Contributors

Masayoshi Akiyama
Kyoto Institute of Technology
Department of Mechanical Engineering
Gosho-Kaido-cho
Matsugasaki
Sakyo-ku
Kyoto 606-8585
Japan

Filipe Xavier Costa Andrade
University of Porto
Faculty of Engineering
Rua Dr. Roberto Frias
4200-465 Porto
Portugal

Irina V. Belova
University of Newcastle
University Centre for Mass and Thermal Transport in Engineering Materials
Priority Research Centre for Geotechnical and Materials Modelling
School of Engineering
Callaghan, NSW 2308
Australia

Eduardo L. Cardoso
State University of Santa Catarina
Department of Mechanical Engineering Centre for Technological Sciences
Campus Universitário Prof. Avelino Marcante
Santa Catarina
89223-100 – Joinville
Brazil

José Manuel de Almeida César de Sá
University of Porto
Faculty of Engineering
Rua Dr. Roberto Frias
4200-465 Porto
Portugal

Guillermo Juan Creus
Federal University of Rio Grande do Sul
Department of Civil Engineering
Centre for Computational and Applied Mechanics
Rua Osvaldo Aranha
90035-190 – Porto Alegre
99, Rio Grande do Sul
Brazil

Advanced Computational Materials Modeling: From Classical to Multi-Scale Techniques.
Edited by Miguel Vaz Júnior, Eduardo A. de Souza Neto, and Pablo A. Munoz-Rojas

ISBN: 978-3-527-32479-8

Luiz Antonio B. da Cunda
Federal University of Rio Grande Foundation
School of Engineering
Rua Alfredo Huch
96201-900 – Rio Grande
475, Rio Grande do Sul
Brazil

Raúl A. Feijóo
Laboratório Nacional de Computação Científica (LNCC/MCT) & Instituto Nacional de Ciência e Tecnologia em Medicina Assistida por Computação Científica (INCT-MACC) Av.
Getúlio Vargas 333
Quitandinha
CEP 25651-070
Petrópolis – RJ
Brazil

Thomas Fiedler
University of Newcastle
University Centre for Mass and Thermal Transport in Engineering Materials
Priority Research Centre for Geotechnical and Materials Modelling
School of Engineering
Callaghan, NSW 2308
Australia

Marc G. D. Geers
Eindhoven University of Technology
Department of Mechanical Engineering
P.O. Box 513
5600 MB Eindhoven
Netherlands

Amir R. Khoei
Sharif University of Technology
Department of Civil Engineering
Center of Excellence in Structures and Earthquake Engineering
P.O. Box. 11365-9313
Tehran
Iran

Shigemitsu Kimura
Pipe & Tube Company
Sumitomo Metal Industries Ltd.
1 Higashi-mukoujima
Amagasaki
Hyogo 660-0856
Japan

Roxane Koeune
University of Liège
Aerospace and Mechanical Engineering Department
1 Chemin des Chevreuils
B4000 Liège
Belgium

Varvara Kouznetsova
Eindhoven University of Technology
Department of Mechanical Engineering
P.O. Box 513
5600 MB
Eindhoven
Netherlands

Kouichi Kuroda
Corporate R & D Laboratories
Sumitomo Metal Industries Ltd.
1-8 Fuso-cho, Amagasaki
Hyogo 660-0891
Japan

Thierry J. Massart
Université Libre de
Bruxelles (ULB)
Building Architecture
and Town Planing
CP 194/2
Av. F.-D. Roosevelt 50
1050 Brussels
Belgium

Pablo Andreś Muñoz-Rojas
Santa Catarina State
University - UDESC
Department of Mechanical
Engineering
Centre for Technological Sciences
Campus Universitário Prof.
Avelino Marcante
89223-100, Joinville
Santa Catarina
Brazil

Graeme E. Murch
University of Newcastle
University Centre for Mass and
Thermal Transport in
Engineering Materials
Priority Research Centre for
Geotechnical and Materials
Modelling
School of Engineering
Callaghan, NSW 2308
Australia

Eduardo A. de Souza Neto
Swansea University
Civil and Computational
Engineering Centre
School of Engineering
Singleton Park
SA2 8PP
Swansea
UK

Gláucio Hermogenes Paulino
University of Illinois at
Urbana-Champaign
Newmark Laboratory
Department of Civil and
Environment Engineering
205 North Mathews Av.
Urbana
IL 61801
USA

Ron H. J. Peerlings
Eindhoven University of
Technology
Department of Mechanical
Engineering
P.O. Box 513
5600 MB Eindhoven
Netherlands

Francisco Manuel Andrade Pires
University of Porto
Faculty of Engineering
Rua Dr. Roberto Frias
4200-465 Porto
Portugal

Jean-Philippe Ponthot
University of Liège
Aerospace and Mechanical
Engineering Department
1 Chemin des Chevreuils
B4000 Liège
Belgium

Wilfredo Montealegre Rubio
National University of Colombia
School of Mechatronic of
the Faculty of Mine
Carrera 80 No. 65-223
bloque M8, oficina 113
Medellin, Antioquia
Colombia

Emílio Carlos Nelli Silva
University of São Paulo
Department of Mechatronics and Mechanical Systems Engineering
Av. Prof. Mello Moraes
2231 - Cidade Universitária
São Paulo
05508-900
Brazil

Kenjiro Terada
Tohoku University
Department of
Civil Engineering
Aza-Aoba 6-6-06
Aramaki
Aoba-ku
Sendai 980-8579
Japan

Sandro Luis Vatanabe
University of São Paulo
Department of Mechatronics and Mechanical Systems Engineering
Av. Prof. Mello Moraes
2231 - Cidade Universitária
São Paulo
05508-900
Brazil

Miguel Vaz Jr.
Santa Catarina State
University - UDESC
Department of Mechanical
Engineering
Centre for Technological Sciences
Campus Universitário Prof.
Avelino Marcante
89223-100 Joinville
Santa Catarina
Brazil

Ikumu Watanabe
National Institute for
Materials Science
Structural Metals Center
Sengen 1-2-1
Tsukuba
Ibaraki 305-0047
Japan

Andreas Öchsner
Technical University of Malaysia
Department of Applied
Mechanics
Faculty of Mechanical
Engineering
Skudai
Johor
81310 UTM
Malaysia

and

University of Newcastle
University Centre for Mass and
Thermal Transport in
Engineering Materials
Priority Research Centre for
Geotechnical and Materials
Modelling
School of Engineering
Callaghan, NSW 2308
Australia

1
Materials Modeling – Challenges and Perspectives

Miguel Vaz Jr., Eduardo A. de Souza Neto, and Pablo Andreś Muñoz-Rojas

1.1
Introduction

The development of materials modeling has experienced a huge growth in the last 10 years. New mathematical approaches (formulations, concepts, etc.), numerical techniques (algorithms, solution strategies, etc.), and computing methods (parallel computing, multigrid techniques, etc.), allied to the ever-increasing computational power, have fostered the research growth observed in recent times. Numerical implementation of some modeling concepts, such as multiscale formulations and optimization procedures, were severally restricted two decades ago due to limitation of computing resources. What were once perspectives of new advancements have become a reality in the last few years and longstanding difficulties have been overcome.

It is important to emphasize that materials modeling is not a recent concept or a new research topic. Some material descriptions widely accepted and used these days were actually proposed in the late eighteenth century. For instance, within the framework of modeling inelastic deformation of metals, the French engineer Henri Tresca (1814–1884), professor at the *Conservatoire National des Arts et Métiers* (CNAM) in Paris, was the first to define distinct rules for the onset of plastic flow in ductile solids [1]. Tresca's groundbreaking studies established a material-dependent critical plastic threshold given by the maximum shear stress. The apparently simple concept gave rise to a completely new approach to studying deformation of solid materials, and, today, his principle is known as *Tresca's yield criterion*. It is interesting to mention that, in spite of many years of proposition, numerical implementation of Tresca's criterion is not straightforward because of the sharp corners of the yield locus and its association with the plastic-normality flow rule [2, 3].

The search for alternate modeling descriptions is also not a new endeavor. For similar problems, Maksymilian Tytus Huber (1872–1950), a Polish engineer, postulated that material strength depends upon the spatial state of stresses and not on a single component of the stress tensor [4]. Independently, the Austrian mathematician and engineer, Richard von Mises (1883–1953), indicated that plastic

Advanced Computational Materials Modeling: From Classical to Multi-Scale Techniques.
Edited by Miguel Vaz Júnior, Eduardo A. de Souza Neto, and Pablo A. Munoz-Rojas

ISBN: 978-3-527-32479-8

deformation of solids is associated with some measure of an equivalent stress state [5]. The assumption indicates that plastic deformation is initiated when the second deviatoric stress invariant reaches a critical value. A few years later, the German engineer, Heinrich Hencky (1885–1952), still within the criterion introduced by Huber and von Mises, suggested that the onset of plastic deformation takes place when the elastic energy of distortion reaches a critical value [6]. An alternate physical interpretation was proposed by Roš and Eichinger, who demonstrated that the critical distortional energy principle is equivalent to defining a critical shear stress on the octahedral plane [7], generally known as *maximum octahedral shear stress criterion*. The aforementioned elastic–plastic modeling assumptions are known today as the *Huber–Mises–Hencky yield criterion*. A brief review of the early works on modeling of plastic deformation of metals illustrates the drive toward understanding the physics of material behavior and its translation into mathematical descriptions.[1)]

Despite the fact that the principles of *plasticity theory* have long been established, application to realistic problems or advanced materials using only mathematical tools is difficult or even impossible. Following the example on deformation of metals, when addressing computational modeling of elastic–plastic deformation at finite strains, the solution requires a physical/material description (e.g., the classical Huber–Mises–Hencky equation), a mathematical formulation able to handle geometrical and material nonlinearity (e.g., multiplicative decomposition of the gradient of deformation tensor into elastic and plastic components), and a computational approximation/discretization of the physical and mathematical problem (e.g., iterative procedures such as the Newton–Raphson and arc-length methods). This class of problems has already been exhaustively investigated in the last 30 years, and the literature shows a wide variety of strategies (see, for instance, Ref. [11] and references therein).

The illustration on the development of physical/mathematical/numerical formulations of elastic–plastic deformation of ductile solids shows that a proper material modeling requires

1) understanding of the physics involved in the problem;
2) comprehensive theoretical and mathematical treatment of the phenomena;
3) sound and consistent numerical approximation/discretization of the governing and constitutive equations; and
4) adequate computing resources.

These principles are extensive to modeling and simulation of any materials-processing operation. In a broader context of materials modeling, the literature has shown an increasing pace in the evolution of each one of the aspects mentioned in items (1–4). Advancements in mathematical and numerical tools have prompted investigation in areas of materials modeling ranging from

1) Further reading on the history and development of yield criteria and concepts of materials behavior can be found in Refs [8–10].

electronic and atomistic level to complex structures within the continuum realm [12]. Despite this considerable progress, there are still pressing challenges to be overcome, mainly those associated with more realistic materials-processing operations or simulation of complex materials structures. This chapter highlights some modeling issues under current and intense scrutiny by researchers and does not intend to be exhaustive. The other chapters of this book present deeper insights into materials modeling and simulation of some class of problems that, in a way, we hope, will serve as a springboard for further realistic applications.

1.2 Modeling Challenges and Perspectives

Materials modeling is as vast as the types of existing materials. For decades, emphasis has been placed on structural (metals, polymers, composites, etc.) and geotechnical (soils and rocks) classical materials. Behavior prediction of such materials subjected to a given load (mechanical or thermal) in process operations or stress–strain/heat transfer analysis has constituted the bulk of numerical approaches available in the literature. The existing solution approaches are comprehensive and provide accurate results for most classical materials subjected to strain paths of reasonable complexity. However, the constant search for technological advances and understanding of some classes of complex materials and processes has posed new challenges, urging scientists to seek new mathematical and computational tools. The following sections discuss general aspects of (i) *the modeling of ductile deformation and mechanical degradation leading to fracture;* (ii) *the modeling of cellular materials;* and (iii) *multiscale approaches.* Many other constitutive modeling issues and material types could have been included in the list; however, the above aspects have attracted substantial attention of academia and industry due to perspectives of realistic applications in a relatively short term.

1.2.1 Mechanical Degradation and Failure of Ductile Materials

In the last few years, numerical simulation of metal-forming operations has been incorporated into the design procedures of many manufacturing processes. Industry is seeking not only to estimate forming loads and energy requirements with higher accuracy but also to predict possible defects and tool life. Forging, extrusion, and deep drawing are some examples of forming processes that are particularly prone to material failure. For instance, a faulty design of extraction angles, tool radius, or workpiece geometry might lead to either external or internal defects. Therefore, aiming at prediction of fracture onset associated with elastic–(visco)plastic deformation, the modeling of mechanical degradation of ductile materials has been extensively studied in the last few years. A brief literature survey shows many research groups engaged in the aspects listed below, which are intrinsically related to ductile failure:

- **Prediction of failure onset**: numerical and experimental investigation of failure criteria for manufacturing processes (e.g., forming limit diagrams for sheet metal forming);
- **Material response to loading**: computation of stress–strain distribution and loads in multistep forming operations (e.g., springback and residual stress evaluation);
- **Multiscale modeling**: approximations for strongly coupled scales, homogenization strategies, and heterogeneous multiscale techniques (e.g., damage modeling, cohesive failure, biomaterials applications, microstructure design, crystal plasticity, and texture evolution);
- **Anisotropic materials**: modeling anisotropic behavior of materials (e.g., complex yield criteria and its interaction with material failure);
- **Nonlocal models**: material modeling including nonlocal effects (e.g., new weighted averages in nonlocal formulations and suitable gradients in gradient approaches. Applications to nonhomogeneous materials);
- **Deformation and failure under complex stress states**: derivation of material models and failure criteria able to describe plastic deformation under complex stress–strain paths (e.g., failure criteria for tensile and compressive-dominant processes);
- **Parameter identification**: identification of material parameters (e.g., identification of elastic–plastic, damage, and fracture parameters using techniques for inverse problems).

The aforementioned topics are not exhaustive and other aspects associated with modeling of the ductile failure process could be added. Furthermore, some topics can (σ_H/σ_{eq}) also be interrelated to each other, for example, deformation and failure under complex stress states using nonlocal damage models. In order to illustrate the challenges faced by researchers and perspectives eagerly awaited by industry, some issues related to deformation and failure under complex stress states are discussed in the following paragraphs.

The literature shows many attempts to describe the mechanical degradation process and failure initiation based on postprocessed criteria owing to the simplicity of modeling. The following summarizes only the most recent advancements in the area. Wierzbicki and coworkers have extensively investigated the failure process for compression, tension, shear, and combined loads [13]. These authors emphasize that the *failure mechanism* plays an important role in failure onset of ductile materials and report that a critical triaxiality factor, given by the ratio between the hydrostatic and von Mises equivalent stresses, defines a limit between shear and void growth fracture modes, as illustrated in Figure 1.1 Furthermore, even restricted to ductile fracture, such differences in the *failure mechanisms* have prevented derivation of a single postprocessed criterion able to successfully predict failure onset for both tensile and compressive-dominant loads [14].

Many other works have attempted to predict ductile failure onset using postprocessed ductile fracture criteria. Most approaches attempt to describe the microscopic phenomena associated with mechanical degradation by either experimental analysis (empirical criteria) or mathematical and/or physical models (e.g., growth of

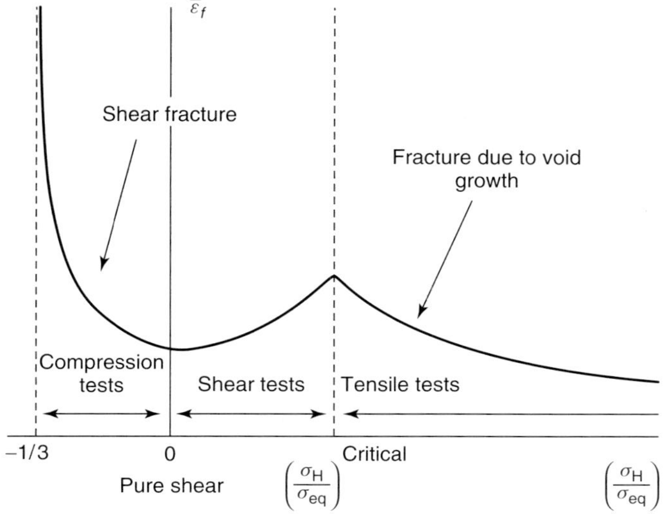

Figure 1.1 Fracture locus in the equivalent strain and stress triaxiality space: Bao and Wierzbicki [10] presented the effect of the stress triaxiality in fracture onset based on three different tests, namely, compression, shear, and tensile tests. The authors also postulate that a change of the fracture mechanism provides a slope discontinuity in the fracture locus. In the range of negative stress triaxiality, the equivalent strain to fracture decreases with the stress triaxiality, reaching a minimum at $\sigma_H/\sigma_{eq} = 0$ (pure shear). The fracture strain increases for low-stress triaxiality factors, reaching a peak at a given (material-dependent) stress triaxiality. For high-stress triaxiality, the shear fracture decreases with the stress triaxiality.

spherical voids, dissipation of plastic energy, etc.). In spite of greater modeling difficulties, a coupled description of elastic–(visco)plastic deformation and mechanical degradation has proved to be the best approach to model the ductile fracture process. In general, such formulations form what is currently known as *continuum damage mechanics* (CDM). However, for years, accurate material degradation and fracture prediction using damage mechanics were restricted to tensile-dominant loading, since the material description was unable to distinguish between tensile and compressive deformation. Its extension to deformation under complex stress–strain paths is one of the most intensively studied topics in recent years. For instance, Vaz *et al.* [15], on the basis of Ladèveze and Lemaitre's [16] and Lemaitre's [17] considerations, proposed a general approach based on the total damage work, W_D,

$$W_D = \int_0^t (-Y)\,\dot{D}\mathrm{d}t = \int_0^{D_c} (-Y)\,\mathrm{d}D \tag{1.1}$$

where D is the damage variable, t is the time, D_c is the critical damage, and $(-Y)$ is the damage strain energy release rate,

$$(-Y) = (-Y)^+ + (-Y)^- - \frac{\nu}{E}\left[\frac{(-f')^{1/2}}{f}\mathrm{tr}\left[\boldsymbol{\sigma}^+\right]\right]\left[\frac{(-f')^{1/2}}{f}\mathrm{tr}\left[\boldsymbol{\sigma}^-\right]\right] \tag{1.2}$$

where $(-Y)^+$ and $(-Y)^-$ are the individual contribution of tensile and compressive stresses,

$$\begin{aligned}(-Y)^+ &= \frac{(-f')}{f^2}\left[\frac{(1+\nu)}{2E}\boldsymbol{\sigma}^+ : \boldsymbol{\sigma}^+ - \frac{\nu}{2E}\left(\mathrm{tr}\left[\boldsymbol{\sigma}^+\right]\right)^2\right] \\ (-Y)^- &= \frac{(-f')}{f^2}\left[\frac{(1+\nu)}{2E}\boldsymbol{\sigma}^- : \boldsymbol{\sigma}^- - \frac{\nu}{2E}\left(\mathrm{tr}\left[\boldsymbol{\sigma}^-\right]\right)^2\right]\end{aligned} \tag{1.3}$$

$f = f(D)$ is the damage function, and $\boldsymbol{\sigma}^+$ and $\boldsymbol{\sigma}^-$ are the individual contributions of tensile and compressive principal stresses to the loss of material stiffness,

$$\boldsymbol{\sigma}^+ = \sum_{i=1}^{3} \langle \sigma_i \rangle\, \mathbf{e}_i \otimes \mathbf{e}_i \quad \text{and} \quad \boldsymbol{\sigma}^- = \sum_{i=1}^{3} \langle -\sigma_i \rangle\, \mathbf{e}_i \otimes \mathbf{e}_i \tag{1.4}$$

in which, mathematically, $\{\boldsymbol{\sigma}_i\}$ and $\{\mathbf{e}_i\}$ denote the eigenvalues and an orthonormal basis of eigenvectors of $\boldsymbol{\sigma}$.

In spite of the apparent success, *local* damage models suffer from dependence on the finite element mesh. In classical plasticity theories, the state of any point in a body depends only on the state of its infinitesimal neighborhood, thereby excluding the internal characteristic lengths of the material from the local field theory [18]. Therefore, in elastic–plastic formulations based on local approaches, the element size defines the minimum dimension within which the plastic deformation takes place, serving as an internal characteristic length of the material. As a consequence, the subsequent mesh refinement in the critical zones causes the damage process to become more concentrated in ever smaller volumes, leading to physical–numerical inconsistencies (e.g., loss of ellipticity of the differential equations in strain localization problems). In order to overcome such difficulties, *nonlocal* formulations have been proposed. It is relevant to mention that nonlocal approximations are not restricted to damage modeling. In general, nonlocal models are formulated using two approaches: integral models and gradient-based formulations. The former builds the nonlocal variable based on weighted and averaged local variables in areas defined by the internal characteristic length. Gradient-based formulations, divided into explicit and implicit schemes, include the gradient of a collection of field variables linked to the inelastic deformation process (e.g., equivalent plastic strain) into the material constitutive equations.

A survey in the recent published literature on the application of damage models to the assessment of failure process shows increasing interest in the use of nonlocal approaches. For instance, Meinders *et al.* [19] reported that, in the area of damage and fracture behavior, a nonlocal damage model provided better predictions of sheet failure than the conventional forming limit diagram. Velde *et al.* [20] presented a nonlocal damage model for viscoelastic materials aiming at time-dependent inelastic behavior of steel structures up to failure. These authors used a nonlocal implicit gradient-based formulation coupled to a hybrid damage approach (Lemaitre and Gurson type damage models), being verified in 3D-structural analysis of Compact Tension (CT) specimens. Failure analysis of CT-specimens was also discussed by Samal *et al.* [21], who presented a nonlocal damage formulation for

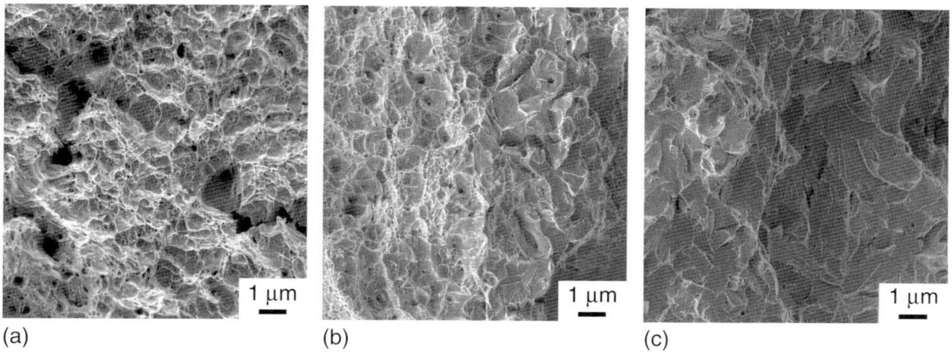

Figure 1.2 Microscopy of the fractured region of a low-carbon V-notched specimen: (a) stress concentration at the root of the V-notch causes failure onset at the external surface of the specimen; the micrography taken at this region shows dimples, which represents the typical texture of ductile fracture. (b) Transitional region: evolution of the stress state caused by a reduction of the resisting area leads to a change in the fracture mechanism from ductile to brittle-type fracture. (c) The end of the failure process exhibits cleavage microplanes, which are typical structures of brittle fracture.

Rousselier's damage model. The model was based on a nonlocal implicit gradient formulation, in which a diffusion-type differential equation correlates the nonlocal damage variable to the local void volume fraction. Finally, it is relevant to emphasize that the considerable potential for model failure analysis in complex materials using nonlocal damage approximations has stimulated research on purely computational issues and modeling approaches, for example, hybrid-displacement finite-element formulations [22], nonhomogeneous elasticity [23], consistent tangent matrix [24], and two-field variational formulation [25].

When addressing the physics of the failure process, using either local or nonlocal formulations as previously mentioned, the *failure mechanism* is fundamentally important when modeling the deformation process leading to failure. For instance, most, if not all, computational models discussed in the literature are not able to properly describe the transition between the initial ductile failure and the brittle-type fracture (catastrophic failure) that takes place at the end of the failure process. Such difficulties are expressed even in the presence of predominantly uniform stress fields. For example, in a tensile test of a low-carbon V-notched specimen, a comprehensive material model should describe the initial ductile failure at the external surface (Figure 1.2a) of the specimen, the transition in the fracture mechanism (Figure 1.2b) as the fracture progresses, and the brittle-type fracture which takes place at the center of the specimen (Figure 1.2c).

1.2.1.1 **Remarks**

Material models able to account for elastic–plastic deformation and change in failure mechanisms are still unavailable even for isotropic materials under one-directional loading. Phenomenological approaches to material degradation and failure progression based on macroscale models seem to lack essential physical tools to properly describe the phenomena involved. Therefore, the natural research direction points to using physics-based failure modeling based upon multiple scales (nano-, micro-, meso-, and macroscales)/nonlocal approximations coupled to a macroscopic scheme of fracture progression.

1.2.2
Modeling of Cellular Structures

Development of manufacturing technologies has instigated investigation on the use of cellular-type materials in many different areas. The literature shows applications ranging from simple filters, as illustrated in Figure 1.3, to flow straighteners, containment matrices and burn-rate enhancers for solid propellants, pneumatic silencers/sound absorbers, catalytic surfaces for chemical reactions, core structures for high-strength panels, crash energy absorbers, flame arresters, heat sinks, and heat exchangers, in general [26–28]. This section summarizes some aspects of the current discussions on modeling strategies and, more importantly, topology design for some classes of problems.

Cellular topology can be generally classified into (i) regular or stochastic cellular structures and (ii) open or closed cellular materials. Most authors agree that stochastic metal foams with open cells have better thermal, acoustic, and energy-absorption properties; however, their load-bearing capacity is significantly inferior to periodic structures with the same weight [30]. Open cell materials have also been regarded as one of the most promising materials for manufacturing heat exchangers due to the high surface area density and strong mixing capability for the fluid [28]. Ultralight structures, energy-absorption systems, and fuel cell and battery subsystems are applications suited to purpose-tailored extruded metal honeycombs or prismatic/periodical cellular materials [31]. A visual summary of the application of cellular metallic materials was presented by Banhart [27], who plotted *purpose* (functional or structural) against the *recommended foam topology* (open, partially open, or closed), as illustrated in Figure 1.4.

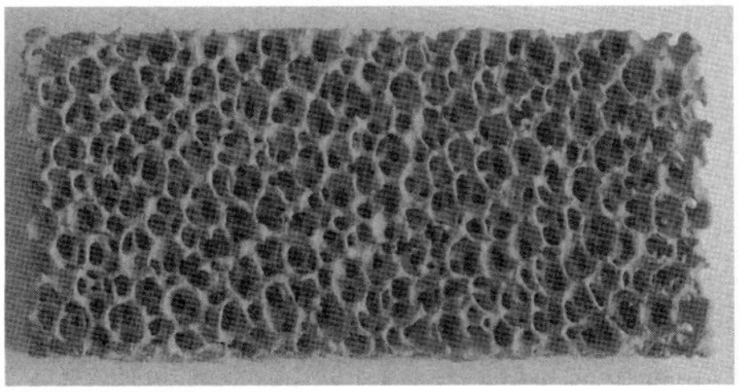

Figure 1.3 Open cell structure: ceramic foam filter used to remove impurities from liquid metals in casting [26]. The materials used in ceramic filters are silicon carbide, alumina, and zirconia. This application requires not only mechanical strength to withstand high-temperature flow but also to yield low pressure loss, erosion resistance, and chemical and thermal stability, to avoid reaction with the molten metal being filtered.

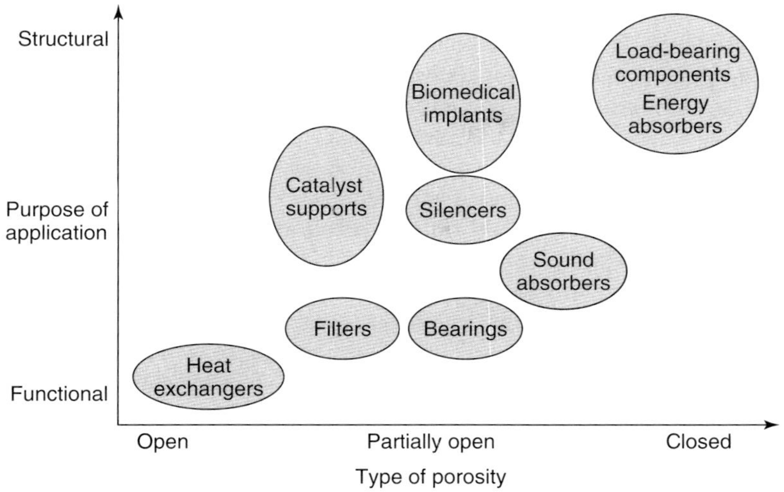

Figure 1.4 Application of porous metals or metal foams [24]. Combination between more than one application poses the greatest challenge when choosing cell topology. For instance, in the automotive industry, the goal is to design components able to combine lightweight, high deformation (plastic) energy absorption, and sound insulation among other desired characteristics. Metal foams, more than other materials, and despite design challenges, have emerged as a possible solution for different types of car parts that require such combinations of functionality.

A brief review on the potential application of cellular materials shows a wide variety of approaches to recommended cellular topology. Furthermore, the modeling strategy also varies according to the analysis of the desired behavior: structural, thermal, or multifunctional formulations. The numerical techniques used to solve the problem are also important when addressing this class of problems. Therefore, modeling and simulation of cellular structures is an enormous research field, still wide open to new developments and modeling strategies. In general, the following aspects have been under intense research in the last few years:

- **Constitutive modeling**: development of global constitutive models for elastic–plastic deformation and failure (e.g., induced anisotropy and yield criteria);
- **Structural and functional behavior**: study of the behavior of a cellular structure under given physical conditions (e.g., deformation and densification and dissipation of thermal energy);
- **Material properties**: material properties of cellular structures (e.g., parameter identification of global mechanical and thermal properties, homogenization, and global properties design);
- **Morphology design**: numerical design of cellular topology (e.g., design optimization, lattice structure modeling, and multiobjective topology design).

The topics listed above represent only the mainstream research on modeling and simulation of cellular materials. One can discern frontline research in each one of those research areas; however, this section is not exhaustive in describing every new approach or modeling technique, but highlights only some issues on topology design of cellular materials using *inverse modeling*.

Cell topology design using *inverse modeling* consists of determining the best configuration of material distribution according to given criteria. Development of techniques using this strategy constitutes one of the greatest challenges in the field of modeling cellular materials. This method, however, does not present solution unicity, making possible to obtain different configurations for the same structural and/or functional requirements. A specific class of *inverse modeling* is concerned with finding the optimum cellular structure from basic known geometries. Note that, in this case, the unit-cell geometrical shape does not change (e.g., triangles and hexagons in 2D configurations), but the individual cell size or other geometrical/material parameters can be determined according to given optimization criteria. Inverse modeling contrasts with *direct approaches*, which are concerned with studying the behavior of cellular structures with known geometrical configuration. Direct strategies can by no means be underestimated, since the target problems and applications may require complex formulations of individual functional characteristics.

Direct approaches are by far the most common strategies used to design cellular structures. The recent work of De Giorgi *et al.* [32] is an example of the use of such modeling techniques to topology design of aluminum foams. The authors aimed at finding the best structural response based on closed-cell configuration using tetrakaidecahedron and ellipsoidal cells of different sizes associated with periodic microstructures. The latter was found to be the best configuration to reproduce the mechanical response of the AlSi10Mg commercial foam produced by Alulight® International GmbH. *Inverse modeling* using cells of predefined geometrical shapes was presented by Kumar and McDowell [33], who used a homogenization-based method to design cellular structures able to maximize heat dissipation and to improve structural performance. Prismatic honeycombs with uniform and graded cell sizes of known topologies (squares, equilateral triangles, and regular hexagons) were used by these authors.

Owing to flexibility and generality, the application of *inverse modeling* techniques to design the unit cell itself has gained increasing attention in the last few years. Sigmund [34] was one of the pioneers in the application of inverse modeling based on *topology optimization* to find the optimum unit-cell configuration. This strategy was subsequently utilized to study different aspects of this class of problems. Recently, application of inverse homogenization for designing two-phase periodic materials under multiple load conditions was introduced by Guedes *et al.* [35]. The technique also used topology optimization to determine the optimum material distribution within a unit cell. The problem considered by the authors corresponded to finding the stiffest microstructure for a minimum compliance problem involving multiple loads (tensile and shear loading conditions). Muñoz-Rojas *et al.* [36], in a general discussion on the application of optimization to heat transfer in cellular

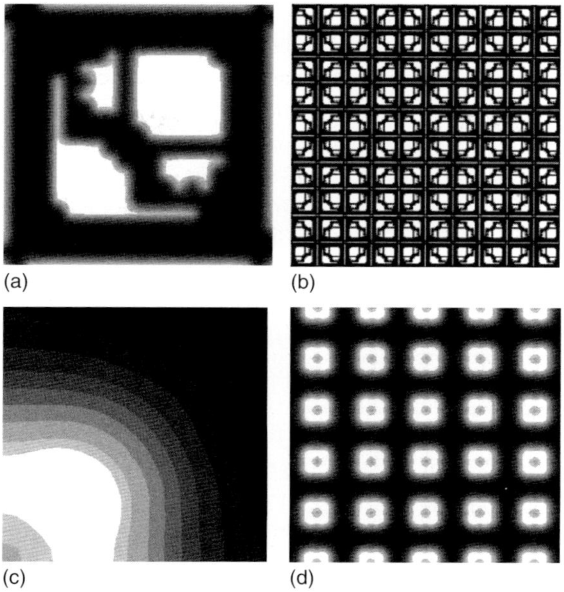

Figure 1.5 One-fourth of a unit-cell pseudodensity distribution and corresponding composite material matrix [33]. The problem consists of achieving the highest possible thermal conductivity for a given volume fraction for an FGM. The figures illustrate results for increasing gradient lengths – from (a)–(b) almost no property gradation to (c)–(d) high gradation. The gradient control makes it possible to address the influence of FGM gradation in the design of purpose-tailored, high-performance materials. Further discussions on this method, including detailed mathematical modeling aiming at structural applications, was presented by Paulino *et al.* [34].

materials, presented some insights into the design of periodic cellular structures using functionally graded materials (FGMs). The strategy uses the concept of the relaxed problem in continuum topology optimization and maximizes the homogenized thermal conductivity for a certain volume fraction. Figure 1.5 illustrates the final material morphology for two different gradation parameters. Paulino *et al.* [37] present a detailed description of the method, in which examples aiming at structural applications are discussed. The great potential of the strategy is illustrated by designing extreme material topologies, such as structures presenting negative Poisson's ratio and near-zero shear modulus. The most recent design techniques use multiscale considerations based upon alternative approaches. For instance, Giusti *et al.* [38] proposed a new numerical strategy that uses the mathematical concept of topological derivative within a variational, multiscale constitutive framework. Application of this method to microstructure design has been recently presented by Giusti *et al.* [39] and Amstutz *et al.* [40], in which the final unit-cell morphology was obtained by reaching the optimality condition defined as the function of a given homogenized property (e.g., Poisson's ratio, bulk modulus, and shear modulus).

A distinct class of cellular materials comprises the lattice/grid structures, which are also known as *lattice-block materials, lattice-truss structures, lattice-block structures, and cellular lattices*. Lattice materials with periodic unit-cell microstructures are trusslike structures mainly conceived to maximize the load-bearing capacity at minimum weight, with potential high energy absorption. Multifunctional applications may also combine heat transfer/thermal dissipation, sound absorption, or other requirements. Industries have shown growing interest in truss microstructures aiming at high-performance applications, motivated by the development of high-precision manufacturing processes, such as rapid prototyping (e.g., selective laser sintering, digital light processing, and microstereolithography [41, 42]). Most research works are concerned with either modeling the global behavior of the lattice materials or determining their homogenized properties, all of which are based on known unit-cell microarchitectures (similar to the *direct approach* discussed in the previous paragraphs). Luxner *et al.* [43], for instance, addressed the macromechanical behavior of six different 3D base-cell geometries presenting cubic material symmetry (simple cubic, Gibson Ashby, reinforced body-centered cubic, body-centered cubic, Kelvin, and Weaire–Phelan structures). The authors' current research is focused on the effect of irregularities on elastic–plastic deformation and localization. A similar approach was also used to address bonelike structures using the simple cubic structure [44]. In all cases, a direct problem was solved, that is, material properties and cell microarchitecture were known in advance.

Conceptually, application of *inverse modeling* to lattice-block materials is similar to that discussed already, that is, (i) use of optimization techniques to design cellular structures based on unit cells of predefined microarchitecture or (ii) designing of the unit-cell lattice microstructure. The former is illustrated by Yan *et al.* [45], who presented an optimization procedure for structural analyses of truss materials. The authors adopted 2D quadrilateral unit cells and used two classes of design variables: relative density and cell-size distribution, under a given total material volume constraint. The technique was able to determine a cell distribution of different sizes and cells with walls of different thickness based only on quadrilateral unit-cell structures. However, the challenge posed to this class of materials is associated with determining the optimum microarchitecture of the unit cell that is capable of achieving maximum performance without predefining its basic geometry.

Studies using homogenization, sensitivity analysis, and optimization are in progress aiming at finding the optimum morphology of lattice materials based on general trusslike unit cells [46]. The combined homogenization–optimization technique has been developed for multifunctional applications, that is, the optimum structure should achieve the best performance for both structural and thermal applications. The method uses homogenization of the elasticity and conductivity tensors, combined with analytical sensitivity analysis based on symbolic computation.

Initially, a general unit cell with arbitrary microarchitecture is defined (which contains an arbitrary number of interlocked struts). The first step corresponds to computing the homogenized elasticity and conductivity tensors, respectively,

$$\mathbf{E}^{\mathrm{H}}(\mathbf{x}) = \frac{1}{|\mathbf{Y}|}\int_{Y} \mathbf{E} \cdot (\boldsymbol{I} - \boldsymbol{\partial}_{y}\boldsymbol{\chi})\mathrm{d}y \quad \text{and} \quad \mathbf{k}^{\mathrm{H}}(\mathbf{x}) = \frac{1}{|\mathbf{Y}|}\int_{Y} \mathbf{k} \cdot (\mathbf{I} - \boldsymbol{\nabla}_{y}\mathbf{R})\mathrm{d}y \qquad (1.5)$$

where $\mathbf{I}$ is the identity tensor, $|\mathbf{Y}|$ is the volume of the unit cell, $\mathbf{E}$ and $\mathbf{k}$ are the elasticity and conductivity tensors, $\boldsymbol{\chi}$ and $\mathbf{R}$ are the characteristic displacements and temperatures of the unit cell, and $\boldsymbol{\partial}_y$ and $\boldsymbol{\nabla}_y$ are the strain and heat flux operators with respect to the unit cell.

The optimization procedure is based on sequential linear programming (SLP) and uses an *objective function* defined according to the desired structural and functional behavior. The initial studies adopt specific components of the homogenized elasticity and conductivity tensors to handle the structural and thermal

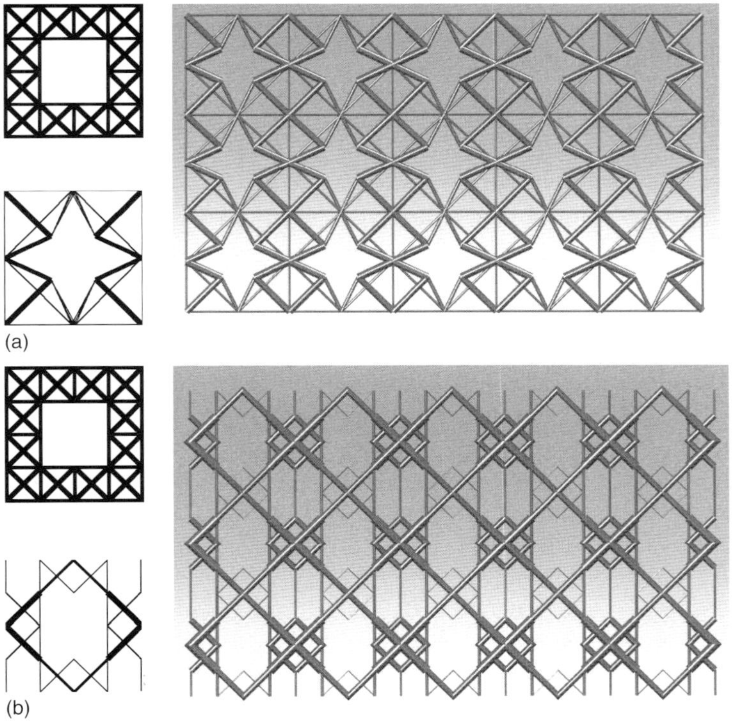

(a)

(b)

Figure 1.6 Two-dimensional lattice structures: initial and final unit-cell microstructures and the corresponding optimized lattice material. Both examples account for structural and functional (thermal) properties. The lattice microarchitectures were obtained by maximizing the ratio between the homogenized shear component of the elasticity tensor and the normal (vertical in the figure) component of the homogenized conductivity tensor for a constant initial volume of the unit cell. The design variables for each case are (a) strut areas and x and y coordinates of the cell internal nodes and (b) only the strut areas. It should be remarked that these results are not necessarily unique, since the optimization problem does not ensure existence of a global minimum.

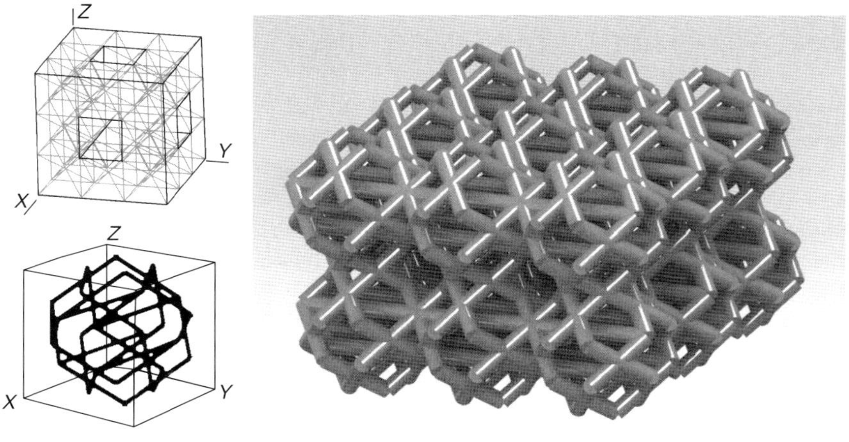

Figure 1.7 Three-dimensional lattice structures: initial and final unit cells and the lattice material. The unit cell is divided into $3 \times 3 \times 3$ subcells so that struts are initially distributed following a pyramidal shape with the vertex located in the center of each subcell. Struts are placed only in the subcells connected to an edge of the unit cell. The design variables are the strut areas. The final lattice material was obtained by maximizing the product between the homogenized shear component, E^{H}_{1212}, of the elasticity tensor, and the homogenized normal component, k^{H}_{11}, of the conductivity tensor.

response, respectively. The multifunctional approach is accounted for an objective function defined as a combination of both homogenized tensors. In the homogenization–optimization process, the strut areas and nodal coordinates inside the unit cell are modified following the direction provided by the structural and thermal requirements. It is interesting to mention that the generality of this strategy makes it possible to recover classical configurations, such as Kagomé and Diamond lattice materials, depending on the structural and functional requirements. The technique was applied to 2D and 3D structures, as illustrated in Figures 1.6 and 1.7. Further aspects of the design technique, including detailed mathematical modeling, is addressed elsewhere in a publication dedicated to lattice materials modeling.

1.2.2.1 **Remarks**

The benefits provided by cellular materials are undisputable. Low specific mass, high energy absorption, and multipurpose thermal behavior are some characteristics of this class of materials. The development of mathematical and numerical tools, allied to the increasing viability of manufacturing complex microstructures, has encouraged investigation on this topic. It is also expected that great advancements in numerical strategies for parameter identification of material properties will be attained in the next few years, especially in the context of hybrid schemes (gradient-based and evolutionary algorithms) and topological derivative-based approaches. Microstructure design using homogenization and optimization techniques, encompassing both topology optimization and lattice-block materials, has also evolved rapidly in the last few years. Although still in the nascent stages, application of multiscale algorithms (e.g., those aiming at strongly coupled scales) to designing the unit-cell microstructure is a welcome event.

1.2.3 Multiscale Constitutive Modeling

Over the last decade or so, the modeling of the dissipative behavior of solids by means of so-called *multiscale theories* has been attracting increasing interest within the applied and computational mechanics communities. The general concept of multiscale modeling extends from quantum mechanics and particle physics, molecular dynamics, and dislocation theory to macroscopic constitutive relations, as illustrated in Figure 1.8. At present it is well accepted that classical, purely phenomenological theories, in which the constitutive response is defined by a set of ordinary differential equations, possess stringent restrictions on the complexity

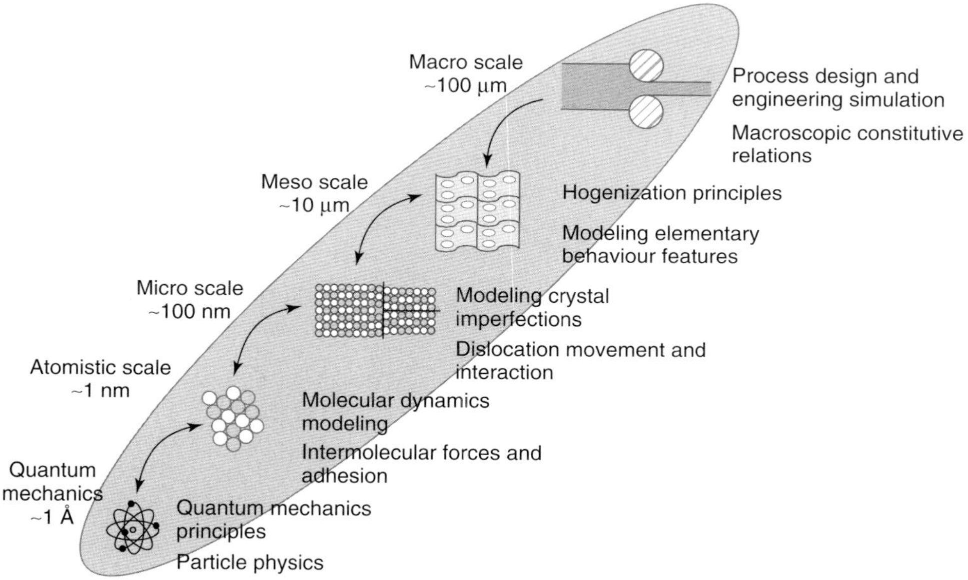

Figure 1.8 Length scales for most metal materials. Classification of physical phenomena in different time and length scales is not an easy task. The concept of scales often differs over the spectrum of nature of materials and microstructures. For instance, biological tissues, such as tendons, frequently present scale classes based upon structural/hierarchical characteristics (from tropocollagen molecules and microfibril structures to the endotendon and tendon itself, the length scale spans from ~1.5 nm to ~3000 μm) [44]. Computational modeling of engineering materials requires also careful considerations owing to different natures and structures: the typical length scales for geotechnical and building materials are significantly different from those characteristic of metal materials, for example, the former can handle macroscales measured in meters, whereas the latter can be modeled using phenomenological approaches with macroscales measured in micrometers.

of strain paths for which reasonable predictions can be obtained. This is particularly true when more intricate micromechanically related phenomena such as damaging, microcracking, or phase debonding are present.

Tackling the problem by the classical approach consists usually in introducing new internal state variables to capture finer details of the phenomenological effects of such mechanisms on the overall response of the material [48]. In many cases, this approach can be very successful but its main side effect is the fact that the greater number of internal variables requires the identification and definition of their corresponding evolution laws with the associated material parameters. Such identification is by no means trivial and may become particularly challenging in situations where phenomena such as, for instance, anisotropy evolution is present.

One possible alternative to address the problem is the adoption of multiscale theories, where the macroscopic stress and strain tensors are defined as volume averages of their microscopic counterparts over a *representative volume element* (RVE) of material. The foundations for this family of constitutive theories in the dissipative range are laid by Germain *et al.* [49]. Owing to their suitability for implementation within nonlinear finite element frameworks, such theories are particularly attractive for the description of complex constitutive response by means of finite element approximations. In such cases, complex macroscopic response can be obtained from the volume averaging of a finite element-discretized RVE containing a relatively accurate representation of the morphology of the microstructure and whose constituents are modeled by simple phenomenological constitutive theories, with possible added nonlinear phase interaction laws. Methodologies of this type are normally used in the following main contexts:

1) determination of the material parameters of an assumed canonical macroscopic constitutive model by fitting the homogenized response produced by finite element solutions of a single RVE under prescribed macroscopic strain paths [50, 51];
2) development of new macroscopic constitutive laws capable of capturing the homogenized response of a finite element-discretized RVE [52–54];
3) fully coupled two-scale finite element analyses where the macroscopic equilibrium problem is solved simultaneously with one RVE equilibrium problem for each Gauss quadrature point of the macroscopic mesh. In this case, the constitutive law at each Gauss point is defined by the homogenized response of the corresponding discretized RVE [55–58].

Items (1) and (2) above are very closely related. In some situations of practical interest, the microstructural features of the material may be such that its macroscopic behavior can be reasonably modeled by means of an existing constitutive law. One such typical situation arises in the modeling of strongly directional fiber-reinforced composites for which macroscopic hyperelastic constitutive models are available to some extent. In Ref. [51], for instance, approach (1) was used to determine the macroscopic material parameters of a suitably chosen hyperelastic model for arterial wall tissue. The idea (generally referred to as *numerical material testing* in Ref. [50]) involves the numerical determination of the homogenized response of the RVE (whose topology and material properties are assumed known) followed by a curve-fitting exercise whereby the material parameters are chosen so as to minimize, in some sense, the error between the homogenized response and the response predicted by the macroscopic model for the range of strains under consideration. The main advantage of such procedures lies in the computing times required for the solution of macroscopic boundary value problems involving only conventional (macroscopic) constitutive models are bound to be far lower than those of similar simulations based on the fully coupled approach of item (3). Hence, whenever it is possible to describe the homogenized behavior of the microstructure by means of an existing macroscopic model with acceptable accuracy, preference should be given to such models. Potential drawbacks of this approach, however, are as follows:

- The set of macroscopic parameters that minimize the errors may not be unique and the selection procedure (based on optimization in Ref. [51]) needs to be sufficiently robust.
- The behavior of the constituents of the RVE needs to be known and appropriate models need to be selected together with their corresponding material parameters. This can often be a problem as, in many realistic situations, it is not possible or practical to test the behavior of the individual constituents of a composite material. Note that this particular issue affects approach (3) equally.

Another important fact here is that, in the presence of dissipative phenomena, the issue of parameter identification in the present context appears to remain largely open, as the fitting of the macroscopic response for a range of strain *paths* is by no means trivial. Experience shows, however, that observations made during *numerical materials testing* often lead to significant insights into the (possibly dissipative) behavior of the tested material. Such insights, in turn, may point to improvements to existing constitutive laws and/or definition of completely new ones – item (2) above. In Refs [52–54], for example, macroscopic yield surfaces are obtained as a result of numerical material testing of elastic, perfectly plastic RVEs. In Ref. [54], macroscopic functional forms of macroscopic yield surfaces are proposed as an alternative to the classical Gurson model [59]. We believe that further studies of the dissipative and nondissipative behavior of materials with a microstructure within the present multiscale framework should lead to the much needed development of new macroscopic models with an ability to capture the material behavior more accurately over a wider range of strain histories of greater complexity. In our view, this is a very interesting research topic with a potential to bring substantial benefits to the field of constitutive modeling of solids.

Finally, in spite of the physical appeal of item (3), this approach remains of relatively limited use in large-scale simulations of industrial problems mainly due to the massive computing costs associated with the fully coupled two-scale analysis. Needless to say, in the case of dissipative RVEs undergoing finite straining, for instance, the solution of one RVE equilibrium problem per macroscopic Gauss quadrature point is a formidable task even for reasonably small macroscopic problems running in high-performance machines. The use of parallel algorithms appears then to be a natural course of action and is discussed by Matsui *et al.* [57] and Kuramae *et al.* [60]. Algorithmic techniques aiming the improvement of solution times [61] are also welcome in this context and should be further pursued in order to make fully coupled analyses a realistic option in the future.

1.3 Concluding Remarks

The role and importance of materials modeling has long been established by the seminal works of Tresca [1], Huber [4], von Mises [5], and Hencky [6], among others, who developed the modern theoretical basis of stress and strain analysis. In the last two decades, *computational* materials modeling has consolidated its importance and has become one of the fastest-growing research areas.

Materials modeling encompasses developments associated with materials as diverse as biological tissues, composite materials for aeronautical and aerospace applications, polymeric materials for technical components and ordinary household objects, and heterogeneous geotechnical materials. Furthermore, physical aspects are also as vast as the nature of the materials: structural and thermal behavior under different physical loading, material degradation and failure, and microstructure design, are some examples of applications under intense investigation. Computational issues have been equally relevant and advancements in a wide range of modeling strategies have been introduced in recent years, among which solution techniques for multiphysics problems, strongly coupled multiscale (or multiple scales) and stable cross-scale formulations, and homogenization and optimization techniques have experienced extraordinary growth. Such variety of materials, physical aspects, and computational issues render the task of reviewing the recent advances almost impossible. However, despite the almost boundless research field, it is possible to discern common directions and challenges faced by scientists across research domains:

- derivation of strong physically based constitutive models;
- numerical and experimental procedures to determine the corresponding material parameters;
- multiphysics and multiple scale formulations, ranging from nano- to meso- and macroscales;
- development of robust solution algorithms able to handle complex materials descriptions;
- Application of homogenization and optimization to new materials design.
- Development of new numerical methods aiming at improving handling of discontinuities and heterogeneities typical of some class of materials.

Most modeling issues are strongly related to one or more of the aforementioned aspects. The brief reviews discussed in Sections 1.2.1–1.2.3 highlight the multidisciplinary character of materials modeling: (i) the failure process under general stress–strain paths requires a multiphysics degradation model based on nonlocal formulation, in which multiscale approximations coupled to macrofracture algorithms are the expected advancements; (ii) modeling of multifunctional cellular materials require cross-scale formulations based on homogenization of local properties and microstructure design using optimization procedures; and (iii) multiscale computational modeling techniques. In particular, advancements are expected in multiple scale formulations coupled to multiphysics approaches able to model individual requirements (structural, plastic energy absorption, thermal dissipation, etc.). Finally, it has been observed that an increasing move toward the application of complex material models in industry or, in cases such as geotechnical and civil materials, more realistic analyses have been used by building companies to improve construction design.

Acknowledgments

Miguel Vaz Jr. gratefully acknowledges the support provided by CNPq (project number 309147/2006-9) in the development of the topics described in Section 1.2.1.

References

1. Tresca, M.H. (1864) Mémoire sur l'écoulement des corps solides soumis à de fortes pressions. *Comptes Rendus Hebdomadaires des Séances de l'Académie des Sciences (Paris)*, **59**, 754–758.
2. Peric, D. and de Souza Neto, E.A. (1999) A new computational model for Tresca plasticity at finite strains with an optimal parameterization in the principal space. *Computer Methods in Applied Mechanics and Engineering*, **171** (3-4), 463–489.
3. He, Q.-C., Vallée, C., and Lerintiu, C. (2005) Explicit expressions for the plastic normality-flow rule associated to the Tresca yield criterion. *Zeitschrift für Angewandte Mathematik und Physik*, **56** (2), 357–366.
4. Huber, M.T. (2004) Specific work of strain as a measure of material effort. *Archives of Mechanics*, **56** (3), 173–190 (Translation from Huber's original paper in Polish: Huber, M.T. (1904) Właściwa praca odkształcenia jako miara wytężnia materiału. *Czasopismo Techniczne – Lwów*, **22**, 38–81).
5. von Mises, R. (1913) Mechanik der festen Körper im plastisch-deformablen Zustand. *Nachrichten von der Koniglichen Gesellschaft der wissenschaften zu Goettinger, Mathematisch-Physikalische Klasse*, **1**, 582–592.
6. Hencky, H. (1924) Zur Theorie plastischer Deformationen und der hierdurch im Material hervorgerufenen Nachspannungen. *Zeitschrift für Angewandte Mathematik und Mechanik*, **4** (4), 323–334.
7. Roš, M. and Eichinger, A. (1926) Versuche zur Klärung der Frage der Bruchgefahr. Sonderdruck aus den Verhandlungen 2. Internationalen Kongress für technische Mechanik, Zürich, pp. 315–327.
8. Bamabic, D. (2000) in *Formability of Metallic Materials: Plastic Anisotropy, Formability Testing, Forming Limits* (ed. D. Bamabic), Springer, Berlin, pp. 119–172.
9. Becchi, A. (1994) I Criteri di Plasticità: Cento Anni di Dibattito (1864–1964), PhD thesis, Università degli Studi di Firenze, Firenze.
10. Lemaitre, J. (ed.) (2001) *Handbook of Materials Behavior Models*, vols. **1–3**, Academic Press, San Diego.

11. de Souza Neto, E.A., Peric, D., and Owen, D.R.J. (2008) *Computational Methods for Plasticity: Theory and Applications*, John Wiley & Sons, Ltd, Chichester.
12. Yip, S. (2005) *Handbook of Materials Modeling*, Springer, Berlin.
13. Bao, Y. and Wierzbicki, T. (2004) On fracture locus in the equivalent strain and stress triaxiality space. *International Journal of Mechanical Sciences*, **46** (1), 81–98.
14. Bao, Y. and Wierzbicki, T. (2004) A comparative study on various ductile crack formulation criteria. *Journal of Engineering Materials and Technology: Transactions of the ASME*, **126** (3), 314–324.
15. Vaz M. Jr., de Santi N. Jr., and de Souza Neto, E.A. (2010) Numerical prediction of ductile failure onset under tensile and compressive stress states. *International Journal of Damage Mechanics*, **19**, 175–195.
16. Ladevèze, P. and Lemaitre, J. (1984) Damage effective stress in quasi unilateral conditions, 16th International Congress of Theoretical and Applied Mechanics, Lyngby, Denmark.
17. Lemaitre, J. (1992) *A Course on Damage Mechanics*, Springer-Verlag, Berlin.
18. Xia, S., Li, G., and Lee, H. (1987) A nonlocal damage theory. *International Journal of Fracture*, **34**, 239–250.
19. Meinders, T., Perdahcıoglu, E.S., van Riel, M., and Wisselink, H.H. (2008) Numerical modeling of advanced materials. *International Journal of Machine Tools and Manufacture*, **48**, 485–498.
20. Velde, J., Kowalsky, U., Zümendorf, T., and Dinkler, D. (2008) 3D-FE-Analysis of CT-specimens including viscoplastic material behavior and nonlocal damage. *Computational Materials Sciences*, **46**, 352–357.
21. Samal, M.K., Seidenfuss, M., and Roos, E. (2009) A new mesh-independent Rousselier's damage model: finite element implementation and experimental verification. *International Journal of Mechanical Sciences*, **51**, 619–630.
22. Silva, C.M. and Castro, L.M.S.S. (2009) Nonlocal damage theory in hybrid-displacement formulations. *International Journal of Solids and Structures*, **46**, 3516–3534.
23. Pisano, A.A., Sofi, A., and Fuschi, P. (2009) Finite element solutions for nonhomogeneous nonlocal elastic problems. *Mechanics Research Communications*, **36**, 755–761.
24. Belnoue, J.P., Nguyen, G.D., and Korsunsky, A.M. (2009) Consistent tangent stiffness for local-nonlocal damage modelling of metals. *Procedia Engineering*, **1**, 177–180.
25. Marotti de Sciarra, F. (2009) A nonlocal model with strain-based damage. *International Journal of Solids and Structures*. doi: 10.1016/j.ijsolstr.2009.08.009.
26. Gibson, L.J. and Ashby, M.F. (1997) *Cellular Solids – Structure and Properties*, 2nd edn, Cambridge University Press, Cambridge.
27. Banhart, J. (2001) Manufacture, characterisation and application of cellular metals and metal foams. *Progress in Materials Science*, **46**, 559–632.
28. Zhao, C.Y., Lu, W., and Tassou, S.A. (2006) Thermal analysis on metal-foam filled heat exchangers. Part II: Tube heat exchangers. *International Journal of Heat and Mass Transfer*, **49** (15-16), 2762–2770.
29. Scheffler, M. and Colombo, P. (2005) *Cellular Ceramics: Structure, Manufacturing, Properties and Applications*, Wiley-VCH Verlag GmbH, Weinheim.
30. Wen, T., Tian, J., Lu, T.J., Queheillalt, D.T., and Wadley, H.N.G. (2006) Forced convection in metallic honeycomb structures. *International Journal of Heat and Mass Transfer*, **49** (19-20), 3313–3324.
31. Seepersad, C.C., Kumar, R.S., Allen, J.K., Mistree, F., and Mcdowell, D.L. (2004) Multifunctional design of prismatic cellular materials. *Journal of Computer-Aided Materials Design*, **11** (2-3), 163–181.
32. De Giorgi, M., Carofalo, A., Dattoma, V., Nobile, R., and Palano, F. (2010) Aluminium foams structural modelling. *Computers and Structures*, **88**, 25–35.
33. Kumar, R.S. and McDowell, D.L. (2009) Multifunctional design of two-dimensional cellular materials

with tailored mesostructure. *International Journal of Solids and Structures*, **46**, 2871–2885.

34. Sigmund, O. (1994) Materials with prescribed constitutive parameters: An inverse homogenization problem. *International Journal of Solids and Structures*, **31**, 2313–2329.
35. Guedes, J.M., Rodrigues, H.C., and Bendsøe, M.P. (2003) A material optimization model to approximate energy bounds for cellular materials under multiload conditions. *Structural and Multidisciplinary Optimization*, **25**, 446–452.
36. Muñoz-Rojas, P.A., Nelli Silva, E.C., Cardoso, E.L., Vaz, M. Jr. (2008) in *On the Application of Optimization Techniques to Heat Transfer in Cellular Materials* (eds A. Öchsner, G.E. Murch, and M.J.S. de Lemos), Wiley-VCH Verlag GmbH, Weinheim, pp. 385–417.
37. Paulino, G.H., Nelli Silva, E.C., and Le, C.H. (2009) Optimal design of periodic functionally graded composites with prescribed properties. *Structural and Multidisciplinary Optimization*, **38**, 469–489.
38. Giusti, S.M., Novotny, A.A., de Souza Neto, E.A., and Feijóo, R.A. (2009) Sensitivity of the macroscopic elasticity tensor to topological microstructural changes. *Journal of the Mechanics and Physics of Solids*, **57**, 555–570.
39. Giusti, S.M., Novotny, A.A., and de Souza Neto, E.A. (2010) Sensitivity of the macroscopic response of elastic microstructures to the insertion of inclusions. *Proceedings of the Royal Society* **A: Mathematical, Physical and Engineering Sciences**, **466**, 1703–1723.
40. Amstutz, S., Giusti, S.M., Novotny, A.A., and de Souza Neto, E.A. (2010) Topological derivative for multi-scale linear elasticity models applied to the synthesis of microstructures. *International Journal of Numerical Methods and Engineering* (in press). DOI: 10.1002/nme.2922.
41. Luxner, M.H., Stampfl, J., and Pettermann, H.E. (2005) Finite element modeling concepts and linear analyses of 3D regular open cell structures. *Journal of Materials Science*, **40**, 5859–5866.
42. McKown, S., Shen, Y., Brookes, W.K., Sutcliffe, C.J., Cantwell, W.J., Langdon, G.S., Nurick, G.N., and Theobald, M.D. (2008) The quasi-static and blast loading response of lattice structures. *International Journal of Impact Engineering*, **35**, 795–810.
43. Luxner, M.H., Stampfl, J., and Pettermann, H.E. (2007) Numerical simulations of 3D open cell structures – influence of structural irregularities on elasto-plasticity and deformation localization. *International Journal of Solids and Structures*, **44**, 2990–3003.
44. Luxner, M.H., Woesz, A., Stampfl, J., Fratzl, P., and Pettermann, H.E. (2009) A finite element study on the effects of disorder in cellular structures. *Acta Biomaterialia*, **5**, 381–390.
45. Yan, J., Cheng, G., Liu, L., and Liu, S. (2008) Concurrent material and structural optimization of hollow plate with truss-like material. *Structural and Multidisciplinary Optimization*, **35**, 153–163.
46. Carniel, T.A., Muñoz-Rojas, P.A., Nelli Silva, E.C., and Öchsner, A. (2009) in *Proceedings of the 5th International Conference on Diffusion in Solids and Liquids* (eds A. Öchsner, G. Murch, A. Shokuhfar, and J. Delgado), Ironix, p. 37.
47. Screen, H.R.C. and Evans, S.L. (2009) Measuring strain distributions in the tendon using confocal microscopy and finite elements. *Journal of Strain Analysis*, **44**, 327–335.
48. Lemaitre, J. and Chaboche, J.-L. (1990) *Mechanics of Solid Materials*, Cambridge University Press, Cambridge.
49. Germain, P., Nguyen, Q., and Suquet, P. (1983) Continuum thermodynamics. *ASME Journal of Applied Mechanics*, **50**, 1010–1020.
50. Terada, K., Inugai, T., and Hirayama, N. (2008) A method of numerical material testing in nonlinear multiscale material analyses. *Transactions of the Japan Society of Mechanical Engineering A*, **74**, 1084–1094 (in Japanese).

51. Speirs, D.C.D., de Souza Neto, E.A., and Perić, D. (2008) An approach to the mechanical constitutive modelling of arterial wall tissue based on homogenization and optimization. *Journal of Biomechanics*, **41**, 2673–2680.
52. Pellegrino, C., Galvanetto, A., and Schrefler, B.A. (1999) Computational techniques for periodic composite materials with non-linear material components. *International Journal of Numerical Methods and Engineering*, **46**, 1609–1637.
53. Michel, J.C., Moulinec, H., and Suquet, P. (1999) Effective properties of composite materials with periodic microstructure: a computational approach. *Computers Methods in Applied Mechanics and Engineering*, **172**, 109–143.
54. Giusti, S.M., Blanco, P., de Souza Neto, E.A., and Feijóo, R.A. (2009) An assessment of the Gurson yield criterion by a computational multi-scale approach. *Engineering Computations*, **26**, 281–301.
55. Miehe, C., Schotte, J., and Schröder, J. (1999) Computational micro-macro transitions and overall moduli in the analysis of polycrystals at large strains. *Computational Materials Science*, **16**, 5477–5502.
56. Terada, K. and Kikuchi, N. (2001) A class of general algorithms for multi-scale analysis of heterogeneous media. *Computer Methods in Applied Mechanics and Engineering*, **190**, 5427–5464.
57. Matsui, K., Terada, K., and Yuge, K. (2004) Two-scale finite element analysis of heterogeneous solids with periodic microstructures. *Computers and Structures*, **82**, 593–606.
58. Kousznetsova, V.G., Geers, M.G.D., and Brekelmans, W.A.M. (2004) Multi-scale second order computational homogenization of multi-phase materials: a nested finite element solution strategy. *Computer Methods in Applied Mechanics and Engineering*, **193**, 5525–5550.
59. Gurson, A.L. (1977) Continuum theory of ductile rupture by void nucleation and growth – part I: yield criteria and flow rules for porous media. *Journal of Engineering Materials and Technology*, **88**, 2–15.
60. Kuramae, H., Ikeya, Y., Sakamoto, H., Morimoto, H., and Nakamachi, E. (2010) Multi-scale parallel finite element analyses of LDH sheet formability tests based on crystallographic homogenization method. *International Journal of Mechanical Sciences*, **52**, 183–197.
61. Somer, D.D., de Souza Neto, E.A., Dettmer, W.G., and Perić, D. (2009) A sub-stepping scheme for multi-scale analysis of solids. *Computer Methods in Applied Mechanics and Engineering*, **198**, 1006–1016.

2
Local and Nonlocal Modeling of Ductile Damage

José Manuel de Almeida César de Sá, Francisco Manuel Andrade Pires, and Filipe Xavier Costa Andrade

2.1
Introduction

The gradual internal deterioration at the microscopic level which may eventually lead to the occurrence of macroscopic failure in ductile metals undergoing plastic deformations, has been subjected to a detail study over the last decades. As experimentally verified for many ductile polycrystalline metals, initially throughout metallographic observations [1–3], the nucleation and growth of voids and microcracks that accompany plastic flow causes considerable reduction of stiffness and strength, and is highly influenced by the triaxiality of the stress state [4]. The understanding of the fundamental mechanisms that govern the material's internal deterioration is growing rapidly and it is becoming feasible to formulate continuum constitutive models capable of accounting for the evolution of internal deterioration. In fact, a considerable body of the literature on applied mechanics has been devoted to the formulation of constitutive models to describe internal degradation of solids within the framework of continuum mechanics.

Pioneering work on the formulation of constitutive equations for ductile metals, which include internal deterioration, was carried out by Lemaitre [5]. Using the method of local state and internal variables, Lemaitre has proposed a phenomenological model that includes the strong coupling between elastoplasticity and damage at the constitutive level. The original isotropic model was then further elaborated [3] and later extended by Lemaitre *et al.* [6] to account for the anisotropy of damage as well as for partial closure of microcracks under compressive stresses. Computational aspects have been addressed by several authors. The numerical integration of the Lemaitre constitutive equations by means of a return mapping-type scheme has been originally proposed by Benallal *et al.* [7] in the infinitesimal strain context and later exploited by De Souza Neto *et al.* [8–10] with finite strain extensions of the model.

Although Lemaitre's ductile damage continuum models have been successful in describing the macroscopic stress and strain fields in the presence of internal degradation, they fall within the category of local formulations. Classical local

Advanced Computational Materials Modeling: From Classical to Multi-Scale Techniques.
Edited by Miguel Vaz Júnior, Eduardo A. de Souza Neto, and Pablo A. Munoz-Rojas

ISBN: 978-3-527-32479-8

continuum theories are based on the assumption that the stress at a given point uniquely depends on the current values (and possibly also the previous history) of deformation and temperature at that point. Moreover, this theory inherently assumes that the material is continuous at any size scale and explicitly ignores the influence of the microstructure on the global behavior. These hypotheses are reasonable for a large number of engineering materials and applications, since the major concern is usually the behavior of the material at the macroscale. Nevertheless, a different approach has to be pursued when deformation mechanisms governed by microscopic phenomena as well as scale effects have to be considered in order to predict certain phenomena. Nonlocal theories have emerged as an alternative constitutive framework on the continuum level, that is, used to bridge the gap between the micromechanical level and the classical continuum level. This is achieved through the incorporation of intrinsic material length parameters in the material constitutive model and by using spatially weighted averages in the evolution equations of the respective internal state variables describing plastic and damage growth.

The first nonlocal developments can be traced back to the 1960s with the pioneering works of Rogula [11] and Eringen [12]. These authors were the first ones trying to incorporate the influence of the microstructure into traditional (local) continuum theory by means of a nonlocal formulation. However, they limited their approach to the elastic domain and the first nonlocal formulations within an elastoplastic framework only appeared in the beginning of the 1980s [13–15]. After these works, significant developments have been made by several authors who addressed theoretical and computational issues on both integral-type and gradient-dependent nonlocal enrichments [16–25].

This chapter is devoted to the constitutive modeling and computational aspects of continuum damage mechanics (CDMs) based on Lemaitre's elastoplastic damage theory. Our intention here is to provide an overview to this promising approach, which has been gaining widespread acceptance over the last three decades. This chapter is organized as follows. In Section 2.2, a brief historical review of CDM is given and particular emphasis is given to the phenomenon of ductile plastic damage. The description of Lemaitre's elastoplastic damage theory [26] is undertaken in Section 2.3. Here, the underlying thermodynamical formulation together with its main assumptions is described in detail. The following section, Section 2.4, describes two modifications of the original constitutive model that yield extremely efficient computational algorithms. In Section 2.5, the topic of modeling the material behavior within a nonlocal framework is addressed in detail. Starting with the description of the main hypothesis behind nonlocal formulations together with an overview of the key approaches, several extensions of Lemaitre's ductile damage model into a nonlocal integral approach are investigated. To illustrate the applicability of the models, three numerical examples are presented in Section 2.6. The final section presents some concluding remarks. Although the text is based on the authors' own research, and inevitably emphasizes the authors' point of view, we attempt to offer a balanced review of the main advances in the field of ductile damage modeling that have taken place within the last two decades or so.

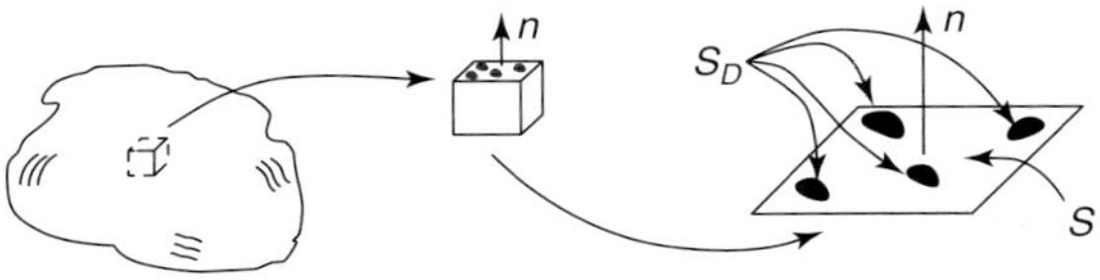

Figure 2.1 Damaged element.

2.2 Continuum Damage Mechanics

Since the pioneering work by Kachanov [27], significant progress has been achieved in the modeling of internal degradation of materials within the theory of continuum mechanics. By incorporating new state variables into the constitutive model, *CDM* has emerged, over the last decades, as a very effective theory to describe material deformability and predict failure onset.

2.2.1 Basic Concepts of CDM

The first modern damage formulations were initially proposed by Kachanov [27] and Rabotnov [28]. Generalizing the concepts proposed by these authors, the damage variable, D, can be physically seen as the corrected area of cracks and cavities per unit surface cut by a plane perpendicular to a normal vector, $\boldsymbol{n}$, as

$$D = \frac{S - \tilde{S}}{S} \tag{2.1}$$

where S is the overall area of the element defined by its normal $\boldsymbol{n}$ and $\tilde{S}$ is the *effective* resisting area (see Figure 2.1).

The definition of damage[1)] by means of an effective resisting area has allowed Kachanov [27] to define the so-called *effective stress*:

$$\tilde{\sigma} = \frac{\sigma}{1 - D} \tag{2.2}$$

The equation above can be straightforwardly obtained by considering the effective area $\tilde{S}$ in the measurement of the uniaxial stress (see Ref. [26] for further details).

These basic concepts, very briefly reviewed above, were the cornerstone for further important developments not only for creep rupture but also for describing the phenomenon of internal degradation in other areas of solid mechanics, including elastic–brittle materials [29, 30], brittle materials [31–33], and fatigue [34, 35], among others. A comprehensive discussion on these models can be found elsewhere [26, 36, 37].

1) A rigorous and more general formulation naturally leads to the definition of a damage tensor [26, 68].

2.2.2 Ductile Plastic Damage

Over the last decades, much effort has been made to model the behavior and predict failure of ductile materials. Two main general approaches have been extensively considered: (i) constitutive theories based on porous material models, such as Gurson plasticity-based models [38–40]; (ii) phenomenological theories motivated by CDM as Lemaitre-based models. In this chapter, we concentrate on the latter approach and, in this section, we shed some light on the basic concepts and assumptions underlying Lemaitre's theory.

The starting point of Lemaitre's original developments was the consideration of a scalar damage variable [5] in the definition of a purely phenomenological model for ductile isotropic damage in metals. By appealing to the hypothesis of strain equivalence[2] which assumes that "the strain associated with a damage state under the applied stress is equivalent to the strain associated with its undamaged state under the effective stress," the model postulates the following elastic constitutive law for a damaged material:

$$\tilde{\sigma} = E\varepsilon \tag{2.3}$$

or, equivalently,

$$\sigma = \tilde{E}\varepsilon \tag{2.4}$$

where E and $\tilde{E} = (1 - D)E$ are the Young's moduli of the virgin (undamaged) and damaged materials, respectively. As a consequence, the standard definition of damage in terms of reduction of the (neither well defined nor easily measurable) load-carrying area is replaced in Lemaitre's model by the reduction of the Young's modulus in the ideally isotropic case. The damage variable from Eq. (2.1) is then redefined as

$$D = \frac{E - \tilde{E}}{E} \tag{2.5}$$

Within the standard local continuum theory, computational aspects have been addressed by several authors where some of these contributions are mentioned below. Simo and Ju [41] used the operator split methodology to derive algorithms for the numerical integration of the elastoplastic damage equations of evolution. Doghri [42], Johansson *et al.* [43], and Lee and Fenves [44] have numerically implemented damage kinetic equations, in a small strain format. De Souza Neto *et al.* [8] presented a comprehensive finite element formulation and error assessment for elastoplastic damage at finite strains, in which isotropic and kinematic hardening and Lemaitre's damage model were accounted for. Steinmann *et al.* [45] formulated Lemaitre's and Gurson's isotropic damage models

2) It is worth to mentioning that other approaches have also been proposed, namely, the hypothesis of equivalent energy [42] and equivalent stress [43].

in the framework of finite multiplicative elastoplasticity, where issues on numerical implementation of the models were specially emphasized. More recently, de Souza Neto [46] proposed an extremely efficient integration algorithm for a simplified version of Lemaitre's model. Moreover, Andrade Pires and *et al.* [47] presented the numerical implementation of an improved damage model, which includes the distinction between tensile and compressive stress states in the evolution of damage.

2.3 Lemaitre's Ductile Damage Model

As mentioned earlier, in this chapter, we focus on Lemaitre-based models for the description and modeling of the ductile damage phenomenon. Therefore, we first present the original constitutive model proposed by Lemaitre where some assumptions and features of the model are discussed. In subsequent sections, we provide the description of modified elastoplastic damage models based on Lemaitre's theory where the details of their numerical implementation are addressed.

2.3.1 Original Model

The original Lemaitre's ductile damage model is based on CDM concepts and in the framework of thermodynamics of irreversible processes. The thermodynamic (or state) potential, ψ, is assumed to be a function of a set of state variables

$$\psi = \psi\left(\boldsymbol{\varepsilon}, T, \boldsymbol{\varepsilon}^{\mathrm{e}}, \boldsymbol{\varepsilon}^{\mathrm{p}}, R, \boldsymbol{\alpha}, D\right) \tag{2.6}$$

where $\boldsymbol{\varepsilon}$ and T respectively represent the strain tensor and the temperature, which are *observable* state variables. The other state variables in Eq. (2.6) are *internal* variables: $\boldsymbol{\varepsilon}^{\mathrm{e}}$ and $\boldsymbol{\varepsilon}^{\mathrm{p}}$ are respectively the elastic and the plastic strain tensors, R and D are scalar variables, associated with isotropic hardening and isotropic damage respectively, and the second-order tensor $\boldsymbol{\alpha}$ is the variable associated with kinematic hardening.

In elastoplasticity, due to the relation $\boldsymbol{\varepsilon} = \boldsymbol{\varepsilon}^{\mathrm{e}} + \boldsymbol{\varepsilon}^{\mathrm{p}}$, it is possible to express the specific free energy as a function of only one strain tensor as internal variable and, therefore,

$$\psi = \psi\left(\boldsymbol{\varepsilon} - \boldsymbol{\varepsilon}^{\mathrm{p}}, T, R, \boldsymbol{\alpha}, D\right) = \psi\left(\boldsymbol{\varepsilon}^{\mathrm{e}}, T, R, \boldsymbol{\alpha}, D\right) \tag{2.7}$$

Assuming a constant density ρ, the state laws are defined, from the state potential, as

$$\boldsymbol{\sigma} = \rho\frac{\partial \psi}{\partial \boldsymbol{\varepsilon}} = \rho\frac{\partial \psi}{\partial \boldsymbol{\varepsilon}^{\mathrm{e}}} = -\rho\frac{\partial \psi}{\partial \boldsymbol{\varepsilon}^{\mathrm{p}}}; \quad S = \frac{\partial \psi}{\partial T} \tag{2.8}$$

where $\boldsymbol{\sigma}$ is the Cauchy stress tensor and S is the entropy density.

The following derivatives define the thermodynamic forces associated with the state variables, that is, with which they are power conjugated, as

$$\chi = \rho\frac{\partial\psi}{\partial R}; \quad \boldsymbol{\beta} = \rho\frac{\partial\psi}{\partial\boldsymbol{\alpha}}; \quad -Y = -\rho\frac{\partial\psi}{\partial D} \tag{2.9}$$

where χ is the thermodynamic force associated with the isotropic hardening variable and $\boldsymbol{\beta}$ is the *backstress tensor*, which is a deviator tensor associated with kinematic hardening. The thermodynamic force associated with damage is denoted as $-Y$ and is called *energy release rate*.

Since elasticity-damage and plasticity-hardening processes are assumed to be decoupled, the state potential is given by the sum

$$\psi = \psi^{\mathrm{ed}}\left(\boldsymbol{\varepsilon}^{\mathrm{e}}, D, T\right) + \psi^{\mathrm{p}}\left(R, \boldsymbol{\alpha}, T\right) \tag{2.10}$$

where ψ^{ed} and ψ^{p} are, respectively, the elastic-damage and plastic contribution to the state potential.

2.3.1.1 The Elastic State Potential

On the basis of the hypothesis of strain equivalence for an isothermal process, Lemaitre's model postulates ψ^{ed} such that

$$\rho\psi^{\mathrm{ed}} = \frac{1}{2}\boldsymbol{\varepsilon}^{\mathrm{e}} : (1-D)\,\mathbf{D}^{\mathrm{e}} : \boldsymbol{\varepsilon}^{\mathrm{e}} \tag{2.11}$$

where $\mathbf{D}^{\mathrm{e}}$ is the standard isotropic elasticity tensor.

For this potential, the elasticity state law is given by

$$\boldsymbol{\sigma} = \rho\frac{\partial\psi^{\mathrm{ed}}}{\partial\boldsymbol{\varepsilon}^{\mathrm{e}}} = (1-D)\,\mathbf{D}^{\mathrm{e}} : \boldsymbol{\varepsilon}^{\mathrm{e}} \tag{2.12}$$

or, equivalently in terms of the effective stress tensor:

$$\tilde{\boldsymbol{\sigma}} = \mathbf{D}^{\mathrm{e}} : \boldsymbol{\varepsilon}^{\mathrm{e}} \tag{2.13}$$

The thermodynamic force associated with damage, $-Y$, is then given by

$$-Y = -\rho\frac{\partial\psi^{\mathrm{ed}}}{\partial D} = \frac{1}{2}\boldsymbol{\varepsilon}^{\mathrm{e}} : \mathbf{D}^{\mathrm{e}} : \boldsymbol{\varepsilon}^{\mathrm{e}} \tag{2.14}$$

which can also be expressed by

$$\begin{aligned} -Y &= -\frac{1}{2(1-D)^2}\boldsymbol{\sigma} : \left[\mathbf{D}^{\mathrm{e}}\right]^{-1} : \boldsymbol{\sigma} \\ &= -\frac{1}{2(1-D)^2}\left[(1+\nu)\,\boldsymbol{\sigma} : \boldsymbol{\sigma} - \nu\,(\mathrm{tr}\boldsymbol{\sigma})^2\right] \\ &= \frac{q^2}{2E(1-D)^2}\left[\frac{2}{3}(1+\nu) + 3(1-2\nu)\left(\frac{p}{q}\right)^2\right] \\ &= \frac{q^2}{6G(1-D)^2} + \frac{p^2}{2K(1-D)^2} \end{aligned} \tag{2.15}$$

where E and ν are the Young's modulus and the Poisson ratio, respectively, G and K are the shear modulus and the bulk modulus, p is the hydrostatic pressure, and q

is the von Mises equivalent stress. It is noteworthy that the damage energy release rate, $-Y$, represents the variation of internal energy density due to damage growth at constant stress [26].

2.3.1.2 The Plastic State Potential

The plastic state potential is selected as a sum of two terms associated, respectively, to isotropic and kinematic hardening. Considering an isothermal process, it may be expressed as

$$\rho\psi^{\mathrm{p}}(R,\boldsymbol{\alpha}) = \rho\psi^{\mathrm{I}}(R) + \frac{a}{2}\boldsymbol{\alpha} : \boldsymbol{\alpha} \tag{2.16}$$

where a is material parameter and $\psi^{\mathrm{I}}(R)$ is the isotropic hardening contribution.

The thermodynamic force related with isotropic hardening is therefore expressed as

$$\chi = \rho\frac{\partial\psi^{\mathrm{p}}}{\partial R} = \rho\frac{\partial\psi^{\mathrm{I}}(R)}{\partial R} = \chi(R) \tag{2.17}$$

while the thermodynamic force associated with kinematic hardening (i.e., the backstress tensor) is expressed as

$$\boldsymbol{\beta} = \rho\frac{\partial\psi^{\mathrm{p}}}{\partial\boldsymbol{\alpha}} = a\boldsymbol{\alpha} \tag{2.18}$$

2.3.1.3 The Dissipation Potential

A second potential, complementary to the state potential, written in terms of the thermodynamic forces, is used to establish evolution laws for the dissipative state variables. Later, we will depart slightly from the original work of Lemaitre and establish these kinematic laws of evolution from a constrained maximization problem and then verify some assumptions and limitations behind Lemaitre's theory.

The complementary potential, ψ^*, here termed as *dissipation potential*, may then be written, for an isothermal process, as

$$\psi^* = \psi^*(\chi, \boldsymbol{\beta}, -Y) = \psi^{*\mathrm{p}}(\chi, \boldsymbol{\beta}) + \psi^{*\mathrm{d}}(-Y) \tag{2.19}$$

which may be separated in two terms, respectively, the contributions of the associated dissipative phenomena, that is, plasticity and damage.

For associated plasticity, the dissipation potential due to plasticity, $\psi^{*\mathrm{p}}$, may be related to the particular yield (limit) function adopted. In the case of the von Mises yield function, coupled with the principle of strain equivalence, the potential of plastic dissipation takes the following form:

$$\psi^{*\mathrm{p}}(\chi, \boldsymbol{\beta}) = F_{\mathrm{p}} = \frac{\sqrt{3J_2(\mathbf{s}-\boldsymbol{\beta})}}{(1-D)} - \left[\sigma_{\mathrm{yo}} + \chi(R)\right] + \frac{b}{2a}\boldsymbol{\beta} : \boldsymbol{\beta} \tag{2.20}$$

where $\tilde{\mathbf{s}}$ is the deviatoric stress tensor, σ_{yo} is the initial yield stress, R is the excess of yield function due to isotropic hardening, and a and b are material parameters associated with kinematic hardening which may be obtained from cyclic loading experiments [48].

Similarly, it is possible to define the dissipation potential for ductile damage in the form of a threshold (limit) function, that is,

$$\psi^{*d}(-Y) = F_d = \frac{(1-D)^s\, r}{(s+1)}\left[\left(\frac{-Y}{(1-D)\,r}\right)^{s+1} - \left(\frac{w^e_{th}}{r}\right)^{s+1}\right] \tag{2.21}$$

where r is a material parameter representing the energy strength of damage, s is a nondimensional material parameter which is a function of the temperature, and w^e_{th} is a threshold value for the elastic strain energy which is required for the development of defects.

2.3.1.4 Evolution of Internal Variables

The behavior of ductile materials is described by Lemaitre's theory by means of the evolution equations of $\dot{\boldsymbol{\varepsilon}}^p$, $\dot{R}$, $\dot{\boldsymbol{\alpha}}$, and $\dot{D}$. In Lemaitre's original model, these equations are obtained through the so-called *normality rule of generalized standard materials*, which states that

$$\dot{\boldsymbol{\varepsilon}}^p = \dot{\gamma}\frac{\partial \psi^*}{\partial \boldsymbol{\sigma}}; \quad \dot{R} = -\dot{\gamma}\frac{\partial \psi^*}{\partial \chi}; \quad \dot{\boldsymbol{\alpha}} = -\dot{\gamma}\frac{\partial \psi^*}{\partial \boldsymbol{\beta}}; \quad \dot{D} = \dot{\gamma}\frac{\partial \psi^*}{\partial (-Y)} \tag{2.22}$$

The derivatives above yield on the following evolution equations

$$\dot{\boldsymbol{\varepsilon}}^p = \sqrt{\frac{3}{2}}\frac{\dot{\gamma}}{(1-D)}\frac{\boldsymbol{s}-\boldsymbol{\beta}}{\|\boldsymbol{s}-\boldsymbol{\beta}\|} \tag{2.23}$$

$$\dot{R} = \dot{\gamma} \tag{2.24}$$

$$\dot{\boldsymbol{\alpha}} = \dot{\gamma}\left[\sqrt{\frac{3}{2}}\frac{\boldsymbol{s}-\boldsymbol{\beta}}{\|\boldsymbol{s}-\boldsymbol{\beta}\|} - \frac{b}{a}\boldsymbol{\beta}\right] \tag{2.25}$$

$$\dot{D} = \frac{\dot{\gamma}}{(1-D)}\left(\frac{-Y}{r}\right)^s H(\varepsilon^p_{ac} - \varepsilon^p_D) \tag{2.26}$$

where H is an added Heaviside function of the equivalent plastic strain ε^p_{ac} and ε^p_D is its critical value which triggers the onset of damage.

It is important to remark that from a numerical implementation point of view, it is more useful to replace the evolution law of the back strain tensor, $\dot{\boldsymbol{\alpha}}$, by the evolution law of its associated variable, the backstress tensor, $\dot{\boldsymbol{\beta}}$. This may be done by taking into account the constitutive equation relating those two tensors and obtained from the state potential, which may be written as [26]

$$\dot{\boldsymbol{\beta}} = a\dot{\boldsymbol{\alpha}} \tag{2.27}$$

Thus, Eq. (2.25) becomes

$$\dot{\boldsymbol{\beta}} = \dot{\gamma}\left[a\sqrt{\frac{3}{2}}\frac{\boldsymbol{s}-\boldsymbol{\beta}}{\|\boldsymbol{s}-\boldsymbol{\beta}\|} - b\boldsymbol{\beta}\right] \tag{2.28}$$

Table 2.1 Lemaitre's ductile damage model.

Strain tensor additive split	$\boldsymbol{\varepsilon} = \boldsymbol{\varepsilon}^{\mathrm{e}} + \boldsymbol{\varepsilon}^{\mathrm{p}}$
Elastic law (coupled with damage)	$\boldsymbol{\sigma} = (1 - D)\, \mathbf{D}^{\mathrm{e}} : \boldsymbol{\varepsilon}^{\mathrm{e}}$
Yield criterion	$F_{\mathrm{p}} = \sqrt{\frac{3}{2}} \frac{\|\mathbf{s}-\boldsymbol{\beta}\|}{1-D} - \left[\sigma_{\mathrm{yo}} + \chi\,(R)\right] + \frac{b}{2a} \boldsymbol{\beta} : \boldsymbol{\beta}$
Plastic flow	$\dot{\boldsymbol{\varepsilon}}^{\mathrm{p}} = \sqrt{\frac{3}{2}} \frac{\dot{\gamma}}{(1-D)} \frac{\mathbf{s}-\boldsymbol{\beta}}{\|\mathbf{s}-\boldsymbol{\beta}\|}$
Evolution of isotropic hardening	$\dot{R} = \dot{\gamma}$
Evolution of kinematic hardening	$\dot{\boldsymbol{\beta}} = \dot{\gamma} \left[a \sqrt{\frac{3}{2}} \frac{\mathbf{s}-\boldsymbol{\beta}}{\|\mathbf{s}-\boldsymbol{\beta}\|} - b\boldsymbol{\beta} \right]$
Evolution of damage	$\dot{D} = \frac{\dot{\gamma}}{(1-D)} \left(\frac{-Y}{r}\right)^{s} H\left(\varepsilon_{\mathrm{ac}}^{\mathrm{p}} - \varepsilon_{\mathrm{D}}^{\mathrm{p}}\right)$
Kuhn–Tucker conditions	$\dot{\gamma} \geq 0; \quad F_{\mathrm{p}} \leq 0; \quad \dot{\gamma} F_{\mathrm{p}} = 0$

Like many other elastoplastic models, the equations above must consider the *Kuhn–Tucker conditions* (also commonly referred to as *loading/unloading conditions*), given by

$$\dot{\gamma} \geq 0; \quad F_{\mathrm{p}} \leq 0; \quad \dot{\gamma} F_{\mathrm{p}} = 0 \tag{2.29}$$

For convenience, the equations for Lemaitre's original ductile damage model are summarized in Table 2.1.

2.3.2 Principle of Maximum Inelastic Dissipation

An alternative manner to obtain the evolution of the internal variables is by generalizing the *principle of maximum plastic dissipation* [49]. In the present case, we apply such a principle for a constitutive model composed of both plastic and damage mechanisms, which is the case in Lemaitre's model. In fact, we consider the assumption that, among all possible stress states that satisfy both the plastic and damage threshold (limit) functions, the solution of the constitutive problem is the one associated with the maximum energy of dissipation. In turn, this energy is directly associated with the inelastic (dissipative) process, characterized by plastic deformation, hardening (both isotropic and kinematic), and damage. Therefore, by defining the energy of the dissipative process as

$$\varphi(\boldsymbol{\sigma}, \chi, \boldsymbol{\beta}, -Y) = \boldsymbol{\sigma} : \dot{\boldsymbol{\varepsilon}}^{\mathrm{p}} - \chi \dot{R} - \boldsymbol{\beta} : \dot{\boldsymbol{\alpha}} - Y\dot{D} \tag{2.30}$$

the solution may be obtained by maximizing $\varphi(\boldsymbol{\sigma}, \chi, \boldsymbol{\beta}, -Y)$ taking as constraints the threshold-limiting functions for plasticity and damage, respectively written in Eqs (2.20) and (2.21). In the present case, we impose these constraints by recurring to Lagrange multipliers; thus,

$$\begin{aligned} L_{\varphi}\left(\boldsymbol{\sigma}, \chi, \boldsymbol{\beta}, -Y, \dot{\gamma}_{\mathrm{p}}, \dot{\gamma}_{\mathrm{d}}\right) &= \boldsymbol{\sigma} : \dot{\boldsymbol{\varepsilon}}^{\mathrm{p}} - \chi \dot{R} - \boldsymbol{\beta} : \dot{\boldsymbol{\alpha}} - Y\dot{D} \\ &\quad - \dot{\gamma}_{\mathrm{p}} F_{\mathrm{p}}\,(\chi, \boldsymbol{\beta}) - \dot{\gamma}_{\mathrm{d}} F_{\mathrm{d}}\,(-Y) \end{aligned} \tag{2.31}$$

The evolution laws for the internal state variables are then obtained at the maximum, that is, the critical point at which the derivatives of the Lagrangian, L_{φ}, with respect to the thermodynamic forces, vanish:

$$\frac{\partial L_\varphi}{\partial \boldsymbol{\sigma}} = 0 \Longleftrightarrow \dot{\boldsymbol{\varepsilon}}^{\mathrm{p}} = \dot{\gamma}_{\mathrm{p}} \frac{\partial F_{\mathrm{p}}}{\partial \boldsymbol{\sigma}} \Longleftrightarrow \dot{\boldsymbol{\varepsilon}}^{\mathrm{p}} = \sqrt{\frac{3}{2}} \frac{\dot{\gamma}_{\mathrm{p}}}{(1-D)} \frac{\boldsymbol{s} - \boldsymbol{\beta}}{\|\boldsymbol{s} - \boldsymbol{\beta}\|} \tag{2.32}$$

$$\frac{\partial L_\varphi}{\partial \chi} = 0 \Longleftrightarrow \dot{R} = -\dot{\gamma}_{\mathrm{p}} \frac{\partial F_{\mathrm{p}}}{\partial R} \Longleftrightarrow \dot{R} = \dot{\gamma}_{\mathrm{p}} \tag{2.33}$$

$$\frac{\partial L_\varphi}{\partial \boldsymbol{\beta}} = 0 \Longleftrightarrow \dot{\boldsymbol{\alpha}} = \dot{\gamma}_{\mathrm{p}} \frac{\partial F_{\mathrm{p}}}{\partial \boldsymbol{\beta}} \Longleftrightarrow \dot{\boldsymbol{\alpha}} = \dot{\gamma}_{\mathrm{p}} \left[\sqrt{\frac{3}{2}} \frac{\boldsymbol{s} - \boldsymbol{\beta}}{\|\boldsymbol{s} - \boldsymbol{\beta}\|} - \frac{b}{a} \boldsymbol{\beta} \right] \tag{2.34}$$

$$\frac{\partial L_\varphi}{\partial (-Y)} = 0 \Longleftrightarrow \dot{D} = \dot{\gamma}_{\mathrm{d}} \frac{\partial F_{\mathrm{d}}}{\partial (-Y)} \Longleftrightarrow \dot{D} = \frac{\dot{\gamma}_{\mathrm{d}}}{(1-D)} \left(\frac{-Y}{r} \right)^{s} \tag{2.35}$$

Moreover, the derivatives of the Lagrangian with respect to the multipliers lead to the plastic and damage limit functions that must be satisfied, that is,

$$\frac{\partial L_\varphi}{\partial \dot{\gamma}_{\mathrm{p}}} = 0 \Longleftrightarrow F_{\mathrm{p}} = 0 \Longleftrightarrow \sqrt{\tfrac{3}{2}} \frac{\|\boldsymbol{s} - \boldsymbol{\beta}\|}{1 - D} - \left[\sigma_{\mathrm{yo}} + \chi(R) \right] + \frac{b}{2a} \boldsymbol{\beta} : \boldsymbol{\beta} = 0 \tag{2.36}$$

$$\frac{\partial L_\varphi}{\partial \dot{\gamma}_{\mathrm{d}}} = 0 \Longleftrightarrow F_{\mathrm{d}} = 0 \Longleftrightarrow -Y = (1 - D)\, w_{\mathrm{th}}^{\mathrm{e}} \tag{2.37}$$

Additionally, the Kuhn–Tucker conditions (or, in this case, the optimality conditions), must hold for any deformation:

$$\begin{array}{lll} \dot{\gamma}_{\mathrm{p}} \geq 0; & F_{\mathrm{p}} \leq 0; & \dot{\gamma}_{\mathrm{p}} F_{\mathrm{p}} = 0; \\ \dot{\gamma}_{\mathrm{d}} \geq 0; & F_{\mathrm{p}} \leq 0; & \dot{\gamma}_{\mathrm{d}} F_{\mathrm{d}} = 0 \end{array} \tag{2.38}$$

Finally, the multipliers may be determined by the consistency conditions:

$$\dot{\gamma}_{\mathrm{p}} F_{\mathrm{p}} = 0; \quad \dot{\gamma}_{\mathrm{d}} F_{\mathrm{d}} = 0 \tag{2.39}$$

2.3.3 Assumptions Behind Lemaitre's Model

In the above section, we have derived the evolution equations for the internal variables from a maximization problem. Plasticity and damage have been considered as uncoupled mechanisms, each one possessing a specific Lagrange multiplier (respectively, $\dot{\gamma}_{\mathrm{p}}$ and $\dot{\gamma}_{\mathrm{d}}$). In turn, these multipliers must be determined by two different consistency conditions, a restriction imposed by Eq. (2.39). However, in the original Lemaitre's model, the evolution of the internal variables considers only one (plastic) multiplier, simply denoted by $\dot{\gamma}$ (with no additional subscript). This corresponds to assuming that in Eq. (2.31) both multipliers, $\dot{\gamma}_{\mathrm{p}}$ and $\dot{\gamma}_{\mathrm{d}}$, are the same. This fact can be, to some extent, seen as a limitation of Lemaitre's model since this assumption automatically implies that damage can take place if, and only if, plasticity occurs. On the other hand, this is a reasonable hypothesis in the case of ductile metals since experimental evidence has already shown that damage is generally induced by plasticity [1–3, 26].

Another important conclusion is the fact that the hydrostatic pressure, p, plays no role in the damage threshold function in Lemaitre's original model. As a matter of fact, the hydrostatic stress only has influence on the rate of damage (i.e., on how much damage will grow), but the mechanism that triggers damage evolution is uniquely dependent of the von Mises equivalent stress, q, which in turn is computed from a purely deviatoric stress measure. In other words, these underlying assumptions state that once damage is triggered, the loading/unloading conditions for damage are considered fulfilled and, therefore, are not checked. As a consequence, the evolution equations in Lemaitre's model take the form previously summarized in Table 2.1.

2.4 Modified Local Damage Models

In this section, two modifications of Lemaitre's original damage model are presented. However, it is important to remark that such modifications are performed within the standard local theory. The improvements achieved by nonlocal formulations of integral type as well as the issues on their numerical implementation are addressed in Section 2.5.

2.4.1 Lemaitre's Simplified Damage Model

One of the possible modifications for Lemaitre's damage model is achieved by disregarding the effects of kinematic hardening. As shown by de Souza Neto [46], the consideration only of isotropic hardening leads to a remarkably simple and efficient numerical algorithm. Therefore, the simplified damage model is, in fact, appealing from the numerical point of view. Nevertheless, one should bear in mind that the use of such a model is exclusively suitable for situations where load reversal is inexistent or negligible, a condition met in several practical applications in metal forming.

2.4.1.1 Constitutive Model

The only difference between the original and the simplified version of Lemaitre's damage model is that the latter disregards kinematic hardening [46, 50, 51]. Thus, the equations are rewritten without the terms related to kinematic hardening effects where the main differences are addressed as follows. First, the yield function is simply given by

$$F_{\mathrm{p}} = \frac{q}{1-D} - \sigma_{\mathrm{y}}(R) \tag{2.40}$$

Table 2.2 Lemaitre's simplified damage model.

Strain tensor additive split	$\boldsymbol{\varepsilon} = \boldsymbol{\varepsilon}^{\mathrm{e}} + \boldsymbol{\varepsilon}^{\mathrm{p}}$
Elastic law (coupled with damage)	$\boldsymbol{\sigma} = (1 - D)\,\mathbf{D}^{\mathrm{e}} : \boldsymbol{\varepsilon}^{\mathrm{e}}$
Yield criterion	$F_{\mathrm{p}} = \tilde{q} - \sigma_{\mathrm{y}}(R)$
Plastic flow	$\dot{\boldsymbol{\varepsilon}}^{\mathrm{p}} = \frac{3}{2}\frac{\dot{\gamma}}{(1-D)}\frac{\mathbf{s}}{q}$
Evolution of isotropic hardening	$\dot{R} = \dot{\gamma}$
Evolution of damage	$\dot{D} = \frac{\dot{\gamma}}{1-D}\left(\frac{-Y}{r}\right)^{s}$
Kuhn–Tucker conditions	$\dot{\gamma} \geq 0;\quad F_{\mathrm{p}} \leq 0;\quad \dot{\gamma} F_{\mathrm{p}} = 0$

or, alternatively,

$$F_{\mathrm{p}} = \tilde{q} - \sigma_{\mathrm{y}}(R) \tag{2.41}$$

where the backstress tensor, $\boldsymbol{\beta}$, has been simply ignored from the original equation. Furthermore, the evolution of the plastic strain tensor is expressed by

$$\dot{\boldsymbol{\varepsilon}}^{\mathrm{p}} = \frac{3}{2}\frac{\dot{\gamma}}{(1 - D)}\frac{\mathbf{s}}{q} \tag{2.42}$$

The model is complete with the definition of the evolution equations for the isotropic hardening and the damage variables, respectively, given by

$$\dot{R} = \dot{\gamma} \tag{2.43a}$$

$$\dot{D} = \frac{\dot{\gamma}}{1 - D}\left(\frac{-Y}{r}\right)^{s} \tag{2.43b}$$

for which the loading/unloading conditions $\dot{\gamma} \geq 0$, $F_{\mathrm{p}} \leq 0$, and $\dot{\gamma} F_{\mathrm{p}} = 0$ must hold. The equations of the simplified constitutive model are grouped in Table 2.2.

2.4.1.2 Numerical Implementation

The numerical implementation of Lemaitre's simplified damage model has been proposed by de Souza Neto [46] and is briefly reviewed in this section. The efficiency is attained by numerically integrating the material model by means of a conventional *fully implicit elastic predictor/return mapping* scheme, typically adopted in a finite element framework (see Ref. [49] for more details on return mapping schemes).

The algorithm for the numerical integration of the constitutive model starts with the definition of the so-called *elastic trial state*. Notice that the constitutive behavior is meant to be locally computed at *every material point* within a generic time interval $[t_n, t_{n+1}]$, where the constitutive variables $\boldsymbol{\varepsilon}_n^{\mathrm{p}}$, $\boldsymbol{\sigma}_n$, R_n, and D_n are known "a prori." The goal of the algorithm is to find the updated values of $\boldsymbol{\varepsilon}_{n+1}^{\mathrm{p}}$, $\boldsymbol{\sigma}_{n+1}$, R_{n+1}, and D_{n+1} for a given strain increment $\Delta\boldsymbol{\varepsilon}$ within $[t_n, t_{n+1}]$. Considering a typical finite element framework, the material points correspond to the Gauss integration points.

In the elastic trial state, the solution of the material problem is assumed to be purely elastic; hence, neither damage nor hardening evolution takes place at this stage. Therefore, the elastic trial strain tensor is given by

$$\boldsymbol{\varepsilon}_{n+1}^{\mathrm{e\ trial}} = \boldsymbol{\varepsilon}_n^{\mathrm{e}} + \Delta\boldsymbol{\varepsilon} \tag{2.44}$$

from which it is possible to compute the *effective* elastic trial stress tensor, written as

$$\tilde{\boldsymbol{\sigma}}_{n+1}^{\mathrm{e\ trial}} = \mathbf{D}^{\mathrm{e}} : \boldsymbol{\varepsilon}_{n+1}^{\mathrm{e\ trial}} \tag{2.45}$$

Considering now the deviatoric/hydrostatic split of the stress tensor, the *undamaged* elastic trial stress tensor is alternatively written as

$$\tilde{\boldsymbol{\sigma}}_{n+1}^{\mathrm{e\ trial}} = \tilde{\mathbf{s}}_{n+1}^{\mathrm{trial}} + \tilde{p}_{n+1}^{\mathrm{trial}} \boldsymbol{I} \tag{2.46}$$

where $\mathbf{s}_{n+1}^{\mathrm{trial}}$ and p_{n+1}^{trial} are the *effective* trial deviatoric and hydrostatic stresses, respectively given by

$$\tilde{\mathbf{s}}_{n+1}^{\mathrm{trial}} = 2G\boldsymbol{\varepsilon}_{\mathrm{d}\ n+1}^{\mathrm{e\ trial}}; \quad \tilde{p}_{n+1}^{\mathrm{trial}} = K\varepsilon_{\mathrm{v}\ n+1}^{\mathrm{e\ trial}} \tag{2.47}$$

where

$$\boldsymbol{\varepsilon}_{\mathrm{d}\,n+1}^{\mathrm{e\,trial}} = \boldsymbol{\varepsilon}_{\mathrm{d}n}^{\mathrm{e}} + \Delta\boldsymbol{\varepsilon}_{\mathrm{d}}; \quad \varepsilon_{\mathrm{v}\,n+1}^{\mathrm{e\,trial}} = \varepsilon_{\mathrm{v}\,n}^{\mathrm{e}} + \Delta\varepsilon_{\mathrm{v}} \tag{2.48}$$

in which the strain deviator and the volumetric strain have been denoted, respectively, by $\boldsymbol{\varepsilon}_{\mathrm{d}}$ and ε_{v}.

Finally, with Eq. (2.47) it is possible to compute the elastic trial von Mises equivalent stress from the *effective* deviatoric stress as

$$\tilde{q}_{n+1}^{\mathrm{trial}} = \sqrt{\frac{3}{2}} \left\| \tilde{\mathbf{s}}_{n+1}^{\mathrm{trial}} \right\| \tag{2.49}$$

where $\tilde{q}_{n+1}^{\mathrm{trial}}$ is necessary for the proper evaluation of the yield function and to check whether the pseudoincrement is elastic or plastic.

Following standard procedures of elastic predictor/return mapping schemes, Lemaitre's constitutive model can be written in its (pseudo)time-discretized version by the following system of equations

$$\begin{cases} \boldsymbol{\varepsilon}_{n+1}^{\mathrm{e}} = \boldsymbol{\varepsilon}_{n+1}^{\mathrm{e\ trial}} - \frac{3}{2} \frac{\Delta\gamma}{(1-D_{n+1})} \frac{\tilde{s}_{n+1}}{q_{n+1}} \\ R_{n+1} = R_n + \Delta\gamma \\ D_{n+1} = D_n + \frac{\Delta\gamma}{1-D_{n+1}} \left(\frac{-Y_{n+1}}{r} \right)^{\mathrm{s}} \\ \frac{q_{n+1}}{1-D_{n+1}} - \sigma_y \left(R_{n+1} \right) = 0 \end{cases} \tag{2.50}$$

where $\boldsymbol{\varepsilon}_{n+1}^{\mathrm{e}}$, R_{n+1}, $\Delta\gamma$, and D_{n+1} are the unknowns of the *incremental initial boundary value constitutive problem*. It is important to remark that the last equation of the system is the *consistency condition* which, in practice, acts as a constraint for the constitutive problem.

The above system of equations is unattractive from the numerical point of view due to the high computational burden if compared to simpler elastoplastic models (e.g., von Mises isotropic plasticity). However, by performing some relatively straightforward operations, it is possible to reduce the system of equations

(Eq. (2.50)) to a *single* scalar nonlinear equation, which is written as

$$F(\Delta\gamma) = \frac{3G\Delta\gamma}{\tilde{q}_{n+1}^{\text{trial}} - \sigma_y(R_n + \Delta\gamma)} - (1 - D_n) + \frac{\left(\tilde{q}_{n+1}^{\text{trial}} - \sigma_y(R_n + \Delta\gamma)\right)}{3G}\left(\frac{-Y(\Delta\gamma)}{r}\right)^{s} = 0 \tag{2.51}$$

where $\Delta\gamma$ is now the only unknown. In fact, the scalar equation above is much simpler to be worked out, significantly reducing the computational cost per Gauss point. The derivation of Eq. (2.51) is omitted here for convenience and the reader is referred to Ref. [46] for further details. The complete stress update algorithm for the efficient numerical integration of Lemaitre's simplified model by means of a fully implicit elastic predictor/return mapping scheme is summarized in Box 2.1 in pseudocode format.

Box 2.1: Stress Update Algorithm for Lemaitre's Simplified Model

(i) Define elastic trial state

$$\boldsymbol{\varepsilon}_{n+1}^{\text{e trial}} = \boldsymbol{\varepsilon}_n^{\text{e}} + \Delta\boldsymbol{\varepsilon}; \qquad \tilde{\boldsymbol{s}}_{n+1}^{\text{trial}} = 2G\boldsymbol{\varepsilon}_{\text{d}\ n+1}^{\text{e trial}}$$

$$R_{n+1}^{\text{trial}} = R_n; \qquad \tilde{q}_{n+1}^{\text{trial}} = \sqrt{\tfrac{3}{2}}\left\|\tilde{\boldsymbol{s}}_{n+1}^{\text{trial}}\right\|$$

$$\tilde{p}_{n+1}^{\text{trial}} = K\varepsilon_{\text{v}\ n+1}^{\text{e trial}};$$

(ii) Check plastic admissibility

IF $\tilde{q}_{n+1}^{\text{trial}} - \sigma_y\left(R_{n+1}^{\text{trial}}\right) \le 0$ THEN

SET $(\cdot)_{n+1} = (\cdot)_{n+1}^{\text{trial}}$ and EXIT

ENDIF

(iii) Solve the single residual equation

$$F(\Delta\gamma) \equiv \frac{3G\Delta\gamma}{\tilde{q}_{n+1}^{\text{trial}} - \sigma_y(R_n + \Delta\gamma)} - (1 - D_n) + \frac{\left(\tilde{q}_{n+1}^{\text{trial}} - \sigma_y(R_n + \Delta\gamma)\right)}{3G}\left(\frac{-Y(\Delta\gamma)}{r}\right)^{s} = 0$$

for $\Delta\gamma$ with the Newton–Raphson method where

$$-Y(\Delta\gamma) = \frac{\left(\sigma_y(R_n + \Delta\gamma)\right)^2}{6G} + \frac{\left(\tilde{p}_{n+1}^{\text{trial}}\right)^2}{2K}$$

(iv) Update stress and internal variables

$$D_{n+1} = 1 - \left(\frac{3G\Delta\gamma}{\tilde{q}_{n+1}^{\text{trial}} - \sigma_y(R_{n+1})}\right); \qquad \boldsymbol{s}_{n+1} = \frac{q_{n+1}}{\tilde{q}_{n+1}^{\text{trial}}}\tilde{\boldsymbol{s}}_{n+1}^{\text{trial}};$$

$$p_{n+1} = (1 - D_{n+1})\,\tilde{p}_{n+1}^{\text{trial}}; \qquad \boldsymbol{\sigma}_{n+1} = \boldsymbol{s}_{n+1} + p_{n+1}\boldsymbol{I};$$

$$q_{n+1} = (1 - D_{n+1})\,\sigma_y(R_{n+1}); \qquad \boldsymbol{\varepsilon}_{n+1}^{\text{e}} = \frac{1}{2G}\boldsymbol{s}_{n+1} + \frac{1}{3}\varepsilon_{\text{v}\ n+1}^{\text{e trial}}\boldsymbol{I}$$

(v) EXIT

2.4.2 Damage Model with Crack Closure Effect

Even though the model presented in the preceding section of this chapter is able to predict damage growth with reasonable accuracy over simple strain paths, increasing deviations from experimental observations should be expected as strain paths become more complex. In fact, this is true not only for damage models but also for inelastic models of continua in general and, at present, it can be said that constitutive refinement in inelasticity remains largely an open issue.

One important feature of Lemaitre's original model is the fact that the state of stress triaxiality has a strong influence on the rate of damage growth. This experimentally observed phenomenon is accounted for through the definition of the damage energy release rate, $-Y$, rewritten here as

$$-Y = \frac{-q^2}{2E(1-D)^2}\left[\frac{2}{3}(1+\nu) + 3(1-2\nu)\left(\frac{p}{q}\right)^2\right] \tag{2.52}$$

which takes part in the damage evolution equation, $\dot{D}$, in Eq. (2.43b).

The inclusion of the hydrostatic component of $\boldsymbol{\sigma}$ in the definition of $-Y$ implies that $\dot{D}$ increases (decreases) with increasing (decreasing) triaxiality ratio. This is in sharp contrast with the standard von Mises plasticity model where only the stress deviator has an influence on the dissipative mechanisms. However, one important aspect of damage growth is not considered by evolution law (Eqs (2.43b) and (2.52)): the clear distinction between rates of damage growth observed for states of stress with identical triaxiality but stresses of opposite sign (tension and compression). This phenomenon can be crucially important in the simulation of forming operations, particularly under extreme strains [47]. It is often the case that, in such operations, the solid (or parts of it) undergoes extreme compressive straining followed by extension or vice versa.

2.4.2.1 Constitutive Model

In order to introduce the distinction between tension and compression, let us first consider a uniaxial stress state. When the stress normal to the crack is compressive, even though the cross section still contains physical cracks, the ability of the material to carry load increases. The crucial point in the definition of the crack closure model is the assumption that the standard elastic relationship $\sigma_1 = (1-D)E\varepsilon$ is valid only under tensile stresses ($\sigma_1 \geq 0$). Under compressive stresses ($\sigma_1 < 0$), the uniaxial stress–strain relation is assumed to take the form

$$\sigma_1 = (1-hD)E\varepsilon \tag{2.53}$$

or

$$\tilde{\sigma}_1 = \frac{\sigma_1}{1-hD} \tag{2.54}$$

where h is an experimentally determined coefficient which satisfies $0 \leq h \leq 1$.

This coefficient characterizes the closure of microcracks and microcavities and depends upon the density and the shape of the defects. It is material dependent and, as a first approximation for simplicity, h is considered as constant. The effect of damage itself on closure is neglected. A value $h \approx 0.2$ is typically observed in many experiments [26]. Note that for $h = 1$, crack closure effects are completely neglected (as in Lemaitre's original model), whereas the other extreme value, $h = 0$, represents full crack closure under compression. Any other value of h describes a partial crack closure effect.

Despite the relative easiness of establishing a piecewise linear-damaged elastic model capable of accounting for crack closure effects in the uniaxial case, the extension of such a simple model to the general three-dimensional situation is, however, not trivial.

In the present model, such a distinction is made on the basis of a tensile/compressive split of the stress tensor. That is, any stress tensor $\boldsymbol{\sigma}$ can be written as

$$\boldsymbol{\sigma} = \sum_{i=1}^{3} \sigma_i \mathbf{e}_i \otimes \mathbf{e}_i \tag{2.55}$$

where σ_i are the principal stresses and $\{\mathbf{e}_1, \mathbf{e}_2, \mathbf{e}_3\}$ is an orthonormal basis of vectors along the principal directions.

The tensile/compressive split of the stress tensor consists in splitting $\boldsymbol{\sigma}$ additively as

$$\boldsymbol{\sigma} = \boldsymbol{\sigma}_+ + \boldsymbol{\sigma}_- \tag{2.56}$$

where $\boldsymbol{\sigma}_+$ and $\boldsymbol{\sigma}_-$ are, respectively, the *tensile* and *compressive component* of $\boldsymbol{\sigma}$ defined as

$$\boldsymbol{\sigma}_+ = \sum_{i=1}^{3} \langle \sigma_i \rangle \mathbf{e}_i \otimes \mathbf{e}_i \tag{2.57}$$

and

$$\boldsymbol{\sigma}_- = \sum_{i=1}^{3} \langle -\sigma_i \rangle \mathbf{e}_i \otimes \mathbf{e}_i \tag{2.58}$$

The symbol $\langle\rangle$ represents the *Macauley bracket*, that is, for any scalar, a,

$$\langle a \rangle = \begin{cases} a & \text{if } a \geq 0 \\ 0 & \text{if } a < 0 \end{cases} \tag{2.59}$$

An interesting alternative to include the crack closure effect was provided by the damage evolution law proposed by Ladevèze *et al.* [52, 53] (described in Ref. [26] in more detail). Their approach consists in modifying the damage energy release rate, $-Y$, of the original Lemaitre ductile damage model by including the tensile/compressive split of the stress tensor. Thus, the original expression (2.52) is replaced by

$$\begin{aligned} -Y = {} & \frac{1}{2E(1-D)^2}\left[(1+\nu)\,\boldsymbol{\sigma}_+ : \boldsymbol{\sigma}_+ - \nu\,\langle \mathrm{tr}\boldsymbol{\sigma}\rangle^2\right] \\ & + \frac{h}{2E(1-hD)^2}\left[(1+\nu)\,\boldsymbol{\sigma}_- : \boldsymbol{\sigma}_- - \nu\,\langle -\mathrm{tr}\boldsymbol{\sigma}\rangle^2\right] \end{aligned} \tag{2.60}$$

and the elastoplastic damage evolution equation keeps the same format as in the original model, that is, as in Eq. (2.43b). The model proposed by Ladevèze and Lemaitre [53] considered the above equation together with the damaged isotropic elasticity law with crack closure effects and the standard plasticity equations of the original Lemaitre model.

However, in the present improved model, elasticity and damage are assumed to be decoupled. This can be justified given the fact that the elastic strain remains truly infinitesimal in many practical processes. Also, kinematic hardening is once again removed in order to obtain a simplified version of the model whose numerical implementation assumes a simpler and more efficient form. Thus, the main differences from Lemaitre's original damage model are described below.

First, the yield function is rephrased to be given by

$$F_{\mathrm{p}} = q - (1 - D)\left[\sigma_{\mathrm{yo}} + \chi(\varepsilon^{\mathrm{p}}_{\mathrm{ac}})\right] \tag{2.61}$$

Here, the material parameter σ_{yo} is the initial uniaxial yield stress of an undamaged and unstrained (or virgin) material, and χ is the isotropic hardening thermodynamical force, now assumed to be a function of the accumulated plastic strain, $\varepsilon^{\mathrm{p}}_{\mathrm{ac}}$. Furthermore, elasticity is assumed to be decoupled from damage; thus, the elastic law is simply given by

$$\boldsymbol{\sigma} = \mathbf{D}^{\mathrm{e}} : \boldsymbol{\varepsilon}^{\mathrm{e}} \tag{2.62}$$

In accordance with the above definitions and assuming associative plasticity, the plastic flow is also decoupled from damage:

$$\dot{\boldsymbol{\varepsilon}}^{\mathrm{p}} = \dot{\gamma}\frac{3}{2}\frac{\mathbf{s}}{q} \tag{2.63}$$

Finally, the evolution equations for the accumulated plastic strain and damage are respectively given by

$$\dot{\varepsilon}^{\mathrm{p}}_{\mathrm{ac}} = \dot{\gamma} \tag{2.64a}$$

$$\dot{D} = \frac{\dot{\gamma}}{1 - D}\left(\frac{-Y}{r}\right)^{s} \tag{2.64b}$$

for which the plastic multiplier, $\dot{\gamma}$, satisfies the Kuhn–Tucker conditions, that is, $\dot{\gamma} \geq 0$, $F_{\mathrm{p}} \leq 0$, $\dot{\gamma} F_{\mathrm{p}} = 0$. The constitutive model is conveniently summarized in Table 2.3.

It is important to remark here, that the choice of the accumulated plastic strain as the isotropic hardening internal variable is at variance with the original model proposed by Lemaitre [54]. In the present case, the original potential structure of the model is lost. However, this choice can be justified on experimental grounds since the measurement of accumulated plastic strain can be carried out experimentally in a straightforward manner allowing the determination of hardening and damage parameters from relatively simple microhardness measurements [55]. A direct consequence of this assumption is the fact that the hardening and damage parameters calibrated for this model will, in general, be different from the ones employed in Lemaitre's original model.

Table 2.3 Lemaitre's simplified damage model with crack closure effect.

Strain tensor additive split	$\boldsymbol{\varepsilon} = \boldsymbol{\varepsilon}^{\mathrm{e}} + \boldsymbol{\varepsilon}^{\mathrm{p}}$
Elastic law (uncoupled from damage)	$\boldsymbol{\sigma} = \mathbf{D}^{\mathrm{e}} : \boldsymbol{\varepsilon}^{\mathrm{e}}$
Yield criterion	$F_{\mathrm{p}} = q - (1 - D)\left[\sigma_{\mathrm{yo}} + \chi\left(\varepsilon_{\mathrm{ac}}^{\mathrm{p}}\right)\right]$
Plastic flow	$\dot{\boldsymbol{\varepsilon}}^{\mathrm{p}} = \dot{\gamma}\frac{3}{2}\frac{\mathbf{s}}{q}$
Evolution of equivalent plastic strain	$\dot{\varepsilon}_{\mathrm{ac}}^{\mathrm{p}} = \dot{\gamma}$
Evolution of damage	$\dot{D} = \frac{\dot{\gamma}}{1-D}\left(\frac{-Y}{r}\right)^{s}$ with $-Y$ given by $-Y = \dfrac{1}{2E(1-D)^2}\left[(1+v)\,\boldsymbol{\sigma}_+ : \boldsymbol{\sigma}_+ - v\,\langle \mathrm{tr}\boldsymbol{\sigma}\rangle^2\right] + \dfrac{h}{2E(1-hD)^2}\left[(1+v)\,\boldsymbol{\sigma}_- : \boldsymbol{\sigma}_- - v\,\langle -\mathrm{tr}\boldsymbol{\sigma}\rangle^2\right]$
Kuhn–Tucker conditions	$\dot{\gamma} \geq 0;\quad F_{\mathrm{p}} \leq 0;\quad \dot{\gamma} F_{\mathrm{p}} = 0$

2.4.2.2 Numerical Implementation

This section proceeds to describe an algorithm for the numerical integration of the elastoplastic damage constitutive equations, including the effect of crack closure. Here, we focus only on the particularization of the fully implicit elastic predictor/return mapping method to the above proposed model.

Let us consider what happens to a typical Gauss point of the finite element mesh within a (pseudo)time interval $[t_n, t_{n+1}]$. Given the incremental strain

$$\Delta\boldsymbol{\varepsilon} = \boldsymbol{\varepsilon}_{n+1} - \boldsymbol{\varepsilon}_n \tag{2.65}$$

and the values $\boldsymbol{\sigma}_n$, $\boldsymbol{\varepsilon}_n^{\mathrm{p}}$, $\varepsilon_{\mathrm{ac}\ n}^{\mathrm{p}}$, and D_n at t_n, the numerical integration algorithm should obtain the updated values at the end of the interval, $\boldsymbol{\sigma}_{n+1}$, $\boldsymbol{\varepsilon}_{n+1}^{\mathrm{p}}$, $\boldsymbol{\varepsilon}_{\mathrm{ac}\,n+1}^{\mathrm{p}}$, and D_{n+1} in a manner consistent with the constitutive equations of the model.

The first step in the algorithm is the evaluation of the elastic trial state where the increment is assumed purely elastic with no evolution of internal variables (internal variables frozen at t_n). The elastic trial strain and trial-accumulated plastic strain are given by

$$\boldsymbol{\varepsilon}_{n+1}^{\mathrm{e\ trial}} = \boldsymbol{\varepsilon}_n^{\mathrm{e}} + \Delta\boldsymbol{\varepsilon}; \qquad \varepsilon_{\mathrm{ac}\ n+1}^{\mathrm{p\ trial}} = \varepsilon_{\mathrm{ac}\ n}^{\mathrm{p}} \tag{2.66}$$

The corresponding elastic trial stress tensor is computed:

$$\boldsymbol{\sigma}_{n+1}^{\mathrm{trial}} = \mathbf{D}^{\mathrm{e}} : \boldsymbol{\varepsilon}_{n+1}^{\mathrm{e\ trial}} \tag{2.67}$$

Equivalently, in terms of stress deviator and hydrostatic pressure, we have

$$\mathbf{s}_{n+1}^{\mathrm{trial}} = 2G\boldsymbol{\varepsilon}_{\mathrm{d}\ n+1}^{\mathrm{e\ trial}}; \qquad p_{n+1}^{\mathrm{trial}} = K\varepsilon_{\mathrm{v}\ n+1}^{\mathrm{e\ trial}} \tag{2.68}$$

The trial yield stress is simply

$$\sigma_{y\ n+1}^{\mathrm{trial}} = \sigma_y\left(\chi_n\right) \tag{2.69}$$

The next step of the algorithm is to check whether $\boldsymbol{\sigma}_{n+1}^{\text{trial}}$ lies inside or outside of the trial yield surface. With variables $\varepsilon_{\text{ac}}^{\text{p}}$ and D frozen at time t_n we compute

$$\begin{aligned} F_{\text{p}}^{\text{trial}} &:= q_{n+1}^{\text{trial}} - (1 - D_n)\,\sigma_y(\chi_n) \\ &= \sqrt{\tfrac{3}{2}}\left\|\mathbf{s}_{n+1}^{\text{trial}}\right\| - (1 - D_n)\left[\sigma_{y\text{o}} + \chi(\varepsilon_{\text{ac}\ n+1}^{\text{p trial}})\right] \end{aligned} \tag{2.70}$$

If $F_{\text{p}}^{\text{trial}} \leq 0$, the process is indeed elastic within the interval and the elastic trial state coincides with the updated state at t_{n+1}. In other words, there is no plastic flow or damage evolution within the interval and

$$\begin{aligned} &\boldsymbol{\varepsilon}_{n+1}^{\text{e}} = \boldsymbol{\varepsilon}_{n+1}^{\text{e trial}}; \quad \boldsymbol{\sigma}_{n+1} = \boldsymbol{\sigma}_{n+1}^{\text{trial}}; \quad \varepsilon_{\text{ac}\ n+1}^{\text{p}} = \varepsilon_{\text{ac}\ n+1}^{\text{p trial}}; \\ &\sigma_{y\ n+1} = \sigma_{y\ n+1}^{\text{trial}}; \quad D_{n+1} = D_{n+1}^{\text{trial}} \end{aligned} \tag{2.71}$$

Otherwise, we apply the *plastic corrector* (or *plastic return mapping* algorithm), that is, we integrate numerically the evolution equations for $\boldsymbol{\varepsilon}^{\text{e}}$, $\varepsilon_{\text{ac}}^{\text{p}}$, and D having the trial state as the initial condition. Using a standard backward Euler approximation, the discrete equations read as follows:

$$\boldsymbol{\varepsilon}_{n+1}^{\text{e}} = \boldsymbol{\varepsilon}_{n+1}^{\text{e trial}} - \Delta\gamma\sqrt{\frac{3}{2}}\frac{\mathbf{s}_{n+1}}{\|\mathbf{s}_{n+1}\|}; \tag{2.72a}$$

$$\varepsilon_{\text{ac}\ n+1}^{\text{p}} = \varepsilon_{\text{ac}\ n}^{\text{p}} + \Delta\gamma; \tag{2.72b}$$

$$D_{n+1} = \begin{cases} 0 & \text{if}\varepsilon_{\text{ac}\ n+1}^{\text{p}} \leq \varepsilon_{\text{D}}^{\text{p}} \\ D_n + \frac{\Delta\gamma}{1-D_{n+1}}\left(\frac{-Y_{n+1}}{r}\right)^{s} & \text{if}\varepsilon_{\text{ac}\ n+1}^{\text{p}} > \varepsilon_{\text{D}}^{\text{p}} \end{cases} \tag{2.72c}$$

The above equations must be complemented by the so-called *consistency condition* that guarantees that the stress state at the end of a plastic step lies on the updated yield surface

$$F_{\text{p}n+1} = q_{n+1} - (1 - D_{n+1})\,\sigma_y(\chi_{n+1}) = 0 \tag{2.73}$$

From standard arguments used in the derivation of return mapping algorithms, the plastic corrector can be reduced to the solution of *single nonlinear scalar equation* for the incremental plastic multiplier $\Delta\gamma$ (the reader is referred to Andrade Pires *et al.* [47] for further details in the derivation):

$$D_{n+1} = D(\Delta\gamma) \equiv 1 - \frac{\sqrt{\frac{3}{2}}\left\|\mathbf{s}_{n+1}^{\text{trial}}\right\| - 3G\Delta\gamma}{\sigma_{y\text{o}} + \chi\left(\varepsilon_{\text{ac}\ n}^{\text{p}} + \Delta\gamma\right)} \tag{2.74}$$

which expresses D_{n+1} as an explicit function of $\Delta\gamma$. Finally, with the introduction of Eq. (2.74) into the discretized damage evolution equation (2.72c), the return mapping algorithm is reduced to the solution of a single equation for the unknown $\Delta\gamma$:

$$F(\Delta\gamma) \equiv \begin{cases} D(\Delta\gamma) = 0 & \text{if} \quad \varepsilon_{\text{ac}\ n+1}^{\text{p}} \leq \varepsilon_{\text{D}}^{\text{p}} \\ D(\Delta\gamma) - D_n - \frac{\Delta\gamma}{1-D(\Delta\gamma)}\left(\frac{-Y(\Delta\gamma)}{r}\right)^{s} = 0 & \text{if} \quad \varepsilon_{\text{ac}\ n+1}^{\text{p}} > \varepsilon_{\text{D}}^{\text{p}} \end{cases} \tag{2.75}$$

The single-equation plastic-damage corrector comprises the solution of the above equation for $\Delta\gamma$ followed by the straightforward update of $\boldsymbol{\varepsilon}^{\text{e}}$, $\varepsilon_{\text{ac}}^{\text{p}}$, and D according to the relevant formulae. For convenience, the resulting algorithm for numerical integration of the elastoplastic damage model with crack closure effects is listed in Box 2.2 in pseudocode format.

Box 2.2: Stress Update Algorithm for Lemaitre's Model with Crack Closure Effect

(i) Define elastic trial state

$$\boldsymbol{\varepsilon}^{\text{e trial}}_{n+1} = \boldsymbol{\varepsilon}^{\text{e}}_{n} + \Delta\boldsymbol{\varepsilon}\ ; \qquad \mathbf{s}^{\text{trial}}_{n+1} = 2G\boldsymbol{\varepsilon}^{\text{e trial}}_{\text{d}\ n+1}\ ;$$

$$\varepsilon^{\text{p trial}}_{\text{ac}\ n+1} = \varepsilon^{\text{p}}_{\text{ac}\ n}\ ; \qquad q^{\text{trial}}_{n+1} = \sqrt{\tfrac{3}{2}}\ \|\mathbf{s}^{\text{trial}}_{n+1}\|$$

$$p^{\text{trial}}_{n+1} = K\varepsilon^{\text{e trial}}_{\text{v}\ n+1}\ ;$$

(ii) Check plastic admissibility

$$\text{IF } q^{\text{trial}}_{n+1} - (1 - D_n)\left\lfloor \sigma_{\text{yo}} + \chi(\varepsilon^{\text{p trial}}_{\text{ac}\ n+1}) \right\rfloor \leq 0 \text{ THEN}$$

$$\text{SET } (\cdot)_{n+1} = (\cdot)^{\text{trial}}_{n+1} \text{ and EXIT}$$

ENDIF

(iii) Solve the nonlinear residual equation

$$F(\Delta\gamma) \equiv \begin{cases} D(\Delta\gamma) = 0 & \text{if } \varepsilon^{\text{p}}_{\text{ac}\ n+1} \leq \varepsilon^{\text{p}}_{\text{D}} \\ D(\Delta\gamma) - D_n - \frac{\Delta\gamma}{1 - D(\Delta\gamma)}\left(\frac{-Y(\Delta\gamma)}{r}\right)^{s} = 0 & \text{if } \varepsilon^{\text{p}}_{\text{ac}\ n+1} > \varepsilon^{\text{p}}_{\text{D}} \end{cases}$$

for $\Delta\gamma$ with the Newton–Raphson method.

(iv) Update stress and internal variables

$$p_{n+1} = p^{\text{trial}}_{n+1}\ ; \qquad \mathbf{s}_{n+1} = \left(1 - \frac{3G\Delta\gamma}{q^{\text{trial}}_{n+1}}\right)\mathbf{s}^{\text{trial}}_{n+1}\ ;$$

$$\boldsymbol{\sigma}_{n+1} = \mathbf{s}_{n+1} + p_{n+1}\mathbf{I}\ ; \qquad \varepsilon^{\text{p}}_{\text{ac}\ n+1} = \varepsilon^{\text{p}}_{\text{ac}\ n} + \Delta\gamma\ ;$$

$$D_{n+1} = D(\Delta\gamma)\ ; \qquad \boldsymbol{\varepsilon}^{\text{e}}_{n+1} = \frac{1}{2G}\mathbf{s}_{n+1} + \frac{1}{3K}\,p_{n+1}\boldsymbol{I}$$

(v) EXIT

In the computational implementation of the model, Eq. (2.75) is solved by the Newton–Raphson algorithm. Therefore, it is necessary to ensure that, throughout the Newton iterations, we remain within the domain where Eq. (2.75) is physically sound, that is, where the damage variable value remains between 0 and 1.

2.5 Nonlocal Formulations

The constitutive models described in the preceding sections have already proved very effective in modeling material internal degradation and predicting failure within a reasonable range of applications. They are, however, based on the standard (local) continuum theory, which inherently assumes that the material is homogeneous and continuous at any size scale.

However, this is not valid when deformation reaches a critical level. At this point, the internal degradation of the material, mainly characterized by the nucleation, growth, and coalescence of microcavities, has an important influence on the macrostructural response to external loads. Moreover, plastic strain tends to concentrate in a localized zone while the body experiences a softening regime. At this phase, heterogeneities in the microstructure play a crucial role, being responsible for the onset of the failure phenomenon that will lead, eventually, to the appearance of a macrocrack.

Since classical local theories disregard important effects of the microstructure of the material, they cannot correctly describe the aforementioned localized failure process. Likewise, the mathematical description of the failure phenomenon using the local theory is inappropriate since it inherently suffers from spurious instabilities. This can be explained as follows. In classical multivariable calculus, the partial differential equilibrium equations that govern static (dynamic) problems are classified as elliptic (hyperbolic). As long as these equations remain elliptic (hyperbolic), the solution of the static (dynamic) IBVP is guaranteed to be unique. However, when the material tangent modulus becomes negative (i.e., under softening regimes), the ellipticity (hyperbolicity) is lost and uniqueness of solution no longer exists. This lack of ellipticity (hyperbolicity) manifests itself through a pathological dependence of solution on the spatial discretization when using numerical methods. Within a typical finite element framework, for instance, the localized zone will have the size of the elements at the critical zone. As the mesh is infinitely refined, plastic strain concentrates in an infinitely small layer of elements and, in this case, the total dissipated energy of the process unrealistically vanishes. In fact, it can be concluded that the problem of the local theory lies on its lack of information about the size of the localized zone. Thus, the mathematical interpretation of the problem implies that this missing information should be, in some manner, incorporated into the continuum theory in order to obtain objective descriptions of the localized failure process. Looking at the problem from the physical point of view, it becomes quite clear that the actual size of the localized zone is related to the heterogeneous microstructure of the material. Therefore, both mathematical and physical interpretations imply that the standard continuum theory must be enriched in order to describe the strain localization correctly.

Among several possible strategies to obtain enriched continua, an attractive theory is the so-called *nonlocal approach*. This incorporates an intrinsic length into the classical continuum by employing spatially weighted averages through an integral operator. As pointed out at the beginning of this chapter, the first nonlocal models were proposed in the elasticity context in the 1960s [11, 12]. Such models aimed to improve the description of microstructural interactions in elastic-wave-dominated problems. The first extension of the theory in the plastic domain has been done by Eringen [13, 14]. In his formulation, the total strain tensor was replaced by its nonlocal average; however, the model was not intended to act as localization limiter.

In the context of CDM, the first nonlocal damage model was proposed by Pijaudier-Cabot and Bažant [16] where they applied a nonlocal averaging operator only to variables that were related to the inelastic process and that could only grow or remain constant. This choice stemmed from the fact that the average of the total strain tensor as in Eringen's model could still lead to spurious instabilities in the IBVP, as previously shown by Bažant and Chang [15]. In the late 1980s, the first nonlocal plasticity model intended to serve as a localization limiter was proposed by Bažant and Lin [56]. After these initial developments, a comprehensive number of relevant contributions have then rapidly emerged (e.g., [19–23, 57–62]).

Another class of nonlocal approaches is the enhanced constitutive theories that incorporate the gradients of one or more variables into originally local models. The *gradient-enhanced models* are the differential counterparts of the nonlocal integral-type theory and have been developed as an alternative to the classical nonlocal theory. Over the last years, several researchers have brought significant developments comprising gradient-enhanced constitutive theories, many of them with detailed contributions on effective and efficient numerical implementations [17, 18, 24, 25, 63, 64]. However, this particular kind of nonlocal theory has not been addressed in this chapter.

2.5.1
Aspects of Nonlocal Averaging

The cornerstone of any nonlocal theory is the definition of a nonlocal variable through a weighted integral average. Basically, any generic scalar field $g(\mathbf{x})$ can be rewritten in its nonlocal form as

$$\bar{g}(\mathbf{x}) = \int_{\mathrm{V}} \beta(\mathbf{x}, \boldsymbol{\xi}) g(\boldsymbol{\xi}) \mathrm{d}V(\boldsymbol{\xi}) \tag{2.76}$$

in which the superimposed bar denotes a nonlocal variable. Equation (2.76) spatially averages the local variable using the weighted averaging operator $\beta(\mathbf{x}, \boldsymbol{\xi})$. The nonlocal average can be interpreted as the incorporation of a diffusive effect into the constitutive model.

2.5.1.1 The Averaging Operator

Any nonlocal enhancement must consider some basic requirements in order to provide physically coherent formulations. Among these, the following normalizing condition

$$\int_{\mathrm{V}} \beta(\mathbf{x}, \boldsymbol{\xi}) \mathrm{d}V(\boldsymbol{\xi}) = 1 \tag{2.77}$$

must hold in order to keep uniform fields. Usually, $\beta(\mathbf{x}, \boldsymbol{\xi})$ is rescaled as in

$$\beta(\mathbf{x}, \boldsymbol{\xi}) = \frac{\alpha(\mathbf{x}, \boldsymbol{\xi})}{\int_{\mathrm{V}} \alpha(\mathbf{x}, \boldsymbol{\xi}) \mathrm{d}V(\boldsymbol{\xi})} \tag{2.78}$$

and $\alpha(\mathbf{x}, \boldsymbol{\xi})$ is a prescribed *weight function*. The disadvantage of this strategy is that the resulting averaging operator is not symmetric (i.e., $\beta(\mathbf{x}, \boldsymbol{\xi}) \neq \beta(\boldsymbol{\xi}, \mathbf{x})$) even for symmetric weight functions $\alpha(\mathbf{x}, \boldsymbol{\xi})$.

2.5.1.2 **Weight Functions**

The definition of the weight function is, to some extent, arbitrary. However, in order to obtain the diffusive effect expected from the nonlocal theory, it should satisfy some basic characteristics. For an instance, the function should have its maximum at the central point and then smoothly decrease around it as the distance of neighboring points increases. Moreover, it must contain at least one length parameter associated with the intrinsic length of the material.

A commonly adopted weight function is the Gaussian distribution, written as

$$\alpha(\mathbf{x}, \boldsymbol{\xi}) = \exp\left(-\frac{\|\mathbf{x}-\boldsymbol{\xi}\|^2}{2\ell_o^2}\right) \tag{2.79}$$

where the parameter ℓ_o is called the *internal length*. A main feature of this function is that it is unbounded, that is, nonlocal interactions can theoretically occur even for infinitely distant points. In practice, the function can be truncated at reasonable finite distances.

Another possibility often adopted is the bell-shaped truncated quartic polynomial function, given by

$$\alpha(\mathbf{x}, \boldsymbol{\xi}) = \begin{cases} \left(1 - \frac{\|\mathbf{x}-\boldsymbol{\xi}\|^2}{\ell_r^2}\right)^2 & \text{if } \|\mathbf{x}-\boldsymbol{\xi}\| \le \ell_r \\ 0 & \text{if } \|\mathbf{x}-\boldsymbol{\xi}\| \ge \ell_r \end{cases} \tag{2.80}$$

where ℓ_r will be referred hereinafter to as *nonlocal characteristic length*. Contrasting with the Gaussian distribution, the bell-shaped function has a bounded support and vanishes for points for which the distance is larger than ℓ_r. Thus, ℓ_r composes a bounded influence area (or volume in 3D) and therefore is also often called *interaction radius* (usually denoted by the letter R).

2.5.2
Classical Nonlocal Models of Integral Type

Classical nonlocal models are the most commonly adopted type of nonlocal approach. The main characteristic of these models is that they are "ad hoc" formulated, that is, this kind of strategy consists in enhancing a previously existing local model by simply replacing one or more constitutive variables by their nonlocal counterparts. Therefore, they are neither based on thermodynamic considerations nor derived from thermodynamic potentials. In fact, these models are merely supported by the physical (and mathematical) evidence that an intrinsic length should be, in some manner, incorporated into the standard (local) continuum theory. However, the nomenclature "classical nonlocal models" is not mentioned in the literature, being probably first suggested here in this publication.

Another class of nonlocal models, fully based on thermodynamic potentials, has recently emerged [20, 21, 23]. The thermodynamical consistency has indeed made such models very appealing from the theoretical point of view and is, in fact, the main motivation for classifying the nonlocal models of the current section as "classical." However, nonlocal models derived from thermodynamic potentials are not addressed here.

In the following subsection, the procedures to extend a local model into a classical nonlocal one are described. For the sake of simplicity, the derivations will be applied to the simplified version of Lemaitre's model only. Nonetheless, the nonlocal extension of other elastoplastic damage models follows the same concepts.

2.5.2.1 **Nonlocal Formulations for Lemaitre's Simplified Model**

The first step of the nonlocal extension is the choice of the nonlocal variable. At first glance, this choice seems to be quite arbitrary. However, some few previous contributions have already shown that some options may lead to inappropriate or spurious formulations under certain conditions [15, 22, 61].

Therefore, in this chapter, Lemaitre's simplified model is enriched by employing three different classical nonlocal formulations. The chosen nonlocal variables are the isotropic hardening variable, R, the damage, D, and the energy release rate, $-Y$. One must bear in mind that each nonlocal formulation leads to a new constitutive model, different from the local one, with the singular particularity that the local model is retrieved if the length parameter is set to zero. Thus, when a variable is chosen to be nonlocal, its nonlocal counterpart is, in fact, the constitutive variable to be actually considered.

Let us first consider the choice of damage as the nonlocal variable. The elastic-damage constitutive relation can then be expressed as

$$\boldsymbol{\sigma}(\mathbf{x}) = \left(1 - \overline{D}(\mathbf{x})\right) \mathbf{D}^{\mathrm{e}}(\mathbf{x}) : \boldsymbol{\varepsilon}^{\mathrm{e}}(\mathbf{x}) \tag{2.81}$$

The yield function, the plastic flow rule, and the evolution of isotropic hardening are also rewritten, respectively, given by

$$F_{\mathrm{p}}(\mathbf{x}) = \frac{q(\mathbf{x})}{1 - \overline{D}(\mathbf{x})} - \sigma_y(R(\mathbf{x})) \tag{2.82}$$

$$\dot{\boldsymbol{\varepsilon}}^{\mathrm{p}}(\mathbf{x}) = \frac{3}{2} \frac{\dot{\gamma}(\mathbf{x})}{\left(1 - \overline{D}(\mathbf{x})\right)} \frac{\mathbf{s}(\mathbf{x})}{q(\mathbf{x})} \tag{2.83}$$

and

$$\dot{R}(\mathbf{x}) = \dot{\gamma}(\mathbf{x}) \tag{2.84}$$

In contrast to the model of Section 2.4.1, all constitutive variables have been explicitly written as a function of their spatial location, generically represented by $\mathbf{x}$ (Table 2.4). This stems from the fact that the nonlocal formulation yields on an enriched continuum theory for which the position of the material point in the body is essential for the determination of its constitutive behavior. Moreover, since nonlocal damage, $\overline{D}(\mathbf{x})$, is now the actual state variable, its rate should play the role of damage evolution. Therefore, it is assumed that

$$\dot{\overline{D}}(\mathbf{x}) = \int_{\mathrm{V}} \beta(\mathbf{x}, \boldsymbol{\xi}) \dot{D}(\boldsymbol{\xi}) \mathrm{d}V(\boldsymbol{\xi}) \tag{2.85}$$

where

$$\dot{D}(\boldsymbol{\xi}) = \frac{\dot{\gamma}(\boldsymbol{\xi})}{1 - \overline{D}(\boldsymbol{\xi})} \left(\frac{-Y(\boldsymbol{\xi})}{r}\right)^{\mathrm{s}} \tag{2.86}$$

Table 2.4 Classical nonlocal damage model with $\overline{D}$.

Strain tensor additive split	$\boldsymbol{\varepsilon}(\mathbf{x}) = \boldsymbol{\varepsilon}^{e}(\mathbf{x}) + \boldsymbol{\varepsilon}^{p}(\mathbf{x})$
Elastic law	$\boldsymbol{\sigma}(\mathbf{x}) = \left(1 - \overline{D}(\mathbf{x})\right) \mathbf{D}^{e}(\mathbf{x}) : \boldsymbol{\varepsilon}^{e}(\mathbf{x})$
Yield criterion	$F_{p}(\mathbf{x}) = \frac{q(\mathbf{x})}{1-\overline{D}(\mathbf{x})} - \sigma_{y}(R(\mathbf{x}))$
Plastic flow	$\dot{\boldsymbol{\varepsilon}}^{p}(\mathbf{x}) = \frac{3}{2}\frac{\dot{\gamma}(\mathbf{x})}{(1-\overline{D}(\mathbf{x}))}\frac{\mathbf{s}(\mathbf{x})}{q(\mathbf{x})}$
Evolution of isotropic hardening	$\dot{R}(\mathbf{x}) = \dot{\gamma}(\mathbf{x})$
Evolution of damage	$\dot{\overline{D}}(\mathbf{x}) = \int_{V} \beta(\mathbf{x}, \boldsymbol{\xi}) \dot{D}(\boldsymbol{\xi}) \mathrm{d}V(\boldsymbol{\xi})$ where $\dot{D}(\boldsymbol{\xi}) = \frac{\dot{\gamma}(\boldsymbol{\xi})}{1-\overline{D}(\boldsymbol{\xi})}\left(\frac{-Y(\boldsymbol{\xi})}{r}\right)^{s}$
Kuhn–Tucker conditions	$\dot{\gamma}(\mathbf{x}) \geq 0; \quad F_{p}(\mathbf{x}) \leq 0; \quad \dot{\gamma}(\mathbf{x})F_{p}(\mathbf{x}) = 0$

Table 2.5 Classical nonlocal damage model with $-\overline{Y}$.

Strain tensor additive split	$\boldsymbol{\varepsilon}(\mathbf{x}) = \boldsymbol{\varepsilon}^{e}(\mathbf{x}) + \boldsymbol{\varepsilon}^{p}(\mathbf{x})$
Elastic law	$\boldsymbol{\sigma}(\mathbf{x}) = (1 - D(\mathbf{x}))\mathbf{D}^{e}(\mathbf{x}) : \boldsymbol{\varepsilon}^{e}(\mathbf{x})$
Yield criterion	$F_{p}(\mathbf{x}) = \frac{q(\mathbf{x})}{1-D(\mathbf{x})} - \sigma_{y}(R(\mathbf{x}))$
Plastic flow	$\dot{\boldsymbol{\varepsilon}}^{p}(\mathbf{x}) = \frac{3}{2}\frac{\dot{\gamma}(\mathbf{x})}{(1-D(\mathbf{x}))}\frac{\mathbf{s}(\mathbf{x})}{q(\mathbf{x})}$
Evolution of isotropic hardening	$\dot{R}(\mathbf{x}) = \dot{\gamma}(\mathbf{x})$
Evolution of damage	$\dot{D}(\mathbf{x}) = \frac{\dot{\gamma}(\mathbf{x})}{1-D(\mathbf{x})}\left(\frac{-\overline{Y}(\mathbf{x})}{r}\right)^{s}$ where $-\overline{Y}(\mathbf{x}) = \int_{V} \beta(\mathbf{x}, \boldsymbol{\xi}) \left(-Y(\boldsymbol{\xi})\right) \mathrm{d}V(\boldsymbol{\xi})$
Kuhn–Tucker conditions	$\dot{\gamma}(\mathbf{x}) \geq 0; \quad F_{p}(\mathbf{x}) \leq 0; \quad \dot{\gamma}(\mathbf{x})F_{p}(\mathbf{x}) = 0$

where ξ corresponds to the global coordinate of a given surrounding point in the vicinity of $\mathbf{x}$.

Finally, the Kuhn–Tucker conditions ($\dot{\gamma}(\mathbf{x}) \geq 0$, $F_{p}(\mathbf{x}) \leq 0$, and $\dot{\gamma}(\mathbf{x})F_{p}(\mathbf{x}) = 0$) must be fulfilled. It is important to remark that, since every material point depends on its neighborhood, the Kuhn–Tucker conditions must, in practice, simultaneously hold for all points of the whole body. As a matter of fact, the integral character of the nonlocal model yields on a substantially more complicated constitutive problem for which analytical solutions are difficult to obtain. Nonetheless, a strategy to numerically solve the nonlocal material problem is addressed in Section 2.5.3. The choices of $-Y$ and R as nonlocal also lead to other different constitutive models that are directly summarized in Tables 2.5 and 2.6 for convenience.

2.5.3
Numerical Implementation of Nonlocal Integral Models

Notwithstanding the relative simplicity in defining a nonlocal elastoplastic damage model, its incorporation into a nonlinear finite element framework is, in general,

Table 2.6 Classical nonlocal damage model with $\overline{R}$.

Strain tensor additive split	$\boldsymbol{\varepsilon}(\mathbf{x}) = \boldsymbol{\varepsilon}^{e}(\mathbf{x}) + \boldsymbol{\varepsilon}^{p}(\mathbf{x})$
Elastic law	$\boldsymbol{\sigma}(\mathbf{x}) = (1 - D(\mathbf{x}))\,\mathbf{D}^{e}(\mathbf{x}) : \boldsymbol{\varepsilon}^{e}(\mathbf{x})$
Yield criterion	$F_{p}(\mathbf{x}) = \frac{q(\mathbf{x})}{1-D(\mathbf{x})} - \sigma_{y}(\overline{R}(\mathbf{x}))$
Plastic flow	$\dot{\boldsymbol{\varepsilon}}^{p}(\mathbf{x}) = \frac{3}{2}\frac{\dot{\gamma}(\mathbf{x})}{(1-D(\mathbf{x}))}\frac{\mathbf{s}(\mathbf{x})}{q(\mathbf{x})}$
Evolution of isotropic hardening	$\dot{\overline{R}}(\mathbf{x}) = \int_{V}\beta(\mathbf{x},\boldsymbol{\xi})\dot{R}(\boldsymbol{\xi})\mathrm{d}V(\boldsymbol{\xi})$
	where $\dot{R}(\boldsymbol{\xi}) = \dot{\gamma}(\boldsymbol{\xi})$
Evolution of damage	$\dot{D}(\mathbf{x}) = \frac{\dot{\gamma}(\mathbf{x})}{1-D(\mathbf{x})}\left(\frac{-Y(\mathbf{x})}{r}\right)^{s}$
Kuhn–Tucker conditions	$\dot{\gamma}(\mathbf{x}) \geq 0;\quad F_{p}(\mathbf{x}) \leq 0;\quad \dot{\gamma}(\mathbf{x})F_{p}(\mathbf{x}) = 0$

not straightforward. The main difficulty is the need to fulfill the consistency condition.

One could think of resolving all material points using a conventional local algorithm and, afterward, performing a nonlocal averaging on the chosen nonlocal variable. However, the new computed stress state may not satisfy the consistency condition, invalidating the solution found. A new constitutive integration procedure should therefore be performed in order to return, if necessary, one or more stress states to the yield surface. Moreover, the nonlocal counterpart of the chosen constitutive variable should be recomputed again. Such procedure should be performed iteratively, using some numerical strategy, until the consistency condition is fulfilled for all material points of the body.

In the present section, a numerical strategy is presented in order to overcome this adversity. As shown in detail, the methodology is based on an extension of the conventional local elastic predictor/return mapping algorithm [49] into a global framework where all material points are integrated simultaneously. The strategy allows the fulfillment of the consistency condition for the whole body.

2.5.3.1 Numerical Evaluation of the Averaging Integral

Before proceeding with the numerical implementation of the nonlocal model, let us first rewrite the averaging integral from Eq. (2.76) in a discrete form. Following the standard procedures employed within the framework of the finite element method (FEM), the evaluation of the averaging integral can be conveniently approximated by the well-established Gaussian quadrature [62]. Therefore,

$$\overline{g}(\mathbf{x}) = \int_{V}\beta(\mathbf{x},\boldsymbol{\xi})g(\boldsymbol{\xi})\mathrm{d}V(\boldsymbol{\xi}) \tag{2.87}$$

is replaced by

$$\overline{g}_{i} = \sum_{j=1}^{ngp_{i}} w_{j}J_{j}\beta_{ij}g_{j} \tag{2.88}$$

where the parameter β_{ij} is the averaging factor, defined in Eq. (2.78), relating the points i and j which represent the Gauss points at the global coordinates $\mathbf{x}$ and $\boldsymbol{\xi}$, respectively. The quantities w_j and J_j are, respectively, the Gaussian weight and the Jacobian at the Gauss point j.

2.5.3.2 Global Version of the Elastic Predictor/Return Mapping Algorithm

As stressed before, the main obstacle in the numerical implementation of nonlocal models is the fulfillment of the consistency condition. This difficulty stems from the fact that the local solution for the constitutive problem is altered by subsequent nonlocal spatial averaging, and hence does not satisfy the Kuhn–Tucker conditions. Thus, the first inspection of the problem suggests that the solution of the material problem should be somehow pursued in a global manner. Moreover, the (pseudo)time discretization of the nonlocal problem using the standard procedures of an elastic predictor/return mapping scheme utterly reveals the global character of the nonlocal constitutive problem.

In the following, we shed some light on the numerical integration of the enhanced constitutive model aiming to clearly illustrate such global characteristics. In order to facilitate comprehension, only the nonlocal formulation for Lemaitre's simplified model with the nonlocal damage variable, $\overline{D}$, is considered. Nevertheless, the concepts described below hold for other nonlocal elastoplastic damage models as well.

To start with, let us consider a set of Gauss points in a generic finite element mesh, as shown in Figure 2.2. As schematically depicted, Gauss point 1 is dependent on a certain set of surrounding points within an interaction radius (representing the intrinsic length). Following now the same procedures used in Section 2.4.2 for the local model, the nonlocal constitutive problem for Gauss point 1 is expressed, in its time-discretized version, as

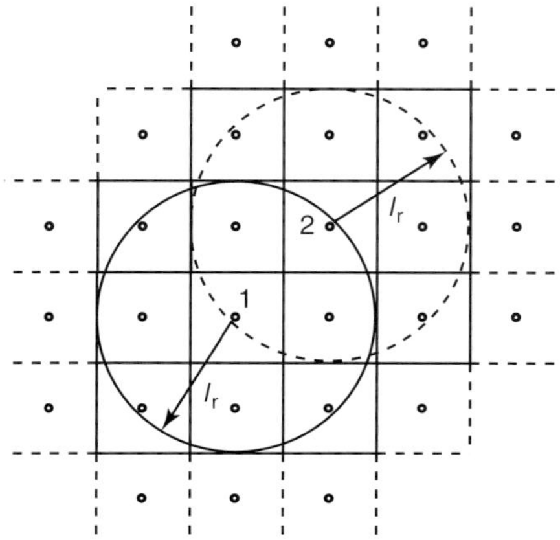

Figure 2.2 Nonlocal influence area in a finite element mesh.

$$\begin{cases} \boldsymbol{\varepsilon}^{\mathrm{e}}_{1_{n+1}} = \boldsymbol{\varepsilon}^{\mathrm{e\ trial}}_{1_{n+1}} - \frac{3}{2}\frac{\Delta\gamma_1}{\left(1-\overline{D}_{1_{n+1}}\right)}\frac{\mathbf{s}_{1_{n+1}}}{q_{1_{n+1}}} & \text{(2.89a)} \\ R_{1_{n+1}} = R_{1_n} + \Delta\gamma_1 & \text{(2.89b)} \\ \overline{D}_{1_{n+1}} = \overline{D}_{1_n} + \sum_{j=1}^{npg_1}\left[w_j J_j \beta_{1j}\frac{\Delta\gamma_j}{1-\overline{D}_{j_{n+1}}}\left(\frac{-Y_{j_{n+1}}}{r}\right)^{s}\right] & \text{(2.89c)} \\ \frac{q_{1_{n+1}}}{1-\overline{D}_{1_{n+1}}} - \sigma_y\left(R_{1_{n+1}}\right) = 0 & \text{(2.89d)} \end{cases}$$

In contrast with the local case, every constitutive variable needs to be written with an additional subscript corresponding to the Gauss point for which it stands for. This is necessary since some quantities in the system belong to other Gauss points of the mesh.

Likewise, the discretized constitutive problem for Gauss point 2 is written as

$$\begin{cases} \boldsymbol{\varepsilon}^{\mathrm{e}}_{2_{n+1}} = \boldsymbol{\varepsilon}^{\mathrm{e\ trial}}_{2_{n+1}} - \frac{3}{2}\frac{\Delta\gamma_2}{\left(1-\overline{D}_{2_{n+1}}\right)}\frac{\mathbf{s}_{2_{n+1}}}{q_{2_{n+1}}} & \text{(2.90a)} \\ R_{2_{n+1}} = R_{2_n} + \Delta\gamma_2 & \text{(2.90b)} \\ \overline{D}_{2_{n+1}} = \overline{D}_{2_n} + \sum_{j=1}^{npg_2}\left[w_j J_j \beta_{2j}\frac{\Delta\gamma_j}{1-\overline{D}_{j_{n+1}}}\left(\frac{-Y_{j_{n+1}}}{r}\right)^{s}\right] & \text{(2.90c)} \\ \frac{q_{2_{n+1}}}{1-\overline{D}_{2_{n+1}}} - \sigma_y\left(R_{2_{n+1}}\right) = 0 & \text{(2.90d)} \end{cases}$$

We can expand some terms of Eqs (2.89c) and (2.90c) to obtain

$$\overline{D}_{1_{n+1}} = \overline{D}_{1_n} + w_1 J_1 \beta_{11}\frac{\Delta\gamma_1}{1-\overline{D}_{1_{n+1}}}\left(\frac{-Y_{1_{n+1}}}{r}\right)^{s} + w_2 J_2 \beta_{12}\frac{\Delta\gamma_2}{1-\overline{D}_{2_{n+1}}}\left(\frac{-Y_{2_{n+1}}}{r}\right)^{s} + \ldots \tag{2.91}$$

and

$$\overline{D}_{2_{n+1}} = \overline{D}_{2_n} + w_1 J_1 \beta_{21}\frac{\Delta\gamma_1}{1-\overline{D}_{1_{n+1}}}\left(\frac{-Y_{1_{n+1}}}{r}\right)^{s} + w_2 J_2 \beta_{22}\frac{\Delta\gamma_2}{1-\overline{D}_{2_{n+1}}}\left(\frac{-Y_{2_{n+1}}}{r}\right)^{s} + \ldots \tag{2.92}$$

Clearly, $\overline{D}_{1_{n+1}}$ depends on the updated values of $\Delta\gamma$, $\overline{D}$, and $-Y$ from all Gauss points that lie inside the circular area defined by ℓ_r from point 1, also including point 2. Moreover, if the nonlocal influence area is drawn from point 2, different Gauss points now affect the pointwise constitutive problem; among them, point 1. However, the contributions of point 1 and point 2 in Eqs (2.89) and (2.90) are distinct due to the different weighted averaging factors β_{11}, β_{12}, β_{21}, and β_{22}.

In fact, the constitutive problem of every Gauss point is coupled with many others of the finite element mesh, each one providing different contributions due to the weighted averaging factors. Thus, the actual problem to be solved is composed of a large system of equations containing several "pointwise systems." This global system could be written as

$$
\left\{
\begin{array}{l}
\left\{
\begin{array}{l}
\boldsymbol{\varepsilon}^{\mathrm{e}}_{1_{n+1}} = \boldsymbol{\varepsilon}^{\mathrm{e\ trial}}_{1_{n+1}} - \frac{3}{2}\frac{\Delta\gamma_1}{\left(1-\overline{D}_{1_{n+1}}\right)}\frac{\mathrm{s}_{1_{n+1}}}{q_{1_{n+1}}} \\
R_{1_{n+1}} = R_{1_n} + \Delta\gamma_1 \\
\overline{D}_{1_{n+1}} = \overline{D}_{1_n} + \sum\limits_{j=1}^{npg_1}\left[w_j J_j \beta_{1j}\frac{\Delta\gamma_j}{1-\overline{D}_{j_{n+1}}}\left(\frac{-Y_{j_{n+1}}}{r}\right)^{s}\right] \\
\frac{q_{1_{n+1}}}{1-\overline{D}_{1_{n+1}}} - \sigma_y\left(R_{1_{n+1}}\right) = 0
\end{array}
\right. \\
\vdots \\
\left\{
\begin{array}{l}
\boldsymbol{\varepsilon}^{\mathrm{e}}_{ngp_{n+1}} = \boldsymbol{\varepsilon}^{\mathrm{e\ trial}}_{ngp_{n+1}} - \frac{3}{2}\frac{\Delta\gamma_{ngp}}{\left(1-\overline{D}_{ngp_{n+1}}\right)}\frac{\mathrm{s}_{ngp_{n+1}}}{q_{ngp_{n+1}}} \\
R_{ngp_{n+1}} = R_{ngp_n} + \Delta\gamma_{ngp} \\
\overline{D}_{ngp_{n+1}} = \overline{D}_{ngp_n} + \sum\limits_{j=1}^{npg_{ngp}}\left[w_j J_j \beta_{ngpj}\frac{\Delta\gamma_j}{1-\overline{D}_{j_{n+1}}}\left(\frac{-Y_{j_{n+1}}}{r}\right)^{s}\right] \\
\frac{q_{ngp_{n+1}}}{1-\overline{D}_{ngp_{n+1}}} - \sigma_y\left(R_{ngp_{n+1}}\right) = 0
\end{array}
\right.
\end{array}
\right.
\tag{2.93}
$$

where ngp is the total number of Gauss points of the body.

Of course, the constitutive problem for every Gauss point in its time-discretized version can, without loss of generality, more shortly (and conveniently) be written as

$$
\left\{
\begin{array}{l}
\boldsymbol{\varepsilon}^{\mathrm{e}}_{i_{n+1}} = \boldsymbol{\varepsilon}^{\mathrm{e\ trial}}_{i_{n+1}} - \frac{3}{2}\frac{\Delta\gamma_i}{\left(1-\overline{D}_{i_{n+1}}\right)}\frac{\mathrm{s}_{i_{n+1}}}{q_{i_{n+1}}} \\
R_{i_{n+1}} = R_{i_n} + \Delta\gamma_i \\
\overline{D}_{i_{n+1}} = \overline{D}_{i_n} + \sum\limits_{j=1}^{npg_i}\left[w_j J_j \beta_{ij}\frac{\Delta\gamma_j}{1-\overline{D}_{j_{n+1}}}\left(\frac{-Y_{j_{n+1}}}{r}\right)^{s}\right] \\
\frac{q_{i_{n+1}}}{1-\overline{D}_{i_{n+1}}} - \sigma_y\left(R_{i_{n+1}}\right) = 0
\end{array}
\right.
\tag{2.94}
$$

where $\boldsymbol{\varepsilon}^{\mathrm{e}}_{i_{n+1}}$, $R_{i_{n+1}}$, $\Delta\gamma_i$, and $D_{i_{n+1}}$ are the unknowns *associated with the generic* ith *Gauss point* of the body. One should bear in mind, however, that the actual unknowns of the nonlocal problem *are not only those associated with the* ith *Gauss point* but also the updated values of $\boldsymbol{\varepsilon}^{\mathrm{e}}$, R, $\Delta\gamma$, and D *corresponding to several other points of the mesh.* This coupling among several pointwise incremental problems makes clear the aforementioned global character of the nonlocal material problem for which the solution must be, somehow, globally found.

Before proceeding with the description of the solution to the global constitutive problem, some algebraic manipulations may be performed in the constitutive equations in order to facilitate the numerical implementation. In Section 2.4.2, it has been mentioned that the simplified version of Lemaitre's model has the particularity of providing a remarkably efficient integration algorithm when the system of equations is reduced to a single scalar nonlinear equation. In a similar manner, it is possible to reduce from *one system of equations* per Gauss point to *one equation* per Gauss point when adopting the nonlocal version of Lemaitre's simplified model. The procedure very closely follows the steps necessary for the system reduction in the local model and is omitted here for convenience. The reader is again referred to Refs [46, 51] for more details in the derivations. Therefore, the single residual scalar equation, associated with the generic Gauss point i, for the nonlocal formulations with $\overline{D}$, $-\overline{Y}$, and $\overline{R}$ are respectively given by

$$\overline{F}_i = 3G\Delta\gamma_i - \left(1 - \overline{D}_{i_n}\right)\left(\tilde{q}^{\text{trial}}_{i_{n+1}} - \sigma^{k+1}_{y_i}\right) + \left(\tilde{q}^{\text{trial}}_{i_{n+1}} - \sigma^{k+1}_{y_i}\right) \times \sum_{j=1}^{npg_i}\left[w_j J_j \beta_{ij} \frac{\left(\tilde{q}^{\text{trial}}_{j_{n+1}} - \sigma^{k+1}_{y_j}\right)}{3G}\left(\frac{-Y^{k+1}_j}{r}\right)^s\right] = 0 \tag{2.95}$$

$$\overline{F}_i = 3G\Delta\gamma_i - \left(1 - D_{i_n}\right)\left(\tilde{q}^{\text{trial}}_{i_{n+1}} - \sigma^{k+1}_{y_i}\right) + \frac{\left(\tilde{q}^{\text{trial}}_{i_{n+1}} - \sigma^{k+1}_{y_i}\right)^2}{3G}\left(\frac{\sum_{j=1}^{npg_i} w_j J_j \beta_{ij}\left(-Y^{k+1}_j\right)}{r}\right)^s = 0 \tag{2.96}$$

and

$$\overline{F}_i = 3G\Delta\gamma_i - \left(1 - D_{i_n}\right)\left(\tilde{q}^{\text{trial}}_{i_{n+1}} - \sigma^{k+1}_{y_i}\right) + \frac{\left(\tilde{q}^{\text{trial}}_{i_{n+1}} - \sigma^{k+1}_{y_i}\right)^2}{3G}\left(\frac{-Y^{k+1}_i}{r}\right)^s = 0 \tag{2.97}$$

where

$$-Y^{k+1}_i = \frac{\left(\sigma^{k+1}_{y_i}\right)^2}{6G} + \frac{\left(\tilde{p}^{\text{trial}}_{i_{n+1}}\right)^2}{2K} \tag{2.98}$$

For the sake of simplicity, the dependency of $\sigma^{k+1}_{y_i}$ on $R^{k+1}_{i_{n+1}}$ has been omitted in the equations above.

Despite the reduction to a single equation per Gauss point, the solution of the nonlocal constitutive problem still must be sought globally. The main advantage of the aforementioned algebraic manipulations is that now one needs to solve an ordinary system of scalar equations instead of a system which contains equations with both scalar and tensorial quantities.

After the aforementioned reduction, only a single equation ($\overline{F}_i$) and a single unknown ($\Delta\gamma_i$) are now associated with every ith Gauss point of the mesh. For notation convenience, one can define $\Delta\boldsymbol{\gamma}$ as a vector that contains the values of the incremental plastic multipliers of all Gauss points of the body. Likewise, the residual function of every point can also be represented by a vector denoted by $\overline{\mathbf{F}}$. With this notation at hand, the nonlinear constitutive problem is solved herein using the Newton–Raphson method, that is,

$$\Delta\boldsymbol{\gamma}^{k+1} = \Delta\boldsymbol{\gamma}^{k} - \left(\left. \frac{\partial \overline{\mathbf{F}}}{\partial \Delta\boldsymbol{\gamma}} \right|_{\Delta\boldsymbol{\gamma}^{K}} \right)^{-1} \overline{\mathbf{F}}\left(\Delta\boldsymbol{\gamma}^{k}\right) \tag{2.99}$$

which can be more conveniently written as

$$\left(\left. \frac{\partial \overline{\mathbf{F}}}{\partial \Delta\boldsymbol{\gamma}} \right|_{\Delta\boldsymbol{\gamma}^{K}} \right) \delta\boldsymbol{\gamma} = -\overline{\mathbf{F}}\left(\Delta\boldsymbol{\gamma}^{k}\right) \tag{2.100}$$

where

$$\delta\boldsymbol{\gamma} = \Delta\boldsymbol{\gamma}^{k+1} - \Delta\boldsymbol{\gamma}^{k} \tag{2.101}$$

and

$$\frac{\partial \overline{\mathbf{F}}}{\partial \Delta\boldsymbol{\gamma}} = \begin{bmatrix} \frac{\partial \overline{F}_1}{\partial \Delta\gamma_1} & \frac{\partial \overline{F}_1}{\partial \Delta\gamma_2} & \cdots & \frac{\partial \overline{F}_1}{\partial \Delta\gamma_{ngp}} \\ \frac{\partial \overline{F}_2}{\partial \Delta\gamma_1} & \frac{\partial \overline{F}_2}{\partial \Delta\gamma_2} & \cdots & \frac{\partial \overline{F}_2}{\partial \Delta\gamma_{ngp}} \\ \vdots & \vdots & \ddots & \vdots \\ \frac{\partial \overline{F}_{ngp}}{\partial \Delta\gamma_1} & \frac{\partial \overline{F}_{ngp}}{\partial \Delta\gamma_2} & \cdots & \frac{\partial \overline{F}_{ngp}}{\partial \Delta\gamma_{ngp}} \end{bmatrix} \tag{2.102}$$

It is important to remark that $\partial\overline{\mathbf{F}}/\partial\Delta\boldsymbol{\gamma}$ is unsymmetric in the general case and that a diagonal matrix is obtained if the local case is considered. In order to avoid significant computational effort, the matrix above can be simplified by disregarding its off-diagonal terms, yielding in a modified Newton–Raphson strategy. Thus, the update at every material point is given by

$$\Delta\gamma_i^{k+1} = \Delta\gamma_i^{k} - \frac{\overline{F}_i^{k}}{\overline{F}_i^{\prime k}} \tag{2.103}$$

where $\overline{F}_i^{\prime}$ is the derivative of the corresponding residual function with respect to $\Delta\gamma_i$, expressed for the nonlocal formulations with $\overline{D}$, $-\overline{Y}$, and $\overline{R}$ respectively as

$$\overline{F}_i^{\prime} = 3G + \left(1 - \overline{D}_{i_n}\right) H_i - H_i \sum_{j=1}^{npg_i} \left[w_j J_j \beta_{ij} \frac{\left(\tilde{q}_{j_{n+1}}^{\text{trial}} - \sigma_{y_j}^{k+1}\right)}{3G} \left(\frac{-Y_j^{k+1}}{r}\right)^{s} \right]$$

$$+ \frac{\left(\tilde{q}_{i_{n+1}}^{\text{trial}} - \sigma_{y_i}^{k+1}\right) w_i J_i \beta_{ii} H_i}{3G}$$

$$\times \left[\frac{\left(\tilde{q}^{\text{trial}}_{i_{n+1}} - \sigma^{k+1}_{y_i}\right)\sigma^{k+1}_{y_i}}{3G} \frac{s}{r}\left(\frac{-Y_i^{k+1}}{r}\right)^{s-1} - \left(\frac{-Y_i^{k+1}}{r}\right)^{s} \right] \tag{2.104}$$

$$\overline{F}'_i = 3G + \left(1 - \overline{D}_{i_n}\right) H_i - \frac{2H_i\left(\tilde{q}^{\text{trial}}_{i_{n+1}} - \sigma^{k+1}_{y_i}\right)}{3G}\left(\frac{\sum\limits_{j=1}^{npg_i} w_j J_j \beta_{ij}\left(-Y_j^{k+1}\right)}{r}\right)^{s}$$
$$+ \frac{\left(\tilde{q}^{\text{trial}}_{i_{n+1}} - \sigma^{k+1}_{y_i}\right)^2 w_i J_i \beta_{ii} H_i \sigma^{k+1}_{y_i}}{9G^2} \frac{s}{r}\left(\frac{\sum\limits_{j=1}^{npg_i} w_j J_j \beta_{ij}\left(-Y_j^{k+1}\right)}{r}\right)^{s-1} \tag{2.105}$$

and

$$\overline{F}'_i = 3G + \left(1 - \overline{D}_{i_n}\right) H_i w_i J_i \beta_{ii} - \frac{2H_i w_i J_i \beta_{ii}\left(\tilde{q}^{\text{trial}}_{i_{n+1}} - \sigma^{k+1}_{y_i}\right)}{3G}\left(\frac{-Y_i^{k+1}}{r}\right)^{s}$$
$$+ \frac{\left(\tilde{q}^{\text{trial}}_{i_{n+1}} - \sigma^{k+1}_{y_i}\right)^2 w_i J_i \beta_{ii} H_i \sigma^{k+1}_{y_i}}{9G^2} \frac{s}{r}\left(\frac{-Y_i^{k+1}}{r}\right)^{s-1} \tag{2.106}$$

A general algorithm for the nonlocal constitutive state update can now be defined (see Box 2.3). However, since a fully implicit elastic predictor/return mapping scheme has been adopted, Eqs (2.95–2.97) are only conceptually valid for plastic points. Otherwise, the material point is elastic and no return mapping is necessary. The question that arises is how to determine which points are elastic and which are plastic, since the solution must be sought globally. This shortcoming can be overcome with a strategy that checks the yield condition (i.e., plastic admissibility) for each material point at every Newton–Raphson iteration.[3] Within this methodology, when $F_p^{k+1} < 0$ and $\Delta\gamma_i^{k+1} = 0$, the point is elastic and $\overline{F}_i$ is set to 0. Otherwise, the point is plastic and the residual function corresponding to the adopted nonlocal formulation is evaluated.

It is important to remark that it is not (usually) possible to know "a priori" which points are elastic and which are plastic. However, with the present algorithm, a point initially elastic can turn plastic and then elastic again as iterations evolve until final convergence is attained.

Finally, the specific algorithms for the solution of the nonlocal problem for the formulations with $\overline{D}$, $-\overline{Y}$, and $\overline{R}$ and are summarized in Boxes 2.4 and 2.5 in pseudocode format.

3) The strategy of checking the yield condition during the solution of the material problem has been firstly proposed by Strömberg and Ristinmaa [19] for the solution of a mixed local/nonlocal problem within the classical elastoplasticity framework.

Box 2.3: General Nonlocal Stress Update Procedure

(i) Define and store the elastic trial state for every Gauss point as

$$\boldsymbol{\varepsilon}^{\text{e trial}}_{i_{n+1}} = \boldsymbol{\varepsilon}^{\text{e}}_{i_n} + \Delta\boldsymbol{\varepsilon}_i \,; \qquad \tilde{\mathbf{s}}^{\text{trial}}_{i_{n+1}} = 2G\boldsymbol{\varepsilon}^{\text{e trial}}_{\text{d}i_{n+1}}$$

$$R^{\text{trial}}_{i_{n+1}} = R_{i_n} \,; \qquad \tilde{q}^{\text{trial}}_{i_{n+1}} = \sqrt{\tfrac{3}{2}}\left\|\tilde{\mathbf{s}}^{\text{trial}}_{i_{n+1}}\right\|$$

$$\tilde{p}^{\text{trial}}_{i_{n+1}} = K\varepsilon^{\text{e trial}}_{\text{v}i_{n+1}} \,;$$

(ii) GOTO Boxes 2.4 or 2.5 to solve

$$\overline{\mathbf{F}}\,(\Delta\boldsymbol{\gamma}) = \mathbf{0}$$

for $\Delta\boldsymbol{\gamma}$ with a modified Newton–Raphson method.

(iii) Update stress state and internal variable for every Gauss point as

$$D_{i_{n+1}} = 1 - \left(\frac{3G\Delta\gamma_i}{\tilde{q}^{trial}_{i_{n+1}} - \sigma_{\text{y}}\left(R_{i_{n+1}}\right)}\right) \,; \qquad \mathbf{s}_{i_{n+1}} = \frac{q_{i_{n+1}}}{\tilde{q}^{\text{trial}}_{i_{n+1}}}\tilde{\mathbf{s}}^{\text{trial}}_{i_{n+1}}$$

$$p_{i_{n+1}} = \left(1 - D_{i_{n+1}}\right)\tilde{p}^{\text{trial}}_{i_{n+1}} \,; \qquad \boldsymbol{\sigma}_{i_{n+1}} = \mathbf{s}_{i_{n+1}} + p_{i_{n+1}}\boldsymbol{I}$$

$$q_{i_{n+1}} = \left(1 - D_{i_{n+1}}\right)\sigma_{\text{y}}\left(R_{i_{n+1}}\right) \,; \qquad \boldsymbol{\varepsilon}^{\text{e}}_{i_{n+1}} = \frac{1}{2G}\mathbf{s}_{i_{n+1}} + \frac{1}{3}\varepsilon^{\text{e trial}}_{\text{v}i_{n+1}}\boldsymbol{I}$$

(iv) EXIT

Box 2.4: Algorithm for the Solution of the Nonlocal Damage Model with $\overline{D}$ or $-\overline{Y}$

(i) Set $k = 0$ and $\Delta\gamma_i^{k+1} = 0$

(ii) Update hardening variable, yield stress, and damage energy release rate for all Gauss points

$$R_i^{k+1} = R_{i_n} + \Delta\gamma_i^{k+1}$$

$$\sigma_{\text{y}_i}^{k+1} = \sigma_{\text{y}}(R_i^{k+1})$$

IF $\tilde{q}_i^{\text{trial}} - \sigma_{\text{y}_i}^{k+1} \leq 0$ THEN

$$-Y_i^{k+1} = 0$$

ELSE

$$-Y_i^{k+1} = \frac{\sigma_{\text{y}_i}^{k+1}}{6G} + \frac{\tilde{p}_i^2}{2K}$$

ENDIF

(iii) Evaluate residual and derivative for every Gauss point as

IF $\tilde{q}_i^{\text{trial}} - \sigma_{y_i}^{k+1} \leq 0$ AND $\Delta\gamma_i^{k+1} = 0$ THEN

$$\overline{F}_i = 0\,; \qquad \overline{F}_i' = 1$$

ELSE

For the nonlocal model with $\overline{D}$

$$\overline{F}_i = \text{Eq. (2.95)}\,; \qquad \overline{F}_i' = \text{Eq. (2.104)}$$

For the nonlocal model with $-\overline{Y}$

$$\overline{F}_i = \text{Eq. (2.96)}\,; \qquad F_i' = \text{Eq. (2.105)}$$

ENDIF

(iv) Check convergence

IF $\left\|\overline{\mathbf{F}}\right\| < TOL$ EXIT

(v) Switch and update incremental plastic multiplier for every Gauss point as

$$\Delta\gamma_i^k = \Delta\gamma_i^{k+1}$$

$$\Delta\gamma_i^{k+1} = \Delta\gamma_i^k - \frac{\overline{F}_i}{\overline{F}_i'}$$

(vi) GOTO (ii)

Box 2.5: Algorithm for the Solution of the Nonlocal Damage Model with $\overline{R}$

(i) Set $k = 0$ and $\Delta\gamma_i^k = 0$

(ii) Evaluate initial residual and derivative for every Gauss point as

IF $q_i^{\text{trial}} - \sigma_{y_i}^k \leq 0$ THEN

$$\overline{F}_i = 0\,; \qquad F_i' = 1$$

ELSE

$$\overline{F}_i = \text{Eq. (2.97)}\,; \qquad \overline{F}_i' = \text{Eq. (2.106)}$$

ENDIF

(iii) Update incremental plastic multiplier for every point as

$$\Delta\gamma_i^{k+1} = \Delta\gamma_i^k - \frac{\overline{F}_i}{\overline{F}_i'}$$

(iv) Update isotropic hardening variable and yield stress for every Gauss point as

$$\overline{R}_i^{k+1} = \overline{R}_i^{(0)} + \sum_{j=1}^{npg_i} w_j J_j \beta_{ij} \Delta\gamma_j^k$$

$$\sigma_{y_i}^{k+1} = \sigma_y(R_i^{k+1})$$

(v) Evaluate residual and derivative for every Gauss point as

IF $\tilde{q}_i^{trial} - \sigma_{y_i}^{k+1} \leq 0$ AND $\Delta\gamma_i^{k+1} = 0$ THEN

$$\overline{F}_i = 0\,; \qquad \overline{F}_i' = 1$$

ELSE

$$\overline{F}_i = \text{Eq. (2.97)}\,; \qquad \overline{F}_i' = \text{Eq. (2.106)}$$

ENDIF

(vi) Check convergence

IF $\left\|\overline{\mathbf{F}}\right\| < TOL$ THEN

EXIT

ELSE

$$\Delta\gamma_i^k = \Delta\gamma_i^{k+1}$$

GOTO (iii)

ENDIF

2.6 Numerical Analysis

2.6.1 Axisymmetric Analysis of a Notched Specimen

This example is based on a similar analysis carried out in Ref. [50]. It consists of the simulation of the fracturing of a cylindrical notched specimen when subjected to a tensile test (see Figure 2.3). Since there is no load reversal, the simplified version of Lemaitre's ductile damage model is perfectly suitable for this analysis. The material properties employed are summarized in Table 2.7. Owing to symmetry, only one-quarter of the specimen has been analyzed[4] where a prescribed vertical displacement $\|u_y\| = 0.43$ mm has been applied at its extremities.

4) All meshes and contour results have been mirrored for the ease of visualization.

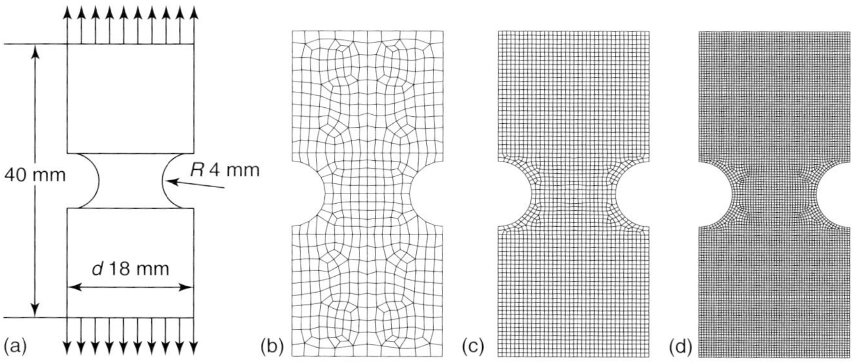

Figure 2.3 Problem geometry and different mesh refinements for the notched specimen.

Table 2.7 Material properties for the notched specimen.

Property	Symbol	Value
Young's modulus	E	69 000 MPa
Poisson's ratio	ν	0.3
Damage exponent	s	1.0
Damage denominator	r	1.25 MPa
Yield stress	σ_y	$589 \cdot (10^{-4} + R)^{0.216}$ MPa
Characteristic length	ℓ_r	0.6325 mm

In order to assess the effectiveness of the nonlocal models presented in Section 2.5 as well as the efficiency of their numerical implementation, different mesh refinements have been considered as shown in Figure 2.3. All meshes contain eight-node quadratic elements with a reduced 2×2 integration scheme.

Figure 2.4 shows the damage contours when the standard local theory is used. If a critical damage value [37, 50], for instance, is used as fracture criterion, it is observed that the spot where failure takes place is correctly predicted by Lemaitre's model. According to the experimental results obtained by Hancock and Mackenzie [1], failure onset occurs at the center of the specimen due to the high triaxiality ratio at that point. As shown by Vaz Jr. and Owen [50], if the criterion of maximum plastic work is adopted, fracture is predicted at the notch root, which is incorrect.

However, due to the softening regime inherently induced by damage in Lemaitre's model, a high mesh dependency is observed in the local solution. Indeed, the damage distribution shrinks at the central region of the specimen upon mesh refinement. As a consequence of the localization effect, the quality of the numerical result dramatically diminishes. In order to alleviate the pathological dependency on spatial discretization, the nonlocal extensions of Lemaitre's model with $\overline{D}$, $-\overline{Y}$, and $\overline{R}$ have been employed. Considering Figures 2.5 and 2.6, the results show that the formulations with $\overline{D}$ and $-\overline{Y}$ have significantly attenuated

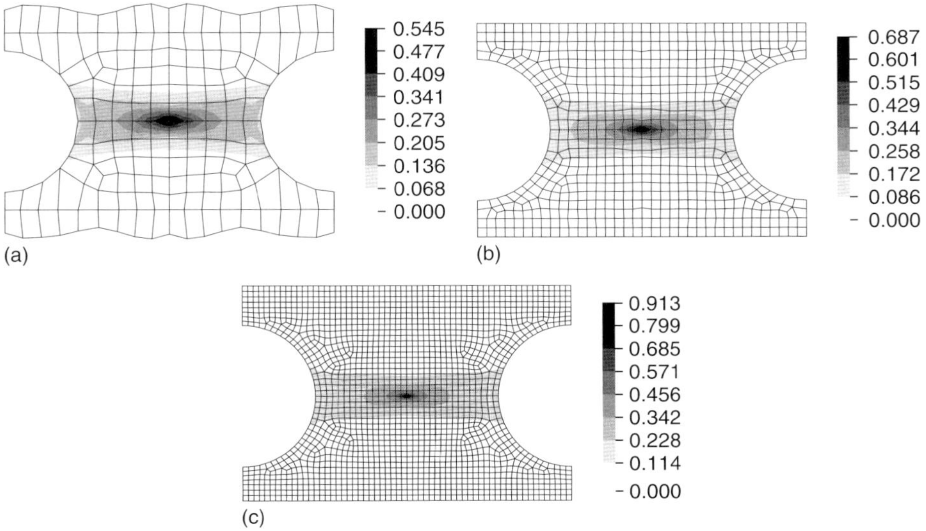

Figure 2.4 Damage contours for the local model.

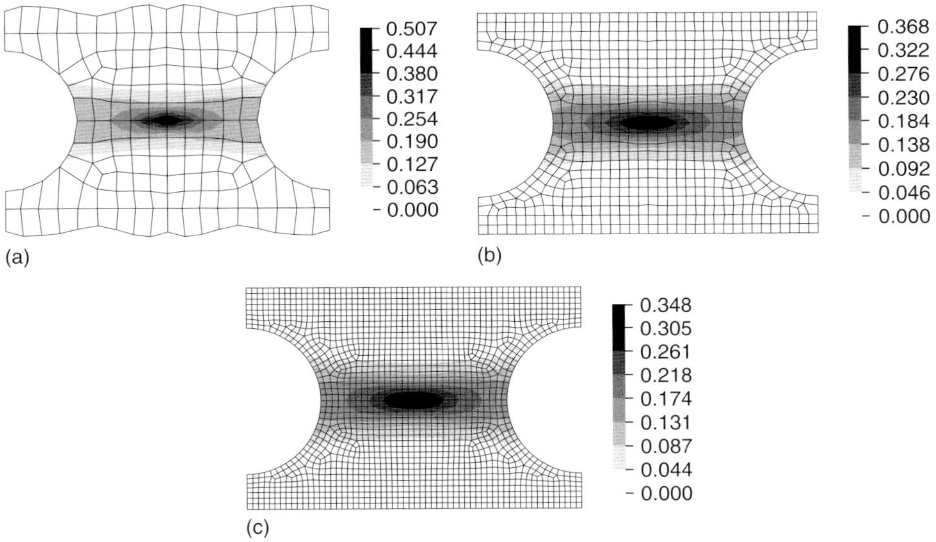

Figure 2.5 Damage contours for the nonlocal model with $\overline{D}$.

mesh dependency, thus improving the reliability on the damage values obtained numerically. On the other hand, the choice of $\overline{R}$ still resulted in a localized solution (see Figure 2.7). This is due to the fact that the isotropic hardening variable does not influence the softening regime since the prescribed curve for the yield stress induces only hardening in the present example.

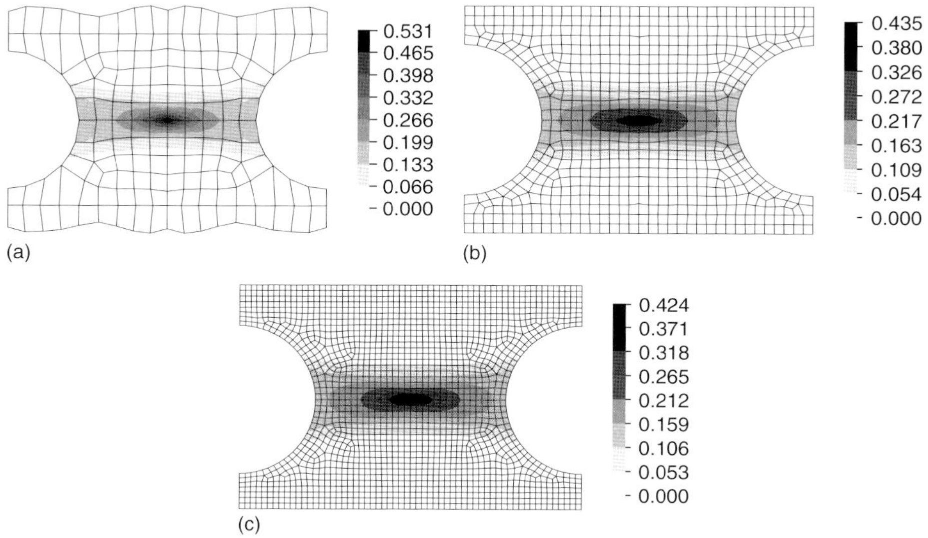

Figure 2.6 Damage contours for the nonlocal model with $-\overline{Y}$.

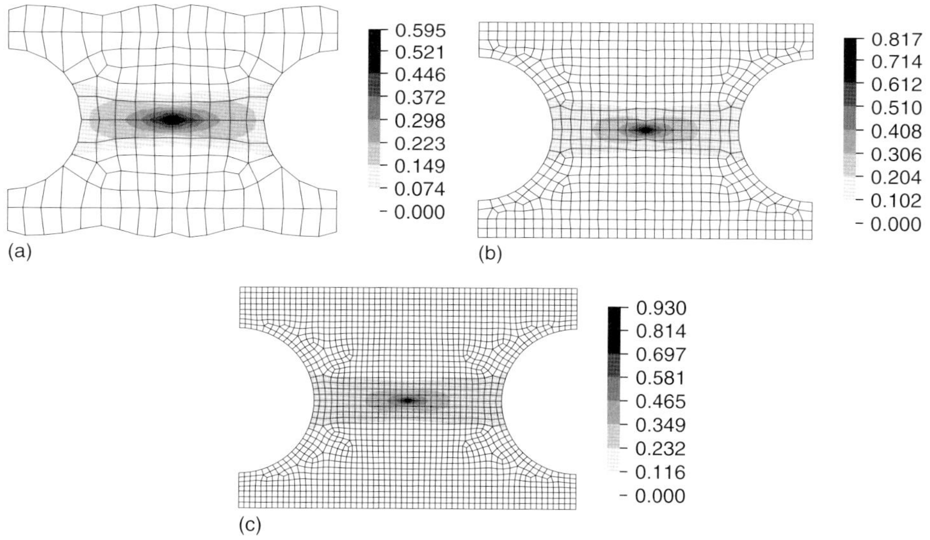

Figure 2.7 Damage contours for the nonlocal model with $\overline{R}$.

Moreover, the damage distribution along the central section of the specimen (on plane X–X) is plotted in Figure 2.8 for the different nonlocal formulations simulated with the finest mesh (Figure 2.3d). It is observed that the choice of $\overline{D}$ yields on a slightly higher diffusive effect than $-\overline{Y}$. Nevertheless, unlike the choice of $\overline{R}$, both nonlocal formulations provided regularized solutions.

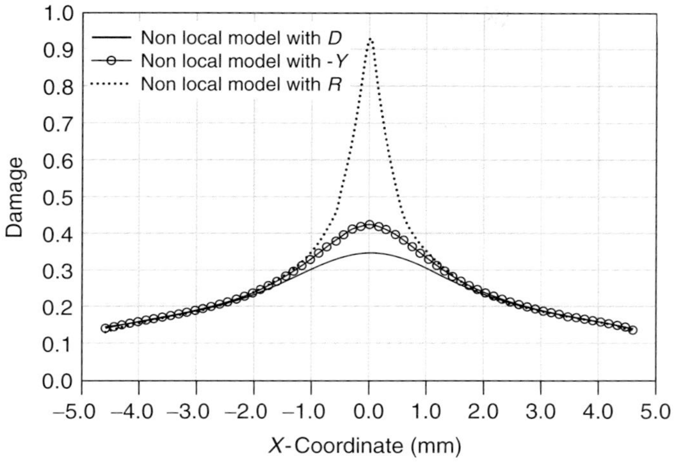

Figure 2.8 Damage distribution along the central section (results for the finest mesh).

The efficiency of the nonlocal stress update algorithms has also been assessed for the current example. Table 2.8 shows typical convergences of the different formulations. Although the off-diagonal terms in Eq. (2.102) have been dropped, both nonlocal formulations with $\overline{D}$ and $-\overline{Y}$ have exhibited remarkably high convergence rates. This was not the case of the nonlocal model with $\overline{R}$ for which convergence was very poor. Nonetheless, as previously pointed out, $\overline{R}$ is not a valid choice to obtain regularized solutions in the present case.

Table 2.8 Typical convergences of the nonlocal stress update algorithms.

	Residual norm		
Iteration	$\overline{D}$	$-\overline{Y}$	$\overline{R}$
1	1.65E+08	3.55E+03	1.08E+02
2	1.53E+04	1.36E+00	5.44E+00
3	1.31E−02	1.39E−04	1.54E+00
4	4.47E−11	2.78E−08	6.62E−01
5	–	6.02E−12	3.19E−01
6	–	–	1.57E−01
⋮	–	–	⋮
30	–	–	1.07E−08
31	–	–	5.40E−09

Table 2.9 Material properties for the flat grooved plate specimen.

Property	Symbol	Value
Young's modulus	E	71150 MPa
Poisson's ratio	ν	0.3
Damage exponent	s	1.0
Damage denominator	r	1.7 MPa
Yield stress	σ_y	$908 \cdot (0.0058 + R)^{0.1742}$ MPa
Characteristic length	ℓ_r	0.35 mm

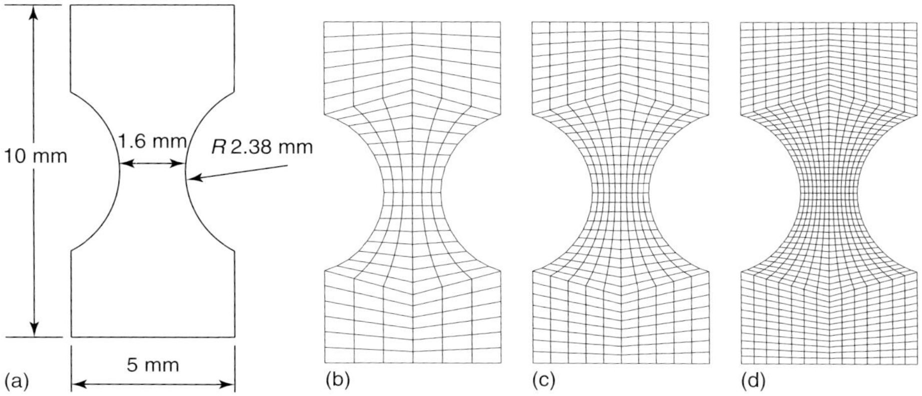

Figure 2.9 Flat grooved specimen's geometry and different mesh refinements.

2.6.2 Flat Grooved Plate in Plane Strain

In this example, a flat grooved plate [65] made of an aluminum alloy (see Table 2.9) is simulated under a tensile loading, in plane strain conditions. Similar to the previous example, the loading is monotonic and therefore Lemaitre's simplified model may be employed. The geometry of the specimen is depicted in Figure 2.9. A prescribed vertical displacement $\|u_y\| = 0.055$ mm has been applied at both ending edges. In order to further illustrate the attenuation of mesh dependency achieved by the nonlocal theory, three different mesh refinements (see Figure 2.9) have been employed in this simulation. Eight-noded quadratic elements with reduced integration have been used in all analyses.

Figure 2.10 shows the results of the local model when the prescribed displacement is applied. Noticeably, the solution obtained is highly mesh dependent since different spatial discretization leads to very different damage distributions and values. Clearly, damage tends to concentrate in narrow bands as the average element size decreases. However, when the nonlocal characteristic length $\ell_r = 0.35$ mm

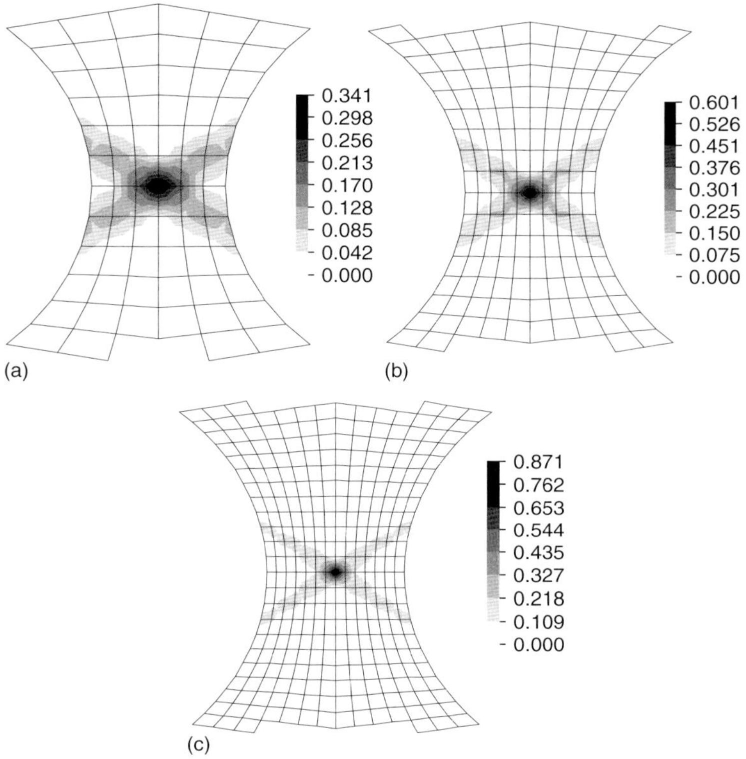

Figure 2.10 Damage contours for the local model.

is considered through the nonlocal formulation with $\overline{D}$, the pathological mesh dependency is eliminated since the damage distribution is almost constant upon mesh refinement (see Figure 2.11).

2.6.3
Upsetting of a Tapered Specimen

The upsetting test of axisymmetric specimens is one of the most used tests to study bulk metal-forming processes, since it is able to reproduce stress/strain states that are similar to the ones observed in these processes. This test presents interesting results regarding the fracture initiation site. In general, fracture initiation occurs either at the specimen center or external surface near the equator.

This example has been drawn from Gouveia *et al.* [66] where the experimental procedure is described: upsetting of a tapered specimen for a UNS L52905 (Unified Numbering System ASTM-SAE) lead alloy. The geometry of the problem, boundary conditions, and the finite element mesh adopted are given in Figure 2.12.

The mesh discretizes one symmetric quarter of the problem with the appropriate symmetric boundary conditions imposed on the relevant edges. The simulation was

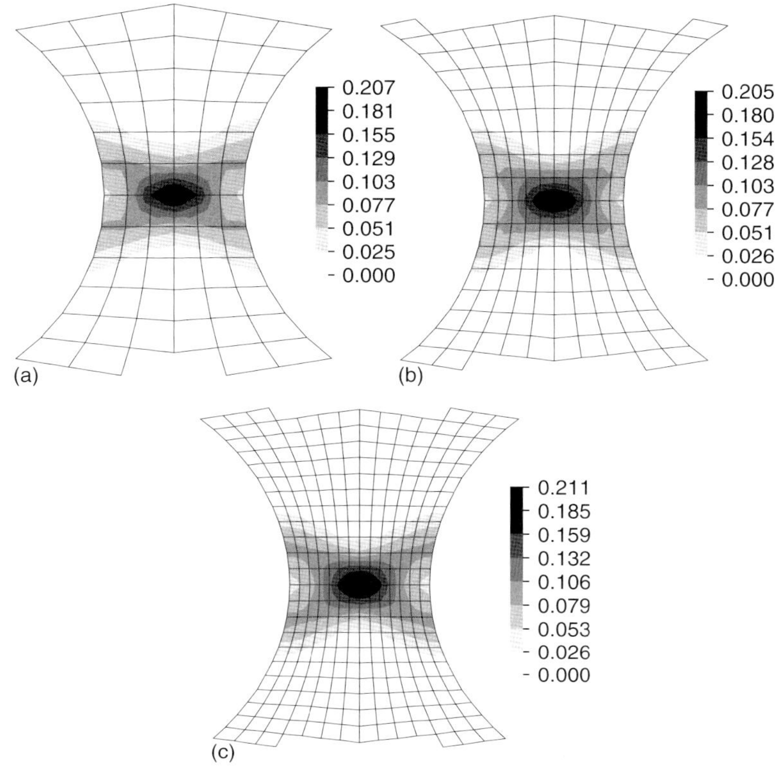

Figure 2.11 Damage contours for the nonlocal model with $\overline{D}$.

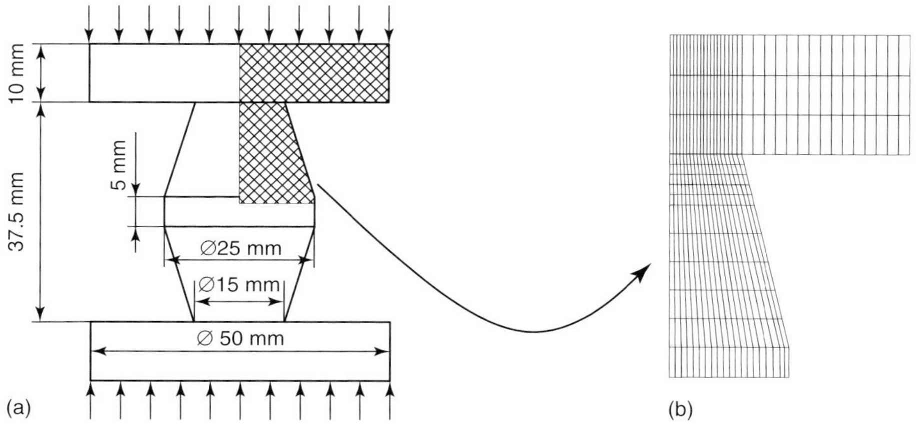

Figure 2.12 Tapered specimen: (a) problem geometry and (b) mesh.

Table 2.10 Material data for UNS I52905 lead alloy and other simulation parameters.

Property	Symbol	Value
Specific mass	ρ	$11\,340\ \text{kg m}^{-3}$
Young's modulus	E	18 000 MPa
Poisson's ratio	ν	0.4
Yield stress	σ_y	$66.656\left(\varepsilon_{\text{acc}}^{\text{p}}\right)^{0.10158}$ MPa
Initial yield stress	σ_{yo}	43 MPa
Friction	m	0.35

executed until a reduction of 65% on the total height was achieved. A total number of 220 four-noded axisymmetric quadrilaterals is used in the discretization of the tapered specimen amounting to a total of 252 nodes. The matrix was discretized with the same type of elements (the total number of elements is 105, resulting in 144 nodes). Boundary frictional conditions are modeled through the utilization of linear friction elements with zero thickness. Friction is characterized by a model that assumes a constant interface friction factor, m (adhesive law type). The material properties adopted are listed in Table 2.10.

2.6.3.1 Damage Prediction Using the Lemaitre's Simplified Model

Since experimental material coefficients associated with the damage evolution are not available, arbitrary values were assigned. As reported by Lemaitre [4], the exponent used in the damage evolution law, $s \approx 1.0$, almost does not change. Therefore, five different damage evolution parameters were initially considered: $r =$ 1.0 MPa; $r = 1.5$ MPa; $r = 3.5$ MPa; $r = 5.0$ MPa; and $r = 10.0$ MPa. The results for $r = \infty$ correspond to the situation where there is no damage evolution, that is, only plasticity is present. In Figures 2.13 and 2.14, it is possible to observe the damage contours plots obtained by the finite element analysis when $r = 1.5$ MPa.

It can be observed that during the initial stages of the upsetting process, maximum damage is detected at the external surface near the contact regions. This is due to the displacement restriction of the contact surfaces. The results are coherent with Zhu *et al.* [67], who simulated the compression of a cylindrical bar under plane strain.

As the compression of the specimen increases (see Figure 2.14), the stress triaxiality ratio "moves" the maximum damage area toward the specimen center and localizes there. This means that fracture initiation should be expected in this region. At the final stage, with a reduction of 65%, damage is highly localized around the center. This result is not in agreement with the experimental observations of Gouveia *et al.* [66], which predicts fracture initiation on the external surface near the equator.

The different simulations executed for the denominator of the damage evolution law, r, did not show significant differences in the way that the damage contours change with the specimen reduction, although the numerical value obtained for

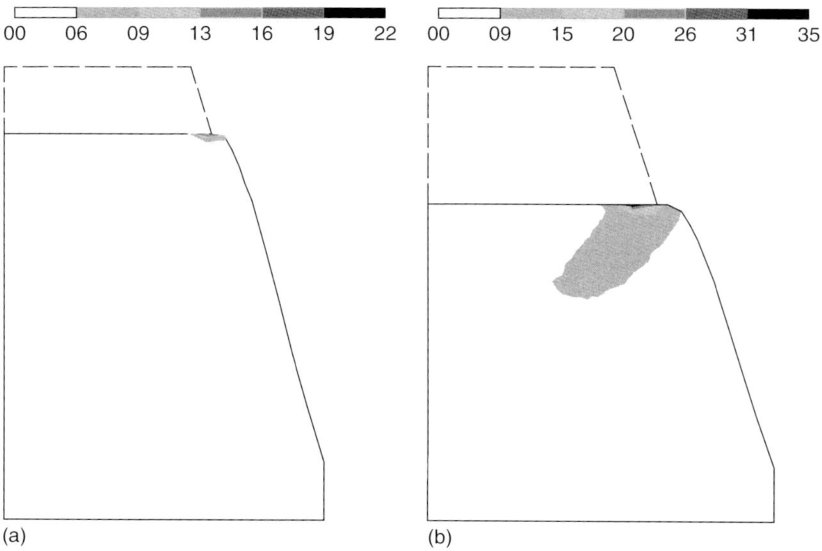

Figure 2.13 Damage contours for the tapered specimen: (a) reduction of 15% and (b) reduction of 30%.

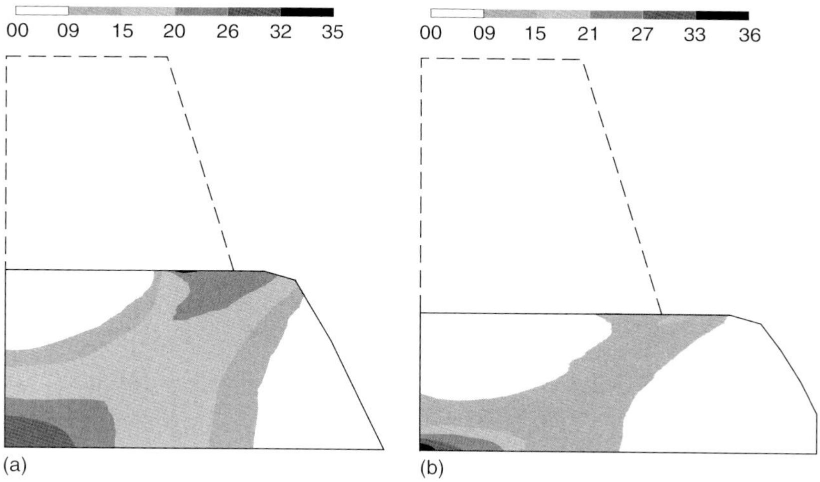

Figure 2.14 Damage contours for the tapered specimen: (a) reduction of 55% and (b) reduction of 65%.

the damage field varies. In Table 2.11, we present the maximum value obtained at the center of the specimen at the final stage (reduction of 65%) for different values of the damage parameter, r.

It can be concluded that Lemaitre's simplified model is not able to correctly predict the fracture site. For any value of the damage parameter, r, the fracture

Table 2.11 Maximum damage value at the center of the specimen.

r	Damage value
1.0	0.75[a]
1.5	0.36
3.5	0.12
5.0	0.074
10.0	0.036

[a]Total reduction not achieved.

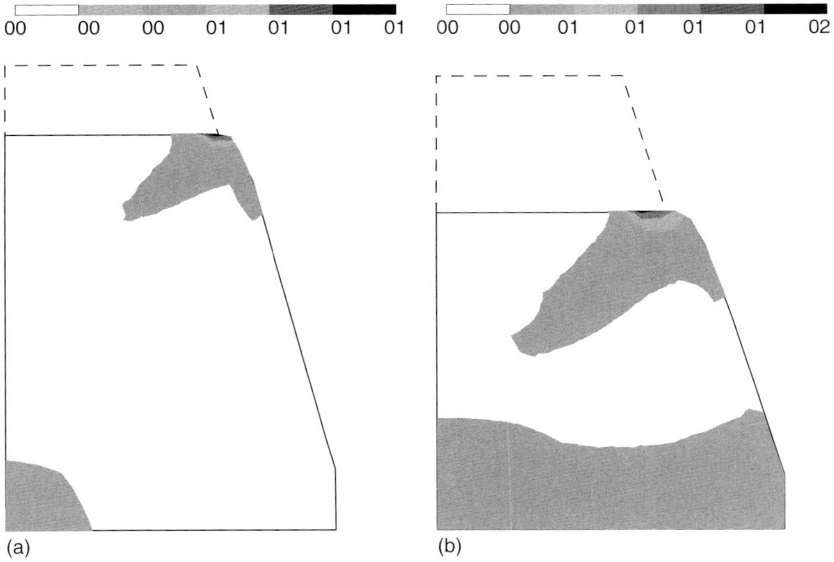

Figure 2.15 Damage contours for the tapered specimen: (a) reduction of 15% and (b) reduction of 30%.

onset always takes place at the specimen's center, which is not in agreement with the experimental observations made by Gouveia *et al.* [66].

2.6.3.2 **Damage Prediction Using the Lemaitre's Model with Crack Closure Effect**

As remarked before, the damage experimental parameters necessary for the simulation are not known. We proceed in a similar way as in the previous section, which means that different values will be attributed for the damage parameters. The only additional parameter that this model needs, when compared with the previous one, is the crack closure effect, h. Again, five different values are initially tested: $h = 0.01$; $h = 0.05$; $h = 0.1$; $h = 0.2$; and $h = 0.5$. The other material parameters for the damage evolution used in this simulation are $s = 1.0$ and $r = 1.5$ MPa (same as in the previous section). In Figures 2.15 and 2.16, it is possible to observe

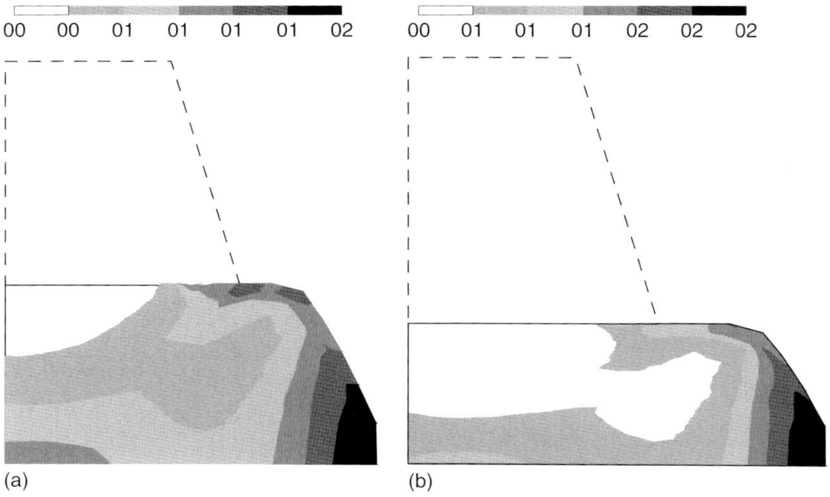

Figure 2.16 Damage contours for the tapered specimen: (a) reduction of 55% and (b) reduction of 65%.

the damage variable field obtained with finite element analysis. In this simulation, the crack closure parameter was set to $h = 0.05$.

In the early stages of the loading process (see Figure 2.15), the maximum value of damage is detected at the external surface near the contact regions because of the displacement restriction of the contact surfaces. This specific area of the specimen was already the one where the standard model had the maximum value of damage. However, as shown in Figure 2.15b, the damage numerical value is much smaller (compare the damage values of Figure 2.13b with Figure 2.15b).

At the final stages of the upsetting process, we can observe a clear distinction between the two damage models. In the damage model with the inclusion of the crack closure effect, the maximum value of damage at the final steps is highly localized at the external surface near the equator, whereas the simplified version of Lemaitre's model predicted the maximum value of damage at the center of the specimen. The introduction of the crack closure effect allows the damage value on the Gauss points, which are under tensile stress state, to grow faster than the points where the stress state is compressive. For this reason, the fracture is predicted at the external surface, near the equator, which is in full accordance to experimental evidence reported by Gouveia *et al.* [66].

2.7 Concluding Remarks

Some advances in the area of continuum constitutive modeling and finite element prediction of ductile plastic damage following Lemaitre's theory have been presented, indicating the progress that has been made both in the theoretical formulation and the associated numerical implementation. In particular, a concise description of Lemaitre's elastoplastic damage theory was carried out and a discussion of the main assumptions of the model, based on the principle of maximum inelastic dissipation, was undertaken. The integration algorithm of two modified models, which results in extremely efficient numerical implementations, was presented. The enhancement of Lemaitre's model into a nonlocal integral constitutive framework was also described and several examples were shown to illustrate the applicability of the models. While the state of knowledge in some areas is relatively mature, considerable further understanding and development is required in others. For example, issues related to the modeling of complex strain paths are far from settled and a more comprehensive treatment of the material intrinsic length may necessitate the integration of micromechanical studies with computational approaches.

Acknowledgments

J. M. A. César de Sá and F. M. Andrade Pires gratefully acknowledge the financial support provided by *Ministério da Ciência e Tecnologia e do Ensino Superior–Fundação para a Ciência e Tecnologia (Portugal)* under the projects PTDC/EME-TME/74589/2006 and PTDC/EME-TME/71325/2006. F. X. C. Andrade was supported by the Programme Alβan, the European Union Programme of High Level Scholarships for Latin America, scholarship no. E07D401751BR.

References

1. Hancock, J.W. and Mackenzie, A.C. (1976) On the mechanism of ductile fracture in high-strength steels subjected to multi-axial stress-states. *Journal of the Mechanics and Physics of Solids*, **24**, 147–169.
2. Le Roy, G., Embury, J.D., and Ashby, M.F. (1981) A model of ductile fracture based on the nucleation and growth of voids. *Acta Metallica*, **29**, 1509–1522.
3. Thomason, P.F. (1990) *Ductile Fracture of Metals*, Pergamon Press, Oxford.
4. Lemaitre, J. (1985) A continuous damage mechanics model for ductile fracture. *Journal of Engineering Materials and Technology*, **107**, 83–89.
5. Lemaitre, J. (1983) A three-dimensional ductile damage model applied to deep-drawing forming limits. ICM 4 Stockholm, vol. 2. pp. 1047–1053.
6. Lemaitre, J., Desmorat, R., and Sauzay, M. (2000) Anisotropic damage law of evolution. *European Journal of Mechanics A: Solids*, **19**, 187–208.
7. Benallal, A., Billardon, R., and Doghri, I. (1988) An integration algorithm and the corresponding consistent tangent operator for fully coupled elastoplastic and damage equations. *Communications in Applied Numerical Methods*, **4**, 731–740.
8. De Souza Neto, E.A., Perić, D., and Owen, D.R.J. (1994) A model for elastoplastic damage at finite strains: algorithm issues and applications. *Engineering Computations*, **11**, 257–281.
9. De Souza Neto, E.A. and Perić, D. (1996) A computational framework for a class of models for fully coupled elastoplastic damage at finite strains with reference to the linearization aspects. *Computer Methods in Applied Mechanics and Engineering*, **130**, 179–193.

10. De Souza Neto, E.A., Perić, D., and Owen, D.R.J. (1998) Continuum modelling and numerical simulation of material damage at finite strains. *Archive for Computer Methods in Engineering*, **5**, 311–384.
11. Rogula D. (1965) Influence of spatial acoustic dispersion on dynamical properties of dislocations. I. *Bulletin de l Academie Polonaise des Sciences-Serie des Sciences Techniques*, **13**, 337–343.
12. Eringen, A.C. (1966) A unified theory of thermomechanical materials. *International Journal of Engineering Science*, **4**, 179–202.
13. Eringen, A.C. (1981) On nonlocal plasticity. *International Journal of Engineering Science*, **19**, 1461–1474.
14. Eringen, A.C. (1983) Theories of nonlocal plasticity. *International Journal of Engineering Science*, **21**, 741–751.
15. Bažant, Z.P. and Chang, T.-P. (1984) Instability of nonlocal continuum and strain averaging. *Journal of Engineering Mechanics, ASCE*, **110**, 1441–1450.
16. Pijaudier-Cabot, G. and Bažant, Z.P. (1987) Nonlocal damage theory. *Journal of Mechanical Engineering*, **113** (10), 1512–1533.
17. De Borst, R. and Mühlhaus, H. (1992) Gradient-dependent plasticity: formulation and algorithmic aspects. *International Journal for Numerical Methods in Engineering*, **35**, 521–539.
18. Peerlings, R.H., De Borst, R., Brekemals, W.A., and De Vree, J.H. (1996) Gradient-enhanced damage for quasi-brittle materials. *International Journal for Numerical Methods in Engineering*, **39**, 1512–1533.
19. Strömberg, L. and Ristinmaa, M. (1996) FE-formulation of a nonlocal plasticity theory. *Computer Methods in Applied Mechanics and Engineering*, **136**, 127–144.
20. Polizzotto, C., Borino, G., and Fuschi, P. (1998) A thermodynamic consistent formulation of nonlocal and gradient plasticity. *Mechanics Research Communications*, **25** (1), 75–82.
21. Borino, G., Fuschi, P., and Polizzotto, C. (1999) A thermodynamic approach to nonlocal plasticity and related variational principles. *Journal of Applied Mechanics*, **66**, 952–963.
22. Bažant, Z.P. and Jirásek, M. (2002) Nonlocal integral formulations of plasticity and damage: survey of progress. *Journal of Engineering Mechanics*, **128** (11), 1119–1149.
23. Borino, G., Failla, B., and Polizzotto, C. (2003) A symmetric nonlocal damage theory. *International Journal of Solids and Structures*, **40**, 3621–3645.
24. Geers, M.G.D., Ubachs, R.L.J.M., and Engelen, R.A.B. (2003) Strongly non-local gradient-enhanced finite strain elastoplasticity. *International Journal for Numerical Methods in Engineering*, **56**, 2039–2068.
25. César de Sá, J.M.A., Areias, P.M.A., and Zheng, C. (2006) Damage modelling in metal forming problems using an implicit non-local gradient model. *Computer Methods in Applied Mechanics and Engineering*, **195**, 6646–6660.
26. Lemaitre, J. (1996) *A Course on Damage Mechanics*, Springer, New York.
27. Kachanov, L.M. (1958) Time of the rupture process under creep condition. *Izvestiya Akademii Nauk Sssr Otdelenie Tekhnicheskikh Nauk*, **8**, 26–31.
28. Rabotnov, Y.N. (1963) *Progress in Applied Mechanics*, Prager Anniversary VOLUME, MacMillan, New York, p. 307.
29. Murakami, S. (1988) Mechanical modeling of material damage. *Journal of Applied Mechanics*, **55**, 280–286.
30. Murakami, S. and Kamiya, K. (1997) Constitutive and damage evolution equations of elastic-brittle materials on irreversible thermodynamics. *International Journal of Mechanical Sciences*, **39**, 473–486.
31. Krajčinović, D. and Fonseka, G.U. (1981) The continuous damage theory of brittle materials – part 1: general theory. *Journal of Applied Mechanics*, **48**, 809–815.
32. Fonseka, G.U. and Krajčinović, D. (1981) The continuous damage theory of brittle materials – part 2: uniaxial and plane response modes. *Journal of Applied Mechanics*, **48**, 816–824.
33. Krajčinović, D. (1983) Constitutive equations for damaging materials. *Journal of Applied Mechanics*, **50**, 355–360.

34. Janson, J. (1978) A continuous damage approach to the fatigue process. *Engineering Fracture Mechanics*, **10**, 651–657.
35. Chaboche, J.L. (1988) Continuum damage mechanics: part I – general concepts and part II – damage growth, crack initiation and crack growth. *Journal of Applied Mechanics*, **55**, 59–72.
36. Lemaitre, J. (1990) Micro-mechanics of crack initiation. *International Journal of Fracture*, **42**, 87–99.
37. Lemaitre, J. and Desmorat, R. (2005) *Engineering Damage Mechanics*, Springer-Verlag.
38. Gurson, A.L. (1977) Continuum theory of ductile rupture by void nucleation and growth – part I: yield criteria and flow rule for porous media. *Journal of Engineering Material Technology*, **99**, 2–15.
39. Tvergaard, V. (1982) Material failure by void coalescence in localized shear bands. *International Journal of Solids and Structures*, **18**, 659–672.
40. Tvergaard, V. and Needleman, A. (1984) Analysis of the cup-cone fracture in a round tensile bar. *Acta Metallurgica*, **32**, 157–169.
41. Simo, J.C. and Ju, J.W. (1987) Strain- and stress-based continuum damage models – I. Formulation and II. Computational aspects. *International Journal of Solids and Structures*, **23**, 821–869.
42. Doghri, I. (1995) Numerical implementation and analysis of a class of metal plasticity models coupled with ductile damage. *International Journal of Numerical Methods in Engineering*, **38**, 3403–3431.
43. Johansson, M., Mahnken, R., and Runesson, K. (1999) Efficient integration technique for generalized viscoplasticity coupled with damage. *International Journal of Numerical Methods in Engineering*, **44**, 1727–1747.
44. Lee, J. and Fenves, G.L. (2001) A return-mapping algorithm for plastic-damage models: 3d and plane stress formulation. *International Journal of Numerical Methods in Engineering*, **50**, 487–506.
45. Steinmann, P., Miehe, C., and Stein, E. (1994) Comparison of different finite deformation inelastic damage models within multiplicative elastoplasticity for ductile metals. *Computational Mechanics*, **13**, 458–474.
46. De Souza Neto, E.A. (2002) A fast, one-equation integration algorithm for the Lemaitre ductile damage model. *Communications in Numerical Methods in Engineering*, **18**, 541–554.
47. Andrade Pires, F.M., de Souza Neto, E.A., and Owen, D.R.J. (2004) On the finite element prediction of damage growth and fracture initiation in finitely deforming ductile materials. *Computer Methods in Applied Mechanics*, **193**, 5223–5256.
48. Lemaitre, J. and Chaboche, J.L. (1990) *Mechanics of Solid Materials*, Cambridge University Press.
49. Simo, J.C. and Hughes, T.J.R. (1998) *Computational Inelasticity*, Springer, New York.
50. Vaz, M. Jr. and Owen, D.R.J. (2001) Aspects of ductile fracture and adaptive mesh refinement in damaged elasto-plastic materials. *International Journal for Numerical Methods in Engineering*, **50**, 29–54.
51. De Souza Neto, E.A., Péric, D., and Owen, D.R.J. (2008) *Computational Methods for Plasticity: Theory and Application*, John Wiley & Sons, Inc.
52. Ladevèze, P. (1983) in *Failure Criteria of Structured Media*, Proceedings of the CNRS International College 351, Villars-de-Lans (ed. J.P. Boehler), Balkema, pp. 355–363.
53. Ladevèze, P. and Lemaitre, J. (1984) Damage effective stress in quasi unilateral conditions. 16th International Congress Theoretical and Applied Mechanics.
54. Lemaitre, J. (1984) How to use damage mechanics. *Nuclear Engineering Design*, **80**, 233–245.
55. Arnold, G., Hubert, O., Dutko, M., and Billardon, R. (2002) Identification of a continuum damage model from micro-hardness measurements. *International Journal of Forming Processes*, **5**, 163–173.
56. Bažant, Z.P. and Lin, F.-B. (1988) Nonlocal yield-limit degradation. *International*

Journal of Numerical Methods in Engineering, **26**, 1805–1823.

57. De Vree, J.H.P., Brekelmans, W.A.M., and van Gils, M.A.J. (1995) Comparison of nonlocal approaches in continuum damage mechanics. *Computers and Structures*, **4**, 581–588.
58. Jirásek, M. (1998) Nonlocal models for damage and fracture: comparison of approaches. *International Journal of Solids and Structures*, **35**, 4133–4145.
59. Jirásek, M. and Patzák, B. (2002) Consistent tangent stiffness for nonlocal damage models. *Computers and Structures*, **80**, 1279–1293.
60. Jirásek, M. and Rolshoven, S. (2003) Comparison of integral-type nonlocal plasticity models for strain-softening materials. *International Journal of Engineering Science*, **41**, 1553–1602.
61. Rolshoven, S. (2003) Nonlocal plasticity models for localized failure. PhD thesis, École Polytechnique Fédérale de Lausanne, Switzerland.
62. Jirásek, M. (2007) Nonlocal damage mechanics. *Revue Européene de Génie Civil*, **11**, 993–1021.
63. Abu Al-Rub, R.K. and Voyiadjis, G.Z. (2005) A direct finite element implementation of the gradient-dependent theory. *International Journal for Numerical Methods in Engineering*, **63**, 603–629.
64. Engelen, R.A.B. (2005) Plasticity-induced damage in metals: nonlocal modelling at finite strain. PhD thesis, Technische Universiteit Eindhoven.
65. Bai, Y. (2008) Effect of loading history on necking and fracture. PhD thesis, Massachusetts Institute of Technology.
66. Gouveia, B.P.P.A., Rodrigues, J.M.C., and Martins, P.A.F. (1996) Fracture predicting in bulk metal forming. *International Journal of Mechanical Sciences*, **38**, 361–372.
67. Zhu, Y.Y., Cescotto, S., and Habraken, A.M. (1992) A fully coupled elastoplastic damage modeling and fracture criteria in metal forming processes. *Journal of Materials Processing Technology*, **32**, 197–204.
68. Chaboche, J.L. (1984) Anisotropic creep damage in the framework of continuum damage mechanics. *Nuclear Engineering and Design*, **79**, 309–319.

3
Recent Advances in the Prediction of the Thermal Properties of Metallic Hollow Sphere Structures

Thomas Fiedler, Irina V. Belova, Graeme E. Murch, and Andreas Öchsner

3.1
Introduction

Porous metals are characterized by high specific stiffness, the ability to absorb high amounts of energy, and the potential for noise control, mechanical damping, and thermal insulation [1]. However, classical cellular metals such as aluminum metallic foams often exhibit inconstant material parameters [2] because of their stochastic geometry. Local density inhomogeneities [3, 4] yield a scattering of macroscopic properties. This problem is decreased in metallic hollow sphere structures (HSSs) that are assembled by spheres with a defined geometry and possess a more homogeneous structure, which results in optimized properties, as illustrated in Figure 3.1. HSSs exhibit a low thermal conductivity in comparison to their metallic sphere wall materials. In particular, adhesively bonded HSSs show very low thermal conductivities, because of the insulating effect of the adhesive matrix between the metallic shells of the spheres. Consequently, HSSs are of interest as thermal insulators.

Earlier research on the thermal properties of cellular metals mainly focused on open-celled structures (e.g., [5, 6]). Owing to their interconnected porosity, open-celled metal foams can be used in heat exchangers [7, 8] or resistance heaters [9]. The investigated HSSs exhibit no or little interconnected porosity and can be considered closed-cell structures. Lu and Chen [10] investigated the thermal transport and fire-inhibiting properties of closed-cell aluminum alloys. They found a strong dependence of the thermal conductivity of the cellular metals on the cell shape, connectivity, and topology. Furthermore, they noted a decrease of the thermal conductivity in the presence of geometrical imperfections. Baumeister *et al.* [11] investigated the thermal properties of syntactic hollow sphere composites. Corundum-based hollow spheres were embedded in an epoxy matrix and the thermal expansion coefficient was determined. It was found that the thermal behavior of these composites was mainly governed by the epoxy resin used. The effective thermal conductivity of adhesively bonded and sintered HSS was numerically investigated in Refs [12, 13]. As a result of these finite element (FE) analyses, a strong dependence of the thermal properties on the joining technology and

Advanced Computational Materials Modeling: From Classical to Multi-Scale Techniques.
Edited by Miguel Vaz Júnior, Eduardo A. de Souza Neto, and Pablo A. Munoz-Rojas

ISBN: 978-3-527-32479-8

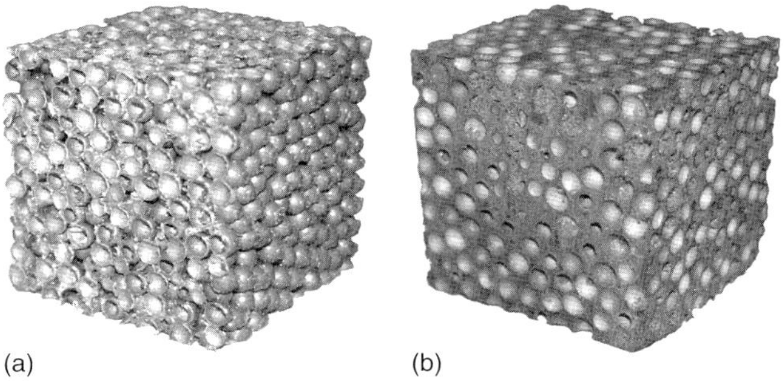

Figure 3.1 Metallic hollow sphere structures: (a) partial morphology and (b) syntactic morphology.

morphology of the structures was found. In Ref. [14], FE and experimental analyses of the effective thermal conductivity of adhesively bonded HSS were conducted.

An alternative to the classical finite element method (FEM) has recently been developed, that is, a particularly flexible one for addressing complex phenomenological mass and thermal diffusion problems. It is based on a lattice model and is addressed using Monte Carlo methods where it is often called the *lattice Monte Carlo* (*LMC*) method (see the recent review [15]). Although Monte Carlo methods traditionally have gained the reputation for being especially demanding of computer time, this is generally no longer a major consideration in their implementation: contemporary PCs can readily and accurately cope with the computational demands often in less than several hours of computational time. In this study, the LMC method is used to determine the effective conductivity in a model of HSS.

In Section 3.2, the methodology of the two numerical approaches is given. Section 3.2.1 addresses the LMC approach, Sections 3.2.2 and 3.2.3 focus on the FEM and FE modeling, respectively. Sections 3.3–3.6 deal with the numerical analysis of various models of HSSs with increasing geometric complexity: Section 3.3 introduces FE analysis on simple porous structures with cubic symmetry. In Section 3.4, these model structures are refined and metallic shells are introduced to account for the contribution of the hollow spheres. The influence of the sphere wall thickness and different cubic-symmetric sphere arrangements on the effective thermal conductivity is investigated. In Section 3.5, the transition to randomly distributed hollow spheres is described. Models of cross sections of HSSs are used for thermal LMC analysis. Section 3.6 addresses numerical analyses of real HSS geometries obtained by computed tomography (CT) scanning. Section 3.7 contains a conclusion of the analysis techniques and obtained results.

3.2
Methodology

Within this chapter, two different numerical methods are used. The FEM is industrial standard and allows the approximate solution of partial differential equations, that is, describing a temperature field. However, FE analysis puts high demands on computer memory and restricts geometric resolution to relatively low values. In contrast, a recently formulated LMC method allows the use of large high-resolution models for thermal analysis.

3.2.1 Lattice Monte Carlo Method

In the following, the use of the LMC method toward the calculation of the effective thermal conductivity (λ_{eff}) of materials will be elucidated. The effective thermal conductivity of a composite is the equivalent of the thermal conductivity of a single phase as long as a sufficiently large volume is considered. The effective thermal conductivity can then, for example, be used in Fourier's law to characterize the thermal behavior of this composite. The LMC method is based on the fact that thermal diffusion is a random process that can be represented by random walks of particles on a discrete lattice. In thermal LMC, these virtual particles represent small energy quantities that increase internal energy and temperature of small volumes represented by lattice sites.

The particle diffusivity $\mathbf{D}$ in an anisotropic material is presented as a tensor:

$$\mathbf{D} = \begin{bmatrix} D_{xx} & D_{xy} & D_{xz} \\ D_{yx} & D_{yy} & D_{yz} \\ D_{zx} & D_{zy} & D_{zz} \end{bmatrix} \tag{3.1}$$

where the Einstein equation describes the diffusivities D_{kl} in the long-time limit:

$$D_{kl} = \frac{\langle R_k R_l \rangle}{2t} \tag{3.2}$$

where R_k, R_l $(k, l = x, y, z)$ are components of the vector displacement $\mathbf{R}$ of a particle in time t and the Dirac brackets refer to a large number of particle histories. In the case of isotropic material and an orthogonal coordinate system, $\mathbf{D}$ can be reduced to

$$\mathbf{D} = D \begin{bmatrix} 1 & 0 & 0 \\ & 1 & 0 \\ \text{sym.} & & 1 \end{bmatrix} \tag{3.3}$$

with one component D given by

$$D_{kk} = D = \frac{\langle \mathbf{R}^2 \rangle}{2dt} \tag{3.4}$$

where d is the geometric dimension of the problem (i.e., $d = 2$ for a two-dimensional analysis or $d = 3$ for a three-dimensional analysis). Equation (3.2) applies to both the thermal diffusivity K and the self particle diffusivity when isolated heat entities and particles are considered. The thermal conductivity λ_i in a phase i is related to the thermal diffusivity K_i in that phase by the expression $K_i = \lambda_i/(\rho_i C_i)$ where ρ_i is the density of phase i and C_i is the specific heat of phase i. It is then possible to make use of the Einstein equation in a Monte Carlo simulation for calculating the *relative* effective thermal conductivity λ_{eff}/λ_i (λ_i is, for convenience, the maximum thermal conductivity of the individual phases present) by simply assigning the densities and the specific heats the value of unity in all phases. Then λ_{eff}/λ_i equals the relative effective thermal diffusivity K_{eff}/K_i. Particles are released, one at a time, from randomly chosen sites within the lattice and each is permitted to explore the lattice on a random walk for the same time t. A time step is an attempt for the particle to jump at a given site. The conductivities (strictly diffusivities) of each phase are represented by different jump frequencies Γ (strictly jump probabilities) noting that the diffusion coefficient can be written for an isotropic material as

$$D = \Gamma s^2/2d \tag{3.5}$$

where s is the jump distance in the lattice. The partitioning of D in Eq. (3.5) contains no correlation effects in the random walks because the particles here are independent. The highest jump frequency is scaled to unity for efficiency. For example, consider the thermal conductivities of three phases designated as the matrix (**M**) $\lambda_M = 0.0214\ \mathrm{W\,m^{-1}\,K^{-1}}$ [14], the inclusions (I) $\lambda_I = 0\ \mathrm{W\,m^{-1}\,K^{-1}}$, and the shells (S), (a shell coats each inclusion) $\lambda_S = 51.9\ \mathrm{W\,m^{-1}\,K^{-1}}$ [16]. The corresponding jump frequencies are $\Gamma_M = 0.00041$, $\Gamma_I = 0$, and $\Gamma_S = 1.0$.

The LMC algorithm follows the following basic steps: prior to the simulation, a large population of thermal particles (N is approximately 10^6) is dispersed in the lattice model. The larger the number of particles N, the higher the accuracy of the calculation (the accuracy can be approximated by $N^{-1/2}$) but the computational load is of course higher at the same time. Accordingly, a balance between accuracy and computation time must be found. Next, a particle is randomly chosen and its position in the lattice is determined. A random jump direction is obtained and based on the thermal conductivities of origin and destination node, a jump frequency Γ is calculated. This jump frequency is compared to a uniformly distributed random number $0 \leq \chi < 1$, and if $\Gamma > \chi$ is valid, the jump attempt is successful, that is, the particle coordinated is updated before the calculation time t is incremented. This procedure is repeated until the final calculation t_f time is reached. Time t_f must be chosen large enough in order to ensure the numerical convergence of the results to the steady-state solution. Time t_f can be approximated using the following random walk upper bound estimate for the average distance traveled by a particle in time t_f : $d = (\lambda_i\ t_f)^{1/2}$; d should be of the order of the linear dimension of the computational box that represents the composite material. The result of the random walk simulation is particle displacements R, which can be used to calculate the effective thermal diffusivity/conductivity according to Eq. (3.4). Further information on thermal LMC simulation and other computationally more efficient algorithms can be found in Ref. [17].

3.2.2
Finite Element Method

3.2.2.1 Basics of Heat Transfer

Heat conduction analysis is based on Fourier's law:

$$\mathbf{q} = -\mathbf{k}\nabla\mathbf{T} \tag{3.6}$$

where $\mathbf{q} = \{q_x \; q_y \; q_z\}^{\mathrm{T}}$ is the heat flux vector and $\nabla T = \{\partial T/\partial x \;\; \partial T/\partial y \;\; \partial T/\partial z\}^{\mathrm{T}}$ is the temperature gradient vector which is generated by the Nabla operator $\nabla T = \{\partial/\partial x \;\; \partial/\partial y \;\; \partial/\partial z\}^{\mathrm{T}}$. The continuum conductivity matrix $\mathbf{k}$ is given for an anisotropic material as

$$\mathbf{k} = \begin{bmatrix} k_{xx} & k_{xy} & k_{xz} \\ k_{yx} & k_{yy} & k_{yz} \\ k_{zx} & k_{zy} & k_{zz} \end{bmatrix} \tag{3.7}$$

which reduces for isotropic materials to

$$\mathbf{k} = k \begin{bmatrix} 1 & 0 & 0 \\ & 1 & 0 \\ \text{sym.} & & 1 \end{bmatrix} \tag{3.8}$$

To solve a heat conduction problem means to determine the temperature field $T = T(x, y, z, t)$ in its spatial (Cartesian coordinates: x, y, z) and temporal (time: t) dependency. Then, the heat flux field $\mathbf{q} = \mathbf{q}(x, y, z, t)$ can be determined according to Fourier's law (Eq. (3.6)).

The unknown temperature field is obtained by solving a partial differential equation, the so-called heat diffusion equation (HDE) [18, 19]

$$\rho c \frac{\partial T}{\partial t} = \nabla^{\mathrm{T}}(\mathbf{k}\nabla T) + \dot{\eta} \tag{3.9}$$

where ρ is the mass density, c is the specific heat, t is the time, and $\dot{\eta}$ is the energy rate per unit volume that accounts for heat sources or sinks.

3.2.2.2 Weighted Residual Method

Let us consider the special case of the steady-state HDE ($\partial T/\partial t = 0$) where no sources or sinks are present ($\dot{\eta} = 0$):

$$\nabla^{\mathrm{T}}(\mathbf{k}\nabla T_0) = 0 \tag{3.10}$$

The basic idea of the weighted residual method [20] consists of multiplying the partial differential Eq. (3.10) with a weighting function w and to demand that the entire integral vanishes over the whole domain. For the true solution T_0, this expression is independent of the weighting function and always fulfilled. Substituting the exact solution T_0 by an approximate solution[1)] T produces a

1) For simplicity, the variable T will be used in the following for the approximate solution. This variable should not be confused with the real continuum temperature T of Section 3.2.2.1.

"residual" function R such that

$$R = \nabla^{\mathrm{T}} (\mathbf{k}\nabla T) \neq 0 \tag{3.11}$$

This error will be distributed according to the scalar weighting function $w = w(x, y, z)$ and the integral over the entire three-dimensional domain $\Omega = (x, y, z)$ will be forced to be zero in a certain *average* sense:

$$\int_{\Omega} w \left(\nabla^{\mathrm{T}} (\mathbf{k}\nabla T)\right) \mathrm{d}\Omega = \int_{\Omega} wR \mathrm{d}\Omega = 0 \tag{3.12}$$

Equation (3.12) is known as the so-called *inner product* while the original statement, that is, Eq. (3.10), is referred to as the *strong formulation or classical form*. The so-called *weak formulation*, that is, where the order[2)] of the differential operator (∇) is the same for w and T, can be derived by the application of the Green–Gauss theorem [21]:

$$\int_{\Omega} w\nabla^{\mathrm{T}} (\mathbf{k}\nabla T) \mathrm{d}\Omega = \int_{\Gamma} w (\mathbf{k}\nabla T)^{\mathrm{T}} \mathbf{n} \mathrm{d}\Gamma - \int_{\Omega} \left(\nabla^{\mathrm{T}} w\right) (\mathbf{k}\nabla T) \mathrm{d}\Omega = 0 \tag{3.13}$$

$$\int_{\Omega} \left(\nabla^{\mathrm{T}} w\right) (\mathbf{k}\nabla T) \mathrm{d}\Omega = \int_{\Gamma} w (\mathbf{k}\nabla T)^{\mathrm{T}} \mathbf{n} \mathrm{d}\Gamma \tag{3.14}$$

The weak formulation (3.14) forms the basis for the derivation of the principal FE equation. Other numerical methods, for example, the boundary element method, require further integration of the weak formulation, which results in the so-called inverse formulation where the differential operator is completely shifted to the weighting function. All classical approximation methods can be derived based on the weighted residual method. The degree of integration and the choice of the weighting function w define finally which method is obtained [20].

3.2.2.3 Discretization and Principal Finite Element Equation

The basic idea of the FEM is to approximate the unknown temperature T not in the entire domain Ω as given in Eq. (3.14) but in a subdomain Ω_{e}, that is, a so-called FE (cf. Figure 3.2), by an expansion:

$$T_{\mathrm{e}} = \mathbf{N}_{\mathrm{e}}^{\mathrm{T}} \mathbf{T}_{\mathrm{e},8} = \{N_1 \ N_2 \ldots N_8\} \begin{Bmatrix} T_{\mathrm{e},1} \\ T_{\mathrm{e},2} \\ \vdots \\ T_{\mathrm{e},8} \end{Bmatrix} \tag{3.15}$$

where $\mathbf{N}_8$ are the shape functions[3)] prescribed in terms of independent variables (such as the special coordinates) and all or most of the nodal temperatures $\mathbf{T}_{\mathrm{e},8}$ are unknown. The index 8 denotes the number of nodes. To derive the FEM, the

2) For k = const., the second-order derivative is obtained in the temperature: $\nabla^2 T$.

3) Let us assume for the derivations a three-dimensional element with 8 nodes.

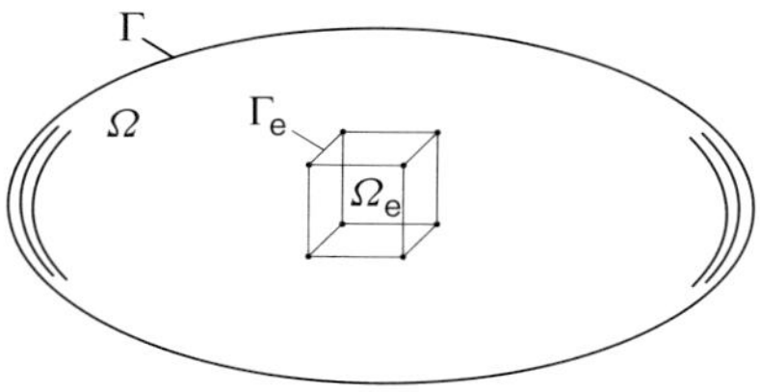

Figure 3.2 Finite element approximation of a spatial domain Ω.

weighting function w is approximated within an element in a similar manner as the temperature (this approach is also known as the *Galerkin method*).

$$w = \delta \mathbf{T}_{e,8}^{T} \mathbf{N}_e = \left\{ \delta T_{e,1}\ \delta T_{e,2} \ldots \delta T_{e,8} \right\} \left\{ \begin{array}{c} N_1 \\ N_2 \\ \vdots \\ N_8 \end{array} \right\} \tag{3.16}$$

where $\delta\mathbf{T}_{e,8}$ are arbitrary temperatures and 8 is the number of unknowns entering the system. Using these two field approximations, the left-hand side (LHS) of Eq. (3.14) can be written as

$$\int_{\Omega_e} \left(\nabla^T \left(\delta \mathbf{T}_{e,8}^T \mathbf{N}_e\right)\right) \left(\mathbf{k}\nabla \left(\mathbf{N}_e^T \mathbf{T}_{e,8}\right)\right) \mathrm{d}\Omega_e \tag{3.17}$$

where the vectors $\delta\mathbf{T}_{e,8}$ and $\mathbf{T}_{e,8}$ are not a function of the spatial coordinates, and thus, can be considered as constants with respect to the Nabla operator ∇. Consequently, these temperature vectors can be taken out of the brackets to give

$$\delta \mathbf{T}_{e,8}^T \int_{\Omega_e} \left(\nabla^T \mathbf{N}_e\right) \left(\mathbf{k} \nabla \mathbf{N}_e^T\right) \mathrm{d}\Omega_e \mathbf{T}_{e,8} = \delta \mathbf{T}_{e,8}^T \mathbf{K}_e \mathbf{T}_{e,8} \tag{3.18}$$

The matrix $\mathbf{K}_e$ is the *thermal conductivity matrix* and is of dimension [8 × 8]. Application of the Nabla operator to the vector of the shape functions gives the following statement:

$$\mathbf{K}_e = \int_{\Omega_e} \left\{ \frac{\partial \mathbf{N}_e}{\partial x}\ \frac{\partial \mathbf{N}_e}{\partial y}\ \frac{\partial \mathbf{N}_e}{\partial z} \right\} \mathbf{k} \left\{ \begin{array}{c} \dfrac{\partial \mathbf{N}_e^T}{\partial x} \\ \dfrac{\partial \mathbf{N}_e^T}{\partial y} \\ \dfrac{\partial \mathbf{N}_e^T}{\partial z} \end{array} \right\} \mathrm{d}\Omega_e \tag{3.19}$$

or with the single shape functions in more detail as

$$\mathbf{K}_e = \int_{\Omega e} \left\{ \left\{ \begin{matrix} \frac{\partial N_1}{\partial x} \\ \frac{\partial N_2}{\partial x} \\ \vdots \\ \frac{\partial N_8}{\partial x} \end{matrix} \right\} \left\{ \begin{matrix} \frac{\partial N_1}{\partial y} \\ \frac{\partial N_2}{\partial y} \\ \vdots \\ \frac{\partial N_8}{\partial y} \end{matrix} \right\} \left\{ \begin{matrix} \frac{\partial N_1}{\partial z} \\ \frac{\partial N_2}{\partial z} \\ \vdots \\ \frac{\partial N_8}{\partial z} \end{matrix} \right\} \right\} \times \mathbf{k} \left\{ \begin{matrix} \left\{ \frac{\partial N_1}{\partial x} \; \frac{\partial N_2}{\partial x} \; \cdots \; \frac{\partial N_8}{\partial x} \right\} \\ \left\{ \frac{\partial N_1}{\partial y} \; \frac{\partial N_2}{\partial y} \; \cdots \; \frac{\partial N_8}{\partial y} \right\} \\ \left\{ \frac{\partial N_1}{\partial z} \; \frac{\partial N_2}{\partial z} \; \cdots \; \frac{\partial N_8}{\partial z} \right\} \end{matrix} \right\} \mathrm{d}\Omega_e \tag{3.20}$$

The integration over the subdomain Ω_e is approximated by numerical integration. To this end, the coordinates (x, y, z) are transformed to the unit space (ξ, η, ζ) where each coordinate ranges from -1 to 1. In the scope of the coordinate transformation, attention must be paid to the derivatives. For example, the derivative of the shape functions with respect to the x-coordinate is transformed in the following way:

$$\frac{\partial N_i}{\partial x} \rightarrow \frac{\partial N_i}{\partial \xi}\frac{\partial \xi}{\partial x} + \frac{\partial N_i}{\partial \eta}\frac{\partial \eta}{\partial x} + \frac{\partial N_i}{\partial \zeta}\frac{\partial \zeta}{\partial x}, i = 1, \ldots, 8 \tag{3.21}$$

Introducing these new derivatives gives the element conductivity matrix:

$$\mathbf{K}_e = \int_{\Omega'_e} \left\{ \frac{\partial \mathbf{N}_e}{\partial \xi} \frac{\partial \mathbf{N}_e}{\partial \eta} \frac{\partial \mathbf{N}_e}{\partial \zeta} \right\} \begin{bmatrix} \frac{\partial \xi}{\partial x} & \frac{\partial \xi}{\partial y} & \frac{\partial \xi}{\partial z} \\ \frac{\partial \eta}{\partial x} & \frac{\partial \eta}{\partial y} & \frac{\partial \eta}{\partial z} \\ \frac{\partial \zeta}{\partial x} & \frac{\partial \zeta}{\partial y} & \frac{\partial \zeta}{\partial z} \end{bmatrix} \times \mathbf{k} \begin{bmatrix} \frac{\partial \xi}{\partial x} & \frac{\partial \xi}{\partial y} & \frac{\partial \xi}{\partial z} \\ \frac{\partial \eta}{\partial x} & \frac{\partial \eta}{\partial y} & \frac{\partial \eta}{\partial z} \\ \frac{\partial \zeta}{\partial x} & \frac{\partial \zeta}{\partial y} & \frac{\partial \zeta}{\partial z} \end{bmatrix} \left\{ \begin{matrix} \frac{\partial \mathbf{N}_e^T}{\partial \xi} \\ \frac{\partial \mathbf{N}_e^T}{\partial \eta} \\ \frac{\partial \mathbf{N}_e^T}{\partial \zeta} \end{matrix} \right\} \mathrm{d}\Omega'_e \tag{3.22}$$

where $\mathrm{d}\Omega'_e = J\mathrm{d}\xi\,\mathrm{d}\eta\,\mathrm{d}\zeta$. The last equation can be written in the following compact form:

$$\mathbf{K}_e = \int_{\Omega'_e} \mathbf{B}\mathbf{k}\mathbf{B}^{\mathrm{T}}\mathrm{d}\Omega'_e \tag{3.23}$$

where $\mathbf{B} = \left\{ \frac{\partial \mathbf{N}_e}{\partial \xi} \; \frac{\partial \mathbf{N}_e}{\partial \eta} \; \frac{\partial \mathbf{N}_e}{\partial \zeta} \right\} \mathbf{J} = \nabla' \mathbf{N}_e^{\mathrm{T}} \mathbf{J}$ is the temperature gradient matrix. Multiplying the gradient vector of the shape functions, $\nabla' \mathbf{N}_e^{\mathrm{T}}$, with the matrix of the geometrical derivatives, $\mathbf{J}$, gives the row vector of dimension $[1 \times 3]$. However, each of the elements of this vector is again a column vector with eight elements and the product can finally be regarded as a matrix of dimension [8 × 3]:

$$\mathbf{B} = \begin{bmatrix} \frac{\partial N_1}{\partial \xi}\frac{\partial \xi}{\partial x} + \frac{\partial N_1}{\partial \eta}\frac{\partial \eta}{\partial x} + \frac{\partial N_1}{\partial \zeta}\frac{\partial \zeta}{\partial x} & \frac{\partial N_1}{\partial \xi}\frac{\partial \xi}{\partial y} + \frac{\partial N_1}{\partial \eta}\frac{\partial \eta}{\partial y} + \frac{\partial N_1}{\partial \zeta}\frac{\partial \zeta}{\partial y} & \frac{\partial N_1}{\partial \xi}\frac{\partial \xi}{\partial z} + \frac{\partial N_1}{\partial \eta}\frac{\partial \eta}{\partial z} + \frac{\partial N_1}{\partial \zeta}\frac{\partial \zeta}{\partial z} \\ \frac{\partial N_2}{\partial \xi}\frac{\partial \xi}{\partial x} + \frac{\partial N_2}{\partial \eta}\frac{\partial \eta}{\partial x} + \frac{\partial N_2}{\partial \zeta}\frac{\partial \zeta}{\partial x} & \frac{\partial N_2}{\partial \xi}\frac{\partial \xi}{\partial y} + \frac{\partial N_2}{\partial \eta}\frac{\partial \eta}{\partial y} + \frac{\partial N_2}{\partial \zeta}\frac{\partial \zeta}{\partial y} & \frac{\partial N_2}{\partial \xi}\frac{\partial \xi}{\partial z} + \frac{\partial N_2}{\partial \eta}\frac{\partial \eta}{\partial z} + \frac{\partial N_2}{\partial \zeta}\frac{\partial \zeta}{\partial z} \\ \vdots & \vdots & \vdots \\ \frac{\partial N_8}{\partial \xi}\frac{\partial \xi}{\partial x} + \frac{\partial N_8}{\partial \eta}\frac{\partial \eta}{\partial x} + \frac{\partial N_8}{\partial \zeta}\frac{\partial \zeta}{\partial x} & \frac{\partial N_8}{\partial \xi}\frac{\partial \xi}{\partial y} + \frac{\partial N_8}{\partial \eta}\frac{\partial \eta}{\partial y} + \frac{\partial N_8}{\partial \zeta}\frac{\partial \zeta}{\partial y} & \frac{\partial N_8}{\partial \xi}\frac{\partial \xi}{\partial z} + \frac{\partial N_8}{\partial \eta}\frac{\partial \eta}{\partial z} + \frac{\partial N_8}{\partial \zeta}\frac{\partial \zeta}{\partial z} \end{bmatrix} \tag{3.24}$$

Multiplying $\mathbf{B}$ with $\mathbf{B}^{\mathrm{T}}$ or $\mathbf{kB}^{\mathrm{T}}$ gives finally the element conductivity matrix, which is of dimension [8 × 8]. Summarizing, we can conclude that the evaluation of the element conductivity matrix $\mathbf{K}_{\mathrm{e}}$ comprises the following steps:

- Determination of the temperature gradient matrix $\mathbf{B}$
- triple matrix product $\mathbf{BkB}^{\mathrm{T}}$ and
- numerical integration.

To evaluate the right-hand side (RHS) of the weak statement according to Eq. (3.14), we introduce the field approximation of the weighting function according to $w = \delta\mathbf{T}_{\mathrm{e},8}^{\mathrm{T}}\mathbf{N}_{\mathrm{e}}$:

$$\int_{\Gamma_{\mathrm{e}}} \left(\delta\mathbf{T}_{\mathrm{e},8}^{T}\mathbf{N}_{\mathrm{e}}\right)\left(k\nabla T\right)^{\mathrm{T}}\mathbf{n}\mathrm{d}\Gamma_{\mathrm{e}} \tag{3.25}$$

Since $\delta\mathbf{T}_{\mathrm{e},8}^{\mathrm{T}}$ can be canceled with the LHS of Eq. (3.18), we get

$$\mathbf{F}_{\mathrm{e}} = \int_{\Gamma_{\mathrm{e}}} \mathbf{N}_{\mathrm{e}}\left(\mathbf{k}\nabla T\right)^{\mathrm{T}}\mathbf{n}\mathrm{d}\Gamma_{\mathrm{e}} \tag{3.26}$$

which needs to be evaluated for each node along the element boundary Γ_{e}. The vector $\mathbf{F}_{\mathrm{e}}$ is the element vector of externally applied equivalent nodal loads. Equations (3.23) and (3.26) can be combined on the elemental level to the so-called principal FE equation:

$$\mathbf{K}_{\mathrm{e}}\mathbf{T}_{\mathrm{e},8} = \mathbf{F}_{\mathrm{e}} \tag{3.27}$$

The simplest representative of a three-dimensional FE is an eight-node hexahedron as shown in Figure 3.2 This element uses trilinear interpolation functions and the thermal gradients tend to be constant throughout the element. Let us derive the element formulation from the assumption that a trilinear temperature field is given in parametric space:

$$T_e\,(\xi,\eta,\zeta) = a_1 + a_2\xi + a_3\eta + a_4\zeta + a_5\xi\eta + a_6\eta\zeta + a_7\xi\zeta + a_8\xi\eta\zeta \qquad (3.28)$$

or in vector notation:

$$T_e\,(\xi,\eta,\zeta) = \chi^{\mathrm{T}}\mathbf{a} = \{1\ \xi\ \eta\ \zeta\ \xi\ \eta\ \eta\ \zeta\ \xi\ \zeta\ \xi\eta\zeta\}\begin{Bmatrix} a_1\\ a_2\\ a_3\\ a_4\\ a_5\\ a_6\\ a_7\\ a_8 \end{Bmatrix} \qquad (3.29)$$

Evaluating Eq. (3.29) for all eight nodes of the quadrilateral element (cf. Figure 3.3) gives

$$\begin{aligned}
&\text{Node 1: } T_{e,1} = T(-1,-1,-1) = a_1 - a_2 - a_3 - a_4 + a_5 + a_6 + a_7 - a_8\\
&\text{Node 2: } T_{e,2} = T(1,-1,-1) = a_1 + a_2 - a_3 - a_4 - a_5 + a_6 - a_7 + a_8\\
&\qquad\vdots \qquad\qquad\qquad\qquad\qquad \vdots\\
&\text{Node 3: } T_{e,8} = T(-1,1,1) = a_1 - a_2 + a_3 + a_4 - a_5 + a_6 - a_7 - a_8
\end{aligned} \qquad (3.30)$$

or in matrix notation:

$$\begin{Bmatrix} T_{e,1}\\ T_{e,2}\\ T_{e,3}\\ T_{e,4}\\ T_{e,5}\\ T_{e,6}\\ T_{e,7}\\ T_{e,8} \end{Bmatrix} = \begin{bmatrix}
1 & -1 & -1 & -1 & 1 & 1 & 1 & -1\\
1 & 1 & -1 & -1 & -1 & 1 & -1 & 1\\
1 & 1 & 1 & -1 & 1 & -1 & -1 & -1\\
1 & -1 & 1 & -1 & -1 & -1 & 1 & 1\\
1 & -1 & -1 & 1 & 1 & -1 & -1 & 1\\
1 & 1 & -1 & 1 & -1 & -1 & 1 & -1\\
1 & 1 & 1 & 1 & 1 & 1 & 1 & 1\\
1 & -1 & 1 & 1 & -1 & 1 & -1 & -1
\end{bmatrix} \begin{Bmatrix} a_1\\ a_2\\ a_3\\ a_4\\ a_5\\ a_6\\ a_7\\ a_8 \end{Bmatrix} \qquad (3.31)$$

Solving for **a** gives

$$\begin{Bmatrix} a_1\\ a_2\\ a_3\\ a_4\\ a_5\\ a_6\\ a_7\\ a_8 \end{Bmatrix} = \frac{1}{8}\begin{bmatrix}
1 & 1 & 1 & 1 & 1 & 1 & 1 & 1\\
-1 & 1 & 1 & -1 & -1 & 1 & 1 & -1\\
-1 & -1 & 1 & 1 & -1 & -1 & 1 & 1\\
-1 & -1 & -1 & -1 & 1 & 1 & 1 & 1\\
1 & -1 & 1 & -1 & 1 & -1 & 1 & -1\\
1 & 1 & -1 & -1 & -1 & -1 & 1 & 1\\
1 & -1 & -1 & 1 & -1 & 1 & 1 & -1\\
-1 & 1 & -1 & 1 & 1 & -1 & 1 & -1
\end{bmatrix} \begin{Bmatrix} T_{e,1}\\ T_{e,2}\\ T_{e,3}\\ T_{e,4}\\ T_{e,5}\\ T_{e,6}\\ T_{e,7}\\ T_{e,8} \end{Bmatrix} \qquad (3.32)$$

or

$$\mathbf{a} = \mathbf{A}\mathbf{T}_{e,8} = \mathbf{X}^{-1}\mathbf{T}_{e,8} \qquad (3.33)$$

The vector of shape functions results as

$$\mathbf{N}_e^T = \{N_1 \quad N_2 \quad N_3 \quad N_4 \quad N_5 \quad N_6 \quad N_7 \quad N_8\} = \chi^T \mathbf{A} \tag{3.34}$$

or

$$N_1 = \frac{1}{8}(1-\xi)\cdot(1-\eta)\cdot(1-\zeta) \tag{3.35}$$

$$N_2 = \frac{1}{8}(1+\xi)\cdot(1-\eta)\cdot(1-\zeta) \tag{3.36}$$

$$N_3 = \frac{1}{8}(1+\xi)\cdot(1+\eta)\cdot(1-\zeta) \tag{3.37}$$

$$N_4 = \frac{1}{8}(1-\xi)\cdot(1+\eta)\cdot(1-\zeta) \tag{3.38}$$

$$N_5 = \frac{1}{8}(-1+\xi)\cdot(-1+\eta)\cdot(1+\zeta) \tag{3.39}$$

$$N_6 = \frac{1}{8}(1+\xi)\cdot(1-\eta)\cdot(1+\zeta) \tag{3.40}$$

$$N_7 = \frac{1}{8}(1+\xi)\cdot(1+\eta)\cdot(1+\zeta) \tag{3.41}$$

$$N_8 = \frac{1}{8}(1-\xi)\cdot(1+\eta)\cdot(1+\zeta) \tag{3.42}$$

or in a more compact form as

$$N_i = \frac{1}{8}(1+\xi\xi_i)\cdot(1+\eta\eta_i)\cdot(1+\zeta\zeta_i) \tag{3.43}$$

where ξ_i, η_i, and ζ_i are the coordinates of the nodes in unit space ($i = 1, \ldots, 8$) (cf. Figure 3.3). The derivatives with respect to the parametric coordinates can easily be obtained as

$$\frac{\partial N_i}{\partial \xi} = \frac{1}{8}\cdot(\xi_i)\cdot(1+\eta\eta_i)\cdot(1+\zeta\zeta_i) \tag{3.44}$$

$$\frac{\partial N_i}{\partial \eta} = \frac{1}{8}\cdot(1+\xi\xi_i)\cdot(\eta_i)\cdot(1+\zeta\zeta_i) \tag{3.45}$$

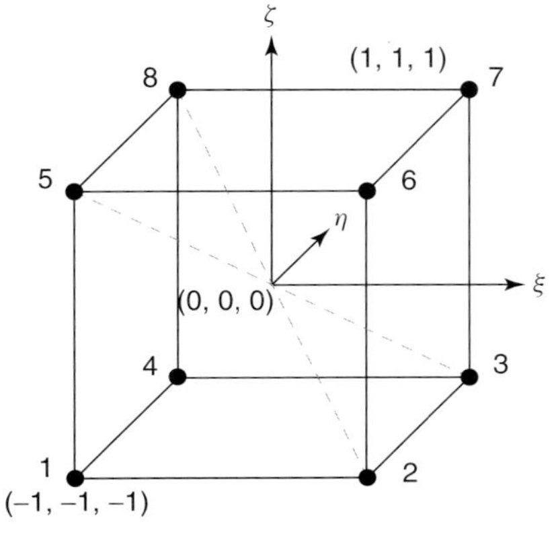

Figure 3.3 Three-dimensional eight-node hexahedron in parametric space.

$$\frac{\partial N_i}{\partial \zeta} = \frac{1}{8} \cdot (1 + \xi \xi_i) \cdot (1 + \eta \eta_i) \cdot (\zeta_i) \tag{3.46}$$

The geometrical derivatives in Eq. (3.22), for example, $\frac{\partial \xi}{\partial x}, \frac{\partial \xi}{\partial y}, \frac{\partial \xi}{\partial y}$, can be calculated on the basis of

$$\begin{bmatrix} \frac{\partial \xi}{\partial x} & \frac{\partial \xi}{\partial y} & \frac{\partial \xi}{\partial z} \\ \frac{\partial \eta}{\partial x} & \frac{\partial \eta}{\partial y} & \frac{\partial \eta}{\partial z} \\ \frac{\partial \zeta}{\partial x} & \frac{\partial \zeta}{\partial y} & \frac{\partial \zeta}{\partial z} \end{bmatrix}
= \frac{1}{J} \cdot \begin{bmatrix} \frac{\partial y}{\partial \eta}\frac{\partial z}{\partial \zeta} - \frac{\partial y}{\partial \zeta}\frac{\partial z}{\partial \eta} & -\frac{\partial x}{\partial \eta}\frac{\partial z}{\partial \zeta} + \frac{\partial x}{\partial \zeta}\frac{\partial z}{\partial \eta} & \frac{\partial x}{\partial \eta}\frac{\partial y}{\partial \zeta} - \frac{\partial x}{\partial \zeta}\frac{\partial y}{\partial \eta} \\ -\frac{\partial y}{\partial \xi}\frac{\partial z}{\partial \zeta} + \frac{\partial y}{\partial \zeta}\frac{\partial z}{\partial \xi} & \frac{\partial x}{\partial \xi}\frac{\partial z}{\partial \zeta} - \frac{\partial x}{\partial \zeta}\frac{\partial z}{\partial \xi} & -\frac{\partial x}{\partial \xi}\frac{\partial y}{\partial \zeta} + \frac{\partial x}{\partial \zeta}\frac{\partial y}{\partial \xi} \\ \frac{\partial y}{\partial \xi}\frac{\partial z}{\partial \eta} - \frac{\partial y}{\partial \eta}\frac{\partial z}{\partial \xi} & -\frac{\partial x}{\partial \xi}\frac{\partial z}{\partial \eta} + \frac{\partial x}{\partial \eta}\frac{\partial z}{\partial \xi} & \frac{\partial x}{\partial \xi}\frac{\partial zy}{\partial \eta} - \frac{\partial x}{\partial \eta}\frac{\partial y}{\partial \xi} \end{bmatrix} \tag{3.47}$$

where the Jacobian J is the determinant as given by

$$J = \left| \frac{\partial (x, y, z)}{\partial (\xi, \eta, \zeta)} \right| = x_\xi y_\eta z_\zeta + x_\eta y_\zeta z_\xi + x_\zeta y_\xi z_\eta - x_\eta y_\xi z_\zeta - x_\zeta y_\eta z_\xi - x_\xi y_\zeta z_\eta \tag{3.48}$$

In Eq. (3.48), abbreviations of the form, for example, $x_\xi = \frac{\partial x}{\partial \xi}$, and so on were used. Let us assume the same interpolation for the global x, y, and z coordinates as for the temperature:

$$x(\xi, \eta, \zeta) = \sum_{i=1}^{8} N_i(\xi, \eta, \zeta) \cdot x_i \tag{3.49}$$

$$y(\xi, \eta, \zeta) = \sum_{i=1}^{8} N_i(\xi, \eta, \zeta) \cdot y_i \tag{3.50}$$

$$z(\xi, \eta, \zeta) = \sum_{i=1}^{8} N_i(\xi, \eta, \zeta) \cdot z_i \tag{3.51}$$

where the global coordinates of the nodes 1, ... , 8 can be used for $x_1, \ldots, x_8$, and so on. Thus, the derivatives can easily be obtained as

$$\frac{\partial x}{\partial \xi} = \sum_{i=1}^{8} \frac{\partial N_i}{\partial \xi} \cdot x_i = \sum_{i=1}^{8} \frac{1}{8} \cdot (\xi_i) \cdot (1 + \eta \eta_i) \cdot (1 + \zeta \zeta_i) \cdot x_i \tag{3.52}$$

$$\frac{\partial x}{\partial \eta} = \sum_{i=1}^{8} \frac{\partial N_i}{\partial \eta} \cdot x_i = \sum_{i=1}^{8} \frac{1}{8} \cdot (1 + \xi \xi_i) \cdot (\eta_i) \cdot (1 + \zeta \zeta_i) \cdot x_i \tag{3.53}$$

$$\frac{\partial x}{\partial \zeta} = \sum_{i=1}^{8} \frac{\partial N_i}{\partial \zeta} \cdot x_i = \sum_{i=1}^{8} \frac{1}{8} \cdot (1 + \xi \xi_i) \cdot (1 + \eta \eta_i) \cdot (\zeta_i) \cdot x_i \tag{3.54}$$

$$\frac{\partial y}{\partial \xi} = \sum_{i=1}^{8} \frac{\partial N_i}{\partial \xi} \cdot y_i = \sum_{i=1}^{8} \frac{1}{8} \cdot (\xi_i) \cdot (1 + \eta\eta_i) \cdot (1 + \zeta\zeta_i) \cdot y_i \tag{3.55}$$

$$\frac{\partial y}{\partial \eta} = \sum_{i=1}^{8} \frac{\partial N_i}{\partial \eta} \cdot y_i = \sum_{i=1}^{8} \frac{1}{8} \cdot (1 + \xi\xi_i) \cdot (\eta_i) \cdot (1 + \zeta\zeta_i) \cdot y_i \tag{3.56}$$

$$\frac{\partial y}{\partial \zeta} = \sum_{i=1}^{8} \frac{\partial N_i}{\partial \zeta} \cdot y_i = \sum_{i=1}^{8} \frac{1}{8} \cdot (1 + \xi\xi_i) \cdot (1 + \eta\eta_i) \cdot (\zeta_i) \cdot y_i \tag{3.57}$$

$$\frac{\partial z}{\partial \xi} = \sum_{i=1}^{8} \frac{\partial N_i}{\partial \xi} \cdot z_i = \sum_{i=1}^{8} \frac{1}{8} \cdot (\xi_i) \cdot (1 + \eta\eta_i) \cdot (1 + \zeta\zeta_i) \cdot z_i \tag{3.58}$$

$$\frac{\partial z}{\partial \eta} = \sum_{i=1}^{8} \frac{\partial N_i}{\partial \eta} \cdot z_i = \sum_{i=1}^{8} \frac{1}{8} \cdot (1 + \xi\xi_i) \cdot (\eta_i) \cdot (1 + \zeta\zeta_i) \cdot z_i \tag{3.59}$$

$$\frac{\partial z}{\partial \zeta} = \sum_{i=1}^{8} \frac{\partial N_i}{\partial \zeta} \cdot z_i = \sum_{i=1}^{8} \frac{1}{8} \cdot (1 + \xi\xi_i) \cdot (1 + \eta\eta_i) \cdot (\zeta_i) \cdot z_i \tag{3.60}$$

The derivatives of the shape functions with respect to the coordinates in parametric space are summarized in Table 3.1. Note that the derivatives (3.52–3.60) are simple constants and independent of ξ, η, and ζ for a cuboid or a parallelepiped (cf. Figure 3.4).

Looking at the example of a cube with edge length $2a$ (cf. Figure 3.4a), one can derive that $\frac{\partial x}{\partial \xi} = \frac{\partial y}{\partial \eta} = \frac{\partial z}{\partial \zeta} = a$, whereas all other geometrical derivatives are zero. For a cuboid with edge lengths $2a$, $2b$, and $2c$ (cf. Figure 3.4b), similar derivation gives $\frac{\partial x}{\partial \xi} = a$, $\frac{\partial y}{\partial \eta} = b$, and $\frac{\partial z}{\partial \zeta} = c$.

Table 3.1 Derivatives of the shape functions in parametric space.

Node	$\frac{\partial N_i}{\partial \xi}$	$\frac{\partial N_i}{\partial \eta}$	$\frac{\partial N_i}{\partial \zeta}$
1	$\frac{1}{8}(-1)(1-\eta)(1-\zeta)$	$\frac{1}{8}(1-\xi)(-1)(1-\zeta)$	$\frac{1}{8}(1-\xi)(1-\eta)(-1)$
2	$\frac{1}{8}(1)(1-\eta)(1-\zeta)$	$\frac{1}{8}(1+\xi)(-1)(1-\zeta)$	$\frac{1}{8}(1+\xi)(1-\eta)(-1)$
3	$\frac{1}{8}(1)(1+\eta)(1-\zeta)$	$\frac{1}{8}(1+\xi)(1)(1-\zeta)$	$\frac{1}{8}(1+\xi)(1+\eta)(-1)$
4	$\frac{1}{8}(-1)(1+\eta)(1-\zeta)$	$\frac{1}{8}(1-\xi)(1)(1-\zeta)$	$\frac{1}{8}(1-\xi)(1+\eta)(-1)$
5	$\frac{1}{8}(-1)(1-\eta)(1+\zeta)$	$\frac{1}{8}(1-\xi)(-1)(1+\zeta)$	$\frac{1}{8}(1-\xi)(1-\eta)(1)$
6	$\frac{1}{8}(1)(1-\eta)(1+\zeta)$	$\frac{1}{8}(1+\xi)(-1)(1+\zeta)$	$\frac{1}{8}(1+\xi)(1-\eta)(1)$
7	$\frac{1}{8}(1)(1+\eta)(1+\zeta)$	$\frac{1}{8}(1+\xi)(1)(1+\zeta)$	$\frac{1}{8}(1+\xi)(1+\eta)(1)$
8	$\frac{1}{8}(-1)(1+\eta)(1+\zeta)$	$\frac{1}{8}(1-\xi)(1)(1+\zeta)$	$\frac{1}{8}(1-\xi)(1+\eta)(1)$

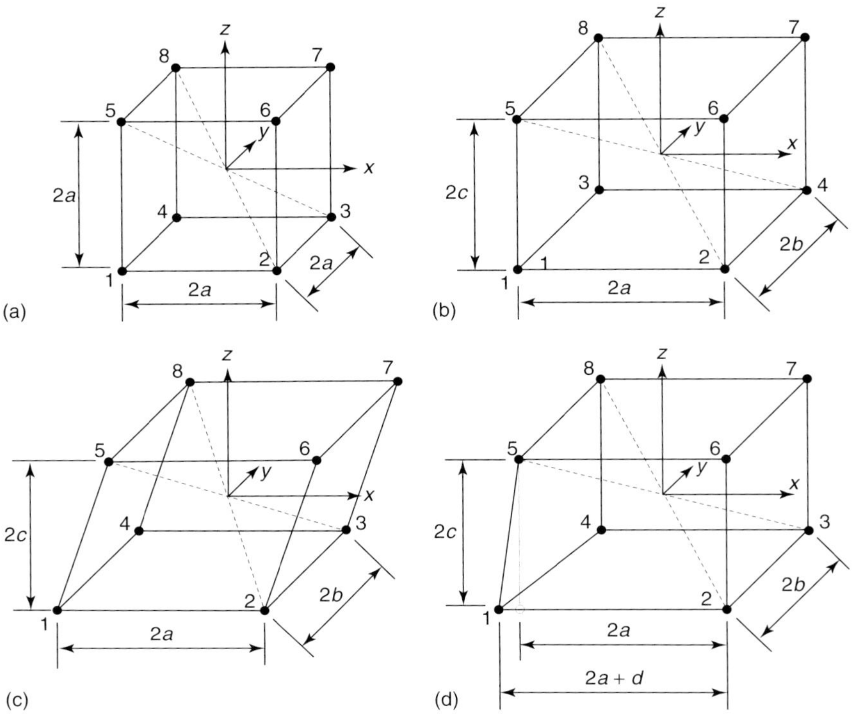

Figure 3.4 Different shapes of an eight-node hexahedron in (x, y, z)-space: (a) cube; (b) cuboid; (c) parallelepiped; and (d) distorted cuboid.

Considering the parallelepiped shown in Figure 3.4c where the lower nodes are moved by $-d$ in the negative x-direction and the upper nodes by $+d$ in the positive x-direction, one obtains $\frac{\partial x}{\partial \xi} = a$, $\frac{\partial y}{\partial \eta} = b$, $\frac{\partial z}{\partial \zeta} = c$, and $\frac{\partial x}{\partial \zeta} = d$ (other derivatives are zero). For other cases, the geometrical derivatives become dependent on the unit coordinates. For example, Figure 3.4d shows a distorted cuboid where node 1 is translated by d along the x-direction. For this case, one obtains

$$\frac{\partial x}{\partial \xi} = a + \frac{1}{8}d - \frac{1}{8}\zeta d - \frac{1}{8}\eta d + \frac{1}{8}\eta\zeta d \tag{3.61}$$

$$\frac{\partial x}{\partial \eta} = \frac{1}{8}d - \frac{1}{8}\zeta d - \frac{1}{8}\xi d + \frac{1}{8}\xi\zeta d \tag{3.62}$$

$$\frac{\partial x}{\partial \zeta} = \frac{1}{8}d - \frac{1}{8}\eta d - \frac{1}{8}\xi d + \frac{1}{8}\xi\eta d \tag{3.63}$$

$$\frac{\partial y}{\partial \eta} = b \tag{3.64}$$

$$\frac{\partial z}{\partial \zeta} = c \tag{3.65}$$

On the basis of the derived equations, the triple matrix product $\mathbf{BkB}^{\mathrm{T}}$ can now be numerically calculated.

Numerical Integration Elementary integration formulas, such as the trapezoidal or Simpson rule, often assume equally spaced data and can become somewhat limited in accuracy and efficiency when used in FE analysis. Gauss quadrature or integration has become the accepted numerical integration scheme in the majority of FE applications. The Gauss–Legendre quadrature locates sampling points and assigns weights so as to minimize integration error when the integrand is a general polynomial. Thus, for a given level of accuracy, Gauss quadrature uses fewer sampling points than other integration rules. Using Gauss–Legendre integration, one can write that

$$\int_{\Omega_e} f(x,y,z)\mathrm{d}\Omega_e = \int_{\Omega_e'} f'(\xi,\eta,\zeta)\mathrm{d}\Omega_e = \int_{-1}^{1}\int_{-1}^{1}\int_{-1}^{1} f'(\xi,\eta,\zeta)J\mathrm{d}\xi\mathrm{d}\eta\mathrm{d}\zeta$$

$$= \sum_{g} f'(\xi,\eta,\zeta)_g J_g W_g \tag{3.66}$$

where the Jacobian J is given in Eq. (3.48) and $(\xi,\eta,\zeta)_g$ are the coordinates of the integration or Gauss points and W_g are the corresponding weighting factors. The locations of the integration points and values of associated weights are given in Table 3.2.

RHS of the Weak Statement The RHS of the weak statement (3.26) needs to be evaluated for each node along the element boundary Γ_e. For node 1, the shape function N_1 is equal to one and identically zero for all other nodes. In addition, all other shape functions are identically zero for node 1. The expression

Table 3.2 Integration rules for hexahedral elements [22].

Points	ξ_g	η_g	ζ_g	Weight W_g	Error
1	0	0	0	8	$O(\xi^2)$
8	$\pm 1/\sqrt{3}$	$\pm 1/\sqrt{3}$	$\pm 1/\sqrt{3}$	1	$O(\xi^4)$
27	0	0	0	$\left(\frac{8}{9}\right)^3$	–
–	$\pm\sqrt{0.6}$	0	0	$\left(\frac{5}{9}\right)\left(\frac{8}{9}\right)^2$	–
–	0	$\pm\sqrt{0.6}$	0	$\left(\frac{5}{9}\right)\left(\frac{8}{9}\right)^2$	$O(\xi^6)$
–	0	0	$\pm\sqrt{0.6}$	$\left(\frac{5}{9}\right)\left(\frac{8}{9}\right)^2$	–
–	$\pm\sqrt{0.6}$	$\pm\sqrt{0.6}$	0	$\left(\frac{8}{9}\right)\left(\frac{5}{9}\right)^2$	–
–	$\pm\sqrt{0.6}$	0	$\pm\sqrt{0.6}$	$\left(\frac{8}{9}\right)\left(\frac{5}{9}\right)^2$	–
–	0	$\pm\sqrt{0.6}$	$\pm\sqrt{0.6}$	$\left(\frac{8}{9}\right)\left(\frac{5}{9}\right)^2$	–
–	$\pm\sqrt{0.6}$	$\pm\sqrt{0.6}$	$\pm\sqrt{0.6}$	$\left(\frac{5}{9}\right)^3$	–

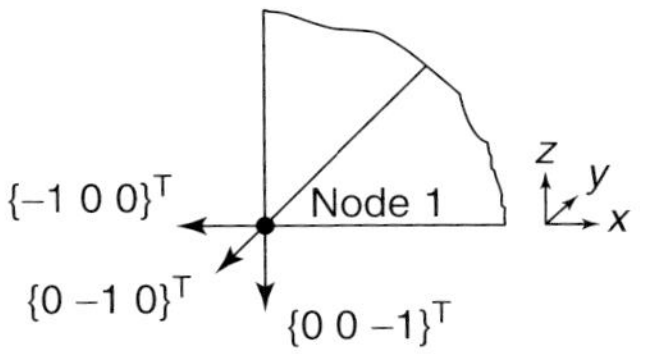

Figure 3.5 Normal vectors for RHS evaluation of Eq. (3.26) at node 1.

$(\nabla T)^{\mathrm{T}}\mathbf{n} = (gradT)^{\mathrm{T}}\mathbf{n} = \partial T/\partial \mathbf{n}$ is equal to the projection of the gradient vector of the temperature in the direction of the boundary normal vector.

Since we have three different normal vectors at each node, one may calculate this expression for node 1 as (cf. Figure 3.5):

$$k\left\{\frac{\partial T}{\partial x}\ \frac{\partial T}{\partial z}\ \frac{\partial T}{\partial z}\right\}\left\{\begin{matrix}-1\\0\\0\end{matrix}\right\} + k\left\{\frac{\partial T}{\partial x}\ \frac{\partial T}{\partial z}\ \frac{\partial T}{\partial z}\right\}\left\{\begin{matrix}0\\-1\\0\end{matrix}\right\}$$
$$+ k\left\{\frac{\partial T}{\partial x}\ \frac{\partial T}{\partial z}\ \frac{\partial T}{\partial z}\right\}\left\{\begin{matrix}0\\0\\-1\end{matrix}\right\} = \left(-k\frac{\partial T}{\partial x} - k\frac{\partial T}{\partial y} - k\frac{\partial T}{\partial z}\right)_{\text{node 1}} \tag{3.67}$$

which is equal to the heat flux entering the element at node 1. Similar results can be obtained for all other nodes and the RHS of the weak statement can be written as

$$\int_{\Gamma_e} \mathbf{N}_e\,(\mathbf{k}\nabla T)^{\mathrm{T}}\,\mathbf{n}\mathrm{d}\Gamma_e = \left\{\begin{matrix}
\left(-k\frac{\partial T}{\partial x} - k\frac{\partial T}{\partial y} - k\frac{\partial T}{\partial z}\right)_{\text{node 1}}\\
\left(+k\frac{\partial T}{\partial x} - k\frac{\partial T}{\partial y} - k\frac{\partial T}{\partial z}\right)_{\text{node 2}}\\
\left(+k\frac{\partial T}{\partial x} + k\frac{\partial T}{\partial y} - k\frac{\partial T}{\partial z}\right)_{\text{node 3}}\\
\left(-k\frac{\partial T}{\partial x} + k\frac{\partial T}{\partial y} - k\frac{\partial T}{\partial z}\right)_{\text{node 4}}\\
\left(-k\frac{\partial T}{\partial x} - k\frac{\partial T}{\partial y} + k\frac{\partial T}{\partial z}\right)_{\text{node 5}}\\
\left(+k\frac{\partial T}{\partial x} - k\frac{\partial T}{\partial y} + k\frac{\partial T}{\partial z}\right)_{\text{node 6}}\\
\left(+k\frac{\partial T}{\partial x} + k\frac{\partial T}{\partial y} + k\frac{\partial T}{\partial z}\right)_{\text{node 7}}\\
\left(-k\frac{\partial T}{\partial x} + k\frac{\partial T}{\partial y} + k\frac{\partial T}{\partial z}\right)_{\text{node 8}}
\end{matrix}\right\} \tag{3.68}$$

From Eq. (3.68), one can see that only the magnitude and the sign of a heat flux can be applied at a node as a boundary condition. However, a component in a specific coordinate direction (e.g., only a heat flux in x-direction) cannot be applied since only the *sum* of all three components is applied at a node.

3.2.3
Numerical Calculation Models

The FE analysis first requires the generation of a calculation model (a so-called FE mesh) that represents the geometry of the investigated structure. In general, HSSs exhibit a quasi-stochastic arrangement of spheres. However, the complexity of FE models required for their geometric characterization fast exceeds the capacity of modern computers. As a solution, simplified cubic-symmetric model structures can be generated (cf. Sections 3.3 and 3.4). The definition of appropriate symmetric boundary conditions allows for a reduction of the model size to one-eighth of a unit cell (UC) while, in fact, simulating an infinite structure. The FE meshes of metallic HSSs in Section 3.4 are based on the geometric dimensions of the experimental samples (outer sphere radius R of 1.5 mm, sphere wall thickness t of 0.1 mm, Figure 3.1). The averaged minimum distance a_{min} between two neighboring spheres varies with the morphology and is 0.09 mm for partial and 0.18 mm for syntactic HSS. The thermal conductivity of the metallic shell (sintered steel) is $\lambda_{\mathrm{St}} = 50\ \mathrm{W\ m^{-1}\ K^{-1}}$ [23] if not mentioned differently, and experimental measurements on the thermal conductivity by the transient plane source (TPS) method for the adhesive epoxy resin gave as result $\lambda_{\mathrm{Ep}} = 0.214\ (\pm 0.001)\ \mathrm{W\ m^{-1}\ K^{-1}}$. The numerical model of the partial HSS incorporating the boundary conditions is shown in Figure 3.6. Two constant temperature boundary conditions T_1 and T_2 are prescribed at the upper and lower surface nodes in order to initiate a heat flux $\dot{Q}$ through the structure. According to the geometry and boundary conditions of the

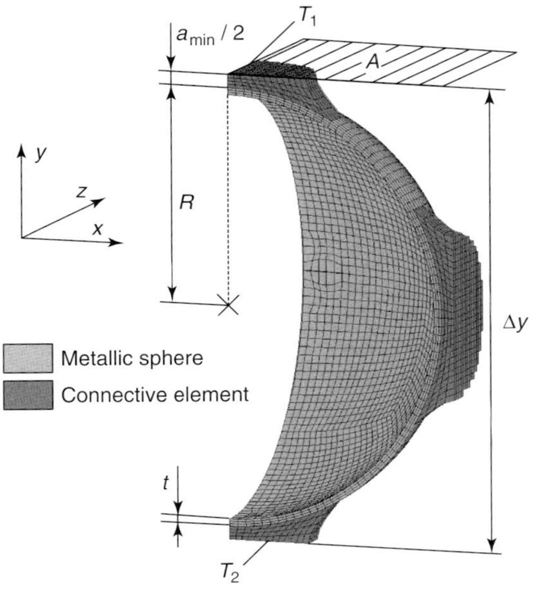

Figure 3.6 Boundary conditions of the finite element calculation.

FE model shown in Figure 3.6, a primitive cubic (pc) arrangement of the spheres is defined. This simplification contradicts the random arrangement of the spheres observed in the experimental samples, but allows for a significant reduction of the complexity of the numerical model.

Within the considered temperature range, the phenomenon of thermal radiation is negligible [24]. Also, owing to either absence (syntactic structure) or low volume fraction of interconnected porosity (partial structure) the contribution of convection to the overall thermal conductivity can be disregarded [25]. On the basis of these assumptions, the effective thermal conductivity λ is defined by Fourier's law:

$$\lambda = \frac{\dot{Q}}{A} \cdot \frac{\Delta y}{\Delta T} \tag{3.69}$$

where the area A and spatial distance Δy are given by the geometry and $\Delta T = T_2 - T_1$ is defined by the boundary conditions (cf. Figure 3.6). Variable $\dot{Q}$ is denoted to be the total heat flux through one of the two surfaces where a temperature boundary condition is prescribed and is the result of the FE analysis. It is obtained by summing up the nodal values of the heat flux of all nodes that lie within one of these surfaces.

In the FE analysis, the real HSS geometry, characterized by a pseudorandom arrangement of hollow spheres, is often substituted by cubic-symmetric model geometries. In contrast, the LMC method allows the utilization of large calculation models with high geometric resolution. In Section 3.5, two-dimensional calculation models representing random cross sections of syntactic HSS were used in thermal LMC analyses. In Section 3.6, this idea is taken one step further by using CT images as geometrical input data. As a consequence, more accurate results can be obtained. Furthermore, FE meshes are used in Section 3.6 on the basis of CT data of *decreased* geometric resolution (cf. Figure 3.7c).

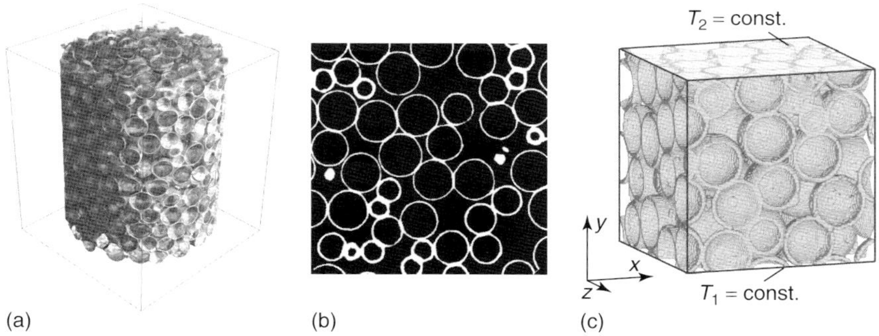

Figure 3.7 Geometry of HSS: (a) three-dimensional reconstruction of CT data; (b) CT cross section used for model generation; and (c) FE calculation model.

3.3 Finite Element Analysis on Regular Structures

First, two- and three-dimensional model structures are addressed. These simplified models do not account for the contribution of the metallic hollow spheres to the effective thermal conductivity but represent porous materials with spherical inclusions.

The numerical investigation of the relative thermal conductivity $\lambda_{rel} = \lambda/\lambda_s$ (λ is the effective conductivity of the porous material and λ_s the conductivity of the solid material) is performed with the commercial FE code MSC.Marc. Within the compass of this study, different porous 2D and 3D structures are investigated (cf. Figure 3.8). As mentioned in Section 3.2.3, the contribution of thermal radiation inside the pores to the overall heat transfer is small and can therefore be disregarded. Furthermore, the thermal conductivity of included media (e.g., air, $\lambda_1 \approx 0.025\ \mathrm{W^{-1}\ m^{-1}K^{-1}}$, [26]) is several orders of magnitude smaller than that of the cell-wall material (e.g., aluminum $\lambda_s = 221\ \mathrm{W^{-1}m^{-1}K^{-1}}$ [26]), and in compliance with Ref. [27] heat transfer due to convection in the small-sized pores of closed-cell foams can also be disregarded. Accordingly, inclusions can be approximated as voids with no contribution to the thermal conductivity of the structure ($\lambda_1 = 0$).

Cellular material can often be characterized on the basis of its relative density ρ_{rel}. This value is defined as the volume of the matrix divided by the total volume (i.e., the volume of the matrix plus the volume of inclusions) of the structure. The relative density ρ_{rel} of the illustrated geometries is varied by modification of the distance of the inclusions, whereas their size and shape remain constant. Since the available computer hardware limits the complexity of the calculation models, not the whole

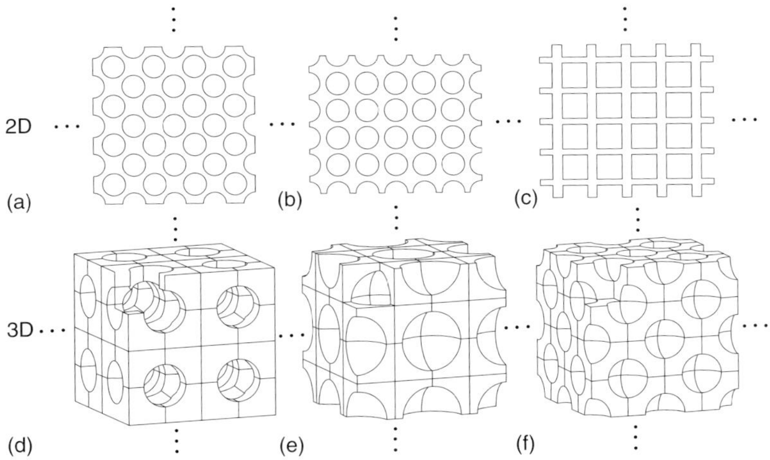

Figure 3.8 Representative areas/volumes: (a) circular 45°, (b) circular 0°, (c) squared, (d) open cell, (e) spherical body-centered cubic (bcc), and (f) spherical face-centered cubic (fcc).

Table 3.3 Mesh density for different configurations.

ρ_{rel}	3D open cell	3D fcc	3D bcc	2D circular (45°)	2D circular (0°)	2D square
0.1	14 329	–	–	–	–	–
0.2	19 197	–	–	–	–	–
0.3	19 960	–	–	1649	–	–
0.4	20 207	–	–	4009	–	–
0.5	17 435	42 385	55 281	5377	26 499	26 499
0.6	28 532	42 385	37 640	6527	23 137	23 137
0.7	28 485	29 797	24 217	1643	14 991	14 991
0.8	35 515	29 797	24 217	1233	9387	9387
0.9	32 249	16 249	14 436	1507	17 013	17 013
0.95	–	16 249	7721	–	22 057	22 057

2D or 3D structure can be meshed. Instead, specific symmetry boundary conditions $\left(\frac{dT}{dx} = \frac{dT}{dy} = \frac{dT}{dz} = 0\right)$ are introduced on the surfaces outside the xz-planes so that only one-fourth of the UC needs to be modeled. Corresponding to these boundary conditions, all FE models describe the thermal behavior of an infinite structure, where the influence of a free boundary is disregarded. This homogenization is accurate for structures that consist of more than 10–15 UCs in every direction. The number of nodes for the different meshes is summarized in Table 3.3.

The boundary conditions and the determination of the effective thermal conductivity using the FEM are elucidated in Figure 3.6, Section 3.2.3.

In addition to the numerical approach, analytical analysis is performed. A composite material with periodically placed cells containing random inclusions is considered. On the basis of a UC representing the whole composite [28, 29], corresponding values of the effective conductivity are approximately derived under the assumption that the inclusions can be treated as empty pores with the conductivity $\lambda_1 \to 0$. For this approach, two-dimensional composite materials with periodic and quasi-periodic structures (i.e., periodically located cells containing a finite number of random nonoverlapping circular inclusions) are considered (cf. Figure 3.9).

Within this approach, conductive properties of composite materials can be described in terms of conjugation conditions $u^+ = u^-, \lambda_s \frac{\partial u^+}{\partial n} = \lambda_1 \frac{\partial u^-}{\partial n}$ on the boundary of inclusions ∂D_k with respect to the function $u(x, y)$ (e.g., the temperature) sectional harmonic in D^+ and D^- [28, 29]. These conjugation conditions correspond to a perfect contact between different materials. Here, D_k are simply connected domains modeling inclusions of conductivity λ_1 in the matrix material D^+ of conductivity λ_s, $D^- := \bigcup_{k=1}^{N} D_k$. If the field is potential, that is, the function $u(x, y)$ satisfies the Laplace equation $\nabla^2 u = 0$ in a domain D, then one can introduce the function $\varphi(z) = u(z) + \mathrm{i}v(z), z = x + \mathrm{i}y, \mathrm{i}^2 = -1$, analytic in D, which is called the *complex potential* and reformulate the problem in terms of this potential.

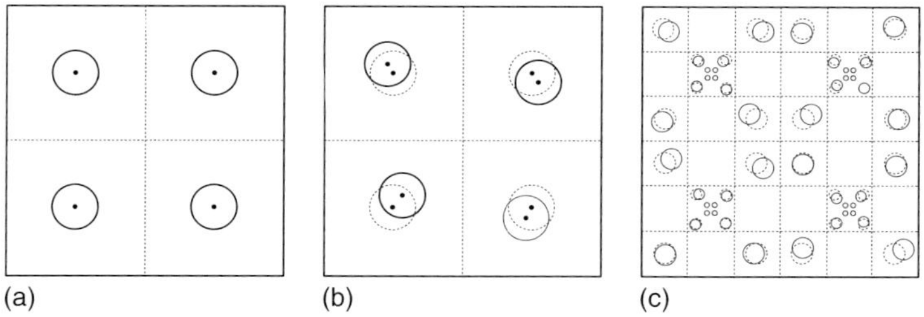

Figure 3.9 (a) Periodic square array of inclusions; (b) perturbed array of inclusions; and (c) periodic and perturbed pseudofractal array of inclusions.

The effective conductivity of macroscopically isotropic composite materials with small concentration ν of nonrandom inclusions is described by the classical Clausius–Mossotti formula (for background and application see [30]):

$$\lambda_{\text{rel}} = \frac{\lambda}{\lambda_s} = \frac{1 + \mu\nu}{1 - \mu\nu} \tag{3.70}$$

where $0 < \lambda_s < +\infty$ is the conductivity of the matrix and $\mu = \frac{\lambda_1 - \lambda_s}{\lambda_1 + \lambda_s} \in [-1, 1]$ is the contrast parameter that expresses the difference between the conductivity of both materials.

The generalization of this formula for quasi-periodic composites with random inclusions of the conductivity $\lambda_1 > 0$ is obtained in Ref. [28] as

$$\frac{\lambda}{\lambda_s} = 1 + \frac{2\mu\nu}{N} \sum_{m=1}^{N} \psi_m(a_m) \tag{3.71}$$

where $\nu = N\pi r^2$ is the concentration of inclusions, which can be expressed in the case of a simple cell of unit size as $\nu = 1 - \rho_{\text{rel}}$. Variable r is the radius of inclusions, N is the number of inclusions in the cell, a_m are the centers of the inclusions, and $\psi_m(z)$ are the derivatives of complex potentials in the cell. The sum on the RHS of Eq. (3.71) is reduced to a power series approximation with respect to ρ_{rel}:

$$\frac{\lambda}{\lambda_s} = 1 + 2\mu(1 - \rho_{\text{rel}})[A_0 + A_1(1 - \rho_{\text{rel}}) + A_2(1 - \rho_{\text{rel}})^2 + A_3(1 - \rho_{\text{rel}})^3 + \ldots] \tag{3.72}$$

The coefficients A_p can be found for square and pseudofractal arrays of inclusions in Ref. [31]. A square array of inclusions refers to their distribution for which the centers of the inclusions form a square (cf. Figure 3.9a). It is shown in Ref. [28] that random perturbation of the inclusions of square (cf. Figure 3.9b) and pseudofractal (cf. Figure 3.9c) arrays increases the effective conductivity of a composite under the requirement that each inclusion is moved in such a way that it cannot touch or cross with others. Thus, periodic square and periodic pseudofractal arrays of circular inclusions provide a minimum for the effective conductivity in the class of periodic macrocells with random microstructure.

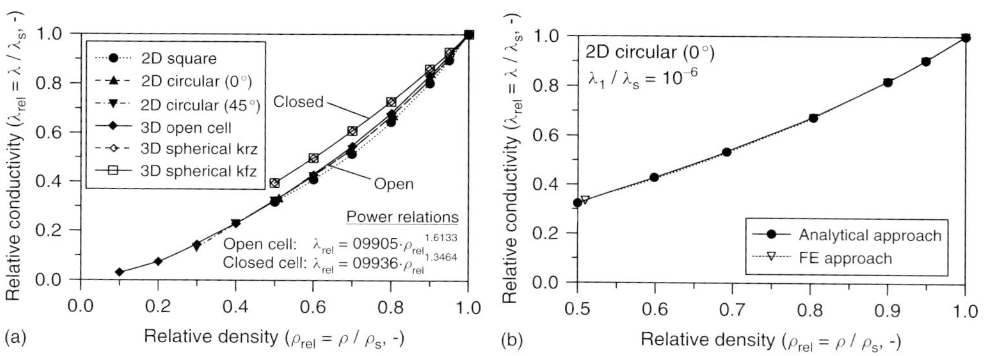

Figure 3.10 The dependence of the relative conductivity λ_{rel} on the relative density ρ_{rel}.

The results of the FE analysis for the relative conductivity λ_{rel} are shown in Figure 3.10a. The relative thermal conductivity λ_{rel} seems to be widely independent of the shape and the topology of the inclusions. Only the morphology of the structure influences the behavior: the results for the two closed-cell structures (face-centered cubic (fcc) and body-centered cubic (bcc)) coincide at slightly higher values compared to the various open-cell geometries. However, the relative density ρ_{rel} seems to be an appropriate generalized parameter for a first description of the relative conductivity of porous materials. Therefore, power relations for open- and closed-cell structures are given in Figure 3.10a. In Figure 3.10b, the results of the analytical approach are compared with the numerical findings in the case of the 2D circular (0°) structure and a very good correlation is found. Furthermore, the effect of an increasing relative conductivity due to small perturbations is analytically verified. The maximum deviations of 0.36 and 0.45% to the results of the unperturbed structures are determined for relative densities of $\rho_{rel} = 0.7$ and $\rho_{rel} = 0.6$, respectively.

3.4 Finite Element Analysis on Cubic-Symmetric Models

In this section, the FE analysis shown in Section 3.3 is refined and spherical pores are replaced by hollow spheres embedded in the matrix. Selected result is presented; however, an encyclopedic study can be found in Ref. [32]. Analogous to the previous section, the quasi-random mesostructure of metallic hollow sphere structure is simplified to cubic-symmetric models. The three different topologies (spatial arrangements) considered are shown in Figure 3.11.

In addition, two different morphologies are considered. Figure 3.1 shows photographs of a partial and a syntactic hollow sphere structure. The corresponding simplified calculation symmetries are shown in Figure 3.12 exemplarily for a primitive cubic arrangement. In the case of the syntactic structure, the external void space between hollow spheres is completely filled by the matrix, whereas in partial

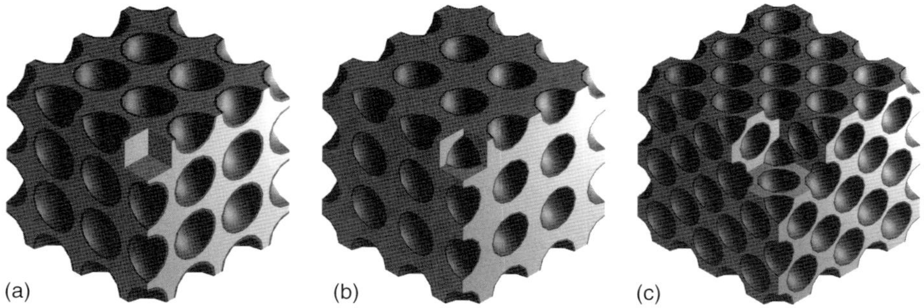

Figure 3.11 Topologies of HSS models: (a) primitive cubic (pc); (b) body-centered cubic (bcc); and (c) face-centered cubic (fcc).

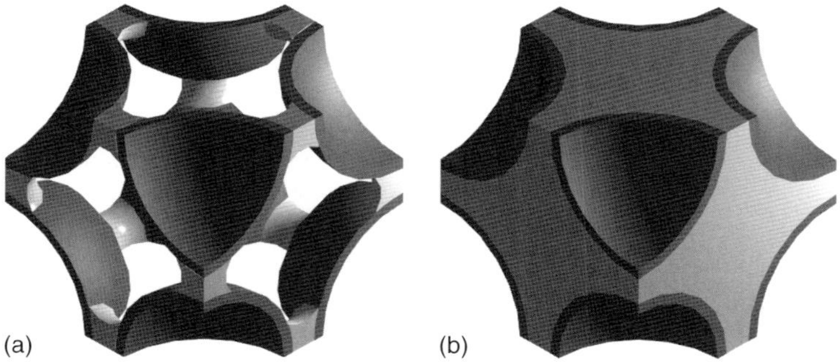

Figure 3.12 Morphologies of primitive cubic HSS models: (a) partial morphology and (b) syntactic morphology.

structures the spheres are only connected close to their contact areas. Accordingly, partial HSSs also have small volume fractions of interconnected porosity.

In the following, selected results of this FE analysis are presented. Figure 3.13 shows the effective thermal conductivity as a function of the sphere wall thickness. The case of primitive cubic sphere arrangement and homogeneous material properties (base material steel, $\lambda = 50\ \mathrm{W\,m^{-1}\,K^{-1}}$) is addressed. The results are compared for the syntactic and partial morphologies. It is obvious that the effective thermal conductivity is much larger for the syntactic case. The explanation is the lower porosity and high volume fraction of steel (cf. Figure 3.12 above). In addition, a linear characteristic is observed for both morphologies.

In Figure 3.14, effective thermal conductivities are plotted for heterogeneous steel (shells) – epoxy (matrix) HSS. Again, only primitive cubic arrangement is considered. The epoxy has a low thermal conductivity ($0.36\ \mathrm{W\,m^{-1}\,K^{-1}}$) and acts as a thermal insulator between the metallic hollow spheres. A comparison of Figures 3.13 and 3.14 shows that the thermal conductivity of adhered (heterogeneous) HSS is much lower in comparison to that of the homogeneous structures.

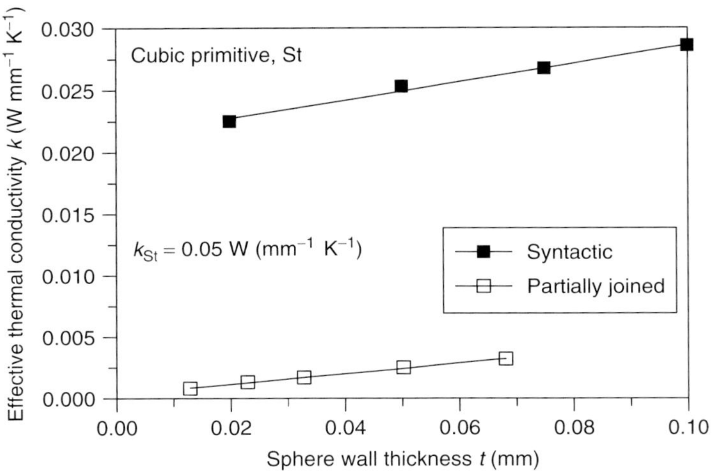

Figure 3.13 Effective thermal conductivity as a function of the sphere wall thickness t: influence of the morphology for homogeneous HSS.

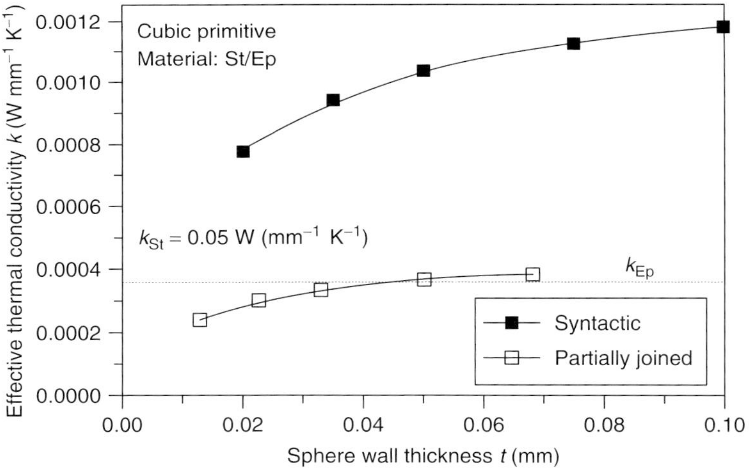

Figure 3.14 Effective thermal conductivity as a function of the sphere wall thickness t: influence of the morphology for heterogeneous HSS.

The thermal conductivity of heterogeneous syntactic HSS is slightly higher than the one of partial structures.

Figure 3.15 shows the effect of different topologies on the thermal properties of HSSs. Primitive cubic sphere arrangement is compared to bcc and fcc arrangements. The effective thermal conductivity is drawn as a function of the sphere wall thickness for homogeneous structures. Independent of the topology, a linear

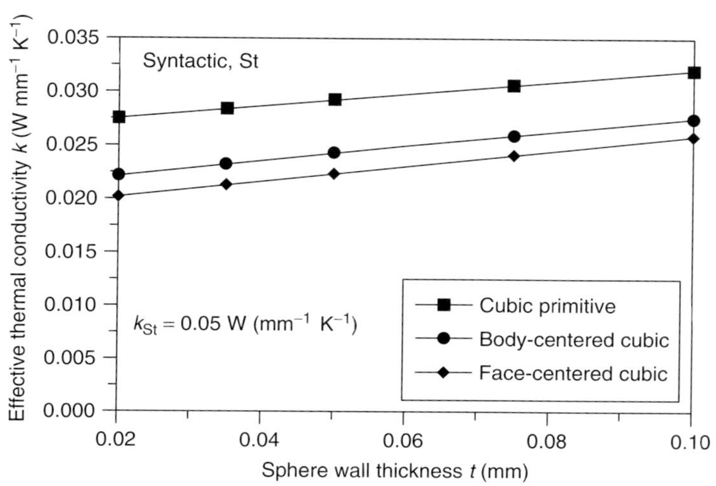

Figure 3.15 Effective thermal conductivity as a function of the sphere wall thickness t: influence of the topology for homogeneous HSS.

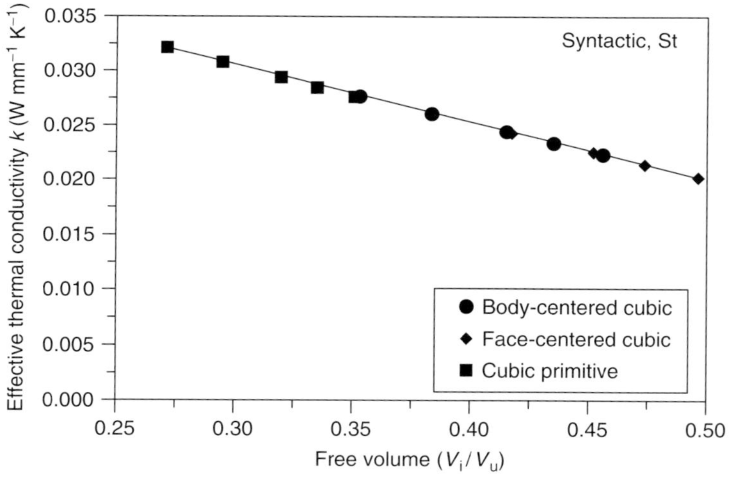

Figure 3.16 Effective thermal conductivity as a function of the porosity: influence of the topology for homogeneous HSS.

characteristic of these functions is observed. In the chosen visualization, primitive cubic topology yields the maximum values.

However, for two-phase materials (metal–void) a plot of the effective thermal conductivity versus porosity (cf. Figure 3.16) reveals a linear dependence. In this case, the effective thermal conductivity is independent of the topology and can be determined solely in dependence of the porosity.

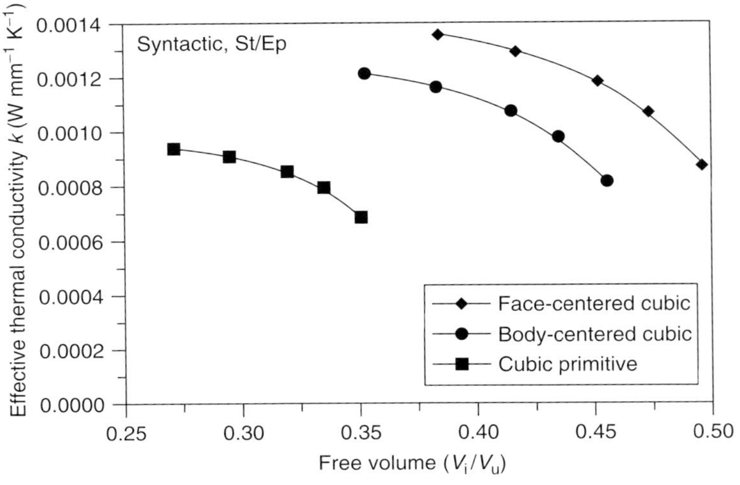

Figure 3.17 Effective thermal conductivity as a function of the sphere wall thickness t: influence of the topology for heterogeneous HSS.

Figure 3.17 shows the thermal properties of heterogeneous HSS for different topologies. A clear dependence on the topology is visible. In contrast to homogeneous structures (cf. Figure 3.16), the values cannot be approximated by a linear fit but strongly depend on topology and porosity. For a given porosity, fcc topology yields the maximum thermal conductivity followed by bcc and cubic primitive topologies.

3.5 LMC Analysis of Models of Cross Sections

Up to this point, only cubic-symmetric models of HSSs were used. This section deals with models of HSS cross sections, where hollow spheres are sequentially added at random coordinates. Accordingly, a more realistic representation of *real* structures can be achieved.

3.5.1 Modeling

Figure 3.18a shows the photography of a syntactic HSS cross section. The outer radii R of the metallic hollow spheres are 1.5 mm and the sphere wall thicknesses are 0.075 mm. The radii r_i (cf. Figure 3.19) of their circular cross sections depend on the distance h_i of their center points to the cutting plane and are given by the relation $r_i(h_i) = \sqrt{R^2 - h_i^2}$. Figure 3.18b shows an example of a two-dimensional periodic model that is used in the Monte Carlo calculation.

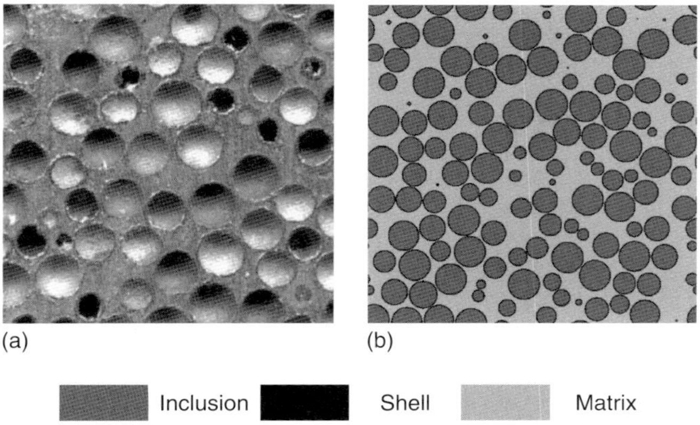

Figure 3.18 Cross section of a hollow sphere structure: (a) photography and (b) calculation model.

The 2D models were obtained as cross sections of the 3D models that were generated using an algorithm, which sequentially fills spheres into a lattice at randomly selected coordinates $C_i(X_i, Y_i, h_i)$. Therefore, the distance h_i must be smaller than the sphere radius R in order to ensure intersection with the cutting plane. A further random number defines the position p_i of the center point C_i relative to the cutting plane (below: $p_i = 1$, above: $p_i = -1$). In the next step of this algorithm, intersection of the last added sphere with preexisting spheres is investigated. In order to do this, the distances (d_i) between the spheres were calculated:

$$d_i = \sqrt{(X_n - X_i)^2 + (Y_n - Y_i)^2 + (p_n \cdot h_n - p_i \cdot h_i)^2}\ .\ (i = 1 \ldots n - 1) \qquad (3.73)$$

If any distance d_i is smaller than two times the sphere radius R, at least two spheres are intersecting and the last added sphere n must be removed. Figure 3.19

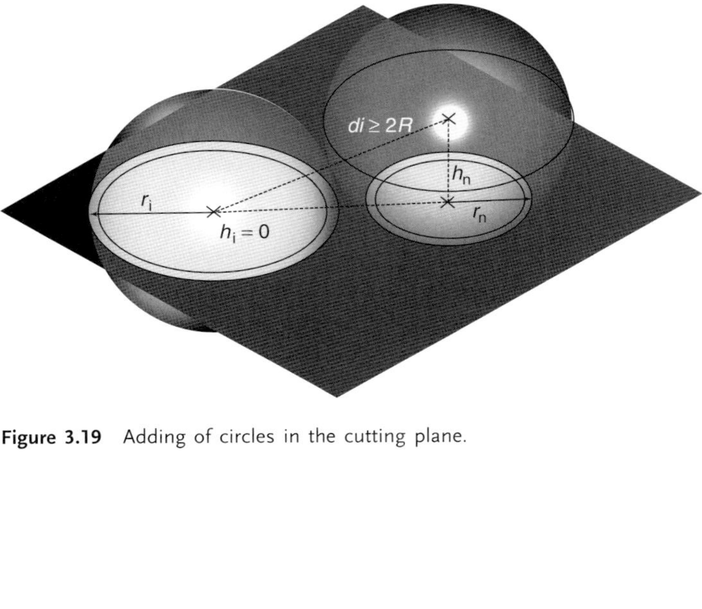

Figure 3.19 Adding of circles in the cutting plane.

Table 3.4 Area fractions and average density of Monte Carlo lattices ($h_i \neq$ const.).

Lattice ID	A_M (%)	A_S (%)	A_I (%)	Average density $\overline{\rho}$ ($g\,cm^{-3}$)
#1	45.43	11.29	43.28	1.22
#2	44.48	11.14	44.38	1.20
#3	45.65	11.29	43.05	1.22
#4	44.06	11.25	44.69	1.20
#5	44.67	11.25	44.08	1.21
#6	45.31	11.35	43.34	1.22
#7	44.43	11.32	44.24	1.21
#8	45.82	11.20	42.98	1.22
#9	45.87	11.12	43.01	1.21
#10	45.61	11.11	43.28	1.21
#11	45.53	11.29	43.18	1.22
#12	43.95	11.11	44.94	1.19

displays this condition for the example $h_i = 0$. Spheres are added in the model until 10^7 repeated attempts fail because of intersection. In this case, it is assumed that no further spheres can be positioned in the gaps between the previously added spheres. In order to analyze a representative cross section of the HSS, each (2D) lattice contains at least 100 circles.

In the final step of the lattice generation, two-dimensional matrices $\mathbf{M}(X, Y)$ with a resolution 500×500 are created. A particular cross section through a HSS can be characterized by its area fractions: the sum of the area fraction of the matrix A_M, the shell A_S, and the inclusion A_I is equal to unity. The area fractions of the lattice models are simply obtained by summing up the lattice points related to each phase and dividing them by the total number of lattice points (500^2). The average density of the HSS can then be calculated:

$$\overline{\rho} = A_S \cdot \rho_S + A_M \cdot \rho_M \qquad (3.74)$$

where $\rho_M = 1.13\,g\,cm^{-3}$ is the density of the epoxy matrix and $\rho_S = 6.2\,g\,cm^{-3}$ (both values were measured at the University of Aveiro, Portugal) is the density of the metallic sphere wall material. Table 3.4 shows the area fractions and average densities of the lattices employed. The calculated average densities are in good agreement with measured average densities of HSS samples ($\overline{\rho} = 1.2\,g\,cm^{-3}$). A second set of lattices (cf. Table 3.5) comprises sequentially filled random distributions of circles with constant heights $h_i = 0.58 \cdot R$. Consequently, these models exhibit constant radii $r \approx 0.815 \cdot R$ of the circular cross sections inside the cutting plane. The comparison of the lattices in Tables 3.4 and 3.5 shows whether the effective thermal conductivity is dependent on particle radii r or is mainly determined by the area fractions and distribution of spheres. Therefore, it is important that all meshes exhibit similar area fractions.

It is important to note that the analysis of two-dimensional cutting planes of three-dimensional HSS introduces a simplification. Only conduction inside the

Table 3.5 Area fractions of lattices ($h_i = 0.58 \cdot R$).

Lattice ID	A_M (%)	A_S (%)	A_I (%)	Average density $\overline{\rho}$ (g cm^{-3})
#13	44.68	11.19	44.13	1.20
#14	44.62	11.19	44.19	1.20
#15	44.68	11.20	44.12	1.20
#16	44.67	11.19	44.14	1.20

cutting plane can be simulated, but out-of-plane thermal conduction, that is, along the curved surface of the metallic shells is neglected.

3.5.2 Results

The effective thermal conductivities λ_{eff} are determined using the Einstein Eq. (3.2). Table 3.6 shows the effective thermal conductivities λ_{eff} of the LMC models with varying radii r ($h \neq$ const.). The average value $\lambda_{eff} = 1.00$ W m^{-1} K^{-1} of the results is in excellent agreement with experimental findings ($\lambda_{eff} = 1.01$ W m^{-1} K^{-1} [14]). Despite the randomly distributed circular cross sections of spheres, the standard deviation of λ_{eff} is only 0.015 W m^{-1} K^{-1}, which corresponds to approximately 1.5% of the average value. Preliminary investigations with a smaller amount of circular cross sections ($\approx$30) showed a higher standard deviation of 3.84%.

Table 3.7 shows the effective thermal conductivities of the lattice models with constant radii r ($h =$ const.). The average value of $\lambda_{eff} = 1.01$ indicates no systematic dependence of the thermal conductivity on the distribution of the radii r in the cutting plane.

Next, the effective thermal conductivity of a primitive cubic arrangement ($h = 0.58 \cdot R$) of circles is investigated. The distance between the spheres is chosen so that the structure exhibits similar area fractions as the lattices in Tables 3.4 and 3.5 ($A_M = 44.67\%$, $A_S = 11.18\%$, $A_I = 44.15\%$). A distinctly lower effective thermal conductivity $\lambda_{eff} = 0.70$ W m^{-1} K^{-1}) is now found. A likely explanation for this deviation is the regular arrangement of spheres that yields the maximum distance between two neighboring spheres. Each sphere is surrounded by the low thermal

Table 3.6 Effective thermal conductivity of the 2D HSS models ($h_i \neq$ const.).

ID	#1	#2	#3	#4	#5	#6	#7	#8	#9	#10	#11	#12
λ_{eff} (W m^{-1} K^{-1})	0.99	1.02	1.00	1.00	1.03	1.02	0.99	0.98	0.98	1.03	0.99	1.00

Table 3.7 Effective thermal conductivity of the 2D HSS models ($h = 0.58 \cdot R$).

ID	#13	#14	#15	#16
λ_{eff} (W m^{-1} K^{-1})	1.02	1.02	0.99	1.00

conducting adhesive matrix that acts as a thermal insulator between the metallic shells.

To investigate the effect of substitution of 3D process by an appropriate 2D one, we can invoke two Maxwell-type analytical solutions for the effective conductivity [33] as upper and lower limits:

$$\lambda_{eff}^{low} = \frac{\lambda_M \left(1 - \dfrac{(d-1)(A_I + A_S)(\lambda_M - \lambda_{SI})}{(d-1)\lambda_M + \lambda_{SI}}\right)}{1 + \dfrac{(A_I + A_S)(\lambda_M - \lambda_{SI})}{(d-1)\lambda_M + \lambda_{SI}}} \quad (3.75a)$$

$$\lambda_{eff}^{up} = \frac{\lambda_{MS} \left(1 - \dfrac{(d-1)A_I(\lambda_{MS} - \lambda_I)}{(d-1)\lambda_{MS} + \lambda_I}\right)}{1 + \dfrac{A_I(\lambda_{MS} - \lambda_I)}{(d-1)\lambda_{MS} + \lambda_I}} \quad (3.75b)$$

with

$$\lambda_{SI} = \frac{\lambda_S \left(1 - \dfrac{(d-1)A_I(\lambda_S - \lambda_I)}{(A_I + A_S)((d-1)\lambda_S + \lambda_I)}\right)}{1 + \dfrac{A_I(\lambda_S - \lambda_I)}{(A_I + A_S)((d-1)\lambda_S + \lambda_I)}} \quad (3.76a)$$

$$\lambda_{MS} = \frac{A_M\lambda_M + A_S\lambda_S}{A_M + A_S} \quad (3.76b)$$

In deriving the lower limit, we accepted the following structural hierarchy: an inclusion and a shell attached considered together as a composite inclusion (with an effective conductivity λ_{SI} given by Eq. (3.76a)); the total effective conductivity can then be calculated using Maxwell relation Eq. (3.75a) with the matrix conductivity λ_M and inclusion conductivity λ_{SI}. For the derivation of the upper limit, we treat the matrix phase as a composite phase consisting of original matrix and shell phases together with the effective conductivity λ_{MS} given by Eq. (3.76b). Then, we can apply Maxwell relation Eq. (3.75b) with matrix conductivity λ_{MS} and inclusion phase conductivity λ_I.

For the set of parameters used ($A_M = 44.67\%$, $A_S = 11.18\%$, $A_I = 44.15\%$, and $\lambda_M = 0.0214$ W m^{-1} K^{-1}, $\lambda_S = 51.9$ W m^{-1} K^{-1}, $\lambda_I = 0$ W m^{-1} K^{-1}), we soon find that according to the 3D version of the analysis $\lambda_{eff}^{low3} = 0.112\,70$ W m^{-1} K^{-1} with $\lambda_{eff}^{up3} = 4.760\,99$ W m^{-1} K^{-1} and according to the 2D version of the analysis $\lambda_{eff}^{low2} = 0.082\,35$ W m^{-1} K^{-1} with $\lambda_{eff}^{up2} = 4.031\,90$ W m^{-1} K^{-1}. We can see that for the given parameters "lowering" the dimensionality should produce about 15–25% lower values of the effective conductivity.

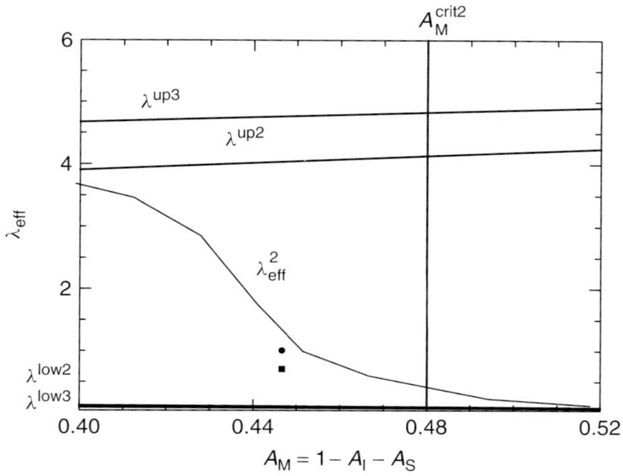

Figure 3.20 Transition between lower and upper limits for the two- and three-dimensional cases and transition between the limits for the case of the lattice model with constant radius.

We need to point out that the lower limit expression (3.75a, 3.76a) will be reasonably accurate when the density of the spheres is not high and they are not touching each other; in other words, when $(A_I + A_S)$ is smaller than corresponding critical percolation point, that is, about 52% for the 2D model with the constant radius and about 67% for the 3D model. On the other hand, when the density of the spheres is high enough so that they all touch their neighbors (the coating phase becomes a network that can produce long-range thermal diffusion), then the upper limits (Eqs. (3.75b, 3.76b)) will be expected to give more accurate results. The transition between the two limits usually occurs smoothly in the vicinity of the critical percolation point (about $\pm$15%); see the schematic representation in Figure 3.20 for the 2D model with the constant radius.

Applying these considerations to our case we can see that $A_I + A_S = 55.33\%$. In Figure 3.20, we plot the values for two 2D models calculated above by means of computer simulations. We can see that the points are in qualitative agreement with the transition between the upper and lower analytical models. Further investigation of the proposed transition behavior is needed.

3.6 Computed Tomography Reconstructions

This section addresses numerical analysis based on CT data. This technique allows an accurate characterization of *real* geometries. Two different numerical analysis methods, namely, the FE and the LMC methods, are used.

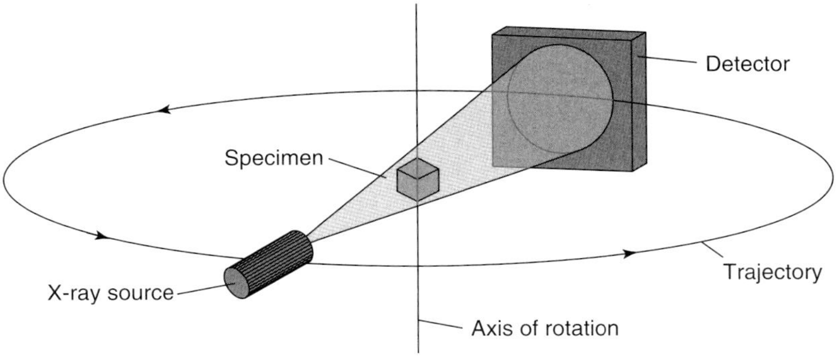

Figure 3.21 Principal scheme of 3D cone-beam tomograph.

3.6.1 Computed Tomography

Modern X-ray tomographic systems produce two-dimensional sectional images (tomograms) that allow the reconstruction of three-dimensional structures with a single 360° rotation. A conical beam from an X-ray source penetrates the entire specimen, cf. Figure 3.21. The attenuated radiation is measured by a large area detector. In order to irradiate the object from all sides, the source and the detector rotate in small steps (<1°) around the object (patient) in medical tomographs, while in industrial applications it is, in most cases, advantageous to rotate the object (specimen).

During rotation a set of projections is measured and stored. The set of projections is then used to numerically reconstruct the 3D structure of the object as 3D voxel data, which is a three-dimensional array with attenuation values. These detected values are proportional to the density of the material and allow the detection of defect inside a sample (nondestructive testing) or the reconstruction of the real sample geometry. The 3D data can be visualized in several ways. It is possible to generate any kind of cuts, slices, or volume parts.

3.6.2 Numerical Analysis

In Section 3.5, two-dimensional calculation models representing random cross sections of syntactic HSS were used in thermal LMC analyses. This section takes this idea one step further by using CT images as geometrical input data. As a consequence, more accurate results can be obtained. In addition to LMC analysis on high-resolution models, FE meshes are generated on the basis of CT data of decreased geometric resolution (cf. Figure 3.7c). In contrast to the adhered structures addressed in previous sections, sintered HSSs are considered. A practical explanation is the difficulty to simultaneously capture two phases of different density, that is, epoxy and steel, in CT scanning.

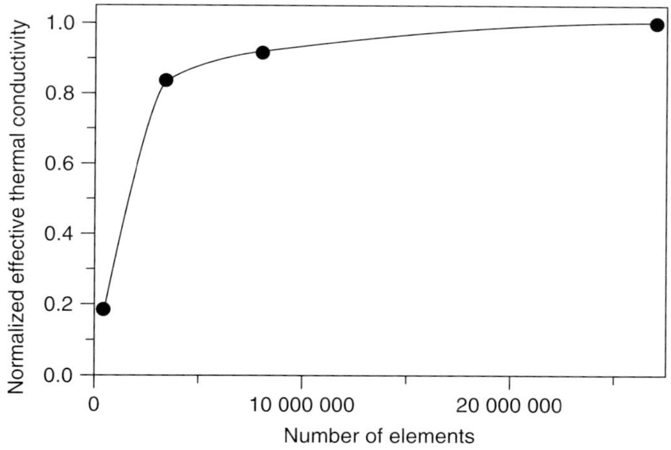

Figure 3.22 Mesh refinement analysis of the finite element model.

For the FE analysis, a meshing algorithm first presented in Ref. [34] is applied. Basically, the FE meshes directly represent the geometry of the material's mesostructure (cf. Figure 3.7). As elucidated in Section 3.2.3, constant temperature boundary conditions are prescribed at two opposing surfaces of the calculation model and the arising heat flux is used to calculate the effective thermal conductivity using Fourier's law (see Ref. [9]). By changing the constant temperature planes, the effective thermal conductivity can be determined in three perpendicular directions (i.e., x, y, and z directions, cf. Figure 3.7). The analysis of anisotropy is currently restricted to the lower resolution FE analysis. A mesh refinement study of the FE meshes used is shown in Figure 3.22.

Four different models with the resolutions 75^3, 150^3, 200^3, and 400^3 elements are considered, where the maximum resolution is limited by computer memory. In addition, the calculation time distinctly increases with increasing mesh density. On the 32-bit system used in the analysis (dual core processor 3.16 GHz, 3.15 GB RAM), the averaged calculation times were 23, 91, 507, and 6800 s for the highest resolution, respectively. It can be observed that accurate convergence is only reached for a large number of elements and accordingly the highest resolution is chosen for subsequent FE analyses.

3.6.2.1 **Microstructure**

The first part of the LMC analyses considers the effective thermal conductivity λ_w of the cell-wall material. This material parameter is governed by the thermal conductivity $\lambda_s = 16.2\ \text{W m}^{-1}\ \text{K}^{-1}$ of the base material steel 316L [35] and its residual porosity. Microstructural characterizations of sintered HSS [36] yielded a microporosity of $p = 5.30\%$. Simple three-dimensional lattice models are generated where nonconducting micropores are randomly distributed. The fraction of nodes related to micropores is thereby identical to the microporosity. The applied model for porosity presumes that all pores are of similar size and homogeneously distributed. As an alternative, the parameter λ_w can be estimated by a Maxwell-type expression [37], valid for the special case of nonconducting inclusions:

$$\lambda_w = \frac{2(1-p)}{2+p}\lambda_m \tag{3.77}$$

3.6.2.2 Mesostructure

The second part of the LMC analyses directly addresses the properties of metallic HSS. The 3D analysis and image acquisition of the hollow structures were carried out using the CT system v|tome|x s of the company phoenix|X-ray. The achieved voxel resolution was 35 μm per voxel. The result of the volume processing was a volume of 650 × 650 × 750 voxels. The 3D visualization and extraction of the STL-iso-surfaces was applied by using the software Volume Graphics 1.2. An optimized gray value was applied to obtain the best segmentation and separation between the metal structure and air. On the basis of that gray value the STL-extraction of the iso-surfaces for the following simulation work was performed. The CT images, representing parallel cross sections of the geometry, are directly translated to LMC calculation models. The jump frequency assigned to a particular node in a lattice is thereby defined on the basis of the gray-scale value of the corresponding voxel. Figure 3.7b shows that the metallic phase can be identified by light pixels and voids by dark pixels, respectively. The gray-scale threshold is empirically chosen to match the relative density 9.7% of the HSS structure [36]. During the LMC model generation the geometry is subdivided into four slightly intersecting subvolumes in order to investigate local fluctuations of the effective thermal conductivity.

The FE analyses use calculation models assembled by uniform cubic FEs [34]. Each FE represents a subset of voxels of the CT images and analogous to the LMC approach, its thermal conductivity is defined in reference to the gray-scale level. However, restrictions in calculation time and computer memory require the bisection of the geometric resolution in the FE models. In a brief comparison, LMC allows for more complex calculation models while FE analyses benefit from distinctly lower calculation times.

3.6.3 Results

The influence of the microporosity on the effective thermal conductivity of the cell-wall material λ_w is shown in Figure 3.23. The drawn line corresponds to the analytical solution (3.74) and is in excellent agreement with corresponding LMC analyses (circles). The porosity of the cell-wall material is 5.30% [36], and accordingly the estimated cell-wall conductivity λ_w is 14.95 W m^{-1} K^{-1}.

Using this result as the thermal conductivity of the metallic phase identified in the CT scans, LMC and FE analyses are performed on sintered HSS. The CT data are subdivided into four subgeometries in order to investigate local fluctuations of the thermal properties. The results of these analyses are shown in Figure 3.24.

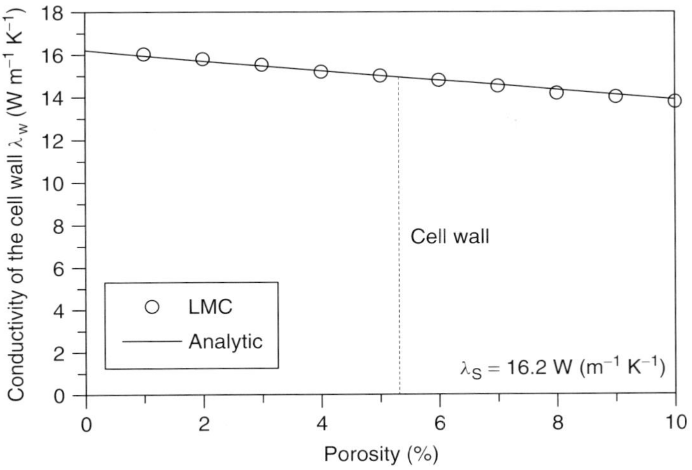

Figure 3.23 Normalized effective thermal conductivity of the microstructure versus microporosity.

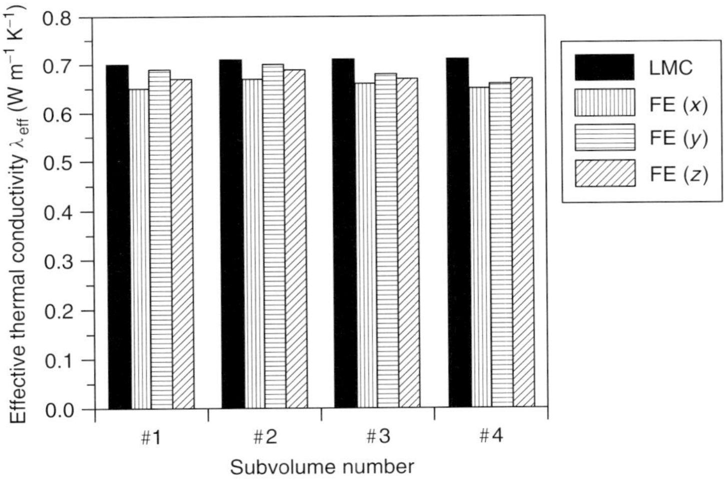

Figure 3.24 Effective thermal conductivity of sintered HSS.

The results of the FE analysis are in good agreement with the LMC findings. The LMC results exhibit circumstantially higher results, which can partially be explained by the superior geometric resolution of the calculation model: the required bisection of the grid resolution (decrease of 650 × 650 × 750 voxels to a total of 325 × 325 × 375 elements) in the FE models might cause the introduction of "artifacts" such as holes in the cell walls that slightly change the FE results. In addition, the results are very similar for all subvolumes. This is remarkable since the analyzed volume elements are relatively small (<100 spheres) and therefore indicates very homogeneous material properties. Furthermore, the FE results indicate quasi-isotropic thermal properties ($\lambda_x \approx \lambda_y \approx \lambda_z$). The LMC results are average values, blurred over all possible directions. The mean effective thermal conductivities are $\lambda_{\text{eff}} = 0.71\ \text{W m}^{-1}\ \text{K}^{-1}$ (LMC) and $\lambda_{\text{eff}} = 0.67\ \text{W m}^{-1}\ \text{K}^{-1}$ (FE). These values only slightly exceed thermal conductivities ($0.57\ \text{W m}^{-1}\ \text{K}^{-1}$) observed for adhesively bonded HSS [14], which can be explained by the relatively low thermal conductivity of the sintered cell-wall material. As a consequence, it can be concluded that sintered HSSs exhibit quasi-isotropic, homogeneous, and low effective thermal conductivities.

3.7 Conclusions

This chapter addressed thermal properties of HSSs. To this end, two numerical analysis techniques, namely, FE and LMC methods, were applied. Various models with increasing geometric complexity were considered. In Section 3.3, cubic-symmetric model structures with spherical and tubular pores were addressed. FE analysis was used and excellent agreement with an analytical solution was achieved. In Section 3.4, improved models were considered that accounted for the contribution of the spherical metal shells to the thermal conduction. Again, FE analysis was performed and a strong dependence of the effective thermal conductivity on the material composition and morphology was found. In particular, a linear dependence of metal-only structures on the free volume was discovered. In Section 3.5, the transition from periodic structures with cubic symmetries to more realistic structures with stochastically distributed spheres was performed. Two-dimensional cross sections of HSSs were considered. Owing to the increased complexity of the geometries only LMC analysis was performed. The obtained results were in good agreement with experimental data. An analytical approach yielded slightly lower values for the effective thermal conductivity, which is probably caused by the two-dimensional modeling of a three-dimensional structure. This shortcoming was overcome in Section 3.6, where the calculation models are directly based on three-dimensional CT data. The geometric complexity of the models only allowed for an FE analysis based on lower resolution models. However, the FE results are in good agreement with the ones obtained by high-resolution LMC analyses.

The comparison of the FE and LMC methods shows that both approaches have particular benefits. Comparative studies on similar geometries have shown that both methods yield identical results. FE analysis is distinguished by the fast solution of calculation models; however, this is currently restricted to models of limited geometric complexity. The LMC method can address extremely complex geometries that are directly based on CT data. The major restriction, large computation time, decreases in importance because of the availability of ever faster and cheaper computer systems.

References

1. Degischer, H. and Kriszt, B. (2002) *Handbook of Cellular Metals*, Wiley-VCH Verlag GmbH.
2. Ramamurty, U. and Paul, A. (2004) *Acta Materialia*, **52**, 869.
3. Solórzano, E., Rodríguez-Pérez, M.A., Reglero, J.A., and de Saja, J.A. (2007) *Journal of Materials Science*, **42**, 2557.
4. Olurin, O.B., Arnold, M., Körner, C., and Singer, R.F. (2002) *Materials Science and Engineering: A*, **328**, 334.
5. Lu, T.J., Stone, H.A., and Ashby, M.F. (1998) *Acta Materialia*, **46**, 3619.
6. Zhao, C.Y., Lua, T.J., Hodson, H.P., and Jackson, J.D. (2004) *Materials Science and Engineering: A*, **367**, 123.
7. Lu, W., Zhao, C.Y., and Tassou, S.A. (2006) *International Journal of Heat and Mass Transfer*, **49**, 2751.
8. Boomsma, K., Poulikakos, D., and Zwick, F. (2003) *Mechanics of Materials*, **35**, 1161.
9. Cookson, E.J., Floyd, D.E., and Shih, A.J. (2006) *International Journal of Mechanical Sciences*, **48**, 1314.
10. Lu, T.J. and Chen, C. (1999) *Acta Materialia*, **47**, 1469.
11. Baumeister, E., Klaeger, S., and Kaldos, A. (2004) *Journals of Materials Processing Technology*, **155–156**, 1839.
12. Fiedler, T. and Öchsner, A. (2007) *Materials Science Forum*, **553**, 39.
13. Fiedler, T. and Öchsner, A. (2007) *Materials Science Forum*, **553**, 45.
14. Fiedler T., Solórzano E., and Öchsner A. (2008) *Materials Letters*, **62**, 1204.
15. Belova, I.V. and Murch, G.E. (2007) *Solid State Phenomena*, **128**, 1.
16. Davis, J.R. (1996) *ASM Specialty Handbook – Carbon and Alloy Steels*, ASM International.
17. Belova, I.V., Murch, G.E., Fiedler, T., and Öchsner, A. (2007) The Lattice Monte Carlo Method for Solving Phenomenological Mass and Heat Transport Problems, Diffusion Fundamentals, published online.
18. Incropera, F.P., DeWitt, D.P., Bergman, T.L., and Lavine, A.S. (2006) *Fundamentals of Heat and Mass Transfer*, John Wiley & Sons, Inc.
19. Rohsenow, W.M., Hartnett, J.P., and Cho, Y.I. (1998) *Handbook of Heat Transfer*, McGraw-Hill.
20. Brebbia, C.A., Felles, J.C.F., and Wrobel, J.C.F. (1984) *Boundary Element Techniques – Theory and Applications*, Springer-Verlag.
21. Buchanan, G.R. (1995) *Finite Element Analysis*, McGraw-Hill.
22. MacNeal, R.H. (1994) *Finite Elements: Their Design and Performance*, Marcel Dekker.
23. Habenicht, G. (2002) *Kleben Grundlagen, Technologien, Anwendung*, Springer-Verlag, New York.
24. Lu, T. and Chen, C. (1999) *Acta Materialia*, **47**, 1469.
25. Öchsner A. and Grácio J., (2004) Proceedings of Applications of Porous Media, Evora, p. 409.
26. Breitz, W. and Grote, K.-H. (1997) *Dubbel Taschenbuch fÄur den Maschinenbau* Springer-Verlag.
27. Gibson, L.J. and Ashby, M.F. (1997) *Cellular Solids Structure and Properties*, 2nd edn, Cambridge University Press.
28. Berlyand, L. and Mityushev, V.V. (2001) *Journal of Statistical Physics*, **102**, 112.
29. Jikov, V.V., Kozlov, S.M., and Olejnik, O.A. (1994) *Homogenization of Di®erential Operators and Integral Functionals*, Springer-Verlag.
30. Milton, G.W. (2000) *Mechanics of Composite*, Cambridge University Press.
31. Pesetskaya, E.V. (2004) *Proceedings of the Institute of Mathematics (Minsk)*, **12**, 117.
32. Öchsner, A., Fiedler, T., Öchsner, A., Murch, G.E., and de Lemos, M.J.S. (2008) *Cellular and Porous Materials – Thermal Properties Simulation and Prediction*, Wiley-VCH Verlag GmbH, Weinheim, pp. 31–71.
33. Belova, I.V. and Murch, G.E. (2005) *Journal of Physics and Chemistry of Solids*, **66**, 722.
34. Fiedler, T., Solórzano, E., Garcia-Moreno, F., Öchsner, A., Belova, I., and Murch, G. (2009)

Materials Science Engineering and Technology, **40**, 139.

35. AK Steel Manufacturer Information (2008) *http://www.aksteel.com/pdf/markets_products/stainless/austenitic/316_316L_Data_Sheet.pdf*.

36. Veyhl, C., Merkel, M., and Öchsner, A. (2008) *Defect and Diffusion Forum*, **280–281**, 105.

37. Maxwell, J.C. (1892) *A Treatise on Electricity and Magnetism*, vol. 1, Clarendon Press, Oxford.

4
Computational Homogenization for Localization and Damage

Thierry J. Massart, Varvara Kouznetsova, Ron H. J. Peerlings, and Marc G. D. Geers

4.1
Introduction

4.1.1
Mechanics Across the Scales

Nowadays, the intrinsic role of different spatial scales in the mechanics of materials is well recognized. At the material level, it is the typical scale at which many heterogeneities can be identified that matters. The mechanics and physics of these – often multiphase – heterogeneous microstructures is recognized as the main driver for the macroscopic engineering response of a material upon mechanical loading, up to the point of failure. A proper understanding of the behavior and evolution of materials at this microscale is the key for predicting and improving their mechanical response. Over time, it has become clear that even smaller scales and thin interfaces may have a pronounced influence on the microscale response. For this reason, multiscale methods have emerged, which link up the relevant phenomena at the microscale to those at smaller and large scales.

The second characteristic of the multidisciplinary field of materials science and engineering is the emphasis that is put on mechanical aspects, covering the role of stress, strain, deformation, and degradation. Generally, the mechanics go hand in hand with material synthesis and microstructure evolution, since internal stress fields are an intrinsic characteristic of heterogeneous microstructures, which cannot be trivially separated from the governing physics. Mechanical aspects generally represent a source of internal (strain) energy, which is an essential ingredient of the underlying thermodynamics. On the other hand, other physical mechanisms (e.g., diffusion, dislocation motion) have a pronounced influence on the relaxation of internal stresses and consequently on the material's mechanical response. In other words, nowadays, multiscale mechanics covers (i) what mechanics contributes to physics across the scales and (ii) how physics at the smallest scales contributes to the mechanics at the macroscale.

Mechanical and physical processes at different spatial scales are particularly interwoven in failure mechanisms. A final, catastrophic fracture penetrates through

Advanced Computational Materials Modeling: From Classical to Multi-Scale Techniques.
Edited by Miguel Vaz Júnior, Eduardo A. de Souza Neto, and Pablo A. Munoz-Rojas

ISBN: 978-3-527-32479-8

all spatial scales, right down to the atomic level. Its effect on relevant processes within the material is so pronounced that it cannot be neglected on any of these scales. However, even the earlier stages of a failure process are generally governed by mechanisms that interact across two or more spatial scales – think, for instance, of diffuse microstructural degradation processes that weaken a material and therefore cause macroscopic stress redistributions, which in turn influence the microscopic damage process.

Inelastic deformation, damage, and failure are perhaps the most relevant mechanisms in engineering practice. Structures, products, and components are usually designed against failure using predictive design methods. Where a certain amount of damage has occurred or is accepted, such methods are used to assess the residual strength and/or ensure a safe operation. And in some particular cases, structures or processes are designed to result in controlled failure. Predictive methods that are necessary for each of these purposes and that go beyond the classical methods, such as fracture mechanics, can only be developed by taking into account the intrinsic multiscale character of failure as discussed above. Constructing multiscale methods for failure processes is therefore extremely relevant, but, at the same time, extremely challenging.

4.1.2 Some Historical Notes on Homogenization

Early steps toward multiscale or microscale modeling were taken long ago, when interest in the mechanics of heterogeneous materials became more pronounced. Preliminary steps go back to the nineteenth century, when the rule of mixtures was first introduced (Voigt, 1887) [1], followed by the Sachs model in 1928 [2], the Reuss estimate (1929), and the frequently used Taylor model (1938) [3]. The Voigt and Reuss models focused more on composite systems, whereas the Taylor and Sachs rules were typically derived for polycrystals. The growing interest in composite materials constituted the main trigger for new developments. The best-known early contribution of this type is probably the work of Eshelby [4]. Even today, these first steps have a pronounced impact, giving rise to alternative continuum mechanics frameworks (Eshelbian mechanics, material forces, etc.). The field of "continuum micromechanics," which was formally established by Hill in 1965 [5], has grown tremendously since then. A survey of activities over the past 40 years is given in Ref. [6].

The 1950s, 1960s, and 1970s were characterized by major progress made in the homogenization of heterogeneous elastic solids. Pioneering work in this time frame was done by, among others, Kröner *et al.* [7], Hill [8], Mori and Tanaka [9], and Willis [10]. Various refinements were developed in the past two decades. First steps into the nonlinear regime of the developed elastic homogenization theories and variational principles were taken by, for example, Kröner [11], Hill [5], and Hutchinson [12], whereas many more papers on the subject appeared in the 1980s and 1990s. Subjects treated covered elastoplasticity (both rate independent and viscoplastic), nonlinear elasticity, and viscoelasticity. Frequently cited contributors in this field are Nemat-Nasser [13, 14], Ponte Castañeda [15], Suquet [16, 17], Willis [18], and Zaoui [19]. Other extensions into the nonlinear range appeared in the late 1970s, for example, the well-known Gurson model for void growth in ductile materials [20], igniting a wealth of follow-up work on the plasticity of porous materials and ductile damage. Since the 1980s, a steady growth of the entire field took place, in which more physics is systematically being integrated. Asymptotic or mathematical homogenization schemes have been frequently used to assess effective properties of elastic heterogeneous materials [21–23]. Extensions toward higher order and nonlocal constitutive equations have been considered, for example, developments including Cosserat media [24], couple stress theory [25], nonlocal effective continua [26], or higher order gradient homogenized elastic materials [27–29]. Homogenization of solids in a geometrically and physically nonlinear regime is clearly more challenging. Interesting contributions are given and cited in Ref. [30]. Another class of hierarchical techniques, also used in the context of localization and damage, is generally known as *variational multiscale methods* [31, 32], showing considerable similarities with applications based on the extended finite element method (X-FEM) [33].

4.1.3
Separation of Scales

Many of the early multiscale methods as discussed above essentially limit themselves to establishing the homogenized response of heterogeneous elastic materials. Homogenization frameworks focus on the equivalent or effective response of a finite volume of material, which is generally assumed to be statistically homogeneous. Characteristic volumes can be identified as unit cells for periodic materials and representative volume elements for statistically heterogeneous media [8, 26]. The response of such a volume is used to calibrate a homogeneous equivalent continuum, which can then be used to solve boundary value problems. This concept critically relies on the principle of separation of scales. This principle states that the scale of the microstructure or microstructural fluctuation ℓ_μ must be much smaller than the size of the representative volume considered ℓ_{m}, which must in turn be smaller than the characteristic fluctuation length in the macroscopic deformation field ℓ_{M}:

$$l_\mu \ll l_{\mathrm{m}} \ll l_{\mathrm{M}} \qquad (4.1)$$

Following this definition, the size of the macrostructure is irrelevant for the scale separation.

The separation of scales as defined by Eq. (4.1) must be satisfied in order to be able to successfully construct a homogeneous equivalent continuum according to the concept of local action. However, it is sometimes violated when either a microstructural length scale tends to be large (e.g., in the presence of long-range correlations or percolation phenomena) or when the macrofluctuation scale tends to be small (e.g., localization of deformation, gradients). Violations of the principle of separation of scales constitute a key difficulty in handling localization and damage in homogenization schemes. If deformations tend to localize, the macroscopic response will ensue from a gradually shrinking volume of material. The term *homogenization* therefore becomes inappropriate. However, this does not imply that it becomes impossible to upscale the local microstructural response to a large scale. If proper scale transition methods are applied, a solution may still be found; this is the main topic of this chapter.

4.1.4 Computational Homogenization and Its Application to Damage and Fracture

Over the last few years, substantial progress has been made in the two-scale computational homogenization of complex multiphase solids. This method is essentially based on the nested solution of two boundary value problems, one at each scale. Although computationally expensive, the procedures developed allow to assess the macroscopic influence of microstructural parameters in a rather straightforward manner. The conventional, first-order computational homogenization method was first addressed in Ref. [34], approximately 20 years ago, even though some elements were already in place earlier [35]. The major developments that have led to the completion of this method took place about 10 years later, through a number of contributions [36–47].

The first-order method has now matured to a standard tool in computational homogenization [48–52]. Several extensions that depart from the first-order method have been addressed in the literature:

- second-order computational homogenization [53–59], which takes into account higher order deformation gradients at the macroscale;
- a continuous–discontinuous multiscale approach for damage, in which the coarse scale is enriched by discrete localization bands (weak discontinuities), whereas the fine scale is modeled using a continuum [60, 61];
- thermomechanically coupled computational homogenization, that is, homogenization of heat conduction, coupling to mechanical homogenization [62, 63];
- computational homogenization of structured thin sheets and shells, based on the application of second-order homogenization principles to through-thickness representative volume elements of thin structures [64];
- computational homogenization of interface problems, by establishing a relationship between a discrete macroscopic interface (e.g., a cohesive zone) and a more refined, often continuous representation of the interface's microstructure [65].

Of the above extensions, scale transitions for damage and fracture are among the most complex. Damage is a phenomenon that develops across all length scales. Since the first-order approach intrinsically violates the principle of separation of scales if localization occurs, alternative solutions are needed. Simplified approaches aim at establishing a rigorous volumetric coupling between a macroscopic finite element size and the damaging representative volume element [66]. However, this solution essentially sidesteps the key ideas and benefits of homogenization and resembles more a domain decomposition approach with an embedded scale refinement.

This chapter focuses on two computational homogenization schemes that have enabled a breakthrough in handling the transition from damage to localization. First, in Section 4.2, a continuous–continuous second-order scheme is briefly reviewed. This scheme allows one to capture moderate localization bands in continua. It inherits all regularization properties of gradient continua of the Mindlin type. The essential limitation of this second-order scheme is that it cannot capture the full transition to a discrete crack.

To treat this transition, a continuous–discontinuous scheme is developed in Section 4.3. This framework handles the gradual transition from homogenization to localization by embedding a discontinuity in the macrodomain once the localization of deformation is detected. The constitutive behavior of the material is still extracted from an underlying microstructural volume element. Several technical issues come into play: detecting the onset of localization, handling the macrovolume with the embedded localization band (compatibility and traction continuity), snapback at the level of finite-size macrodomain, path-following within the microscale volume element, and so on, each of which is discussed in detail. The approach is applied to masonry, since its regular structure considerably simplifies the interpretation and implementation of such a continuous–discontinuous approach.

Recent developments related to the continuous–discontinuous homogenization scheme can be found in Refs [67–69], where advantage is taken of an X-FEM approach to incorporate the discontinuity at the macroscale. Generalized schemes, where localization bands can evolve steadily, are presently under construction and will be published in forthcoming work.

4.2 Continuous–Continuous Scale Transitions

4.2.1 First-Order Computational Homogenization

At the macroscopic scale, consider a general quasi-static boundary value problem, governed by the equilibrium equation, which is, in the absence of body forces, expressed as

$$\nabla_{0\mathrm{M}} \cdot \mathbf{P}_{\mathrm{M}}^{T} = \mathbf{0} \tag{4.2}$$

supplemented by natural or essential boundary conditions. In Eq. (4.2), $\mathbf{P}_{\mathrm{M}}$ is the first Piola–Kirchhoff stress tensor; $\nabla_{0\mathrm{M}}$ is the gradient operator with respect to the reference configuration; the subscript "M" refers to a macroscale quantity, while the subscript "m" denotes a microscale quantity.

To complete this boundary value problem, a constitutive relation between the stress and kinematical quantities needs to be added. Instead of assuming a constitutive equation in a closed-form, computational homogenization techniques extract the constitutive response numerically from the detailed computational analysis of a microstructural representative volume element (RVE).

First-order computational homogenization departs from the classical linearization of the macroscopic nonlinear deformation map, $\mathbf{x} = \phi(\mathbf{X})$, with $\mathbf{x}$ and $\mathbf{X}$ corresponding position vectors in the deformed and reference state, respectively. When applied to a material vector $\Delta\mathbf{x}$ in the deformed state, this linearization reads

$$\Delta\mathbf{x} = \mathbf{F}_\mathrm{M} \cdot \Delta\mathbf{X} + \mathbf{w} \tag{4.3}$$

where $\Delta\mathbf{x}$ and $\Delta\mathbf{X}$ are relative position vectors with respect to an arbitrary reference point; $\mathbf{F}_\mathrm{M} = (\boldsymbol{\nabla}_{0\mathrm{M}}\mathbf{x})^T$ is the macroscale deformation gradient tensor. The microfluctuation field $\mathbf{w}$ is identified as the local fine-scale contribution superimposed on to the macroscale deformation. From Eq. (4.3), the microscale deformation gradient tensor $\mathbf{F}_\mathrm{M}$ is determined as

$$\mathbf{F}_\mathrm{m} = (\boldsymbol{\nabla}_{0\mathrm{m}}\Delta\mathbf{x})^T = \mathbf{F}_\mathrm{M} + (\boldsymbol{\nabla}_{0\mathrm{m}}\mathbf{w})^T \tag{4.4}$$

with $\boldsymbol{\nabla}_{0\mathrm{m}}$ as the gradient operator with respect to the reference configuration of the RVE.

Relations (4.3) and (4.4) are valid for every point at the microscale, with the first terms readily known for a given macroscale deformation tensor $\mathbf{F}_\mathrm{M}$, while the microfluctuation $\mathbf{w}$ will follow from the solution of the microscale boundary value problem.

The microscale boundary value problem is also a standard problem in quasi-static continuum solid mechanics. In the absence of body forces, the equilibrium equation for the microstructural RVE in terms of the microscale first Piola–Kirchhoff stress tensor $\mathbf{P}_\mathrm{m}$ takes the form

$$\boldsymbol{\nabla}_{0\mathrm{m}} \cdot \mathbf{P}_\mathrm{m}^T = \mathbf{0} \tag{4.5}$$

The material behavior of each microstructural constituent α (e.g., matrix, inclusion, interface, etc.) is assumed to be known and described by constitutive laws, specifying a time- and history-dependent stress–strain relationship, possibly involving microstructural evolution,

$$\mathbf{P}_\mathrm{m}^{(\alpha)}(t) = \mathcal{F}^{(\alpha)}\left\{\mathbf{F}_\mathrm{m}^{(\alpha)}(\tau), \tau \in [0, t]\right\} \tag{4.6}$$

The above problem (4.5) has to be completed by boundary conditions. The essential step in the computational homogenization methodology is the derivation of RVE boundary conditions from postulated scale transition relations. For example, the kinematical averaging relation is one of the most commonly used. It requires the volume average of the microscale deformation gradient tensor $\mathbf{F}_\mathrm{m}$ to be equal to the corresponding macroscale deformation gradient tensor $\mathbf{F}_\mathrm{M}$:

$$\mathbf{F}_\mathrm{M} = \frac{1}{V_0}\int_{V_0} \mathbf{F}_\mathrm{m}\,\mathrm{d}V_0 \tag{4.7}$$

where V_0 is the RVE volume in the reference configuration. Substituting Eq. (4.4) into the right-hand side of the scale transition relation (4.7) yields

$$\begin{aligned}\frac{1}{V_0}\int_{V_0} \mathbf{F}_{\mathrm{m}}\, \mathrm{d}V_0 &= \mathbf{F}_{\mathrm{M}} + \frac{1}{V_0}\int_{V_0} (\nabla_{0\mathrm{m}}\mathbf{w})^T\, \mathrm{d}V_0 \\ &= \mathbf{F}_{\mathrm{M}} + \frac{1}{V_0}\int_{S_0} \mathbf{w}\otimes\mathbf{N}_{\mathrm{m}}\, \mathrm{d}S_0 \end{aligned} \tag{4.8}$$

where the divergence theorem has been used to transform the volume integral to an integral over the undeformed boundary S_0 of the RVE, with outward normal $\mathbf{N}_{\mathrm{m}}$.

From Eq. (4.8) it is immediately clear that in order for the kinematical scale transition relation (4.7) to be satisfied a priori, the boundary conditions imposed on the RVE should be such that the contribution of the microfluctuation field $\mathbf{w}$ in Eq. (4.8) vanishes. This can be achieved in many alternative ways. Some of the possibilities proposed and used in the literature are listed below.

1) Do not allow for any microstructural fluctuations in the RVE, that is,

$$\mathbf{w} = \mathbf{0} \quad \forall \mathbf{X} \in V_0 \tag{4.9}$$

forcing the entire volume to deform according to the prescribed $\mathbf{F}_{\mathrm{M}}$. In the literature, this is usually referred to as the *Taylor* (or *Voigt*) *assumption*.

2) Suppress the microfluctuation only at the RVE boundary

$$\mathbf{w} = \mathbf{0} \quad \forall \mathbf{X} \in S_0 \tag{4.10}$$

while leaving the microstructural fluctuations inside the volume yet undetermined. With this condition, the displacements of the RVE boundary are fully prescribed according to the given $\mathbf{F}_{\mathrm{M}}$. These are often termed *uniform displacement boundary conditions.*

3) For an RVE with an initially geometrically periodic boundary (i.e., the boundary can be split in "+" and "−" parts defined by the opposite outward normal vectors at the corresponding points, $\mathbf{N}_{\mathrm{m}}^{+} = -\mathbf{N}_{\mathrm{m}}^{-}$), the so-called periodic boundary conditions can be imposed by requiring periodicity of the microfluctuation field:

$$\mathbf{w}^{+} = \mathbf{w}^{-} \tag{4.11}$$

4) The weakest possible constraint is to require the boundary integral to vanish as a whole:

$$\int_{S_0} \mathbf{w}\otimes\mathbf{N}_{\mathrm{m}}\, \mathrm{d}S_0 = \mathbf{0} \tag{4.12}$$

In the literature, this constraint is sometimes called *minimal kinematic boundary conditions* [70] and it is, in fact, equivalent to uniform traction boundary conditions [71].

Of the above choices, the Taylor (Voigt) assumption is computationally the most efficient, since it does not require detailed modeling of microstructural geometry.

Accordingly, only a rough estimate of the overall material response is obtained, usually significantly overestimating the stiffness. Nevertheless, the Taylor assumption is often used in crystal plasticity modeling with a reasonable accuracy, but it performs poorly for a general complex nonlinear microstructure. Some special cases, such as strain localization and damage, cannot be treated at all with the Taylor assumption. The other above-mentioned alternatives to enforce the kinematical scale transition all require the solution of the RVE boundary value problem, at the same time allowing the incorporation of local microstructural details. The apparent overall properties obtained by application of uniform displacement boundary conditions on a microstructural cell usually overestimate the real effective properties, while the minimal kinematic boundary conditions lead to an underestimation and usually are sensitive to microstructural details near the RVE boundary. For a given microstructural cell size, the periodic boundary conditions are known to provide a better estimation of the overall properties than the other mentioned alternatives [42, 72–74]. The periodic boundary conditions are the most frequently used in practice, although the uniform displacement boundary conditions are also often used, mostly owing to the simplicity of their implementation. The elaboration of the periodic boundary conditions (4.11) toward a format suitable for the implementation in a general finite element code is beyond the scope of this chapter and can be found elsewhere [45, 75].

After the solution of the RVE boundary value problem, the obtained microscale stress state is to be homogenized toward the macroscopic stress response. For this, another scale transition is typically used, for example, the Hill–Mandel macrohomogeneity condition [8, 35]. This condition requires the volume average of the microscale virtual work to be equal to the corresponding local macroscale virtual work:

$$\frac{1}{V_0}\int_{V_0} \mathbf{P}_{\mathrm{m}} : \delta\mathbf{F}_{\mathrm{m}}^T \,\mathrm{d}V_0 = \mathbf{P}_{\mathrm{M}} : \delta\mathbf{F}_{\mathrm{M}}^T \tag{4.13}$$

Using the divergence theorem, with incorporation of microstructural equilibrium (4.5), the volume average of the microstructural virtual work may be expressed in terms of RVE surface quantities as

$$\delta W_{\mathrm{m}} = \frac{1}{V_0}\int_{V_0} \mathbf{P}_{\mathrm{m}} : \delta\mathbf{F}_{\mathrm{m}}^T \,\mathrm{d}V_0 = \frac{1}{V_0}\int_{S_0} \mathbf{p}_{\mathrm{m}} \cdot \delta\mathbf{x} \,\mathrm{d}S_0 \tag{4.14}$$

where $\mathbf{p}_{\mathrm{m}} = \mathbf{N}_{\mathrm{m}} \cdot \mathbf{P}_{\mathrm{m}}^T$ is the first Piola–Kirchhoff stress vector. Substitution of the variation of relation (4.3) leads to

$$\delta W_{\mathrm{m}} = \left(\frac{1}{V_0}\int_{S_0} \mathbf{p}_{\mathrm{m}} \otimes \mathbf{X} \,\mathrm{d}S_0\right) : \delta\mathbf{F}_{\mathrm{M}}^T + \frac{1}{V_0}\int_{S_0} \mathbf{p}_{\mathrm{m}} \cdot \delta\mathbf{w} \,\mathrm{d}S_0 \tag{4.15}$$

The last term in Eq. (4.15), involving the average work done by the microfluctuations, can be shown to vanish for each of the possible RVE boundary conditions listed above. Then, comparing Eq. (4.15) and the right-hand side of Eq. (4.13), the micro-to-macro stress scale transition is obtained as

$$\mathbf{P}_{\mathrm{M}} = \frac{1}{V_0}\int_{S_0} \mathbf{p}_{\mathrm{m}} \otimes \mathbf{X} \,\mathrm{d}S_0 = \frac{1}{V_0}\int_{V_0} \mathbf{P}_{\mathrm{m}} \,\mathrm{d}V_0 \tag{4.16}$$

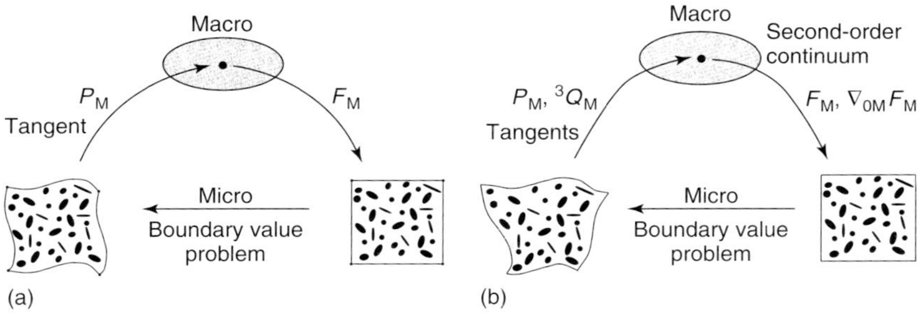

Figure 4.1 (a) First-order and (b) second-order continuum–continuum computational homogenization schemes.

Note that sometimes in first-order computational homogenization, the stress volume averaging relation (4.16) is postulated, together with the kinematics scale transition (4.7), leading to a selection of the boundary conditions, and then the validity of the Hill–Mandel condition is shown. Obviously, the obtained formulation is exactly the same.

Thus, for a given macroscale deformation gradient tensor $\mathbf{F}_M$, the corresponding stress response $\mathbf{P}_M$ is obtained from the analysis of the microscale RVE, combined with the consistent scale transition relations. The first-order computational homogenization loop is schematically illustrated in Figure 4.1(a). This procedure can be repeated for each macroscale (integration) point for each macroscale loading increment. If an implicit procedure is used to solve the macroscale nonlinear problem, the multiscale loop should also be performed for each iteration. The tangent operator needed in this case at each integration point can be efficiently extracted by static condensation of the RVE tangent stiffness matrix [45, 75].

4.2.2
Second-Order Computational Homogenization

Second-order computational homogenization establishes the scale transition between a microstructural RVE treated as a standard first-order continuum and a full second-gradient continuum at the macroscale. In the second-gradient continuum theory [76], the internal work is assumed to be determined by the deformation gradient tensor $\mathbf{F}_M$ and its gradient $^3\mathbf{G}_M = \boldsymbol{\nabla}_{0M}\mathbf{F}_M$ combined with their respective work-conjugated quantities: the first Piola–Kirchhoff stress tensor $\mathbf{P}_M$ and the higher order stress tensor $^3\mathbf{Q}_M$. Starting from this assumption, after some mathematical manipulations, the strong form of local equilibrium for a second-gradient continuum can be derived as (in the absence of the body forces)

$$\boldsymbol{\nabla}_{0M} \cdot (\mathbf{P}_M - \boldsymbol{\nabla}_{0M} \cdot {}^3\mathbf{Q}_M)^T = \mathbf{0} \tag{4.17}$$

The associated boundary conditions consist of tractions and double stress tractions, displacements and normal gradients of displacements, or combinations thereof.

To close the second-gradient boundary value problem, constitutive relations between the stress quantities ($\mathbf{P}_M$ and $^3\mathbf{Q}_M$) and the history of the kinematical variables

($\mathbf{F}_M$ and ${}^3\mathbf{G}_M$) should be added. Again, rather than assuming the constitutive laws in a closed form, the second-order computational homogenization technique is used to numerically obtain the macroscale constitutive response from the detailed microscale analysis, as schematically shown in Figure 4.1(b).

The second-order computational homogenization scale transition departs from the Taylor series expansion of the nonlinear deformation map, truncated after the second term (cf. (4.3)):

$$\Delta\mathbf{x} = \mathbf{F}_M \cdot \Delta\mathbf{X} + \frac{1}{2}\Delta\mathbf{X} \cdot {}^3\mathbf{G}_M \cdot \Delta\mathbf{X} + \mathbf{w} \tag{4.18}$$

where $\mathbf{w}$ is, as before, the microfluctuation field. The microscale deformation gradient tensor $\mathbf{F}_m$ follows as

$$\mathbf{F}_m = (\nabla_{0m}\Delta\mathbf{x})^T = \mathbf{F}_M + {}^3\mathbf{G}_M \cdot \Delta\mathbf{X} + (\nabla_{0m}\mathbf{w})^T \tag{4.19}$$

Applying the earlier introduced kinematical scale transition relation (4.7) to Eq. (4.19) leads to two kinematical constraints:

$$\frac{1}{V_0}\int_{V_0} \Delta\mathbf{X}\, \mathrm{d}V_0 = \mathbf{0} \tag{4.20}$$

$$\frac{1}{V_0}\int_{V_0} (\nabla_{0m}\mathbf{w})^T\, \mathrm{d}V_0 = \frac{1}{V_0}\int_{S_0} \mathbf{w} \otimes \mathbf{N}_m\, \mathrm{d}S_0 = \mathbf{0} \tag{4.21}$$

where the Gauss theorem has been used to obtain the latter relation. Equation (4.20) is clearly satisfied if the Taylor series (4.18) has been expanded with respect to the RVE's geometrical center. This appears to be a necessary condition in the second-order framework, which deviates from the first-order scheme, where essentially any point could have been used to develop Eq. (4.3), leading to the same result [56]. The second constraint (4.21) is the same as in the first-order approach and can be satisfied in various ways as discussed previously in Section 4.2.1. Note that the macroscopic deformation is, in general, not periodic in the second-order case; thus, the application of the periodic boundary conditions ensures the periodicity of the microfluctuation field only and not of the deformed RVE geometry.

As has been shown in Refs [53, 54], to fully impose the macroscale second-order kinematics on a microstructural RVE, an additional kinematical scale transition is required. Perhaps the most straightforward relation, equating ${}^3\mathbf{G}_M$ and the volume average of ${}^3\mathbf{G}_m = \nabla_{0m}\mathbf{F}_m$ would, however, lead to higher order boundary conditions on the microfluctuation field, which would require a second-gradient problem on the RVE level as well. To be able to retain a classical continuum problem at the RVE level, an alternative scale transition relation between ${}^3\mathbf{G}_M$ and microstructural quantities has been elaborated in Ref. [54], ultimately leading to the following additional constraint for the microfluctuation field:

$$\int_{S_0} (\mathbf{N}_m \otimes \mathbf{w} \otimes \Delta\mathbf{X} + \Delta\mathbf{X} \otimes \mathbf{w} \otimes \mathbf{N}_m)\, \mathrm{d}S_0 = {}^3\mathbf{0} \tag{4.22}$$

which can be imposed in different alternative ways as well.

To extract the macroscopic stress quantities upon the microscale analysis, the Hill–Mandel macrohomogeneity is again employed, with the variation of the macroscale work involving two terms in the case of the second-gradient continuum:

$$\frac{1}{V_0}\int_{V_0} \mathbf{P}_{\mathrm{m}} : \delta\mathbf{F}_{\mathrm{m}}^T \,\mathrm{d}V_0 = \mathbf{P}_{\mathrm{M}} : \delta\mathbf{F}_{\mathrm{M}}^T + {}^3\mathbf{Q}_{\mathrm{M}} : \delta^3\mathbf{G}_{\mathrm{M}} \tag{4.23}$$

Elaboration of this equation along the same steps as done previously for the first-order case leads to the following expressions of the macroscale stress quantities in terms of RVE surface or volume integrals [54]:

$$\mathbf{P}_{\mathrm{M}} = \frac{1}{V_0}\int_{S_0} \mathbf{p}_{\mathrm{m}} \otimes \mathbf{X}\,\mathrm{d}S_0 = \frac{1}{V_0}\int_{V_0} \mathbf{P}_{\mathrm{m}}\,\mathrm{d}V_0 \tag{4.24}$$

$${}^3\mathbf{Q}_{\mathrm{M}} = \frac{1}{2V_0}\int_{S_0} \mathbf{X}\otimes\mathbf{p}_{\mathrm{m}} \otimes \mathbf{X}\,\mathrm{d}S_0 = \frac{1}{2V_0}\int_{V_0} (\mathbf{P}_{\mathrm{m}}^T\otimes\mathbf{X} + \mathbf{X}\otimes\mathbf{P}_{\mathrm{m}})\,\mathrm{d}V_0 \tag{4.25}$$

The consistent tangent operators at the macroscale integration point level can also be extracted through the static condensation of the RVE tangent matrix [58].

4.2.3
Application of the Continuous–Continuous Homogenization Schemes to Ductile Damage

To scrutinize the applicability of the first- and second-order continuum–continuum computational homogenization schemes to problems of strain localization and damage, an academic benchmark problem has been set up, mimicking ductile damage development across the two scales.

At the microscale, void nucleation by matrix debonding from hard inclusions has been simulated using a simple unit-cell model, shown in Figure 4.2(a). The size of the two-dimensional (plane-strain) square unit cell is $h = 3.3\,\mu\mathrm{m}$, with the diameter of the circular inclusion $d = 1.4\,\mu\mathrm{m}$. The matrix material is modeled as an elastoplastic material with $E = 210\,\mathrm{GPa}$, $\nu = 0.3$, the initial yield stress

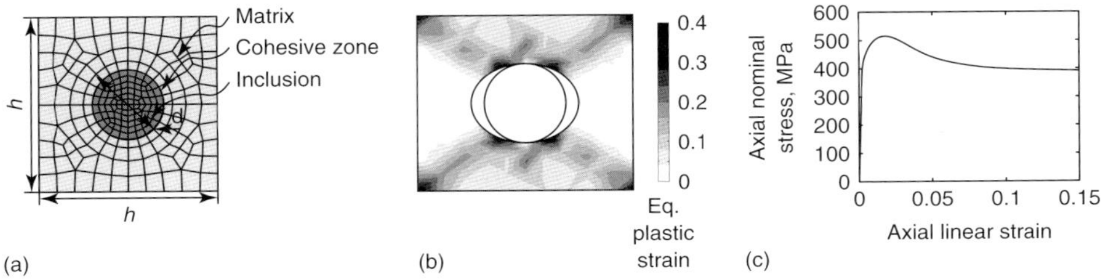

Figure 4.2 (a) The microstructural unit cell with the finite element mesh. (b) The deformed geometry and distribution of the equivalent plastic strain within the unit cell after uniaxial tensile loading, and (c) the corresponding overall stress–strain response.

$\sigma_y^0 = 507\,\text{MPa}$ and the linear hardening modulus $h = 200\,\text{MPa}$. The inclusion is assumed elastic with $E = 500\,\text{GPa}$ and $\nu = 0.3$. Interface debonding is modeled using the cohesive zone approach, as proposed by Xu and Needleman [77], with an exponential form of traction–separation law. The work of separation in both normal and tangential direction is taken $\phi = 45\,\text{J/m}^2$ and the characteristic length $\delta = 0.03\,\mu\text{m}$. Application of a tensile load on such a unit cell at a certain stage leads to the opening of the interface and further void growth (Figure 4.2(b)). Additionally, local bands of localized deformation appear within the unit cell. The combination of these two effects results in a strain softening overall unit-cell response (Figure 4.2(c)). The presence of the RVE's global softening and local localization bands will prove crucial where the applicability limits of various computational homogenization schemes are concerned.

For this benchmark problem, at the macroscale, a plate consisting of a periodic array of the unit cells is considered, as shown in Figure 4.3. To trigger the appearance of a macroscale localization band, an imperfection, that is, a reduction of the interface work of separation by 10%, is introduced in the left bottom corner of the plate. At the bottom and left edges of the plate, the symmetry boundary conditions are applied, at the top edge, a uniform vertical displacement is applied, whereas the right edge is traction free. Additionally, for the second-gradient continuum zero normal gradients of the tangential displacement components at

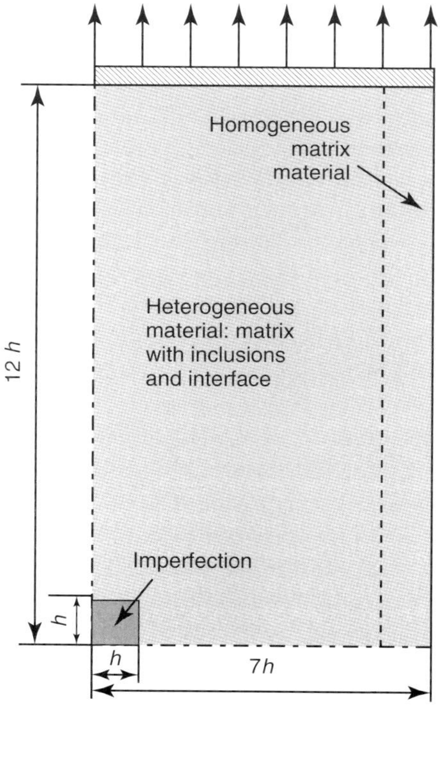

Figure 4.3 Macroscopic problem for the benchmark homogenization analysis.

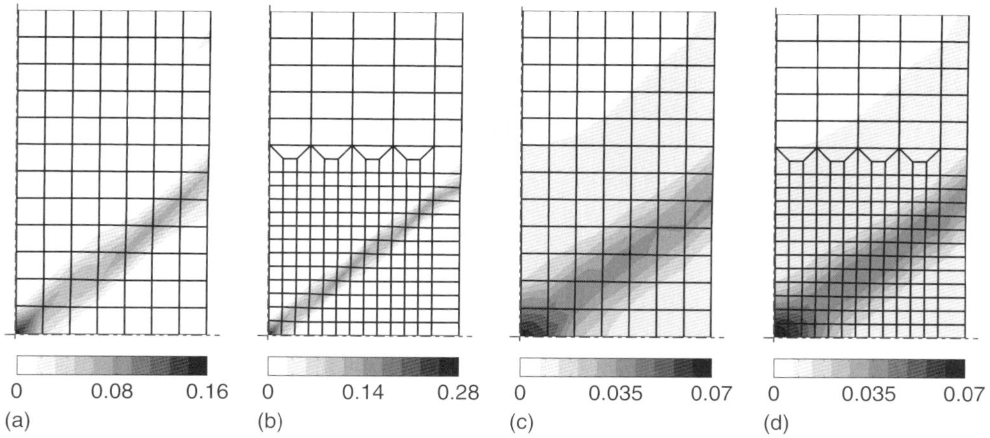

Figure 4.4 Localized equivalent total strain resulting from (a,b) the first-order computational homogenization for the (a) coarse and (b) fine finite element discretizations; (c,d) the second-order computational homogenization for the (c) coarse and (d) fine finite element discretisations.

the bottom, left, and top edges are prescribed, while the remaining higher order boundary conditions are zero double stress tractions.

Before the onset of overall unit-cell softening both the first- and second-order schemes lead to identical solutions, which can also be shown to be the same as the direct numerical simulation (results not shown here). After the onset of microscale softening, however, the inadequacy of the first-order scheme to deal with this type of problems becomes apparent. Figures 4.4(a) and 4.4(b) show the equivalent total strain distribution at the macroscale obtained with the first-order scheme using two different finite element discretizations of the macroscale problem. Clearly, the solution fully localizes according to the size of the elements used. This property of the first-order scheme is inherently linked to the principle of local action combined with the full-scale separation. It associates a microscale volume of an infinitesimal size with each macroscopic point, such that the assumption of a constant macroscale deformation gradient over the RVE, used to write the relation (4.3), holds. Upon further refinement of the macroscale mesh, the energy dissipated in the softening RVE at the microscale is in fact dissipated in a shrinking macroscale volume, which is one of the main manifestations of the ill-posedness of the macroscale boundary value problem to be solved. This is, in fact, the same shortcoming as has been extensively investigated by many authors for local closed-form constitutive softening models.

The second-order scheme homogenizes toward a macroscale second-gradient continuum, which is known to possess certain regularizing properties due to the incorporation of a length scale, making the numerical solution independent of the mesh size (i.e., the width of the localization band converges to a finite value upon mesh refinement). In the second-order computational homogenization context, the length scale of the macroscopic homogenized continuum has been shown to be intrinsically related to the RVE size and the unit processes therein (e.g., the development of percolation paths) [57]. Indeed, the solution of the benchmark problem considered here with the second-order scheme proves to be mesh independent, as shown in Figure 4.4(c,d).

The two-scale results obtained with the second-order approach are shown in Figure 4.5, showing the macroscale localized deformation together with the deformed geometries of several unit cells corresponding to different macroscale integration points, thus providing insight into the interaction between the deformation processes taking place at the two scales.

As demonstrated by this example, the first-order computational homogenization scheme is not suited for the analysis of localization problems (beyond the point of the macroscale localization onset). The second-order extension, on the other hand, can well tackle moderate localization bands, that is, when the macroscopic strain fields vary only linearly (quadratic in displacements) over the microscale RVE. The method cannot properly resolve macroscopic localization bands beyond this linear variation. Sharper localization regions, for example, a localization band within the RVE itself, will naturally violate this condition. For instance, in ductile damage development, the stage of necking between the voids and void coalescence, leading to the macroscale discrete crack, cannot be upscaled by continuous–continuous computational homogenization schemes. This calls for the development of continuous–discontinuous approaches, one of which is discussed in detail in the next section.

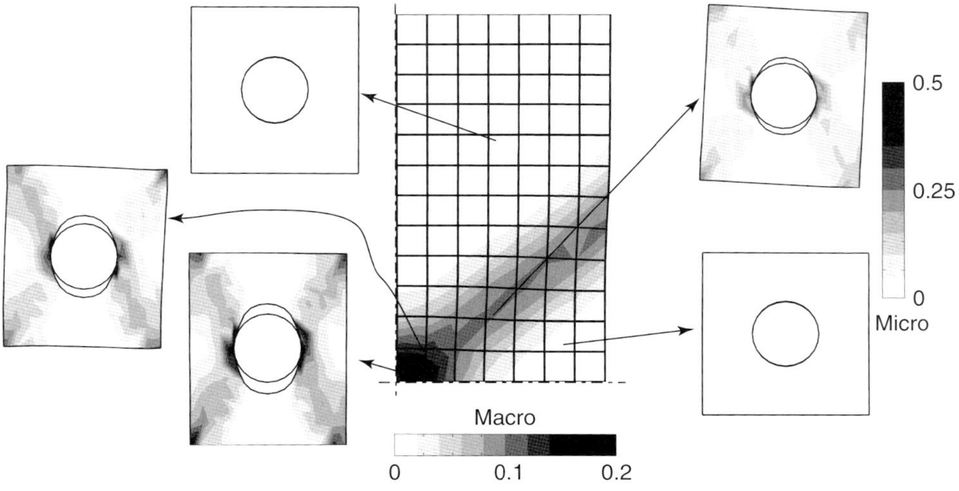

Figure 4.5 Distribution of the equivalent total strain in the deformed macrostructure and several microstructural unit cells obtained by the second-order computational homogenization analysis.

4.3 Continuous–Discontinuous Scale Transitions

Problems in which the degradation process leads to localization of damage and deformation at the scale of the microstructure can no longer be dealt with efficiently by the continuous–continuous scale transitions discussed in Section 4.2 – even those of second order. Many examples of this class of problems may be found in failure processes of quasi-brittle materials used in civil engineering, for example, of concrete or masonry. The complex failure behavior exhibited by these materials, nevertheless, requires a multiscale modeling approach. A strategy that explicitly accounts for the localization process is called for in such cases and a natural approach is to incorporate discrete localization bands, that is, strong or weak displacement discontinuities, at the coarse scale.

At the underlying scale of material heterogeneity, either a discrete or a continuous description of the failure process may be employed. As quasi-brittle materials generally exhibit a diffuse degradation at the scale of the microstructure, we use a continuum damage model at this scale. To regularize the microstructural localization of deformation and damage, a nonlocal (implicit gradient) formulation is adopted.

To establish a relationship between the continuous micromodel and the discontinuous macroscopic model, proper continuous–discontinuous scale transitions must be formulated. In this section, we demonstrate how such transitions can be constructed by appropriately modifying the first-order computational homogenization framework as presented in Section 4.2.1. This is a challenge particularly for quasi-brittle materials due to their relative brittleness and the instabilities associated with this brittleness. A number of issues must, therefore, be addressed, which do not arise in conventional computational homogenization or for more ductile materials and which can be summarized as follows:

- The appearance and the orientation of discrete localization zones at the coarse scale, starting from a more uniform response, should result from the average behavior extracted computationally from the fine scale.
- An RVE size must be identified to extract the average material behavior. The average softening response of an RVE intrinsically depends on its size. A multiscale scheme should, therefore, properly transfer to the coarse scale the amount of energy dissipated by the fine-scale damage process. When a discrete band is used at the coarse scale, this can be ensured by incorporating the finite volume on which the fine-scale damage is sampled. For initially periodic materials such as masonry, a characteristic size of the localization band can be identified relatively easily, as is shown below.
- Because of the heterogeneous nature of quasi-brittle materials, elastic unloading may occur in stronger phases upon failure of the weaker phase(s), resulting in snapback-type instabilities in the coarse-scale average response. This snapback can originate from unloading inside the RVE itself or from unloading of the material surrounding the macroscopic localization band. The boundary value problem on the RVE and the coarse-scale path-following technique should

be constructed in such a way that these complex meso- and macrostructural instabilities can be captured.

- Finally, coarse-scale discontinuities are used to represent the energy dissipation of a cluster of cracks, which successively evolves into a single macroscopic crack. The orientation of these (average) localization zones should, therefore, be updated according to the microstructural damage evolution during the computation.

Each of the above issues is discussed in detail in the following for an initially periodic quasi-brittle material. The methods developed are illustrated using a model material for which all these adaptations have to be addressed, that is, masonry. Extensive use is made of the first-order computational homogenization concepts of Section 4.2.1. However, given the quasi-brittle nature of the materials considered, we can limit ourselves to an infinitesimal strain description. Furthermore, we denote the relevant scale of heterogeneity, that is, that of individual bricks and mortar joints, as mesoscale, reserving the term *microscale* for the internal structure of each of these two phases. Throughout this section, uppercase symbols denote macroscopic quantities, whereas lowercase symbols are used for the mesoscale.

4.3.1
Scale Transitions and RVE for Initially Periodic Materials

4.3.1.1 First-Order Scale Transitions

As explained in Section 4.2.1, a macro–meso scale transition consists in applying the macroscopic strain to an RVE of the meso- (or micro-) structure in an average sense. Boundary conditions must be defined for the RVE, which ensure that the volume average of the mesoscopic strain on the RVE equals the imposed macroscopic strain.

In terms of the infinitesimal strain setting used in this section, the mesoscopic displacement field can be written as (cf. Eq. (4.3))

$$\mathbf{u} = \mathbf{E} \cdot \mathbf{x} + \mathbf{w} \tag{4.26}$$

where $\mathbf{E}$ is the macroscopic infinitesimal strain tensor, $\mathbf{x}$ is the position vector within the RVE, and $\mathbf{w}$ is a mesoscopic displacement fluctuation originating from the heterogeneity of the material. The periodicity of the masonry material considered here suggests the use of periodicity conditions for $\mathbf{w}$, similar to those defined in Eq. (4.11). As a result, averaging of the mesoscopic strain field ε ensuing from Eq. (4.26) results in

$$\langle \varepsilon \rangle = \frac{1}{V} \int_V [\nabla_{\mathrm{m}}(\mathbf{E} \cdot \mathbf{x} + \mathbf{w})]^{\mathrm{sym}} \, \mathrm{d}V = \mathbf{E} + \frac{1}{V} \int_S (\mathbf{w} \otimes \mathbf{n})^{\mathrm{sym}} \, \mathrm{d}S = \mathbf{E} \tag{4.27}$$

where the boundary integral vanishes because of the periodicity of $\mathbf{w}$ (cf. Section 4.2.1).

The equivalence between macroscopic and mesoscopic virtual work (Hill–Mandel condition; see Section 4.2.1) reads

$$\boldsymbol{\Sigma} : \delta\mathbf{E} = \frac{1}{V} \int_V \boldsymbol{\sigma} : \delta\varepsilon \, \mathrm{d}V \quad \forall \delta\varepsilon \tag{4.28}$$

where $\delta\mathbf{E}$ and $\delta\boldsymbol{\varepsilon}$ are related as defined above. Using mesoscopic equilibrium and Gauss's divergence theorem, the displacement field (4.26) may be introduced in Eq. (4.28), yielding

$$\boldsymbol{\Sigma} : \delta\mathbf{E} = \frac{1}{V}\int_S (\mathbf{t} \otimes \mathbf{x})^{\mathrm{sym}}\,\mathrm{d}S : \delta\mathbf{E} + \frac{1}{V}\int_S (\mathbf{t} \otimes \mathbf{w})^{\mathrm{sym}}\,\mathrm{d}S \tag{4.29}$$

where $\mathbf{t}$ is the traction vector acting on the boundary S of the RVE. Taking into account the periodicity of $\mathbf{w}$ and the resulting antiperiodicity of $\mathbf{t}$ at the boundary, the second integral vanishes and the macroscopic stress tensor is obtained as (cf. Eq. (4.16))

$$\boldsymbol{\Sigma} = \frac{1}{V}\int_S (\mathbf{t} \otimes \mathbf{x})^{\mathrm{sym}}\,\mathrm{d}S \tag{4.30}$$

4.3.1.2 **Choice of the Mesoscopic Representative Volume Element**

The selection of the RVE should be performed with care, in order to limit the computational effort at the mesoscopic level, at the same time still capturing all possible failure mechanisms. Different RVE shapes and sizes have been used in the literature for masonry [78–80]. Note that any periodic RVE delivers the same results as long as the average behavior remains unique, that is, before localization. On the basis of periodicity of the initial mesostructure of masonry, the RVE is therefore chosen as the smallest possible periodic unit cell of the mesostructure. This choice is based on the assumption that given its initially periodic structure, the average stiffness degradation of the material is mainly due to the geometrical arrangement of the constituents and this degradation – including damage-induced anisotropy effects – is correctly captured using any unit cell. The unit cell used here is illustrated for the case of running bond masonry as shown in Figure 4.6 and consists of a brick surrounded by half a joint. The running bond periodicity conditions have been indicated on the cell by arrows linking mutually tied boundary segments. The proposed unit cell was used in Refs [81, 82], together with an isotropic damage model at the mesoscopic scale. It was shown to deliver realistic average responses for nonperforated bricks, both in terms of load-bearing capacity and observed failure modes, including nonorthotropic damaging effects.

It is important that one realizes that using this periodic cell throughout the degradation process would imply a failure mechanism which satisfies the periodicity conditions, since periodicity implies that the damage pattern in a unit cell

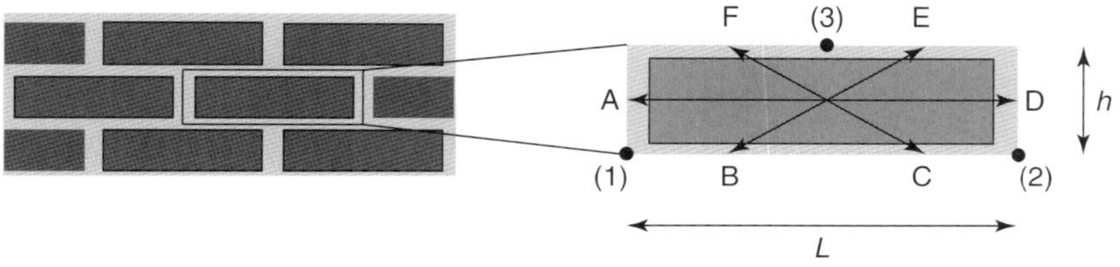

Figure 4.6 Unit cell and periodicity tyings for running bond masonry.

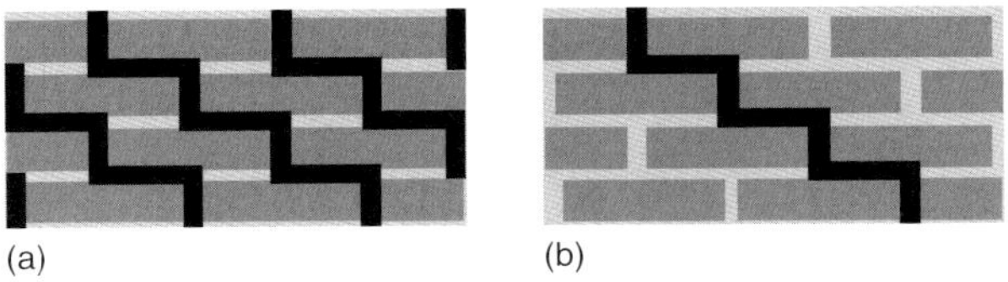

Figure 4.7 Influence of the periodicity assumption on the representation of failure patterns in the vicinity of a macroscopic material point: periodicity implies the failure pattern (a), whereas the pattern (b) is more realistic.

attributed to a macroscopic material point repeats itself in the vicinity of this point. An illustration of such a periodic failure pattern is shown in Figure 4.7(a). Actual damaged configurations usually exhibit strongly localized crack patterns at the macroscopic scale, as sketched in Figure 4.7(b). A methodology to represent macroscopically localized states is therefore developed in the next sections, which is based on relaxed periodicity assumptions.

4.3.1.3 Boundary Conditions for the Unit Cell

Three macroscopic strain components are imposed on the unit cell by prescribing the displacements of three controlling points, indicated in Figure 4.6 by (1), (2), and (3). These displacements may be expressed in terms of the components of the macroscopic infinitesimal strain tensor as

$$\begin{aligned}(u_1, v_1) &= (0, 0)\\ (u_2, v_2) &= (LE_{xx}, 0)\\ (u_3, v_3) &= \left(2hE_{xy} + \frac{1}{2}E_{xx}, hE_{yy}\right)\end{aligned} \tag{4.31}$$

where L and h are the horizontal and vertical sizes of the unit cell, respectively.

Periodicity of the fluctuation field w is imposed through the relations

$$\begin{aligned}\mathbf{u}(\mathbf{x}_D) &= \mathbf{u}(\mathbf{x}_A) + \mathbf{u}_2 - \mathbf{u}_1\\ \mathbf{u}(\mathbf{x}_E) &= \mathbf{u}(\mathbf{x}_B) + \mathbf{u}_3 - \mathbf{u}_1\\ \mathbf{u}(\mathbf{x}_F) &= \mathbf{u}(\mathbf{x}_C) + \mathbf{u}_3 - \mathbf{u}_2\end{aligned} \tag{4.32}$$

where $\mathbf{x}_A \ldots \mathbf{x}_F$ are the position vectors of points on the boundary segments $A \ldots F$.

Because of periodicity, the macroscopic stress can be obtained directly from the external forces $\mathbf{f}^{(i)}$ acting only on the three controlling points [45]:

$$\boldsymbol{\Sigma} = \frac{1}{V}\sum_{i=1}^{3} \mathbf{x}^{(i)} \otimes \mathbf{f}^{(i)} \tag{4.33}$$

where $\mathbf{x}^{(i)}$ is the position of point (i). A macroscopic consistent tangent operator can be obtained by static condensation of the mesostructural tangent stiffness matrix toward the controlling points [45]. This condensed tangent relation, applied to a mesoscopically converged state, may be written as

$$\sum_{p=1}^{3} \mathbf{K}_{\mathrm{M}}^{(np)} \cdot \delta \mathbf{u}^{(p)} = \delta \mathbf{f}^{(n)}, \qquad n = 1, 2, 3 \tag{4.34}$$

where $\mathbf{K}_{\mathrm{M}}^{(np)}$ is a second-order tensor relating the variations of the displacement vector of the controlling point p to the variations of the mesoscopic force vector at controlling point n. Combining relations (4.26), (4.33), and (4.34) allows us to define the macroscopic constitutive tangent ${}^4\mathbf{L}_{\mathrm{M}}$ according to

$$\delta \boldsymbol{\Sigma} = \underbrace{\left(\sum_{n=1}^{3} \sum_{p=1}^{3} \mathbf{x}^{(n)} \otimes \mathbf{K}_{\mathrm{M}}^{(np)} \otimes \mathbf{x}^{(p)} \right)^{(\mathrm{rs})}}_{{}^4\mathbf{L}_{\mathrm{M}}} : \delta \mathbf{E} \tag{4.35}$$

where $(.)^{(\mathrm{rs})}$ denotes condensation with right symmetrization to relate the symmetric Cauchy stresses to the symmetric infinitesimal strain tensor.

4.3.2 Localization of Damage at the Fine and Coarse Scales

An important issue in modeling failure by continuum methods is the localization of deformation and degradation. Classical continuum theories notoriously suffer from pathological localization when used to model damage and fracture [83]. At a certain stage of the degradation process, further deformation tends to localize in a surface, while the remaining volume unloads elastically. As a consequence, no energy dissipation is predicted anymore. This deficiency is intimately linked to a loss of well-posedness of the underlying boundary value problem. In numerical analyses, this results in a pathological sensitivity of the analysis results with respect to the spatial discretization. Damage localization thus calls for a specific treatment, in the present multiscale case at both the mesoscopic and macroscopic scales.

4.3.2.1 Fine-Scale Localization – Implicit Gradient Damage

A number of approaches have been proposed in order to enhance continuum formulations by including an intrinsic length parameter to avoid ill-posedness [83–85]. For the present study, simple nonperforated bricks are assumed. An isotropic damage model is used at the mesoscopic scale, assuming that damage-induced anisotropy at the macroscopic scale essentially originates from the geometrical arrangement of the constituents. It is, however, emphasized that the principles presented in the remainder of this chapter are mostly independent of the modeling choice at the mesoscopic scale.

Pathological mesoscopic localization is prevented here by the use of a gradient damage model [84]. The internal length scale included in this model implicitly sets the width of damage bands and thus prohibits bands of zero width. A scalar, isotropic damage field is introduced for each constituent via the stress–strain relationship:

$$\boldsymbol{\sigma} = (1 - D)\,{}^4\mathbf{L}_{\mathrm{m}} : \boldsymbol{\varepsilon} \tag{4.36}$$

where ${}^4\mathbf{L}_{\mathrm{m}}$ is the undamaged elastic stiffness tensor of the respective phases. The value of the damage variable D is deduced from a damage evolution law based on a variable κ, which represents the most severe strain state experienced so far by the material:

$$D = D(\kappa) \tag{4.37}$$

To determine whether a strain state change is accompanied by further damage growth, a damage loading function is expressed in terms of a nonlocal or averaged scalar measure of the strain state: $\overline{\varepsilon}_{\mathrm{eq}}$ and of the variable κ as

$$f(\overline{\varepsilon}_{\mathrm{eq}}, \kappa) = \overline{\varepsilon}_{\mathrm{eq}} - \kappa \tag{4.38}$$

The damage loading function f enters the Kuhn–Tucker relations

$$f \le 0 \qquad \dot{\kappa} \ge 0 \qquad f\dot{\kappa} = 0 \tag{4.39}$$

and an initial condition is defined for κ as

$$\kappa(t = 0) = \kappa_{\mathrm{i}} \tag{4.40}$$

The nonlocal $\overline{\varepsilon}_{\mathrm{eq}}$ field is introduced as the solution of an averaging partial differential equation incorporating a material intrinsic length scale l_{c} in the mesoscale description:

$$\overline{\varepsilon}_{\mathrm{eq}} - l_{\mathrm{c}}^2 \nabla_{\mathrm{m}}^2 \overline{\varepsilon}_{\mathrm{eq}} = \varepsilon_{\mathrm{eq}} \tag{4.41}$$

This partial differential equation is complemented by a boundary condition of the Neumann type applied at the interface between materials:

$$\boldsymbol{\nabla}_{\mathrm{m}} \overline{\varepsilon}_{\mathrm{eq}} \cdot \mathbf{n} = 0 \tag{4.42}$$

Details related to the damage criteria used for the constituents and the application of this framework for masonry mesostructures are discussed in Refs [81, 82], where it is shown that realistic results are obtained using this damage model.

4.3.2.2 **Detection of Coarse-Scale Localization as a Bifurcation into an Inhomogeneous Deformation Pattern**

The macroscopic continuum used in a first-order multiscale approach remains local, and a pathologically localized response may therefore be expected at the macroscale. The characteristic spatial variation of the macroscopic solution would localize in an infinitesimal volume, whereas that of the underlying mesostructural analysis remains finite, thus violating the scale separation that was assumed a priori between the meso- and macroscale. For disordered random heterogeneous materials, the framework presented in Section 4.2.2 provides a solution to the above macroscopic localization issue. In this second-order framework, the physical size of the mesostructural RVE enters the macroscopic equilibrium problem, providing a length scale at which the macroscopic deformation field tends to localize. This

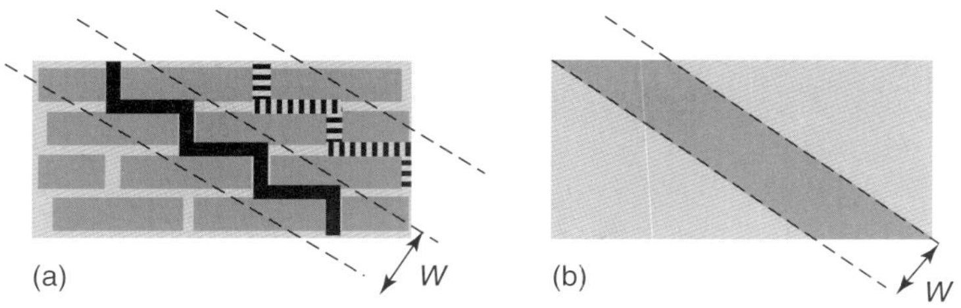

Figure 4.8 Extraction of a localization bandwidth based on the periodicity of the material. The dashed crack pattern (a) indicates the periodicity bandwidth for the considered pattern. A localization band of the same width is then introduced in the macroscopic continuum description (b).

second-order description, however, requires a sufficiently fine discretization of the localized zone at the macroscopic scale to capture its kinematics. In the case of quasi-brittle materials, damage appears at the lowest scale, below the unit-cell size. As a consequence, the use of a macroscopic gradient-enhanced description would not have any benefit, as the required macroscopic discretization would be of the same resolution as the mesoscopic discretization.

In the case of masonry, macroscopic higher order continua can be avoided by further exploiting the periodicity of the mesostructure. For each possible failure mechanism, a localization bandwidth can be determined in a natural way by identifying the smallest possible period over which the damage pattern may be repeated. For most crack patterns, the bandwidth equals the unit-cell size perpendicular to the considered average crack orientation. The crux of the method that we develop here is now that a localization band is introduced in the macroscopic description via a couple of weak displacement discontinuities, which have a distance equal to the bandwidth extracted from the unit cell. This is illustrated in Figure 4.8 for the case of a staircase crack pattern. The inclusion of such localization bands requires the following ingredients: (i) a criterion to decide when to introduce a localization band and (ii) a criterion to determine its orientation.

The method used to retrieve the above localization characteristics relies on pointwise concepts as defined for closed-form constitutive relations. In homogeneous problems, the appearance of a band in which further growth of damage localizes may be considered as the bifurcation of a secondary solution from the homogeneous fundamental solution. This approach is used in classical continuum mechanics to derive the conditions for the onset of localization [86], and is related to the mathematical concept of loss of ellipticity [83, 85]. For continuous bifurcation – that is, with the same material tangent ${}^4\mathbf{L}_M$ at both sides of the discontinuity – macroscopic localization in an orientation $\mathbf{n}$ may occur whenever the condition

$$\det(\mathbf{A}) = \det(\mathbf{n} \cdot {}^4\mathbf{L}_M \cdot \mathbf{n}) = 0 \tag{4.43}$$

is fulfilled. Under the assumption of a linear comparison solid, meeting this criterion is a sufficient condition for the boundary value problem to become ill-posed. Note that this condition was later extended to discontinuous bifurcations – where the tangent $^4\mathbf{L}_M$ is no longer identical on both sides of the discontinuity [87]. The orientations $\mathbf{n}$ for which localization may appear once Eq. (4.43) is satisfied are determined by the inequality:

$$\det\left(\mathbf{n} \cdot {}^4\mathbf{L}_M \cdot \mathbf{n}\right) \leq 0 \tag{4.44}$$

In a multiscale setting, the loss of uniqueness in the underlying mesostructural problem may also be used to detect macroscopic localization. A necessary condition for it is the loss of positive-definiteness of the underlying mesostructural discretized stiffness [83]. Since the homogenized material stiffness is obtained from this mesostructural stiffness, loss of positive-definiteness can be detected based on the eigenvalues of the homogenized stiffness: $\lambda_i^{\mathbf{L}_M}$ via

$$\lambda_{i,\min}^{\mathbf{L}_M} \leq 0 \tag{4.45}$$

which corresponds to a limit point in the homogenized stress–strain response.

4.3.2.3 Illustration of the Localization Analysis

The use of the above localization conditions in the computational homogenization scheme for running bond masonry is now illustrated. An analysis is performed on the eigenspectrum of the acoustic tensor associated with the macroscopic material tangent $^4\mathbf{L}_M$ obtained from Eq. (4.35). An eigenvalue analysis of the homogenized tangent stiffness itself is also performed. These analyses are done for the entire deformation path in a uniformly and proportionally stressed "structure". The macroscopic "structure" consists of a single finite element that has the precise dimensions of a single unit cell, in order to obtain the same load–displacement curves at the two scales.

The considered brick has dimensions $L \times h \times e$ equal to $165 \times 52 \times 100\,\text{mm}^3$, with mortar joints that are 10 mm thick. The mesostructure is represented by means of a finite element mesh containing 396 eight-noded elements and 1440 nodes. A plane-stress behavior is assumed since the failure patterns for the considered loading cases are not strongly affected by three-dimensional effects [82]. The damage criteria used for the brick and mortar materials are the same as in Ref. [82], namely, a maximum principal stress criterion for the brick and a Drucker–Prager-like criterion with a compressive cap for mortar. The material parameters used in this illustration are listed in Table 4.1. The internal length parameter of the nonlocal damage model is denoted by l_c, while f_t and f_c denote the uniaxial tensile and the uniaxial compressive strengths of the materials, respectively. The compressive cap used for the mortar criterion is controlled by the triaxial strength parameters f_b and f_h. Advanced path-following techniques [88] are used to follow the solution path up to complete failure.

The loading case investigated here consists in vertical compression in addition to shear. It leads to a typical failure pattern for masonry: a staircase crack. Figure 4.9 shows the final damage pattern obtained for this loading case. If complete failure

Table 4.1 Mesoscopic material parameters.

Material	E (MPa)	ν	l_c (mm)	f_t (MPa)	G_f (J/m^2)	f_c (MPa)	f_b (MPa)	f_h (MPa)
Brick	16700	0.15	1.73	0.75	16.2	15	–	–
Mortar	3900	0.20	1.73	0.13	9	5.6	8.72	7.45

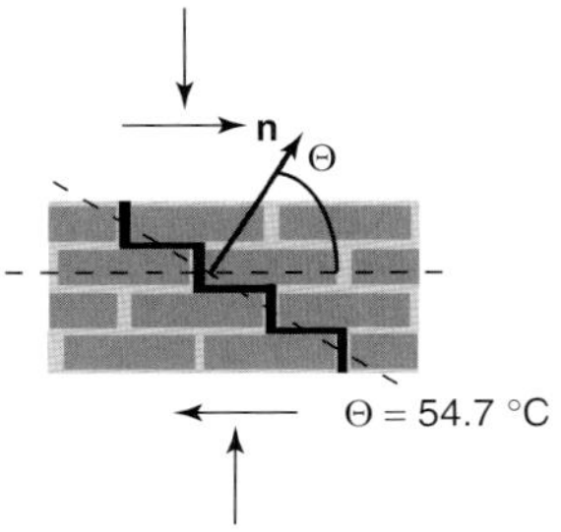

Figure 4.9 Average crack orientation defined by the normal **n** for the considered failure pattern.

is assumed, an overall crack direction can be identified, based only on geometric arguments, as indicated by the normal vector **n** in Figure 4.9. For the geometry used here, an orientation of $\Theta = 54.7^\circ$ is thus found. This is the direction that one would like to retrieve from the analysis at the mesostructural level, without any a priori knowledge of the final failure pattern.

The eigenvalue data for the case of vertical compression combined with an equal amount of shear are shown in Figures (4.10) and (4.11). Figure 4.10 shows the load–displacement response as well as the lowest eigenvalue of the tangent stiffness, $\lambda_{i,\min}^{L_M}$, for each possible orientation **n**. Figure 4.11 shows the lowest eigenvalue of

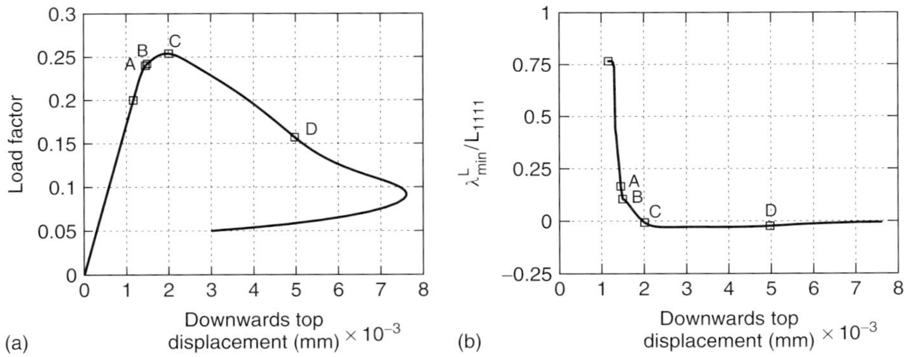

Figure 4.10 (a) Load factor versus vertical displacement of the top controlling node of the cell for vertical compression combined with shear. (b) Evolution of the lowest eigenvalue of the homogenized tangent stiffness.

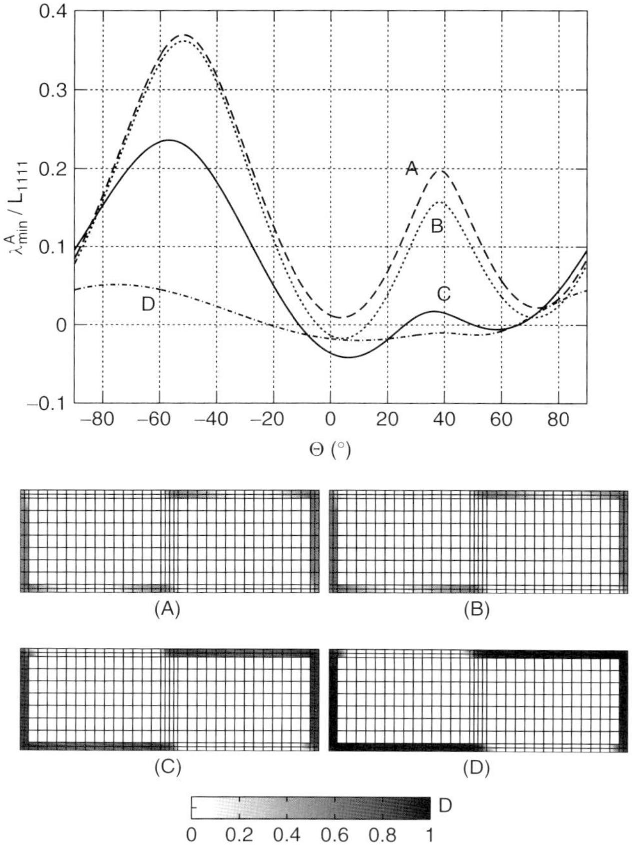

Figure 4.11 Acoustic tensor eigenspectrum at different stages of the compression plus shear loading and corresponding damage distributions.

the acoustic tensor, λ^A_{min}, as a function of the orientation or $\mathbf{n}$ at various stages of the damage process, along with the corresponding damage distributions. Loss of ellipticity occurs when λ^A_{min} vanishes for a certain direction $\mathbf{n}$.

Initially, the head joint is partially degraded and damage is initiated in parts of the bed joints, as a precursor to the expected staircase crack pattern (Figure 4.11). However, the damaged segment of the bed joint still contains some undamaged material. This evolution is reflected in the corresponding acoustic tensor's eigenspectrum, in which the values decrease faster for orientations between $\Theta = 0^\circ$ and $\Theta = 90^\circ$ compared to orientations between $\Theta = -90^\circ$ and $\Theta = 0^\circ$. The lowest eigenvalue of the homogenized tangent stiffness remains positive.

The crack pattern continues to evolve, as shown in Figure 4.11, state B, and a first minimum of the acoustic tensor eigenspectrum becomes negative for an orientation of $\Theta = 4.1^\circ$. Note that the damaged bed joint segment still contains some undamaged mortar as shown in Figure 4.11, state B. The acoustic tensor criterion (4.44) detects loss of ellipticity at this stage. However, the lowest eigenvalue of the tangent stiffness remains positive as can be seen in Figure 4.10, point B. The limit point of the homogenized stress–strain response has not yet been reached, and the load can still be increased since no continuous crack path is formed and periodic boundary conditions are applied (Section 4.3.1.3).

Upon further loading, damage spreads over the complete bed joint segment, as illustrated for state C, leading to the formation of a staircase crack pattern. This happens when the macroscopic limit point is reached. Simultaneously, a second minimum of the acoustic tensor's eigenspectrum becomes negative for an orientation of approximately $\Theta = 58.3^\circ$, (Figure 4.11, state C). This orientation is close to the averaged crack orientation specified above, the difference being linked to the nonuniform damage distribution inside the damage zone at this stage. As from this point, the lowest eigenvalue of the homogenized tangent stiffness becomes negative as well (Figure 4.10), and both the loss-of-ellipticity criterion (4.44) and the limit-point condition (4.45) are now satisfied.

The orientation corresponding to this new minimum evolves toward $\Theta = 53.8^\circ$, which differs by less than 1° from the angle of $\Theta = 54.7^\circ$ predicted by geometric arguments. Localization, therefore, seems to be best detected from the homogenized tangent via criterion (4.45), as the average damage pattern orientation appears to be linked to the second bifurcation point detected by the loss of ellipticity criterion (4.44). This bifurcation point is reached at the limit point in the homogenized response. The minima of the acoustic tensor eigenspectrum first pick up an orientation other than the expected one. Clearly, the first negative minimum of the acoustic tensor's eigenspectrum, in the rising part of the averaged stress–strain response, is linked to the nonsymmetry of the homogenized tangent stiffness.

4.3.2.4 Identification and Selection of the Localization Orientation

On the basis of the above observations, the limit-point criterion is used as an indicator of localization in the following. For the shearing compression case presented above, this means that the first bifurcation indication obtained from the acoustic tensor is ignored. Although this choice is in principle pragmatic, it can be supported by energy considerations. It is justified if the fundamental homogeneous solution path is more critical than the localized solution at this first bifurcation, which is indeed demonstrated in Section 4.3.6.1.

Once localization is detected, the orientation of the strain discontinuity can be determined as the one corresponding to the latest minimum of the acoustic tensor eigenspectrum to become negative.

4.3.3 Localization Band Enhanced Multiscale Solution Scheme

Once localization is detected, the first-order multiscale technique presented in Section 4.2.2 has to be amended in order to include localization bands as outlined in Section 4.3.2.2. Localization bands in the form of a pair of weak discontinuities embedded in an element were used in Ref. [89] to simulate strain localization in isotropic materials. The proper representation of a true discontinuity, that is, with complete separation of the crack faces for mode I or mode II opening and proper energy dissipation, requires the use of kinematically enhanced strain fields [90, 91]. This improvement is, however, only clearly motivated for constant strain elements and does not completely solve the mesh alignment sensitivity; see Ref. [92] for a detailed discussion. More advanced discretization techniques forcing crack-path continuity are now available for closed-form constitutive frameworks [92], but their adaptation to multiscale techniques is not considered here.

In the present approach, strain discontinuities defining a localization band are introduced to represent the behavior of the physical volume associated with a macroscopic sample point (Gauss point) when macroscopic localization is detected. The localization detection is based on the criterion presented in Section 4.3.2.2, whereas the localization width is extracted from the unit-cell dimensions along the direction **n** given by the localization detection. The material surrounding the localization band is assumed to unload elastically, thus resulting in a discontinuous bifurcation.

4.3.3.1 Introduction of the Localization Band

To include a discontinuous bifurcation in a macroscopic point, an approximate embedded band model is used, based on a relaxed Taylor assumption [93]. The Gauss point volume is split into a localized band (b), and its surrounding, unloading volume (s) (Figure 4.12). The volume fractions of the band and surrounding material are denoted by $f^{\rm b}$ and $f^{\rm s}$, and the strain jump between the band and surrounding material is defined by the normal to the band **n** and the strain jump mode **m**. The relaxed Taylor model consists in imposing a constant strain in each subregion according to

$$\begin{aligned} \mathbf{E}^{\rm b} &= \mathbf{E} + f^{\rm s}(\mathbf{m}\otimes\mathbf{n})^{\rm sym} \\ \mathbf{E}^{\rm s} &= \mathbf{E} - f^{\rm b}(\mathbf{m}\otimes\mathbf{n})^{\rm sym} \end{aligned} \tag{4.46}$$

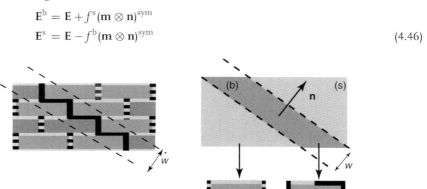

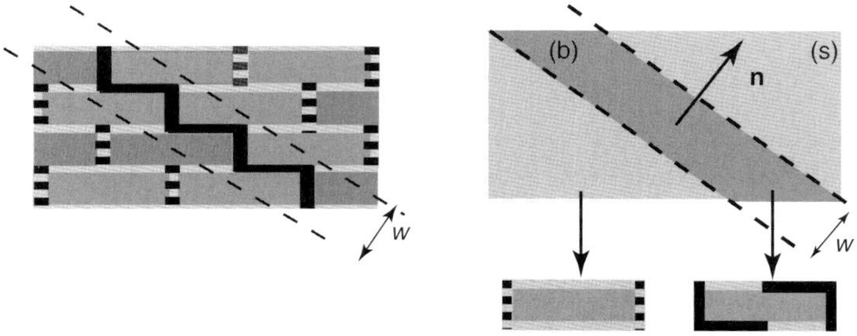

Figure 4.12 Idealization of the constitutive response for a macroscopic material point using a relaxed Taylor model. The responses of the material within and outside the localization band are extracted from separate unit cells.

With these assumptions, the volume-averaged strain is identical to the macroscopically applied strain:

$$\mathbf{E} = f^{\mathrm{b}}\mathbf{E}^{\mathrm{b}} + f^{\mathrm{s}}\mathbf{E}^{\mathrm{s}} \tag{4.47}$$

The stress variations in the two phases are obtained using their respective tangent stiffnesses as

$$\begin{aligned} \delta\boldsymbol{\Sigma}^{\mathrm{b}} &= {}^4\mathbf{L}_{\mathrm{M}}^{\mathrm{b}} : \delta\mathbf{E}^{\mathrm{b}} \\ \delta\boldsymbol{\Sigma}^{\mathrm{s}} &= {}^4\mathbf{L}_{\mathrm{M}}^{\mathrm{s}} : \delta\mathbf{E}^{\mathrm{s}} \end{aligned} \tag{4.48}$$

and the macroscopic stress variation $\delta\Sigma$ is obtained by volume averaging via

$$\delta\boldsymbol{\Sigma} = f^{\mathrm{b}}\delta\boldsymbol{\Sigma}^{\mathrm{b}} + f^{\mathrm{s}}\delta\boldsymbol{\Sigma}^{\mathrm{s}} \tag{4.49}$$

As is clarified in Figure 4.12, the assumption is made that the material responses of the band and of its surrounding volume can be deduced from the mesostructure according to the same scale transition as in the initial first-order framework. This clearly introduces an approximation, since this scale transition is based on a (local) periodicity assumption. According to this assumption, a cell within the band behaves as if it were surrounded by identically behaving cells, whereas in the real mesostructure the adjacent cells may be unloading. Similarly, the behavior of the material outside the band does not take into account the presence of weaker, localizing cells within the band. This assumption may, therefore, have some influence on the predicted postpeak response, for example, on the resulting energy dissipation. An improvement could be obtained by using larger periodic cells that include several periods of the mesostructure together with weaker (integral) boundary conditions at their boundaries, but these options clearly require a more extensive treatment. Another limitation of the assumed periodicity is the inability to correctly represent the energy dissipation for failure patterns which do not match the initial periodicity of the material. For example, for the case of localized vertical cracking in running bond masonry (single vertical crack through the head joints and brick), the development of two failure zones inside the unit cell may result in an overestimation of the energy dissipation.

4.3.3.2 **Coupled Multiscale Scheme for Localization**

On the basis of the assumptions made above, the first-order multiscale solution scheme can now be enhanced in order to capture macroscopic localization. Before localization, the first-order multiscale scheme is applied. Upon the detection of localization, a band is inserted and the amended scheme illustrated in Figure 4.13 is initiated.

On the basis of the localization orientation $\mathbf{n}$ and the related volume fractions, the macroscopic strain is decomposed into strains within the band and in the surrounding volume according to Eq. (4.46). The material response of the band is then evaluated through the first-order multiscale scheme using a unit-cell computation. Since the material surrounding the band is assumed to unload elastically, its secant stiffness ${}^4\mathbf{L}_{\mathrm{M}}^{\mathrm{s}}$ is evaluated only once. The macroscopic stress is obtained from the averaging relation (4.49). The determination of the complete

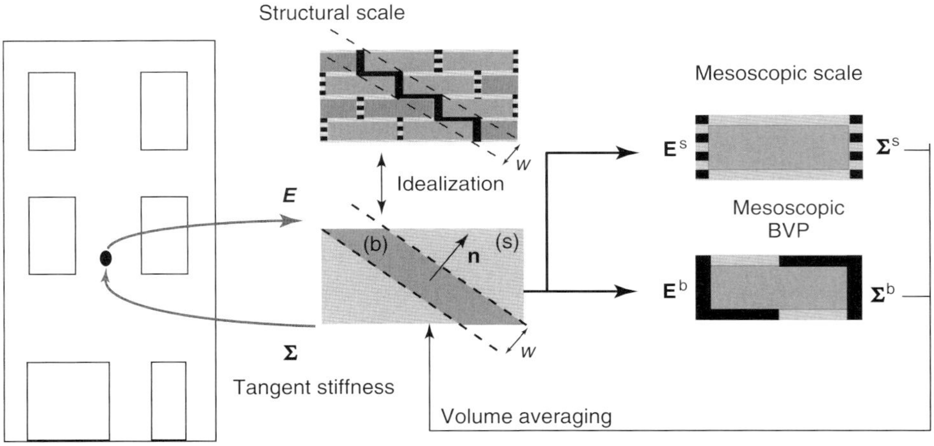

Figure 4.13 Sketch of the enhanced first-order multiscale scheme with embedded strain discontinuity for localized behavior.

macroscopic material response for a given macroscopic load step now consists in solving the following problem: given a macroscopic strain increment $\Delta\mathbf{E}$ and the orientation of the localization band $\mathbf{n}$, determine (i) the macroscopic stress increment $\Delta\boldsymbol{\Sigma}$ (three independent unknowns in two dimensions) and (ii) the increment of the strain jump $\Delta\mathbf{m}$ (two unknowns). These unknowns may be determined from the traction continuity requirement at the interface between the band and its surrounding material (two equations)

$$\mathbf{n}\cdot(\boldsymbol{\Sigma}^{\mathrm{b}}-\boldsymbol{\Sigma}^{\mathrm{s}})=\mathbf{0} \tag{4.50}$$

and the stress averaging relation (three equations)

$$\boldsymbol{\Sigma}=f^{\mathrm{b}}\boldsymbol{\Sigma}^{\mathrm{b}}+f^{\mathrm{s}}\boldsymbol{\Sigma}^{\mathrm{s}} \tag{4.51}$$

It is emphasized that the above problem (Eqs (4.50) and (4.51)) is nonlinear and thus requires an additional iterative solution procedure at the level of a macroscopic sampling point. This becomes apparent by expressing the traction continuity in Eq. (4.50) in terms of the strain increments in the band and the surrounding material as defined in relation (4.46):

$$\mathbf{n}\cdot({}^4\mathbf{L}_{\mathrm{M}}^{\mathrm{b}}-{}^4\mathbf{L}_{\mathrm{M}}^{\mathrm{s}}):\Delta\mathbf{E}+\mathbf{n}\cdot(f^{\mathrm{s}\,4}\mathbf{L}_{\mathrm{M}}^{\mathrm{b}}+f^{\mathrm{b}\,4}\mathbf{L}_{\mathrm{M}}^{\mathrm{s}}):(\Delta\mathbf{m}\otimes\mathbf{n})^{\mathrm{sym}}=\mathbf{0} \tag{4.52}$$

In relation (4.52), the tangent material stiffness in the band, ${}^4\mathbf{L}_{\mathrm{M}}^{\mathrm{b}}$, is a nonlinear function of the strain increment in the band $\Delta\mathbf{E}^{\mathrm{b}}$ and thus also of the strain jump increment $\Delta\mathbf{m}$. As a consequence, three nested solution procedures are required in this enhanced framework: (i) a solution procedure for equilibrium at the macroscopic scale, (ii) a solution procedure to solve the nonlinear set of equations of the localization band model, and (iii) a solution procedure at the unit-cell level for the solution of the mesostructural boundary value problem.

The orientation of the localization band is not considered as an unknown, but is determined by the localization analysis. For some fundamental equilibrium paths, a rotation of the localization orientation may occur as a result of mesoscopic damage evolution inside the unit cell. This effect must be taken into account to avoid stress locking effects at the macroscopic scale. The acoustic tensor associated to the band material tangent is, therefore, analyzed at each converged configuration and the band orientation is updated accordingly. No crack closure effects are included at the macroscopic scale. A localization band cannot be deactivated, and a material point in which a localization band has been inserted is not allowed switch back to a distributed damage pattern.

4.3.4 Scale Transition Procedure for Localized Behavior

4.3.4.1 Multiscale Solution Procedure

The multiscale scheme involves the numerical solution of equilibrium problems at two scales: the macroscopic or structural scale and the mesostructural scale. When localization is triggered, three sets of equations need to be solved. At the macroscopic scale, the global equilibrium problem can be written in the usual way as

$$\boldsymbol{\nabla}_{\mathrm{M}} \cdot \boldsymbol{\Sigma} + \mathbf{F} = \mathbf{0} \tag{4.53}$$

$$\boldsymbol{\Sigma} = \mathcal{F}_{\Sigma}(\mathbf{E}) \tag{4.54}$$

where $\mathcal{F}_{\Sigma}$ denotes the numerically obtained response of the embedded band model. The decomposition of the material behavior in a band and its surrounding material is characterized by Eqs (4.46, 4.50, 4.51), and

$$\begin{aligned} \boldsymbol{\Sigma}^{\mathrm{b}} &= \mathcal{F}_{\Sigma^{\mathrm{b}}}(\mathbf{E}^{\mathrm{b}}) \\ \boldsymbol{\Sigma}^{\mathrm{s}} &= \mathcal{F}_{\Sigma^{\mathrm{s}}}(\mathbf{E}^{\mathrm{s}}) \end{aligned} \tag{4.55}$$

where $\mathcal{F}_{\Sigma^{\mathrm{b}}}$ and $\mathcal{F}_{\Sigma^{\mathrm{s}}}$ denote the numerical response of the respective unit-cells.

At the mesoscopic scale, the governing partial differential equations are the equilibrium equation

$$\boldsymbol{\nabla}_{\mathrm{m}} \cdot \boldsymbol{\sigma} = \mathbf{0} \tag{4.56}$$

and the nonlocal averaging Eq. (4.41). Equations (4.56) and (4.41) are solved iteratively for each iteration of the macroscopic solution procedure.

The multiscale solution procedure as discussed so far implicitly assumes that a solution of the mesostructural problem and of the embedded band model always exists for a given macroscopic strain $\mathbf{E}$. This condition is not satisfied when the average response exhibits a snapback. The possibility that snapbacks occur, therefore, requires adaptations in the scale transition procedures.

4.3.4.2 Causes of Snapback in the Averaged Material Response

Snapback is a structural phenomenon appearing when localization of damage occurs in zones that are narrow with respect to the structural size. Energetically, it occurs when the rate at which the elastic zone releases elastically stored energy becomes larger than the rate at which energy is dissipated by the degradation of the material in a process zone. This calls for adequate path-following techniques [88]. Such methods use a generalized load control, in which a scalar load factor constitutes an additional unknown of the problem. To again close the set of equations, a constraint equation is added in terms of the conventional degrees of freedom of the structure and the load factor. A monotonically increasing quantity is used to define the constraint and hence to control the computation. For problems involving damage localization in narrow zones, local constraints (involving only degrees of freedom related to the damage process zone itself) have been shown to perform better than global constraints [88].

The macroscopic material point response is obtained here from a mesostructural computation. This implies that the average response may also show snapback, since relatively brittle mesostructural materials are considered and damage growth tends to localize in the weaker and relatively narrow mortar joints. If classical nonlinear displacement-based finite element procedures are used at the macroscopic scale, it is implicitly assumed that for a given strain increment a resulting stress can always be determined, that is, that the mesostructural boundary value problem always possesses a solution for the prescribed macroscopic strain increment. This is clearly not true for strains beyond a snapback point, and a macroscopic displacement-based multiscale scheme may hence fail in capturing the macroscopic equilibrium path.

A second potential source of snapback of the macroscopic material response is formed by the averaging relations of the localization band model, that is, Eqs (4.46) and (4.51). These relations take into account localization bands with a characteristic volume that is smaller than the volume associated with the Gauss point where the macroscopic material behavior is sampled. Since this band behaves in a quasi-brittle manner and the remaining material is assumed to unload elastically, the overall macroscopic material response may show snapback, even if the average behavior of the band unit cell does not. This type of snapback is more likely to occur for relatively coarse meshes, in which the damage band is thin compared to the volume associated with the Gauss points.

To be able to handle mesostructural snapbacks in our multiscale solution scheme, two difficulties need to be solved. First, the macroscopic solution procedure must be able to predict a decreasing macroscopic strain increment as from the strain control limit point. Secondly, in the snapback regime, following the elastic unloading equilibrium path should be avoided for a given negative strain increment. For this reason, the scale transition procedure should apply a decreasing macroscopic strain to the mesostructure, along with an additional condition that pushes the solution toward a dissipating path.

4.3.4.3 Strain Jump Control for Embedded Band Snapback

Let us assume that at a certain stage of a computation, a localization band has already been introduced in the considered material volume and that the homogenized unit-cell response does not exhibit snapback. Given the small width of the band and the brittleness of its behavior, snapback may appear in the averaged response of the finite volume associated with the considered macroscopic material point. The kinematic variable that characterizes the strain jump between the band and the surrounding material is the vector **m**. If the band material follows a softening branch while the surrounding material unloads, the strain jump $(\mathbf{m} \otimes \mathbf{n})^{\mathrm{sym}}$ should increase, even if the overall material response exhibits a snapback. To enforce further dissipation in the material, the macroscopic solution procedure should thus enforce the growth of **m**.

A way to enforce this condition is to impose on the localization band model a positive increment of the strain jump vector **m**. The vector **m** is, therefore, defined as a variable at the macroscopic level in each macroscopic Gauss point. Since Eq. (4.50) is nonlinear, simultaneously imposing a strain jump **m** and a macroscopic strain **E** does not automatically lead to an equilibrium configuration. An equivalent number of equations conjugate with these unknowns must, therefore, be formulated at the macroscopic level and solved at this scale. The traction continuity requirement between the band and surrounding material, Eq. (4.50), may be used for this purpose. This equation couples the strain jump to the macroscopic displacement field through the components of the macroscopic strain tensor, as shown by relation (4.52). Because of the nonlinear nature of Eq. (4.52), the prescribed values of the strain jump **m** and the applied overall strain **E** will generally not lead to an equilibrium state immediately, meaning that traction continuity will not be directly satisfied for the prescribed **m**. Equation (4.50) therefore has to be solved iteratively, together with the discrete macroscopic equilibrium equations that result from Eq. (4.53). This means that the traction continuity between the band and the surrounding phases (Eq. (4.50)) is only satisfied in macroscopically converged configurations and not in the intermediate iterative configurations.

4.3.4.4 Dissipation Control for Unit-Cell Snapback

As illustrated in Ref. [94], averaging of the unit-cell behavior may cause mesostructural snapback in the response of the localization band itself. To follow the dissipative equilibrium path of the cell, an additional condition related to mesoscopic dissipation inside the cell must be added to the data provided to the mesostructural problem by the macroscopic solution procedure. The principle of this technique is briefly recalled for the sake of clarity. Assuming a discontinuous bifurcation upon appearance of the band, the surrounding material unloads elastically, and an enhancement is only needed for the unit-cell computation associated with the localization band. A variable related to the dissipating damage process zone – here, a nonlocal equivalent strain field degree of freedom that drives the mesoscopic damage growth – can be imposed by the macroscopic solution procedure to the cell to force the solution onto a dissipative path. This nonlocal degree of freedom must, therefore, be transferred to the macroscopic scale, and a conjugate equation has to be solved iteratively at this scale.

The "external" prescription of a nonlocal strain in the cell gives rise to a conjugate "reaction force" or conjugate residual $f_{\bar{\varepsilon}}$. This residual can be interpreted as the external action needed to enforce the prescribed value of the nonlocal strain increment $\Delta\bar{\varepsilon}$ in addition to the imposed average strain **E**. Since in reality at equilibrium no such external action exists, it should vanish in the equilibrium state:

$$f_{\overline{\varepsilon}} = 0 \tag{4.57}$$

Rather than using this condition at the mesoscopic scale – which would be equivalent with not prescribing $\Delta\overline{\varepsilon}$ – it is transferred to the macroscopic solution procedure. This allows one to have more control on the mesoscopic equilibrium iterations, thereby still satisfying the original equilibrium equations in the final macroscopically converged solution. This feature allows to pass strain control limit points of the homogenized stress–strain behavior and, since it selects the corresponding tangent stiffness, to follow the dissipative snapback solution.

For details on the practical implementation of the enhanced control of the mesoscopic dissipation, the reader is referred to Ref. [94]. Note that the nonlocal degree of freedom may already be introduced in the macroscopic description at the start of the computation, as it does not have any effect in the prelocalized regime.

An essential difference with the treatment of snapback due to the macroscopic localization band discussed in Section 4.3.4.3 resides in the fact that the selection of the nonlocal degree of freedom, which is used to enforce dissipation at the cell level, should evolve as a result of mesoscopic damage evolution. The nonlocal degree of freedom corresponding to the largest incremental damage growth is selected at the end of each macroscopic increment as the controlling variable of a given unit cell in the subsequent increment. Note that the procedure described here is rather general. If another type of model is used at the mesoscale, for example, a cohesive zone approach, the solution control can also be based on a relative displacement between two nodes or even other quantities such as the dissipated energy, as suggested in Ref. [95].

4.3.5 Solution Strategy and Computational Aspects

4.3.5.1 Governing Equations for the Macroscopic and Mesoscopic Solution Procedures

As a result of the enhancements introduced in order to handle mesostructural snapbacks, mesostructural equations have been incorporated in the macroscopic solution procedure. The governing equations and the level at which they are solved are summarized in Table 4.2, and the corresponding scale transition procedure is represented graphically in Figure 4.14. The essential difference with the scheme presented in Figure 4.13 resides in the presence of a set of mesoscopic nonlocal degrees of freedom in the macroscopic solution procedure. The band orientation is adapted at the start of each step, based on the evaluation of the tangent operator in the previous converged state. In case the band unit-cell response undergoes mesostructural snapback, its tangent stiffness becomes positive-definite, and no localization direction can be determined anymore, in which case the orientation is fixed to that previously determined.

Table 4.2 Distribution of the field-governing equations among the different solution procedures in the enhanced first-order scheme.

Solution level	Equations	Physical meaning	Unknowns
Macroscopic	(4.53)	Macro-equilibrium	Displacement field **u** (macroscopic variation)
	(4.50)	Traction continuity	Strain discontinuities **m**
	(4.57)	Non-local residual	Mesoscopically prescribed $\bar{\varepsilon}_{eq}$
Mesoscopic	(4.56)	Mesoscopic equilibrium	Displacement field **u** (mesoscopic variation)
	(4.41)	Nonlocal averaging	Nonlocal field $\bar{\varepsilon}_{eq}$

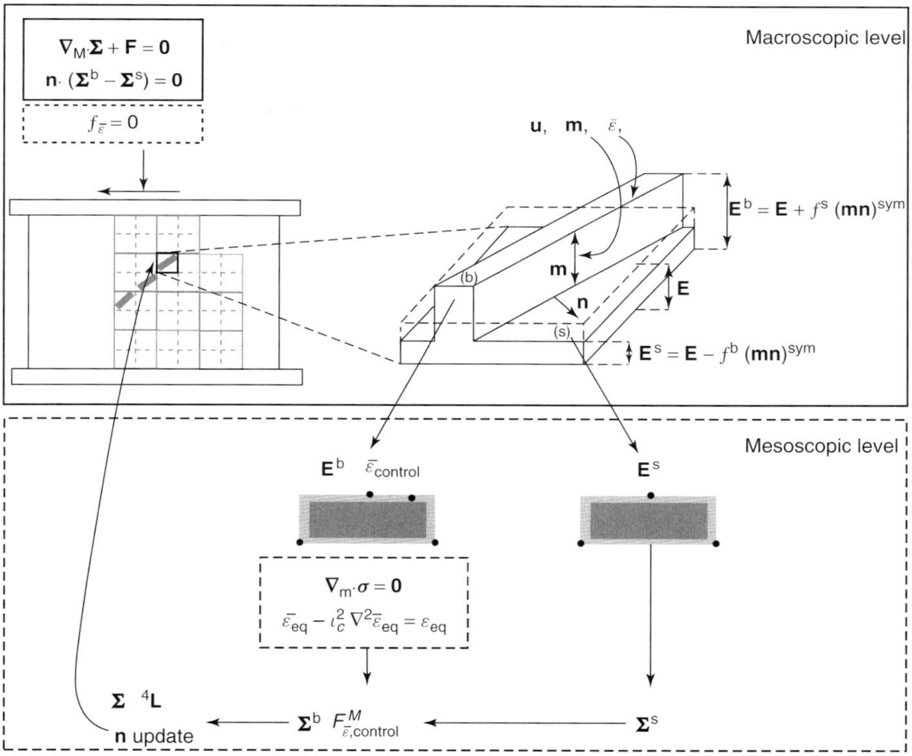

Figure 4.14 Principles of the localization enhanced first-order scheme with nested macroscopic and mesoscopic solution procedures and with mesostructural snapback handling. (a) The structural scale, including localization bands at the level of Gauss point volumes. (b) The mesoscopic unit-cell computations.

4.3.5.2 Extraction of Consistent Tangent Stiffness for Unit-Cell Snapback Control

The fact that a mesoscopic degree of freedom is now included in the set of unknowns at the macroscopic level means that the extraction of the macroscopic tangent operator presented in Ref. [55] must be adapted as well. This tangent now consists of four tensors that relate variations of the stress tensor Σ and the nonlocal residual $f_{\bar{\varepsilon}}$ to variations of the strain tensor $\mathbf{E}$ and the nonlocal variable $\bar{\varepsilon}$ according to

$$\delta\boldsymbol{\Sigma} = {}^4\mathbf{C}_{\mathrm{M}}^{uu} : \delta\mathbf{E} + \mathbf{C}_{\mathrm{M}}^{ue}\delta\bar{\varepsilon}, \quad \delta f_{\bar{\varepsilon}} = \mathbf{C}_{\mathrm{M}}^{eu} : \delta\mathbf{E} + : C_{\mathrm{M}}^{ee}\delta\bar{\varepsilon} \tag{4.58}$$

The tangent stiffness relation of the mesostructural problem, condensed toward the control displacements and the controlling nonlocal degree of freedom, reads in matrix format:

$$\begin{bmatrix} [\mathbf{K}_{\mathrm{M}}^{uu}] & \{\mathbf{K}_{\mathrm{M}}^{ue}\} \\ \langle \mathbf{K}_{\mathrm{M}}^{eu} \rangle & K_{\mathrm{M}}^{ee} \end{bmatrix} \begin{Bmatrix} \{\delta u_{\mathrm{M}}\} \\ \delta\bar{\varepsilon} \end{Bmatrix} = \begin{Bmatrix} \{\delta f_{\mathrm{M}}\} \\ \delta f_{\bar{\varepsilon}} \end{Bmatrix} \tag{4.59}$$

When expressed in tensorial format, this relation can be reworked to obtain the following expressions for the consistent tangent [94]:

$$\begin{aligned} {}^4\mathbf{C}_{\mathrm{M}}^{uu} &= \frac{1}{V}\left(\sum_{n=1}^{3}\sum_{p=1}^{3}\mathbf{x}^{(n)} \otimes \mathbf{K}_{\mathrm{M}}^{uu(np)} \otimes \mathbf{x}^{(p)}\right)^{(rs)}, \quad \mathbf{C}_{\mathrm{M}}^{ue} = \frac{1}{V}\left(\sum_{n=1}^{3}\mathbf{x}^{(n)} \otimes \mathbf{K}_{\mathrm{M}}^{ue(n)}\right) \\ \mathbf{C}_{\mathrm{M}}^{eu} &= \frac{1}{V}\left(\sum_{p=1}^{3}\mathbf{K}_{\mathrm{M}}^{eu(p)} \otimes \mathbf{x}^{(p)}\right)^{(rs)}, \quad C_{\mathrm{M}}^{ee} = K_{\mathrm{M}}^{ee} \end{aligned} \tag{4.60}$$

The consistent tangent operator ${}^4\mathbf{C}_{\mathrm{M}}^{uu}$ defined by relations (4.60) is not the material tangent in the usual sense, due to the presence of additional information related to the mesoscopic nonlocal degree of freedom $\bar{\varepsilon}$. Upon macroscopic convergence, the unit cell is in equilibrium under the prescribed macroscopic strain increment and the residual $f_{\bar{\varepsilon}}$ vanishes, and the material tangent operator may then be recovered as

$${}^4\mathbf{L}_{\mathrm{M}} = {}^4\mathbf{C}_{\mathrm{M}}^{uu} - \frac{1}{C_{\mathrm{M}}^{ee}}\mathbf{C}_{\mathrm{M}}^{ue}\mathbf{C}_{\mathrm{M}}^{eu} \tag{4.61}$$

for the detection or update of the localization orientation.

4.3.5.3 Discretization and Linearization of the Macroscopic Solution Procedure

A nested finite element strategy is used in order to solve the governing equations according to the scheme given in Table (4.2). To achieve this, unknowns related to the following kinematic variables have to be defined in the macroscopic solution procedure: (i) the displacement field $\mathbf{u}$, (ii) the strain jump vector $\mathbf{m}$ in each quadrature point where bifurcation has occurred, and (iii) a mesoscopic nonlocal strain degree of freedom $\bar{\varepsilon}$ in each quadrature point. Two unknowns related to the strain jump $\mathbf{m}$ and one related to $\bar{\varepsilon}$ are thus defined in each integration point. The displacement field is interpolated by conventional finite element shape functions. Note that the piecewise uniform strain state assumed in the localization band model is only consistent with the finite element interpolation of the displacements in case constant strain triangle elements are used. This restriction is also present in "classical" embedded discontinuity elements available in the literature, the formulation of which is usually well motivated only for constant strain elements [92].

Using the finite element displacement interpolation, the discretized form of the governing equations in iteration k is given in matrix format by

$$\int_V [\mathrm{B}]^T \left\{\Sigma^{(k)}\right\} \mathrm{d}V = \left\{\mathrm{f}_{\mathrm{ext}}^{(k)}\right\} \tag{4.62}$$

$$[n]\left(\left\{\Sigma^{b,(k)}\right\} - \left\{\Sigma^{s,(k)}\right\}\right) = \{0\} \tag{4.63}$$

$$f_{\overline{\varepsilon}}^{(k)} = 0 \tag{4.64}$$

where [B] is the matrix that links the interpolated macroscopic strain field to the macroscopic nodal displacements and $[n]$ is a matrix containing the components of the normal to the localisation band. The integral in Eq. (4.62) is defined on the entire structure. Equations (4.63) and (4.64) are formulated for each quadrature point where the macroscopic material behavior is sampled. Relations (4.62)–(4.64) are linearized below

$$\left\{\Sigma^{(k)}\right\} = \left\{\Sigma^{(k-1)}\right\} + \{\delta\Sigma\} \tag{4.65}$$

$$\left\{m^{(k)}\right\} = \left\{m^{(k-1)}\right\} + \{\delta m\} \tag{4.66}$$

$$\overline{\varepsilon}^{(k)} = \overline{\varepsilon}^{(k-1)} + \delta\overline{\varepsilon} \tag{4.67}$$

where δ indicates the change of a quantity between two successive iterations.

The macroscopic stress variation in a given quadrature point may be obtained using the stress averaging on the localization band and the remaining material according to Eq. (4.51), in which the band and the surrounding responses (4.55) are substituted and in which Eq. (4.58) is used for the band. If $[\mathrm{L}^{\mathrm{s}}]$ contains the components of the secant stiffness of the surrounding material, one obtains

$$\{\delta\Sigma\} = f^{\mathrm{b}}\left(\left[{}^4\mathrm{C}_{\mathrm{M}}^{uu}\right]\left\{\delta\mathrm{E}^{\mathrm{b}}\right\} + \left[\mathrm{C}_{\mathrm{M}}^{ue}\right]\delta\overline{\varepsilon}\right) + f^{\mathrm{s}}\left[{}^4\mathrm{L}_{\mathrm{M}}^{\mathrm{s}}\right]\left\{\delta\mathrm{E}^{\mathrm{s}}\right\} \tag{4.68}$$

Expressing the strains in the band and in the surrounding material in terms of the strain discontinuity via Eq. (4.46) and introducing the finite element interpolation for the macroscopic strain **E**, the variation of stress can be written as

$$\begin{aligned}\{\delta\Sigma\} = &\left(f^{\mathrm{b}}\left[{}^4\mathrm{C}_{\mathrm{M}}^{uu}\right] + f^{\mathrm{s}}\left[{}^4\mathrm{L}_{\mathrm{M}}^{\mathrm{s}}\right]\right)[\mathrm{B}]\{\delta u\} \\ &+ f^{\mathrm{b}}f^{\mathrm{s}}\left(\left[{}^4\mathrm{C}_{\mathrm{M}}^{uu}\right] - \left[{}^4\mathrm{L}_{\mathrm{M}}^{\mathrm{s}}\right]\right)[n]\{\delta m\} + f^{\mathrm{b}}\left[\mathrm{C}_{\mathrm{M}}^{ue}\right]\delta\overline{\varepsilon}\end{aligned} \tag{4.69}$$

where the matrix $[n]$ is defined such that its multiplication with $\{\delta m\}$ yields the symmetric part of $\mathbf{n} \otimes \delta\mathbf{m}$. Note that in this expression the first term connects the macroscopic displacement degrees of freedom of the considered macroscopic finite element, whereas the remaining terms are related to variations of variables that are local to the considered quadrature point.

Substituting Eqs (4.65) and (4.69) in the discretized macroscopic equilibrium, Eq. (4.62) yields

$$[\mathrm{K}_{uu}]\{\delta u\} + \sum_{(i)}\left([\mathrm{K}_{um,(i)}]\left\{\delta m_{(i)}\right\} + \left\{\mathrm{K}_{ue,(i)}\right\}\delta\overline{\varepsilon}_{(i)}\right) = \left\{\mathrm{f}_{\mathrm{ext}}^{(k)}\right\} - \left\{\mathrm{f}_{\mathrm{int}}^{(k-1)}\right\} \tag{4.70}$$

where

$$[\mathrm{K}_{uu}] = \int_V [\mathrm{B}]^T (f^{\mathrm{b}}\ [{}^4\mathrm{C}_{\mathrm{M}}^{uu}] + f^{\mathrm{s}}\ [{}^4\mathrm{L}_{\mathrm{M}}^{\mathrm{s}}])\ [\mathrm{B}]\mathrm{d}V \tag{4.71}$$

$$[\mathrm{K}_{um,(i)}] = V_{(i)} f^{\mathrm{b}} f^{\mathrm{s}}\ [\mathrm{B}]^T ([{}^4\mathrm{C}_{\mathrm{M}}^{uu}] - [{}^4\mathrm{L}_{\mathrm{M}}^{\mathrm{s}}])\ [n] \tag{4.72}$$

$$\left\{\mathrm{K}_{ue,(i)}\right\} = V_{(i)} f^{\mathrm{b}}\ [\mathrm{B}]^T\ \left\{\mathrm{C}_{\mathrm{M}}^{ue}\right\} \tag{4.73}$$

and $\left\{\mathbf{f}_{\mathrm{ext}}^{(k)}\right\}$ are the external nodal forces in iteration k. The sum over the Gauss points, indicated by the index i in the two latter relations, reflects that the strain jump unknowns and the mesoscopic nonlocal strain unknowns are associated with these points. All terms in Eqs (4.72) and (4.73) are evaluated at the considered Gauss points and $V_{(i)}$ denotes the volume associated with the considered Gauss point.

Traction continuity across the boundary of the localization band in a given Gauss point volume, as expressed by Eq. (4.63), can be linearized using similar substitutions as above. This leads to the following linearized equation for iteration k:

$$[\mathrm{K}_{mu}]\,\{\delta u\} + [\mathrm{K}_{mm}]\,\{\delta m\} + \{\mathrm{K}_{me}\}\,\delta\overline{\varepsilon} = -[n]\left(\left\{\Sigma^{b,(k-1)}\right\} - \left\{\Sigma^{s,(k-1)}\right\}\right) \tag{4.74}$$

with the corresponding stiffness matrices defined as

$$[\mathrm{K}_{mu}] = [n]\ ([{}^4\mathrm{C}_{\mathrm{M}}^{uu}] - [{}^4\mathrm{L}_{\mathrm{M}}^{\mathrm{s}}])\ [\mathrm{B}] \tag{4.75}$$

$$[\mathrm{K}_{mm}] = [n]\ (f^{\mathrm{s}}[{}^4\mathrm{C}_{\mathrm{M}}^{uu}] - f^{\mathrm{b}}[{}^4\mathrm{L}_{\mathrm{M}}^{\mathrm{s}}])\ [n] \tag{4.76}$$

$$\{\mathrm{K}_{me}\} = [n]\left\{\mathrm{C}_{\mathrm{M}}^{ue}\right\} \tag{4.77}$$

Finally, the linearization of Eq. (4.64) is obtained by substituting the band strain given by Eq. (4.46) into the expression of the nonlocal residual variation (4.58), yielding for iteration k:

$$\langle \mathrm{K}_{eu}\rangle\,\{\delta u\} + \langle \mathrm{K}_{em}\rangle\,\{\delta m\} + \mathrm{K}_{ee}\delta\overline{\varepsilon} = -f_{\overline{\varepsilon}}^{(k-1)} \tag{4.78}$$

with

$$\langle \mathrm{K}_{eu}\rangle = \langle \mathrm{C}_{\mathrm{M}}^{eu}\rangle[\mathrm{B}] \tag{4.79}$$

$$\langle \mathrm{K}_{em}\rangle = f^{\mathrm{s}}\langle \mathrm{C}_{\mathrm{M}}^{eu}\rangle[n] \tag{4.80}$$

$$\mathrm{K}_{ee} = \mathrm{C}_{\mathrm{M}}^{ee} \tag{4.81}$$

4.3.5.4 Introduction of Localization Bands upon Material Bifurcation

In the first increment after localization detection in a certain macroscopic material point, separate responses of the band and surrounding material are evaluated in the considered point. On the basis of the detected localization orientation $\mathbf{n}$, a branch switching procedure is applied to force the appearance of a strain jump $(\mathbf{m} \otimes \mathbf{n})^{\mathrm{sym}}$ between the band and the surrounding material. The vector $\mathbf{m}$ is unknown, and no estimate for it is available from the previous increment.

However, a prediction of $\mathbf{m}$ can be obtained from the discontinuous bifurcation assumption. The rate of the vector $\mathbf{m}$ must satisfy the linearized form of the traction continuity (4.50)

$$(\mathbf{n} \cdot {}^4\mathbf{L}_{\mathrm{M}}^{\mathrm{b}} \cdot \mathbf{n}) \cdot \dot{\mathbf{m}} = \mathbf{n} \cdot ({}^4\mathbf{L}_{\mathrm{M}}^{\mathrm{s}} - {}^4\mathbf{L}_{\mathrm{M}}^{\mathrm{b}}) : \dot{\mathbf{E}} \tag{4.82}$$

where ${}^4\mathbf{L}_M^b$ denotes the tangent stiffness in the band and ${}^4\mathbf{L}_M^s$ the tangent stiffness in the remaining material. Initially, the latter is taken equal to the secant stiffness, whereas the tangent stiffness before bifurcation is used as an initial estimate for ${}^4\mathbf{L}_M^b$. Since the direction $\mathbf{n}$ has been determined by the localization condition, Eq. (4.82) can be used to estimate the finite increment $\Delta\mathbf{m}$, which is consistent with the finite increment of the overall strain $\Delta\mathbf{E}$. This estimate is only an approximation since finite increments are estimated from the linearized equation and since the tangent stiffness before the bifurcation is used as an estimate for the band stiffness. With this strain jump prediction $\Delta\mathbf{m}$, the strain increment in the localization band may be predicted as

$$\Delta\mathbf{E}^b = \Delta\mathbf{E} + f^s(\mathbf{n} \otimes \Delta\mathbf{m})^{sym} \tag{4.83}$$

So far, the prediction of the nonlocal degree of freedom increment $\Delta\bar{\varepsilon}$ available from the macroscopic solution procedure does not take into account the branching at the bifurcation point, since it still relates to the overall strain increment $\Delta\mathbf{E}$. A new prediction of $\Delta\bar{\varepsilon}$ thus has to be obtained, related to the strain increment $\Delta\mathbf{E}^b$ applied to the band rather than to $\Delta\mathbf{E}$. This new prediction of the nonlocal degree of freedom may be obtained from relation (4.78) by setting the nonlocal residual increment $\Delta f_{\bar{\varepsilon}}$ to zero. A correction for $\Delta\bar{\varepsilon}$ that takes into account the appearance of the strain jump is thus obtained as

$$\Delta\bar{\varepsilon} = \frac{-1}{\mathrm{K}_{ee}}[\langle \mathrm{K}_{eu}\rangle\,\{\Delta u\} + \langle \mathrm{K}_{em}\rangle\,\{\Delta m\}] \tag{4.84}$$

4.3.6 Applications and Discussion

In this section, the numerical multiscale framework developed above is applied in a number of relevant test problems. The selection of bifurcated solutions suggested in Section 4.3.2.4 is first scrutinized for the case of homogeneous compressive shearing loading. Two simple test cases with homogeneous loading at the macroscopic scale are then considered in order to illustrate the capability of the algorithm to treat mesostructural snapback and its sensitivity to the absolute size of the mesostructure. Finally, an example of a structural computation is given.

4.3.6.1 Selection of Localized Solutions

As illustrated and discussed in Section 4.3.2.3, the macroscopic localization criterion detects loss of ellipticity in the rising part of the load–displacement curve, prior to the satisfaction of the limit-point criterion. The question thus arises whether the proposed algorithm for the selection of the solution path detects the most critical path. To verify this, the enhanced multiscale technique presented in Section 4.3.3 is applied to a "structure" made of a single finite element subjected to homogeneous vertical compression combined with shear. The same unit cell is used as in Section 4.3.2.3 and in Ref. [82]. The macroscopic element dimensions are $800 \times 800\,\text{mm}^2$. The computation is performed for the fundamental, homogeneous path, as well

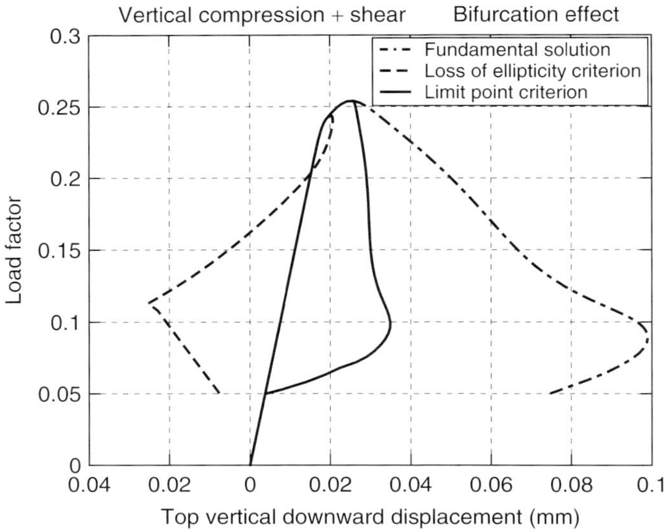

Figure 4.15 Load–displacement curve for vertical compression combined with shear for fundamental and bifurcated solutions based on loss-of-ellipticity and limit-point criteria.

as for all bifurcated paths detected by the loss-of-ellipticity criterion and by the limit-point criterion.

The load–displacement curves obtained are depicted in Figure 4.15. It can be noticed that an equilibrium path can indeed be traced for both detected bifurcation points. When the localized solution is followed at the first bifurcation, a new path is found, which shows rather large rotations of the localization band. Localization is first detected for an orientation of approximately $\Theta = 4^\circ$, and the last equilibrium configuration for this bifurcated solution shows an orientation of $\Theta = 35^\circ$. The final damage pattern is a staircase crack pattern, as for the fundamental solution. If the homogeneous path is followed at the first bifurcation, another bifurcation point is detected when the limit point is reached. A secondary solution is also available in this case as illustrated in Figure 4.15. The band orientation associated with this bifurcation is consistent with the average orientation of the staircase crack pattern.

The most critical solution at each bifurcation point is that for which the smallest amount of incremental work must be supplied to the material. As no closed-form material law is available here, the supplied work is evaluated numerically, on the basis of the computed overall stress $\mathbf{\Sigma}$ and strain increment $\Delta\mathbf{E}$ for the obtained solution paths. The equilibrium paths are traced with small loading steps near the bifurcation points. The supplied energy density variation versus load factor is depicted for both solutions at the two bifurcation points in Figure 4.16. Figure 4.16a indicates that for the first bifurcation, the material needs less energy to proceed along the fundamental path compared to the localized solution. This suggests

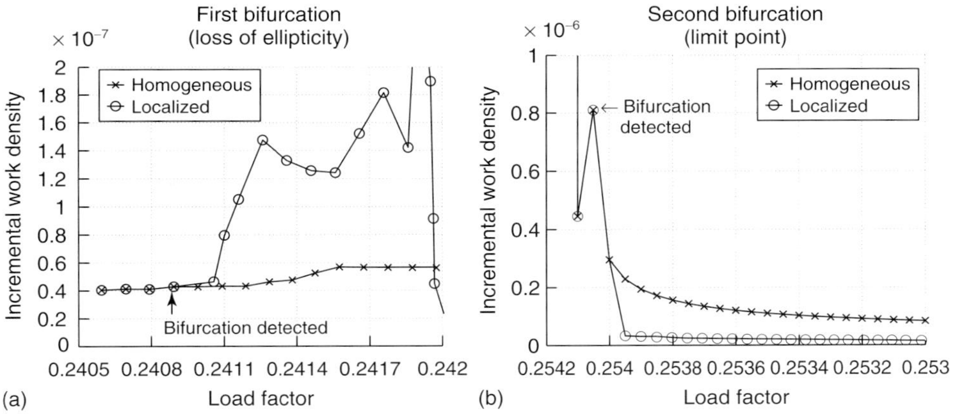

Figure 4.16 Supplied incremental work density for the alternative solution paths in the compression–shear test: (a) first bifurcation detected by the loss-of-loss-of-ellipticity criterion and (b) second bifurcation detected by limit-point criterion. At the second bifurcation, the load factor is decreasing.

that the first bifurcation point should indeed be passed without introducing a localization band, as predicted by the limit point criterion. At the bifurcation detected by the limit-point criterion, on the contrary, the localized solution is more critical than the fundamental path, as it minimizes the incremental energy supply to the material (Figure 4.16b).

This illustration thus gives a heuristic justification for the procedure defined for localization detection in Section 4.3.2.3. It is, however, emphasized that this verification is difficult to implement in real structural computations since it requires knowledge of all bifurcated solutions to decide on which path to follow.

4.3.6.2 **Mesostructural Snapback in a Tension–Compression Test**

As shown in Ref. [94], the multiscale framework can be applied to trace the homogenized response where mesostructural unit-cell snapback occurs. If a unit cell is considered to be a part of a structure within the nested multiscale scheme, the unit-cell computation is strain driven and the snapback control technique discussed in Section 4.3.4 is required. To illustrate this, a macroscopic "structure" is tested, which consists of a single, quadratic (serendipity) quadrilateral finite element under homogeneous macroscopic loading. As a result, all four Gauss points of the element exhibit the same response. The bifurcation detection and branching are not activated in this computation to ensure that the unit-cell response is the only possible cause of snapback in the homogenized material response. The structure is proportionally stressed with vertical compression combined with horizontal tension along the stress path $(\Sigma_{xx}, \Sigma_{yy}, \Sigma_{xy}) = (0.2, -1, 0)$. The macroscopic "structural" dimensions are taken such that the volume associated with each Gauss point by the quadrature scheme is identical to that of the unit cell, which has the same

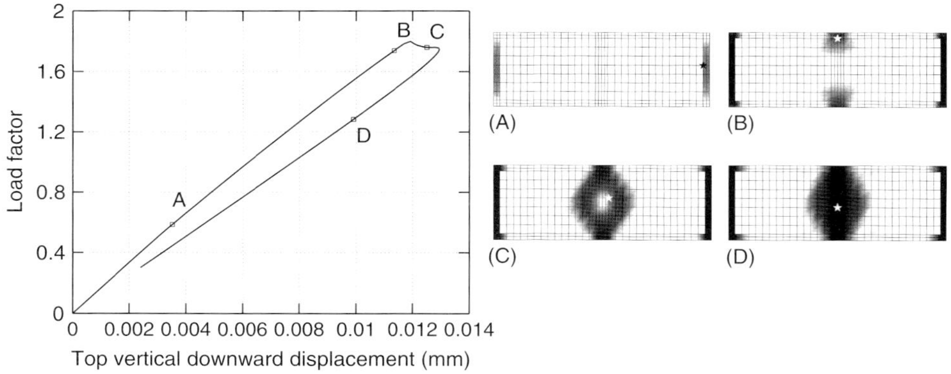

Figure 4.17 Load–displacement curve and damage distribution obtained by the multiscale modeling for homogeneous macroscopic tension–compression loading. Each damage distribution corresponds to the point indicated in the load–displacement curve. The stars in the damage distributions indicate the degree of freedom that is selected for the snapback control in the next increment.

dimensions as in Section 4.3.2.3. The damage criteria and material properties used for the constituents are taken from Ref. [82].

The load factor evolution obtained in Ref. [94] is shown in Figure 4.17 as a function of the top vertical displacement of the structure divided by the number of unit cells along the height of the structure. This curve shows that unit-cell snapback occurs and it is correctly dealt with by the dissipation-enhanced scale transition. For each point of the load–displacement curve marked by a capital letter, the damage state inside the unit cell is also depicted in the figure. The node used for the dissipation control in each increment is identified by a star. At each stage, these selected nodes are clearly positioned where the incremental damage growth is highest.

To investigate the effect of mesoscopic bifurcation on this result, the analysis is repeated with a macroscopic "structure" of modified dimensions $600 \times 300\,\mathrm{mm}^2$. Only one macroscopic finite element is again used, which means that the volume associated with each Gauss point by the quadrature scheme is now larger than the unit-cell dimensions. The macroscopic loading is kept homogeneous such that all Gauss points exhibit the same behavior. For this test, the homogeneous structural solution without localization is compared with that in which the material response is allowed to localize simultaneously in all Gauss points, that is, the solution in which part of every Gauss point volume is allowed to unload once the bifurcation condition is satisfied.

The results of this analysis are presented in Figure 4.18. The predicted localization bands have a vertical orientation, corresponding to the final damage pattern. The effect of the localization band clearly appears in the load–displacement curve, in which a sharper snapback is found for the localized solution. An asymmetric damage pattern is observed in the cell for the localized solution, due to the use of the nonlocal strain control in the multiscale procedure. As illustrated in

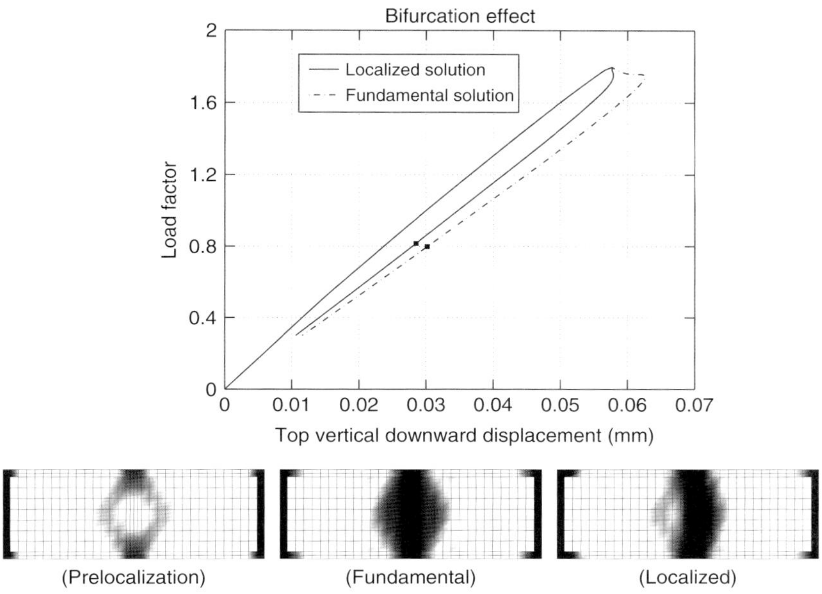

Figure 4.18 Load–displacement curves and damage distribution obtained by multiscale modeling for macroscopic tension–compression loading for fundamental and localized solutions.

Figure 4.18, joint and brick damage is initiated in a symmetric fashion prior to localization. This symmetry causes a bifurcation to appear at the mesoscopic level. This mesoscale bifurcation is not detected in the stress-controlled, direct unit-cell computation, which therefore shows a symmetric response. If the enhanced multiscale technique is used, together with the branching procedure for the band introduction, the unit-cell dissipation control selects one nonlocal strain degree of freedom for controlling the unit cell even if its symmetric counterpart would be equally suitable. As a consequence, the asymmetric solution in which the selected node continues to be damaged is followed. The sharper snapback observed in Figure 4.18 is, thus, caused not only by the presence of the localization band in each Gauss point but also by the occurrence of a different mesoscopic damage pattern inside the unit cell.

4.3.6.3 Size Effect in a Shear–Compression Test

As a second elementary test, a single macroscopic element is subjected to vertical compression combined with shear, similar to that discussed in Section 4.3.2.3. The influence of the ratio of the structural and the mesostructural dimensions is examined. Since the mesostructural size enters the macroscopic computation via the localization band width, a size effect is observed at the macroscopic scale. For a higher ratio of the structural dimension and the unit-cell dimension, a

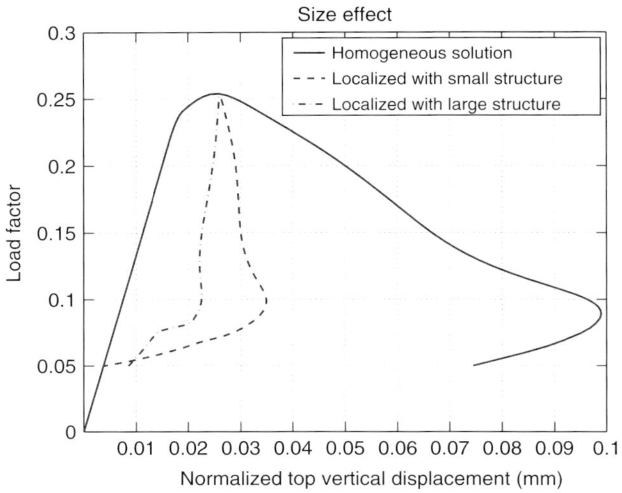

Figure 4.19 Load–displacement curve for vertical compression combined with shear for two different structural dimensions and a fixed unit-cell size.

more brittle structural postpeak response is expected, since the volume in which energy is dissipated is smaller compared to the structural size. To show this effect, the vertical compression–shearing test is repeated with the same unit-cell dimensions, but with two different structural dimensions, namely, $800 \times 800\,mm^2$ and $1600 \times 1600\,mm^2$.

Figure 4.19 shows the load factor versus the displacement, normalized by the structural size, for both structural dimensions, together with the structural fundamental paths. For a fixed unit-cell size, a larger structure indeed clearly leads to a more brittle structural response.

4.3.6.4 **Masonry Shear Wall Test**

Data related to large-scale tests on masonry structures are scarce in the literature because of the practical difficulties associated with such experiments. Confined shear wall tests are the most commonly used structural experiments [96]. Similar experimental results were already used in Ref. [97] for the validation of a discrete mesoscopic masonry model. For practical reasons, the tested structures usually consist of only a few bricks. Since the scale separation between the structural and mesoscopic scales remains intrinsically small in such panels, the corresponding results cannot be used for the quantitative validation of homogenization based models. However, given the lack of experimental data on large-scale structures, the small-scale test data are nevertheless used here for a qualitative assessment of the multiscale modeling. The type of test reported in Ref. [96] for shear walls with openings are considered for this purpose, with altered mesoscopic and macroscopic dimensions for the multiscale approach to remain applicable. The dimensions of the opening are also modified with respect to the dimensions of the wall.

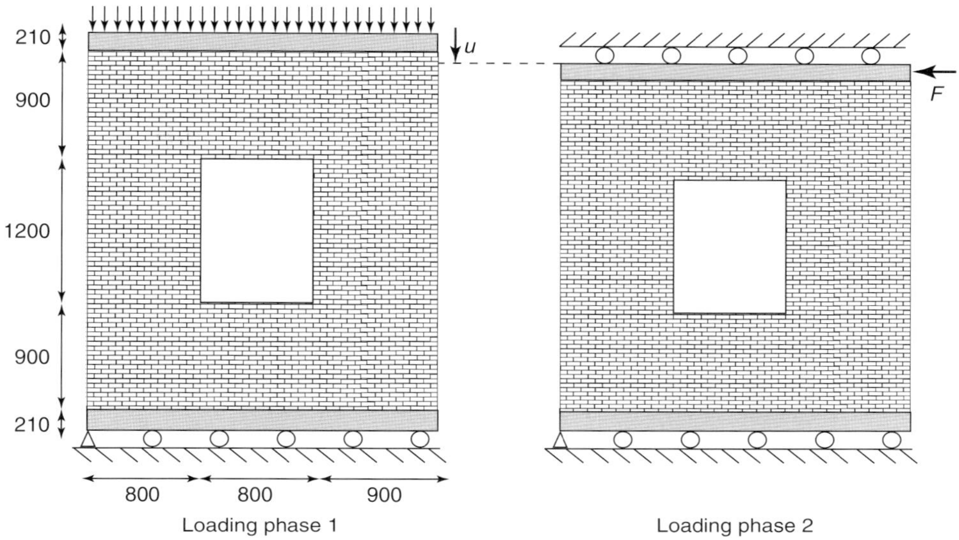

Figure 4.20 Shear test on a wall with opening: two-phase loading and dimensions.

The wall geometry considered in the simulation is shown in Figure 4.20. It consists of a planar masonry wall of dimensions $2500 \times 3000 \times 100\,\mathrm{mm}^3$. An opening with dimensions $800 \times 1200\,\mathrm{mm}^2$ is present in the wall. The geometry is slightly asymmetric in order to trigger an asymmetric response. The assumed size of the bricks relative to the wall is shown in Figure 4.20 in order to emphasize the costly character of a complete, fine-scale modeling of this structure. In the experiments reported in Ref. [96], the top and bottom boundaries of the wall were "clamped" in steel beams. The loading was applied in two phases (Figure 4.20). In a first phase, the wall was compressed by a vertical distributed load applied at the top of the wall, resulting in a uniform vertical displacement. In the second loading phase, the vertical displacement of the top was fixed and a horizontal shearing force was applied. Here, the effect of the loading setup is represented by two bands of elements with elastic behavior. The vertical displacement of the top boundary is forced to remain uniform to simulate the presence of a steel beam. To represent the (imperfect) clamping of the wall in the steel beam, a lower stiffness is assumed for the two rows of elastic elements. The shearing load is applied as a distributed load on the right-hand side of this row of elements.

Experimentally, a complex crack evolution pattern was obtained. The observed crack pattern stages are illustrated in Figure 4.21, after Ref. [96]. Damage was first initiated in the form of diagonal cracks, starting at corners of the opening and in the middle of its top border (Figure 4.21, state A). These specific damage initiation locations were linked to the small number of bricks in the specimen (the opening in the wall had a width of approximately one brick). Soon after their initiation, these diagonal cracks became inactive, upon the appearance of two tensile horizontal cracks at the free boundaries (Figure 4.21, state B). Later, these horizontal cracks

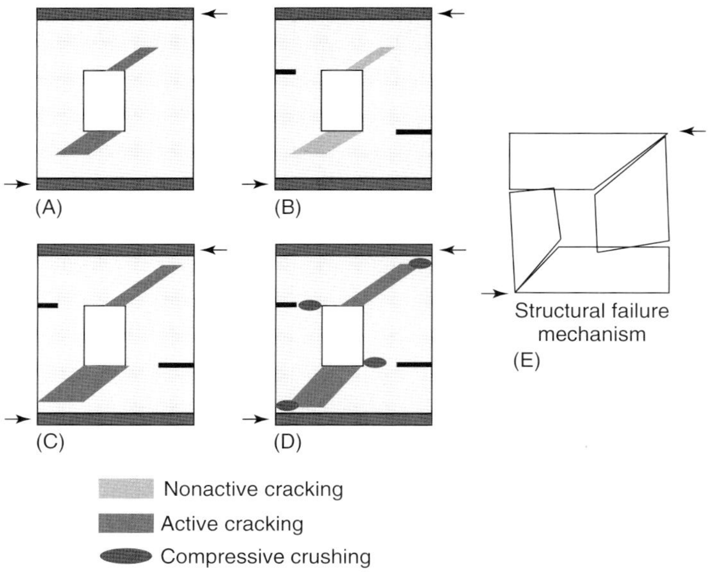

Figure 4.21 Shear test on a wall with opening – successive cracking states observed in experiments [96].

were arrested and two of the diagonal cracks were reactivated and propagated toward the top-right and bottom-left corners of the wall (Figure 4.21, state C). Finally, a structural failure mechanism was observed, with compressive crushing of two corners of the wall and of the two corners of the opening ahead of the tensile cracks (Figure 4.21, state D). The created cracks define four distinct parts of the wall, which rotate with respect to each other in an almost rigid manner, as sketched in Figure 4.21, state E. Note that the diagonal cracking present in this test causes nonorthotropic damage states, which are difficult to model with closed-form models.

For the numerical simulation, the mesostructure of the material is made of bricks of dimensions $90 \times 30 \times 100\ \text{mm}^3$ with 10 mm thick mortar joints. The unit cell is discretized with a rather coarse mesh of 120 elements, since the objective here is to illustrate the application of the scale transition. Two elements are used on the width of a joint. A biquadratic displacement field and a bilinear nonlocal strain field are used at the mesoscopic scale, together with plane-stress assumption. Note that the compressive crushing observed in the last cracking stage of the experiment cannot be captured by this plane-stress computation. A generalized plane-state description would be useful in this respect, but would require a considerably finer mesoscopic discretization to simulate brick cracking [79]. A maximum principal stress criterion is used for the brick material and a Drucker–Prager criterion with a compressive cap is used for mortar. The same mesoscopic material parameters as reported in Table (4.1) are used, except for the intrinsic length scale, which is taken here as $l_c = 2.2$ mm in order to have a sufficient number of elements within the localizing zone for the coarse discretization used. At the macroscopic scale, a mesh of 48 elements is used with a biquadratic displacement interpolation and a four-point integration scheme. Each iteration of the macroscopic solution procedure thus requires the solution of 192 mesostructural boundary value problems. A resultant compressive vertical load of 37.5 kN is considered for the first loading phase. The computation is continued until loss of convergence, that is, the postpeak response is not traced.

The evolution of the macroscopic localization during the computation is illustrated in Figures (4.22)–(4.27) for each of the states A, B, and C indicated in Figure 4.21. In these figures, the embedded localization bands are represented by their respective orientations for the Gauss points in which localization has appeared. For each of these states, the mesoscopic damage field in typical unit cells is shown. Unless stated otherwise, the unit-cell damage patterns are related to the localization band of the Gauss point where a band is present. The macroscopic stress distribution is shown on the deformed shape of the structure – displacements have been magnified by a factor of 1000.

As depicted in Figure 4.22, cracking initiates at the top-right and bottom-left corners of the opening. This crack initiation is due to horizontal tension combined with shear. At this stage, the orientation of the localization bands clearly reflects the staircase damage pattern obtained at the mesoscopic scale, which involves nonorthotropic damage effects. Some further damage evolution is already present in nonlocalized Gauss points, indicating that the diagonal cracking propagates in the direction of the corners of the wall. For the first localization band at the top of the opening, further damage evolution already influenced the staircase pattern in the situation represented in Figure 4.22. Fully anisotropic effects are, however, still present due to unloading of the material surrounding the localization band,

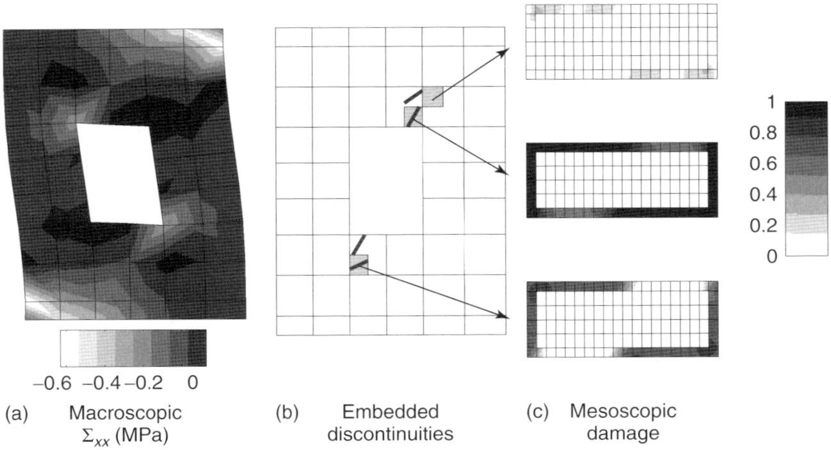

Figure 4.22 State A – initiation of diagonal cracking: (a) macroscopic horizontal stress distribution, (b) embedded discontinuities, and (c) mesoscopic damage distributions.

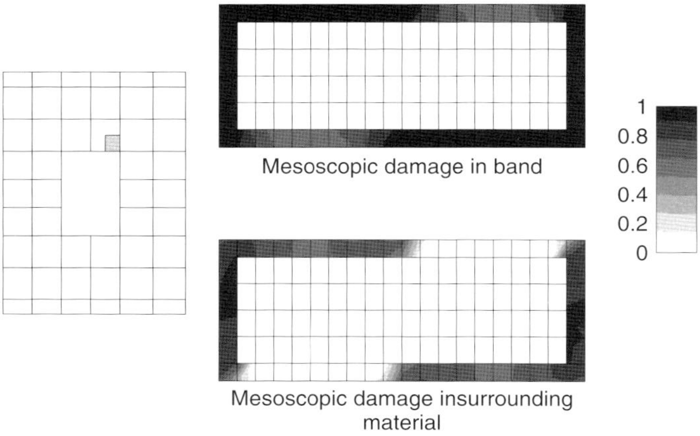

Figure 4.23 Damage distribution in state A with diagonal cracking: comparison between damage states inside and outside the localization band of the first localized Gauss point above the opening.

which exhibits a staircase pattern as illustrated in Figure 4.23. The rather high level of damage reached in the surrounding material has a considerable impact on the aggregate response of this Gauss point, since the band occupies only a limited volume fraction of the associated material volume.

Upon further shearing, tensile damage zones localize at the edges of the wall as illustrated in Figure 4.24. During this stage, the existing diagonal cracks at the top and bottom of the opening do not evolve strongly (top unit cell in

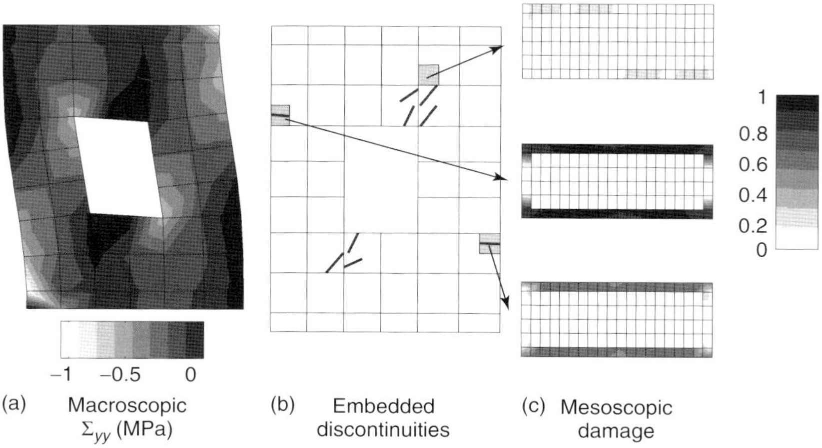

Figure 4.24 State B – appearance of lateral horizontal cracks: (a) macroscopic vertical stress distribution, (b) embedded discontinuities, and (c) mesoscopic damage distributions.

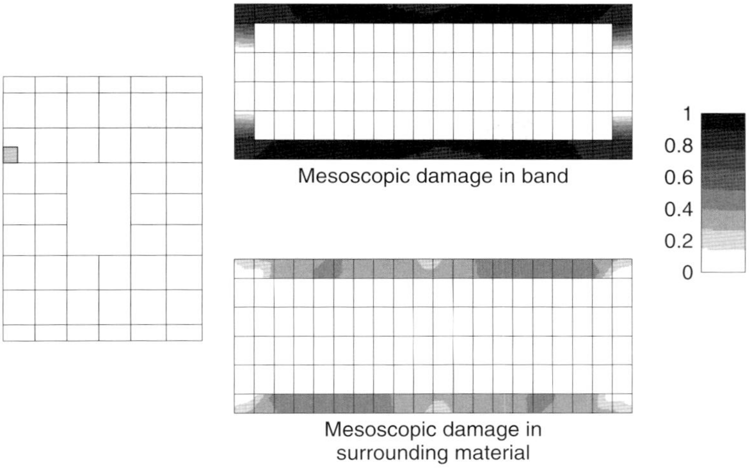

Figure 4.25 Damage distribution in state B with horizontal tensile cracking within the localization band and in the surrounding material.

Figure 4.24). The multiscale approach is thus able to represent the transition from the diagonal cracking to the tensile lateral cracking as observed in experiments on small structures of a similar shape (Figure 4.21). The comparison of the damage distribution within the band and in the surrounding material for the tensile cracks is presented in Figure 4.25. A rather low damage level is obtained in the surrounding material as a result of the pure mode I opening of the bed joint, for which the behavior becomes unstable early in the damaging process.

After the opening of the horizontal tensile cracks, the growth of the diagonal cracks near the opening is reactivated as observed in the experiments. This further propagation is accompanied by a reduced progression of the lateral tensile cracks. It is also observed that the embedded discontinuities near the corners of the wall tend to be inclined toward a horizontal direction, as shown in Figure 4.26 (top and bottom unit cells). The comparison between the damage distribution within the band and in the surrounding material for a point located just below the top beam is shown in Figure 4.27. Although damage is concentrated in the bed joints, as in the case of the horizontal tensile cracks (Figure 4.25), the level of damage in the surrounding material is considerably higher here. This reflects the fact that under the compressive stress path followed by this point, damage may grow substantially before triggering localization.

Finally, the damage continues to grow in the direction of the corners of the wall, simultaneous with a reduced evolution of the lateral horizontal cracks (Figure 4.26). The result of this damage evolution is that strong redistributions of stresses occur, which lead to high compressive vertical stresses at four locations as observed in the experiments (Figure 4.26a). Note also that the vertical tensile stress field is altered around the horizontal cracks by the presence of the damage. Figure 4.26 also confirms the presence of four highly compressed regions. The compressive failure

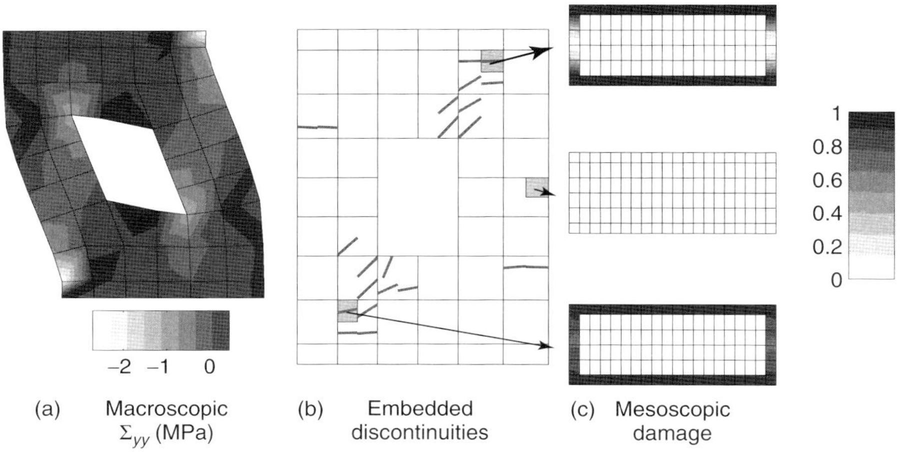

Figure 4.26 State C – propagation of diagonal cracking: (a) macroscopic vertical stress distribution, (b) embedded discontinuities, and (c) mesoscopic damage distributions.

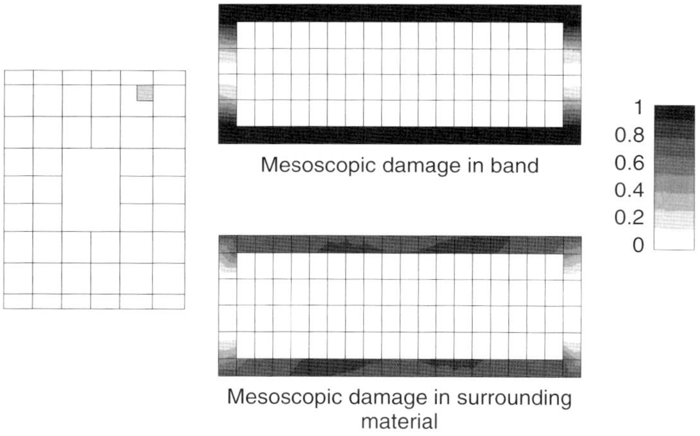

Figure 4.27 Damage distribution in state C with compressive bed joint damage within the localization band and in the surrounding material.

at these locations represents the last stage of the cracking process as observed in experiments. The use of a generalized plane state would have an impact at this stage of the computation. Noteworthy is the fact that already at this stage, a good indication of the structural failure mechanism is found. The deformed shape of the wall depicted in Figure 4.26 shows that a mechanism is formed with the four parts of the cracked wall, which rotate with respect to each other.

It is emphasized that the last stage, represented in Figure 4.26, corresponds to a stage in which the final failure pattern has not yet been reached, since the postpeak branch of the structural response was not traced. Nevertheless, a qualitatively correct prediction of the evolution of damage is obtained by the multiscale approach. In particular, the complex evolution of local cracking mechanisms, coupled with the appearance of mesoscopic damaging zones, is captured in the same order as observed experimentally.

4.4 Closing Remarks

The examples shown in the previous sections demonstrate the capability of the computational homogenization methodology in general and in the context of localization of deformation and damage in particular. The major benefit of the method is that it requires no a priori assumptions on the macroscopic constitutive response. As a result, it can represent extremely complex material behavior due to evolving microstructures in a natural manner, as exemplified by the simulations of masonry cracking shown in the previous section.

Although the conventional, first-order computational homogenization method is well established, it is limited to problems in which a clear separation of scales exists (Section 4.1). The extensions summarized in this chapter allow one to push the concept beyond the conventional realm of homogenization, toward problems in which damage and degradation lead to macroscopically localized solutions and the scale separation principle is thus violated. As discussed in Section 4.2, continuous–continuous scale transitions can be extended to include higher order deformation gradients. The resulting framework can be used for problems with mild localization phenomena. Where the localization is more pronounced, perhaps culminating in final fracture, a transition must be made to a discontinuous description. One possible approach here is to employ the continuous– discontinuous scale transition discussed in Section 4.3; similar and alternative routes have been suggested, for instance, in Refs [67–69].

Unfortunately, the increasing complexity of the material behavior dealt with, as well as of the interaction between the different spatial scales of a problem, at the same time, requires an increasingly complex computational treatment. Quasi-brittle materials such as the brick masonry considered in Section 4.3 are perhaps the most challenging in this respect, as they present the added difficulty of snapbacks and other instabilities at several spatial scales. Although the cornerstones of a general framework for them are now in place, many of the remaining bricks are presently under development – see, for example, Refs [98, 99] for recent developments. However, the specialization to other application domains may also present further challenges, perhaps requiring the combination of the methods developed here with other extensions of the first-order framework as listed in Section 4.1.4. We believe that we are still far from employing the full potential of these computational scale transitions and that the subject will therefore remain an active field of research in the years to come.

References

1. Voigt, W. (1887) Theoretische Studien über die Elasticitätsverhältnisse der Krystalle. *Abhandlungen der Königlichen Gesellschaft der Wissenschaften Göttingen, Math. Kl*, **34**, 3–51.
2. Sachs, G. (1928) Zur ableitung einer fliessbedingung. *Zeitschrift des Vereins Deutscher Ingenieure*, **72**, 734–736.
3. Taylor, G.I. (1938) Plastic strain in metals. *Journal of the Institute of Metals*, **62**, 307–324.
4. Eshelby, J.D. (1957) The determination of the elastic field of an ellipsoidal inclusion. *Proceedings of the Royal Society of London*, **A241**, 376–396.
5. Hill, R. (1965) Continuum micromechanics of elastoplastic polycrystals. *Journal of the Mechanics and Physics of Solids*, **13**, 89–101.
6. Zaoui, A. (2002) Continuum micromechanics: Survey. *Journal of Engineering Mechanics*, **128**(8), 808–816.
7. Hashin, Z. and Shtrikman, S. (1963) A variational approach to the theory of the elastic behaviour of multiphase materials. *Journal of the Mechanics and Physics of Solids*, **11**, 127–140.
8. Hill, R. (1963) Elastic properties of reinforced solids: some theoretical principles. *Journal of the Mechanics and Physics of Solids*, **11**, 357–372.
9. Mori, T. and Tanaka, K. (1973) Average stress in the matrix and average elastic energy of materials with misfitting inclusions. *Acta Metallurgica*, **21**, 571–574.
10. Willis, J.R. (1977) Bounds on self-consistent estimates for the overall properties of anisotropic composites. *Journal of the Mechanics and Physics of Solids*, **25**, 185–202.
11. Kröner, E. (1961) Zur plastischen verformung des vielkristalls. *Acta Metallurgica*, **9**, 155–161.
12. Hutchinson, J.W. (1976) Bounds and self-consistent estimates for creep of polycrystalline metals. *Proceedings of the Royal Society. London A*, **394**, 87–119.
13. Nemat-Nasser, S. and Obata, M. (1986) Rate-dependent finite elasto-plastic deformation of polycrystals. *Proceedings of the Royal Society. London*, **A407**, 343–375.
14. Nemat-Nasser, S. and Hori, M. (1993) *Micromechanics: Overall Properties of Heterogeneous Materials*, Elsevier, Amsterdam.
15. Ponte Castañeda, P. (1991) The effective mechanical properties of nonlinear isotropic composites. *Journal of the Mechanics and Physics of Solids*, **39**, 45–71.
16. Suquet, P. (1993) Overall potentials and extremal surfaces of power law or ideally plastic materials. *Journal of the Mechanics and Physics of Solids*, **41**, 981–1002.
17. Suquet, P. (1997) *Continuum Micromechanics*, vol. 377 of CISM Courses and Lectures, Springer.
18. Willis, J.R. (1994) Upper and lower bounds for nonlinear composite behaviour. *Materials Science and Engineering: A*, **175**, 7–14.
19. Zaoui, A. and Masson, R. (2000) Micromechanics based modelling of plastic polycrystals: an affine formulation. *Materials Science and Engineering: A*, **285**, 8–4.
20. Gurson, A.L. (1977) Continuum theory of ductile rupture by void nucleation and growth: Part I yield criteria and flow rules for porous ductile media. *Journal of Engineering Materials and Technology-Transactions of the ASME*, **99**, 2–15.
21. Chung, P.W., Tamma, K.K., and Namburu, R.R. (2001) Asymptotic expansion homogenization for heterogeneous media: computational issues and applications. *Composites part A: Applied Science and Manufacturing*, **32**, 1291–1301.
22. Fish, J., Shek, K., Pandheeradi, M., and Shephard, M.S. (1997) Computational plasticity for composite structures based on mathematical homogenization: Theory and practice. *Computer Methods in Applied Mechanics and Engineering*, **148**, 53–73.
23. Fish, J. and Chen, W. (2001) Higher-order homogenization of initial/

boundary-value problem. *Journal of Engineering Mechanics*, **127**, 1223–1230.
24. Forest, S., Pradel, F., and Sab, K. (2001) Asymptotic analysis of heterogeneous Cosserat media. *International Journal of Solids and Structures*, **38**, 4585–4608.
25. Smyshlyaev, V.P. and Fleck, N.A. (1994) Bounds and estimates for linear composites with strain gradient effects. *Journal of the Mechanics and Physics of Solids*, **42**, 1851–1882.
26. Drugan, W.J. and Willis, J.R. (1996) A micromechanics-based nonlocal constitutive equation and estimates of representative volume element size for elastic composites. *Journal of the Mechanics and Physics of Solids*, **44**, 7–524.
27. Triantafyllidis, N. and Bardenhagen, S. (1996) The influence of scale size on the stability of periodic solids and the role of associated higher order gradient continuum models. *Journal of the Mechanics and Physics of Solids*, **44**, 1891–1928.
28. Smyshlyaev, V.P. and Cherednichenko, K.D. (2000) On rigorous derivation of strain gradient effects in the overall behaviour of periodic heterogeneous media. *Journal of the Mechanics and Physics of Solids*, **48**, 1325–1357.
29. Peerlings, R.H.J. and Fleck, N.A. (2001) Numerical analysis of strain gradient effects in periodic media. *Journal de Physique IV*, **11**, 153–160.
30. Doghri, I. and Friebel, C. (2005) Effective elasto-plastic properties of inclusion-reinforced composites. Study of shape, orientation and cyclic response. *Mechanics of Materials*, **37**, 45–68.
31. Hughes, T.J.R., Feijóo, G.R., Mazzei, L., and Quincy, J. (1998) The variational multiscale method – a paradigm for computational mechanics. *Computer Methods in Applied Mechanics and Engineering*, **166**, 3–24.
32. Garikipati, K. and Hughes, T.J.R. (2000) A variational multiscale approach to strain localization – formulation for multidimensional problems. *Computer Methods in Applied Mechanics and Engineering*, **188**, 39–60.
33. Moës, N. and Belytschko, T. (2002) Extended finite element method for cohesive crack growth. *Engineering Fracture Mechanics*, **69**, 813–833.
34. Renard, J. and Marmonier, M.-F. (1987) Etude de l'initiation de l'endommagement dans la matrice d'un matériau composite par une méthode d'homogénéisation. *Aerospace Science and Technology*, **6**, 37–51.
35. Suquet, P.M. (1985) Local and global aspects in the mathematical theory of plasticity, in *Plasticity Today: Modelling, Methods and Applications* (eds A. Sawczuk and G. Bianchi), Elsevier, London, pp. 279–310.
36. Guedes, J.-M. and Kikuchi, N. (1990) Preprocessing and postprocessing for materials based on the homogenisation method with adaptative finite element methods. *Computer Methods in Applied Mechanics and Engineering*, **83**, 143–198.
37. Ghosh, S., Lee, K., and Moorthy, S. (1995) Multiple scale analysis of heterogeneous elastic structures using homogenisation theory and voronoi cell finite element method. *International Journal of Solids Structures*, **32**, 27–62.
38. Smit, R.J.M., Brekelmans, W.A.M., and Meijer, H.E.H. (1998) Prediction of the mechanical behaviour of non-linear systems by multi-level finite element modeling. *Computer Methods in Applied Mechanics and Engineering*, **155**, 181–192.
39. Moulinec, H. and Suquet, P. (1998) A numerical method for computing the overall response of non-linear composites with complex microstructure. *Computer Methods in Applied Mechanics and Engineering*, **157**, 69–94.
40. Miehe, C., Schröder, J., and Schotte, J. (1999) Computational homogenization analysis in finite plasticity. Simulation of texture development in polycrystalline materials. *Computer Methods in Applied Mechanics and Engineering*, **171**, 387–418.
41. Miehe, C., Schotte, J., and Schröder, J. (1999) Computational micro-macro transitions and overall moduli in the analysis of polycrystals at large strains. *Computational Materials Science*, **16**, 372–382.

42. Terada, K., Hori, M., Kyoya, T., and Kikuchi, N. (2000) Simulation of the multi-scale convergence in computational homogenization approach. *International Journals of Solids and Structures*, **37**, 2285–2311.
43. Feyel, F. and Chaboche, J.-L. (2000) FE^2 multiscale approach for modelling the elasto-viscoplastic behaviour of long fibre SiC/Ti composite materials. *Computer Methods in Applied Mechanics and Engineering*, **183**, 309–330.
44. Terada, K. and Kikuchi, N. (2001) A class of general algorithms for multi-scale analyses of heterogeneous media. *Computer Methods in Applied Mechanics and Engineering*, **190**, 5247–5464.
45. Kouznetsova, V., Brekelmans, W.A.M., and Baaijens, F.P.T. (2001) An approach to micro-macro modeling of heterogeneous materials. *Computational Mechanics*, **27**, 37–48.
46. Miehe, C. and Koch, A. (2002) Computational micro-to-macro transition of discretized microstructures undergoing small strain. *Archieve of Applied Mechanics*, **72**, 300–317.
47. Miehe, C. and Bayreuther, C.G. (2007) On multiscale FE analyses of heterogeneous structures: from homogenization to multigrid solvers. *International Journal for Numerical Methods in Engineering*, **71**, 1135–1180.
48. Matsui, K., Terada, K., and Yuge, K. (2004) Two-scale finite element analysis of heterogeneous solids with periodic microstructures. *Computers and Structures*, **82**, 593–606.
49. McVeigh, C., Vernerey, F., Liu, W.K., and Brinson, L.C. (2006) Multiresolution analysis for material design. *Computer Methods in Applied Mechanics and Engineering*, **195**, 5053–5076.
50. Temizer, I. and Zohdi, T.I. (2007) A numerical method for homogenization in non-linear elasticity. *Computational Mechanics*, **40**, 281–298.
51. Hain, M. and Wriggers, P. (2008) Computational homogenization of micro-structural damage due to frost in hardened cement paste. *Finite Elements in Analysis and Design*, **44**, 233–244.
52. Yuan, Z. and Fish, J. (2008) Toward realization of computational homogenization in practice. *International Journal for Numerical Methods in Engineering*, **73**, 361–380.
53. Geers, M.G.D., Kouznetsova, V.G., and Brekelmans, W.A.M. (2001) Gradient-enhanced computational homogenization for the micro-macro scale transition. *Journal de Physique IV*, **11**, 145–152.
54. Kouznetsova, V., Geers, M.G.D., and Brekelmans, W.A.M. (2002) Multi-scale constitutive modelling of heterogeneous materials with a gradient-enhanced computational homogenization scheme. *International Journal for Numerical Methods in Engineering*, **54**, 1235–1260.
55. Kouznetsova, V.G. (2002) Computational homogenization for the multi-scale analysis of multi-phase materials. PhD thesis, Eindhoven University of Technology.
56. Geers, M.G.D., Kouznetsova, V.G., and Brekelmans, W.A.M. (2003) Multi-scale second-order computational homogenization of microstructures towards continua. *International Journal for Multiscale Computational Engineering*, **1**, 371–386.
57. Kouznetsova, V.G., Geers, M.G.D., and Brekelmans, W.A.M. (2004) Size of a representative volume element in a second-order computational homogenization framework. *International Journal for Multiscale Computational Engineering*, **2**, 575–598.
58. Kouznetsova, V.G., Geers, M.G.D., and Brekelmans, W.A.M. (2004) Multi-scale second-order computational homogenization of multi-phase materials: a nested finite element solution strategy. *Computer Methods in Applied Mechanics and Engineering*, **193**, 5525–5550.
59. Kaczmarczyk, L., Pearce, C.J., and Bicanic, N. (2008) Scale transition and enforcement of RVE boundary conditions in second-order computational homogenization. *International Journal for Numerical Methods in Engineering*, **74**, 506–522.
60. Massart, T.J., Peerlings, R.H.J., and Geers, M.G.D. (2007) Structural damage analysis of masonry walls using computational homogenization. *International*

Journal of Damage Mechanics, **16**, 199–226.

61. Massart, T.J., Peerlings, R.H.J., and Geers, M.G.D. (2007) An enhanced multi-scale approach for masonry wall computations with localization of damage. *International Journal for Numerical Methods in Engineering*, **69**, 1022–1059.
62. Özdemir, I., Brekelmans, W.A.M., and Geers, M.G.D. (2008) Computational homogenization for heat conduction in heterogeneous solids. *International Journal for Numerical Methods in Engineering*, **73**, 185–204.
63. Özdemir, I., Brekelmans, W.A.M., and Geers, M.G.D. (2008) Fe^2 computational homogenization for the thermo-mechanical analysis of heterogeneous solids. *Computer Methods in Applied Mechanics and Engineering*, **198**, 602–613.
64. Geers, M.G.D., Coenen, E.W.C., and Kouznetsova, V.G. (2007) Multi-scale computational homogenization of structured thin sheets. *Modelling and Simulation in Material Science Engineering*, **15**, S393–S404.
65. Matouš, K., Kulkarni, M.G., and Geubelle, P.H. (2008) Multiscale cohesive failure modeling of heterogeneous adhesives. *Journal of the Mechanics and Physics of Solids*, **56**, 1511–1533.
66. Gitman, I.M., Askes, H., and Sluys, L.J. (2008) Coupled-volume multi-scale modelling of quasi-brittle material. *European Journal of Mechanics. A Solids*, **27**, 302–327.
67. Loehnert, S. and Belytschko, T. (2007) A multiscale projection method for macro/microcrack simulations. *International Journal for Numerical Methods in Engineering*, **71**, 1466–1482.
68. Belytschko, T., Loehnert, S., and Song, J.H. (2008) Multiscale aggregating discontinuities: a method for circumventing loss of material stability. *International Journal for Numerical Methods in Engineering*, **73**, 869–894.
69. Hettich, T., Hund, A., and Ramm, E. (2008) Modeling of failure in composites by X-FEM and level sets within a multiscale framework. *Computer Methods in Applied Mechanics and Engineering*, **197**, 414–424.
70. Mesarovic, S.D. and Padbidri, J. (2005) Minimal kinematic boundary conditions for simulations of disordered microstructures. *Philosophical Magazine*, **85**, 65–76.
71. Miehe, C. (2002) Strain-driven homogenization of inelastic microstructures and composites based on an incremental variational formulation. *International Journal for Numerical Methods in Engineering*, **55**, 1285–1322.
72. van der Sluis, O., Schreurs, P.J.G., Brekelmans, W.A.M., and Meijer, H.E.H. (2000) Overall behaviour of heterogeneous elastoviscoplastic materials: effect of microstructural modelling. *Mechanics of Materials*, **32**, 449–462.
73. Kanit, T., Forest, S., Galliet, I. *et al.* (2003) Determination of the size of the representative volume element for random composites: statistical and numerical approach. *International Journal of Solids and Structures*, **40**, 3647–3679.
74. Kanit, T., N'Guyen, F., Forest, S. *et al.* (2006) Apparent and effective physical properties of heterogeneous materials: representativity of samples of two materials from food industry. *Computer Methods in Applied Mechanics and Engineering*, **195**, 3960–3982.
75. Kouznetsova, V.G., Geers, M.G.D., and Brekelmans, W.A.M. (2009) Computational homogenization for non-linear heterogeneous solids, in *Multiscale modeling in solid mechanics: computational approaches* (eds U. Galvanetto and M.H. Aliabadi), chapter 1. Imperial College Press.
76. Mindlin, R.D. (1964) Micro-structure in linear elasticity. *Archieve for Rational Mechanics and Analysis*, **16**, 51–78.
77. Xu, X.P. and Needleman, A. (1993) Void nucleation by inclusion debonding in crystal matrix. *Modelling and Simulation in Materials Science and Engineering*, **1**, 111–132.
78. Anthoine, A. (1995) Derivation of the in-plane elastic characteristics of masonry through homogenization theory. *International Journal of Solids and Structures*, **32**, 137–163.
79. Pegon, P. and Anthoine, A. (1997) Numerical strategies for solving continuum damage problems with softening:

application to the homogenization of masonry. *Computers and Structures*, **64**, 623–642.

80. Luciano, R. and Sacco, E. (1997) Homogenization technique and damage model for old masonry material. *International Journal of Solids and Structures*, **34**, 3191–3208.
81. Massart, T.J., Peerlings, R.H.J., and Geers, M.G.D. (2004) Mesoscopic modeling of damage-induced anisotropy in brick masonry. *European Journal of Mechanics. A Solids*, **23**, 719–735.
82. Massart, T.J., Peerlings, R.H.J., Geers, M.G.D., and Gottcheiner, S. (2005) Mesoscopic modeling of failure in brick masonry accounting for three-dimensional effects. *Engineering Fracture Mechanics*, **72**, 1238–1253.
83. de Borst, R., Sluys, L.J., Muhlhaus, H.-B., and Pamin, J. (1993) Fundamental issues in finite element analyses of localization of deformation. *Engineering Computation*, **10**, 99–121.
84. Peerlings, R.H.J., de Borst, R., Brekelmans, W.A.M., and de Vree, J.H.P. (1996) Gradient-enhanced damage for quasi-brittle materials. *International Journal for Numerical Methods in Engineering*, **39**, 3391–3403.
85. Pijaudier-Cabot, G. (1991) Rupture et calculs à la ruine. Thèse d'habilitation, Université Paris VI Pierre et Marie Curie – LMT ENS Cachan.
86. Rice, J.R. (1976) *The localization of plastic deformations*, in *Theoretical and Applied Mechanics* (ed. W.T. Koiter), North-Holland.
87. Rice, J.R. and Rudnicki, J.W. (1980) A note on some features of the theory of localization of deformation. *International Journal of Solids and Structures*, **16**, 597–605.
88. Geers, M.G.D. (1999) Enhanced solution control for physically and geometrically non-linear problems. Part I – the subplane control approach. *International Journal for Numerical Methods in Engineering*, **46**, 177–204.
89. Sluys, L.J. and Berends, A.H. (1998) Discontinuous failure analysis for mode-I and mode-II localization problems. *International Journal of Solids and Structres*, **35**, 4257–4274.
90. Jirásek, M. (2000) Comparative study on finite elements with embedded discontinuities. *Computer Methods in Applied Mechanics and Engineering*, **188**, 307–330.
91. de Borst, R., Wells, G.N., and Sluys, L.J. (2001) Some observations on embedded discontinuity models. *Engineering Computation*, **18**, 241–254.
92. Wells, G.N. (2001) Discontinuous modelling of strain localization and failure. PhD thesis, Delft University of Technology.
93. Evers, L.P., Parks, D.M., Brekelmans, W.A.M., and Geers, M.G.D. (2002) Crystal plasticity model with enhanced hardening by geometrically necessary dislocation accumulation. *Journal of the Mechanics and Physics of Solids*, **50**, 2403–2424.
94. Massart, T.J., Peerlings, R.H.J., and Geers, M.G.D. (2005) A dissipation-based control method for the multi-scale modelling of quasi-brittle materials. *Comptes Rendu Mecanique*, **333**, 521–527.
95. Gutiérrez, M.A. (2003) Path-following constraint based on fracture energy control, in *Proceedings of the VII International Conference on Computational Plasticity - COMPLAS 2003* (eds E. Oñate and D.R.J. Owen), CIMNE, Barcelona.
96. Raijmakers, T.M.J. and Vermeltfoort, A.T. (1992) Deformation controlled tests in masonry shear walls. Technical Report B-92-1156, TNO Bouw, Delft, The Netherlands.
97. Lourenço, P.B. (1996) Computational Strategies for Masonry Structures. PhD thesis, Delft University of Technology.
98. Mercatoris, B.C.N. and Massart, T.J. (2009) Assessment of periodic homogenization-based multiscale computational schemes for quasi-brittle structural failure. *International Journal for Multiscale Computational Engineering*, **7**, 153–170.
99. Mercatoris, B.C.N., Bouillard, Ph., and Massart, T.J. (2009) Multi-scale detection of failure in planar masonry thin shells using computational homogenisation. *Engineering Fracture Mechanics*, **76**, 479–499.

5
A Mixed Optimization Approach for Parameter Identification Applied to the Gurson Damage Model

Pablo Andreś Muñoz-Rojas, Luiz Antonio B. da Cunda, Eduardo L. Cardoso, Miguel Vaz Jr., and Guillermo Juan Creus

5.1
Introduction

Numerical simulation of metal-forming operations has experienced a steady growth in the last decade, mainly due to the development of robust elastic–plastic finite element commercial packages. Large strain plasticity, contact and friction modeling, meshing and remeshing procedures, and complex material constitutive relations are some areas that experienced growth in the last years. In spite of the widening spectrum of successful modeling strategies, the determination of material parameters still constitutes a challenge. The most common strategies to determine material parameters are based on standard mechanical tests assuming uniform stress–strain fields. Such strategies are well known and generally easy to be implemented in industries. However, it is worth noting that material parameters should be determined on the basis of experiments that reproduce stress–strain states close to the intended application of the material.

Many different strategies have already been conceived and proposed for the identification of constitutive parameters of a wide range of materials, most of which are based on the solution of so-called inverse problems. Techniques to solve inverse problems (e.g., genetic algorithms (GAs) and gradient-based optimization techniques, among others) provide adequate tools to address parameter identification. Higher complexity of material constitutive relations would require more elaborate and robust identification techniques (the number of parameters and the corresponding degree of interdependence play a fundamental role in the success or failure of a given identification strategy). For instance, a fully coupled elastic–plastic and damage material model requires not only a reliable finite element approximation but also an optimization strategy able to avoid the many local minima typical of this class of materials.

In this chapter, the major concern is the identification of the material parameters of the Gurson model for ductile damage evolution. Section 5.2 reviews the damage model and highlights some important aspects through particular examples, especially the role played by its material parameters. Section 5.3 presents the general

Advanced Computational Materials Modeling: From Classical to Multi-Scale Techniques.
Edited by Miguel Vaz Júnior, Eduardo A. de Souza Neto, and Pablo A. Munoz-Rojas

ISBN: 978-3-527-32479-8

concept used for parameter identification based on an optimization approach. In order to skip local minima, the strategy uses a combination of evolutionary and gradient-based optimization algorithms. General issues and individual modeling aspects of both optimization techniques are discussed in Section 5.4. The approach to sensitivity analysis used in conjunction with the gradient-based optimization procedure is summarized in Section 5.5, emphasizing applications for path-dependent problems. Special attention is devoted to a modified finite difference method and its particularization to a semianalytical approach. Section 5.6 summarizes the proposed mixed optimization strategy. Section 5.7 presents applications for identification of plastic and damage parameters of low carbon steel and aluminum alloys. Finally, highlights of the proposed strategy are given in Section 5.8.

5.2
Gurson Damage Model

The purpose of this theory (originally presented by A. L. Gurson [1, 2] in a PhD dissertation advised by J. R. Rice) was to develop approximate yield criteria and flow rules for porous materials, focusing the effect of void nucleation, and growth, as observed in ductile fracture. Porosity may be assumed as a measure of isotropic damage according to the concept of Kachanov [3, 4].

Figure 5.1, taken from the original work [2], defines macroscopic and microscopic stress and strain and the spherical model of a unit cell. The isotropic damage

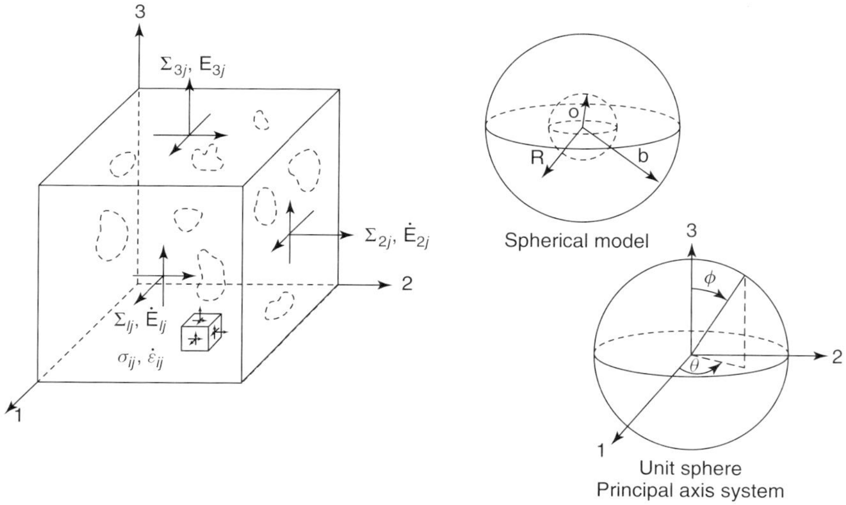

Figure 5.1 Void-matrix aggregate, with random void shapes and orientations, evidencing macroscopic and microscopic tensor quantities, and also the unit cell model studied by Gurson [2].

variable is the volumetric void fraction or porosity $f = V_v/V$, where V_v is the volume of voids in a representative small volume V, corrected for effects as stress concentration, and so on; f is defined at each point of the continuum.

The yield criterion is approximated through an upper bound approach. Simplified physical models for ductile porous materials (as aggregates of voids and ductile matrix) are employed, with the matrix material idealized as rigid–perfectly plastic obeying von Mises yield criterion. Using a distribution of macroscopic flow fields and working with a dissipation integral, the upper bound for the macroscopic stress fields required for yield is determined [2]. The locus in stress space determines the yield surface. It is also shown that normality holds for this yield surface.

The yield surface originally proposed by Gurson had the form

$$\Phi = T_{\mathrm{eqv}}^2 + 2f\cosh\left(\frac{T_{\mathrm{nn}}}{2}\right) - 1 - f^2 = 0 \tag{5.1}$$

with

$$T_{\mathrm{eqv}} = \left(\frac{3s_{ij}s_{ij}}{2\sigma_y^2}\right)^{1/2}, \quad s_{ij} = \sigma_{ij} - p\delta_{ij}, \quad p = {}^1\!/_3\sigma_{ij}\delta_{ij}, \quad T_{\mathrm{nn}} = \frac{3p}{\sigma_y} \tag{5.2}$$

where σ_{ij} are the Cauchy stresses and σ_y is the yield stress in simple tension.

A prior model, the Drucker–Prager theory [5], also proposed a yield criterion dependent on hydrostatic stress in the general form

$$\sqrt{1/2 s_{ij}s_{ij}} = a_1 + a_2 p \tag{5.3}$$

The importance of Gurson theory comes from the establishment of a direct relationship between the effect of the hydrostatic stress p and the porosity f.

The equation usually employed in computational damage analyses, the so-called Gurson–Tvergaard model [2, 6], differs from Eq. (5.1) and considers a yield surface defined by

$$\Phi = \sqrt{\frac{3}{2}s_{ij}s_{ij}} - \overline{\omega}\sigma_y = 0 \tag{5.4}$$

where

$$\overline{\omega} = \left[1 - 2\alpha_1 f\cosh\left(\frac{\alpha_2 3p}{2\sigma_y}\right) + \alpha_3 f^2\right]^{1/2} \tag{5.5}$$

In the expressions above, α_i are material parameters. The original Gurson's model [2] does not include the parameters α_i. The introduction of these parameters was proposed by Tvergaard [6, 7] to improve the approximation of experimental results for slip bands at low values of f. The first parameter (α_1) is a coefficient multiplying the porosity f, to be adjusted by comparing numerical simulations of the representative volume element (RVE) aggregates and the predictions of the model [6]. As the element studied by Gurson was a single hollow sphere, disregarding interaction among voids, this coefficient attempts to introduce the void interaction effect. The values proposed by Tvergaard are

$$\alpha_1 = 1.5, \quad \alpha_2 = 1.0, \quad \alpha_3 = \alpha_1^2 \tag{5.6}$$

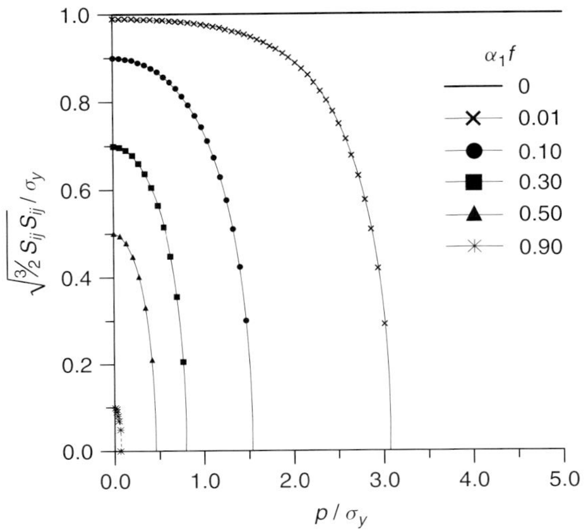

Figure 5.2 Gurson yield surface for a porous material, showing the influence of pressure and volumetric void fraction.

According to Thomason [8], α_1 allows to adjust the influence of f on the yield surface, while α_2 allows to represent the pressure effect. The parameter α_1 may also be interpreted as the inverse of f_U, which in turn is the volumetric void fraction that corresponds to rupture in the absence of hydrostatic pressure $\overline{\omega}(p = 0, \sigma_y, f) = 0$,

$$f_U = \frac{1}{\alpha_1} \tag{5.7}$$

In Figure 5.2, yield surfaces for different levels of void contents are shown in a plot of normalized deviatoric stress versus normalized pressure. It can be seen that the elastic domain depends on the hydrostatic pressure. When the volumetric void fraction f decreases, the influence of pressure also decreases, leading to a larger elastic domain. For $f = 0$, the model reduces to the classical von Mises model, which is independent of pressure. It should be noted here that, in the absence of hydrostatic pressure, the coefficient $\overline{\omega}$ reduces to

$$\overline{\omega} = 1 - \alpha_1 f = 1 - \frac{f}{f_U} \tag{5.8}$$

It must be emphasized that rupture only occurs at a porosity level f_U in the absence of pressure. In the presence of pressure, rupture takes place for a porosity value lower than f_U. The combinations of pressure and porosity that lead to rupture can be determined from the equation

$$\frac{2}{3\alpha_2} \operatorname{arc\,cosh}\left(\frac{1 + (f/f_U)^2}{2(f/f_U)}\right) - \frac{p}{\sigma_y} = 0 \tag{5.9}$$

derived by imposing the condition $\overline{\omega} = 0$. Figure 5.3 shows combinations of pressure and porosity corresponding to Eq. (5.9) for $\alpha_2 = 1.0$ and $\alpha_2 = 0.7$.

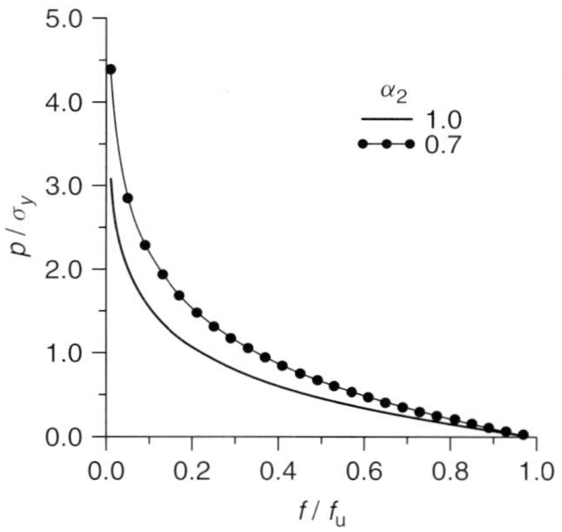

Figure 5.3 Combinations of pressure and porosity values that correspond to loss of the load-bearing capacity ($\overline{\omega} = 0$).

The plastic strain rate tensor is given by the normality rule

$$D_{ij}^{\mathrm{p}} = \dot{\lambda} \frac{\partial \Phi}{\partial \sigma_{ij}} \tag{5.10}$$

and the equivalent plastic strain rate is defined as

$$\dot{\varepsilon}^{\mathrm{p}} = \sqrt{2/3 D_{ij}^{\mathrm{p}} D_{ij}^{\mathrm{p}}} \tag{5.11}$$

In damage theory, in addition to the yield criteria and flow rule, a damage evolution law (porosity) is required. The basic mechanisms of damage evolution considered in the Gurson model are *nucleation*, *growth*, and *coalescence* of voids. The onset of *nucleation* is mainly due to material defects in the presence of tension. *Growth* takes place when voids (preexistent or nucleated) change their size according to the volume change in the continuum. *Coalescence* is related to the fast rupture process that occurs after the volumetric void fraction reaches a limit, usually indicated by f_{C}. Coalescence consists in the union of neighboring voids due to the rupture of the ligaments among them.

The equations that govern damage evolution are modeled in a simplified form as follows. First, it is assumed that the total void rate is given by

$$\dot{f} = \begin{cases} \dot{f}_{\mathrm{n}} + \dot{f}_{\mathrm{g}} & f \le f_{\mathrm{C}} \\ \dot{f}_{\mathrm{c}} & f > f_{\mathrm{C}} \end{cases} \tag{5.12}$$

where $\dot{f}_{\mathrm{n}}$ is the void nucleation rate, $\dot{f}_{\mathrm{g}}$ is the void growth rate, and $\dot{f}_{\mathrm{c}}$ is the void coalescence rate. Thus, as long as f is smaller than a characteristic value f_{C}, only nucleation and growth develop. Above f_{C}, only coalescence takes place.

The nucleation rate is proportional to the equivalent plastic strain rate

$$\dot{f}_{\mathrm{n}} = A(\varepsilon^{\mathrm{p}})\dot{\varepsilon}^{\mathrm{p}} \tag{5.13}$$

Chu and Needleman [9] propose a statistical distribution for $A(\varepsilon^{\mathrm{p}})$ as

$$A(\varepsilon^{\mathrm{p}}) = \frac{f_{\mathrm{N}}}{s_{\mathrm{N}}\sqrt{2\pi}} \exp\left[-\frac{1}{2}\left(\frac{\varepsilon^{\mathrm{p}} - \varepsilon_{\mathrm{N}}}{s_{\mathrm{N}}}\right)^2\right] \tag{5.14}$$

where f_{N} is the void nucleation volumetric fraction, ε_{N} is the mean plastic strain for nucleation, and s_{N} is the standard deviation for the distribution. Sometimes it is assumed that nucleation does not take place when the material is compressed [10]. The compression state is indicated by a negative pressure p, so that

$$A(\varepsilon^{\mathrm{p}}) = 0 \quad \text{if} \quad p < 0 \tag{5.15}$$

The void growth rate is controlled by mass conservation through the expression

$$\dot{f}_{\mathrm{g}} = (1 - f) D_{ii}^{\mathrm{p}} \tag{5.16}$$

Voids increase or decrease their volume according to the volume variation in the continuum. Coalescence is governed [11] by the relation

$$\dot{f}_{\mathrm{c}} = \frac{f_{\mathrm{U}} - f_{\mathrm{C}}}{\Delta\varepsilon} \dot{\varepsilon}^{\mathrm{p}} \tag{5.17}$$

where $\Delta\varepsilon$ is a material parameter that controls the coalescence rate.

An alternative strategy to account for void coalescence [12] replaces the volumetric void fraction, f, in the Gurson yield surface (Eq. (5.6)) by a corrected volumetric void fraction f^* given by

$$f^* = \begin{cases} f & f < f_{\mathrm{C}} \\ f_{\mathrm{C}} + \frac{(1.0 - f_{\mathrm{C}})}{(f_{\mathrm{F}} - f_{\mathrm{C}})}(f - f_{\mathrm{C}}) & f > f_{\mathrm{C}} \end{cases} \tag{5.18}$$

where f_{F} is a material parameter. In this case, only nucleation and growth are considered in Eq. (5.12).

The three mechanisms of evolution of voids described in the previous paragraphs can be classified into reversible and irreversible. The nucleation and coalescence mechanisms are irreversible. Thus, it is natural to assume that the phenomena Eqs (5.13) and (5.17) are governed by the equivalent plastic strain, since this measure is always positive. The growth mechanism is reversible, making it possible to assume its dependence on the volume change (Eq. 5.16): if the volume reduces, growth rate is negative, and, conversely, if the volume increases, growth rate is positive.

The presence of embedded voids in a metallic matrix changes also the elastic behavior. This effect is usually accounted for by adopting the Mori–Tanaka [13] relations,

$$K = \frac{4K_0 G_0 (1 - f)}{4G_0 + 3K_0 f} \tag{5.19}$$

and

$$G = \frac{G_0(1 - f)}{1 + \frac{6K_0 + 12G_0}{9K_0 + 8G_0} f} \tag{5.20}$$

in which K_0 and G_0 are the undamaged values of compressibility and shear modulus, respectively. There are other proposals to include the effect of porosity

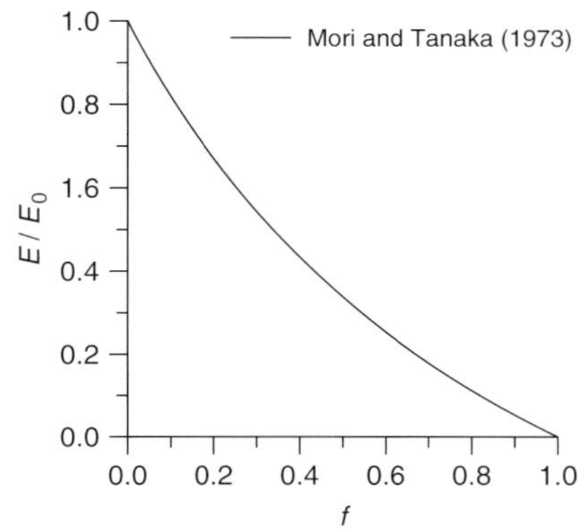

Figure 5.4 Dependence of Young modulus on the void volumetric fraction as introduced by the Mori–Tanaka equations.

on elastic constants [14–16], leading to similar results. Figure 5.4 shows the dependence of damaged Young modulus on the porosity evaluated from Eqs (5.19) and (5.20).

5.2.1 Influence of the Parameter Values on Behavior of the Damage Model

The material parameters present in the Gurson damage model can be grouped [17] into three classes:

1) **Constitutive parameters**, related to the Gurson yield surface (α_1, α_2 e α_3);
2) **Initial parameters**, associated with the origin of the porosity, if present in the virgin material (f_0) or nucleated by plastic straining (f_N, s_N, and ε_N); and
3) **Critical parameters**, related to the interaction between neighboring voids, describing the coalescence stage and the final rupture of the material.

The constitutive parameters α_1 and α_2 act as multipliers on the volumetric void fraction and on the pressure, respectively, thereby increasing the possibility to adjust the Gurson yield surface to available experimental or numerical data. Larger values of α_1 and α_2 correspond to a smaller elastic domain. Figure 5.5 shows the dependence of Gurson yield surface on the α_2 parameter.

The second group of parameters is related to the origin of the voids. Some choices must be made with relation to the origin of voids. A nonnull initial porosity, f_0, can be employed in two situations: (i) when the material actually contains preexisting voids and (ii) when voids are developed from inclusions that break or debond from the matrix in a very low strain level. Otherwise, it is usual to employ a strain-governed nucleation equation, as proposed by Chu and Needleman [9], and described by Eqs (5.13) and (5.14).

The value of f_N determines the level of nucleated voids. The parameter ε_N corresponds to the mean equivalent plastic strain at which nucleation is developed.

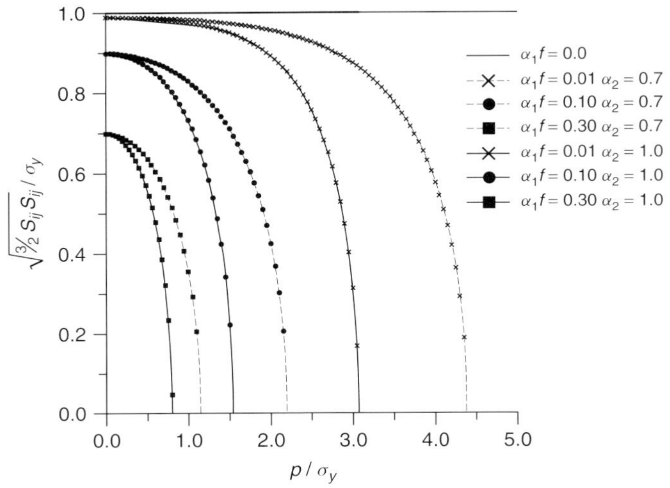

Figure 5.5 Dependence of Gurson yield surface on α_2. Results presented for $\alpha_2 = 0.7$ and $\alpha_2 = 1.0$.

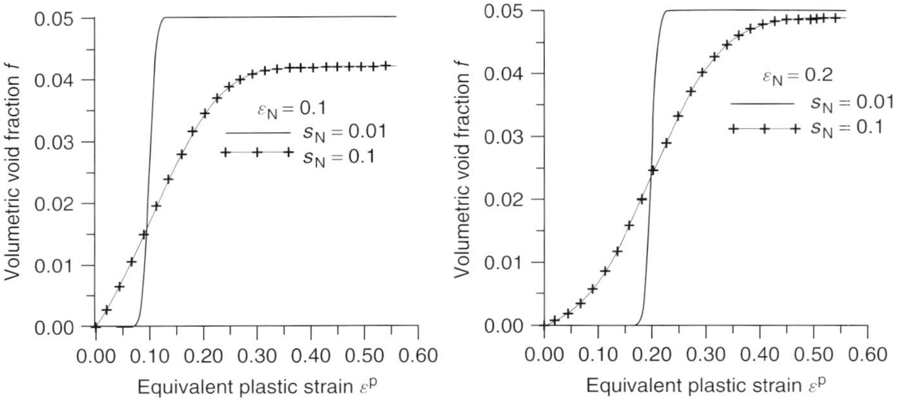

Figure 5.6 Evolution of porosity in the nucleation stage for different values for ε_N and s_N. The simulations show localized nucleation around ε_N if a smaller value of s_N is employed.

The nucleation standard deviation, s_N, controls the localization of nucleation around ε_N. Figure 5.6 shows two nucleation processes for different ε_N, in which smaller values indicate an earlier nucleation. The same figures show results for different s_N, so that smaller values of s_N correspond to a faster nucleation, with nucleation localized around the mean equivalent plastic strain nucleation ε_N.

Figure 5.6 also indicates that the imposed nucleation void volumetric fraction, f_N, is not reached. This is because the porosity evolution law in the nucleation stage, given in a rate form, $\dot{f}_n$, must be integrated as the material is plastically

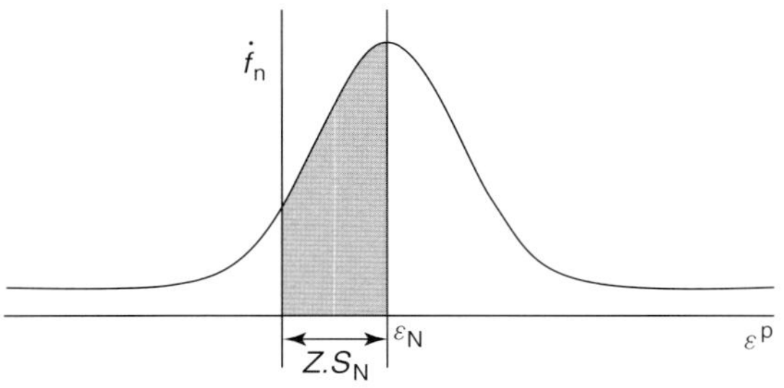

Figure 5.7 Dependence of the nucleation evolution rate on the equivalent plastic strain, indicating the undesired possibility of part of the area under the curve be located in the "fictitious negative" domain of the equivalent plastic strain. Such characteristics are directly affected by the relation between ε_N and s_N.

deformed. If an inadequate relation between s_N and ε_N is employed, a significant part of the porosity evolution rate $\dot{f}_n$ will take place in a "fictitious negative" part of the equivalent plastic strain domain (Figure 5.7). In this case, no integration is performed due to the positive nature of the equivalent plastic strain, leading to an incomplete nucleation.

In order to avoid this problem, a relation between s_N and ε_N in the form

$$\varepsilon_N \geq z.\, s_N \tag{5.21}$$

must be taken into consideration. Therefore, each value of z corresponds to a different level of nucleation. For instance, to ensure a nucleation level of at least 95% of f_N, it is necessary to employ $z = 1.645$; a nucleation level of 97% of f_N requires $z = 1.882$ and, to attain a nucleation level of 99% of f_N, $z = 2.337$ must be employed.

An important issue to be addressed concerns the influence of the pressure sign on the nucleation of voids. There are some possible strategies to account for this effect. The first approach considers void nucleation completely independent of the pressure sign [18, 19]. For a material without initial porosity, this strategy leads to a Gurson yield surface symmetric in relation to pressure, as indicated in Figure 5.8. On the other hand, nucleation can be associated with debonding between inclusions and metallic matrix, which, in this case, will be reduced if the region is submitted to negative pressure p (compression).

To avoid the aforementioned contradiction, there are proposals that consider nucleation only for $p > 0$ (tensile stress states). The first consequence can be described as follows: for an initially void-free material, the deformation behavior changes to a mixed form depending on pressure, from Gurson behavior in tension to von Mises behavior in compression, as indicated in Figure 5.9. This approach has also its drawbacks. When considering a material initially compressed and plastically deformed to an equivalent plastic strain level higher than ε_N, if loading is reversed and the hydrostatic stress state in the material changes to tension, nucleation will not take place. If there are no nucleated voids, voids will obviously

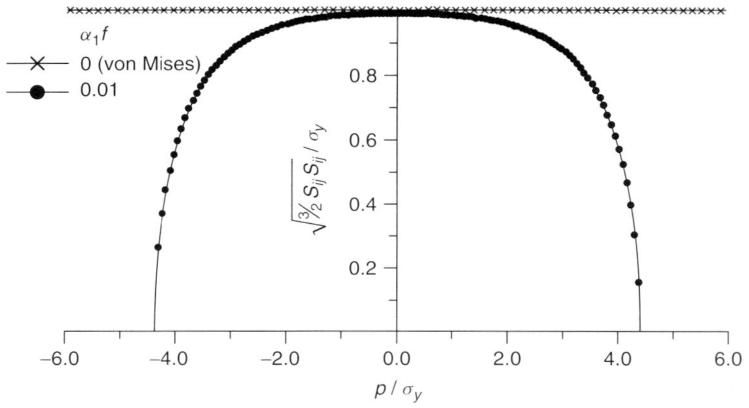

Figure 5.8 Gurson yield surface associated with a virgin material free of voids with nucleation law independent of the pressure sign: the yield locus is symmetric with respect to pressure.

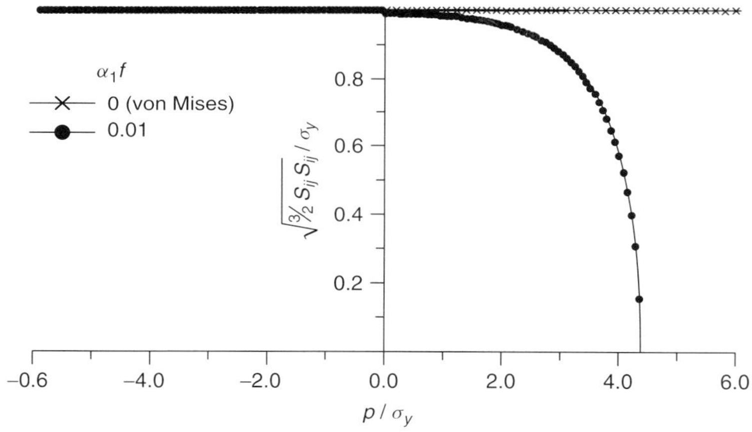

Figure 5.9 Gurson yield surface associated with a virgin material free of voids with nucleation inhibited in compression: von Mises behavior in compression and Gurson behavior in tension.

not grow and the material will continue obeying von Mises yield surface even for a higher level of plastic straining and positive hydrostatic tension. Finally, it should be remarked that both described approaches give the same response to monotonic positive hydrostatic pressure.

The third class of parameters includes those related to the coalescence of voids. The void coalescence rate defined by Eqs (5.11) and (5.16) presents two material parameters: f_C and $\Delta\varepsilon$. The former indicates the initial level of voids at which coalescence onsets, whereas the latter indicates how fast coalescence takes

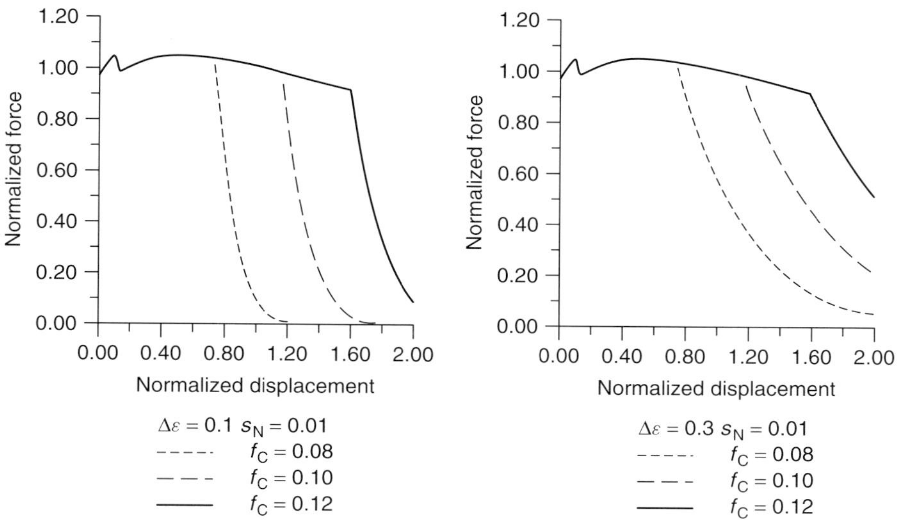

Figure 5.10 Normalized force versus displacement for a tensioned material obeying Gurson behavior. It can be seen that after yielding there is decay in tension corresponding to nucleation, and that the final branch corresponding to rupture starts earlier or later depending on f_C and with its slope depending on $\Delta\varepsilon$.

place. Considering a material submitted to uniform tension, it can be seen that coalescence parameters controls the final branch of the load–displacement relation presented in Figure 5.10: f_C controls the start of the branch and $\Delta\varepsilon$ controls its slope. Smaller values of f_C cause the final branch to start earlier and smaller values of $\Delta\varepsilon$ correspond to a steeper slope. Figure 5.11 illustrates the evolution of porosity for a tensioned material showing the three mechanisms of void evolution: nucleation, growth, and coalescence.

5.2.2
Recent Developments and New Trends in the Gurson Model

The Gurson model has been modified by several authors, particularly in respect of its parameters. There are proposals to make those parameters functions of porosity [20, 21], stress triaxiality [22], and void shape [23]. Extensions of the model to kinematic hardening have also been proposed [24–26]. Further enhancements include thermomechanical coupling [27], making it possible to account for temperature-dependent parameters.

Several authors addressed special issues related to void size and shape. Thomason [8, 28] introduced a model incorporating Rice and Tracey equations [29], which is able to represent both void growth and its change of shape. For this model, the α_i parameters proposed by Tvergaard [6, 7] would be unnecessary. Klöcker

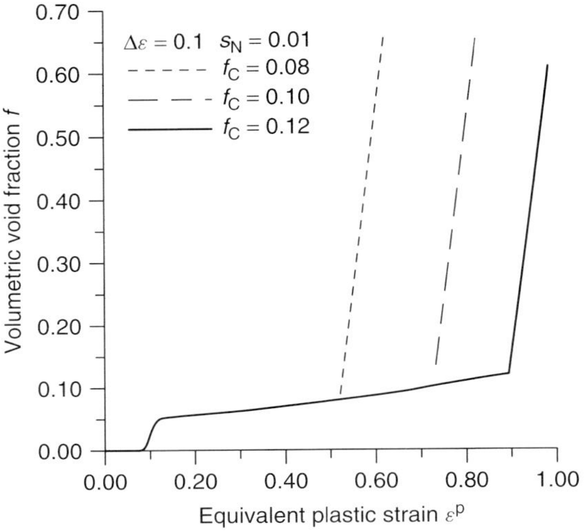

Figure 5.11 Evolution of porosity for a tensioned material obeying Gurson behavior showing the three mechanisms of void evolution: nucleation, growth, and coalescence. The parameter $\Delta\varepsilon$ is the equivalent plastic strain increment from the beginning of coalescence until the final rupture of the material.

and Tvergaard [23] also proposed a modification of the Gurson model to account for changes in void shape and the coalescence process itself. Zhang and Niemi [30] presented a coalescence mechanism that avoids the need to experimentally determine the critical value of the coalescence onset, f_C. In this model, voids are assumed to remain spherical during growth with coalescence onset controlled by the stress triaxiality level. The void size effect was also addressed by Wen *et al.* [31], who demonstrated that the yield surface is larger for materials with very small voids. This effect becomes important for high porosities. Voyiadjis and Kattan [21] proposed a formulation that introduces damage through a damage tensor. The authors extended the Gurson model to account for a yield surface with porosity-dependent parameters.

Gurson-type models have also been used in conjunction with typical fracture mechanics approaches. For instance, Aravas and McMeeking [32] used the crack tip opening displacement (CTOD) theory coupled to a Gurson-type model to study fracture initiation. Kikuchi *et al.* [33] studied the interaction between the macrocrack and microvoids near the crack tip using the J_2 corner theory. The dynamic crack growth and mesh sensitivity for blunt and sharp cracks were analyzed by Needleman and Tvergard [34]. Ductile crack extension in precracked Charpy specimens subjected to static and impact loading was discussed by Koppenhoefer and Dodds [35]. The cyclic J-integral combined to a modified Gurson–Tvergaard model was also used to approximate low-cycle fatigue [36].

Important research effort is also being dedicated to the numerical problems observed in the integration of the nonlinear equations in finite element elastic–plastic damage codes. The main difficulty is the lack of objectivity that is observed in the results of Gurson damage problems as well as in all solutions that involve softening. A subject that requires particular attention is the influence of mesh size in finite element analysis using Gurson model, as well as in all situations that involve softening. In such cases, results become strongly mesh dependent, unless special procedures are employed. Strategies, including nonlocal models and gradient plasticity, have been used to avoid mesh sensitivity. This class of models introduces a characteristic length associated with the material structure but unrelated to the mesh size (e.g., Refs [37–43] for general materials). Nonlocal strategies for the Gurson model have been proposed by Leblond *et al.* [44], Tvergaard and Needleman [45], and Reusch *et al.* [46, 47]. Gradient plasticity formulations consider that the yield surface depends not only on internal variables but also on their gradients. As discussed by Gologanu *et al.* [48] and Ramaswamy and Aravas [49, 50] within the framework of the Gurson model, this dependence leads to a behavior similar to nonlocal approaches. An extension of Leblond *et al.* [44] to an elastic–viscoplastic version of the Gurson model was introduced by Tvergaard and Needleman [45]. A viscoplastic formulation introducing a characteristic length in the context of Gurson damage was also used by Stainier [19].

5.3 Parameter Identification

As seen in Section 5.2, phenomenological models can be applied to consider the effects of *nucleation*, *growth*, and *coalescence* of voids in the yield surface of ductile materials. Therefore, in order to have a fairly complete description of the material behavior upon yielding, two aspects of the material laws must be accounted for: (i) the yield surface definition and its evolution, which is given by the Gurson model defined in Eqs (5.4) and (5.12); and (ii) the work-hardening, generically defined by the yield stress and its dependence with respect to the equivalent plastic strain, which, in this work is, represented as

$$\sigma_y = (P_2 - P_1)[1 - \exp(-P_3\,\varepsilon^p)] + P_4(\varepsilon^p)^{P_5} + P_1(1 + P_6\,\varepsilon^p)^{P_7} + P_8\,\varepsilon^p \quad (5.22)$$

Hence, by adjusting the set of damage parameters, given by $S_d = \{f_U, f_{lim}, f_0, f_N, f_C, s_N, \varepsilon_N, \Delta\varepsilon\}$, and the set of elastic–plastic parameters, represented by $S_{ep} = \{P_1, \ldots, P_8\}$ (or a subset of S_{ep}), it is possible to obtain an approximation to an experimental load–displacement evolution. A numerical load–displacement curve can be obtained by selecting arbitrary values for the parameters in the sets S_d and S_{ep}, and by simulating the experimental test using a finite element model. In general, and most certainly, the curves will not match, as illustrated in Figure 5.12.

Figure 5.12 shows experimental and numerical load–displacement curves of a tensile test with discrete values sampled at different deformation stages. The present work takes experimental points as reference and interpolates their numerical

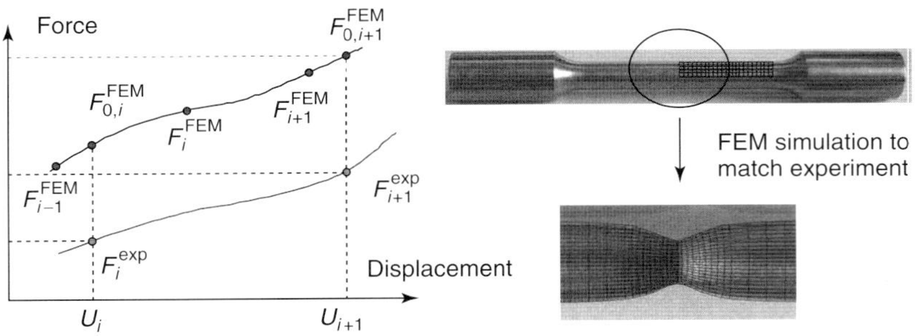

Figure 5.12 Sketch of experimental and numerical curves for parameter identification. Specimen used in tensile tests and the corresponding finite element mesh.

counterparts. The sum of the gaps between the experimental and the interpolated numerical points is computed, which is used to quantify the difference between both curves as

$$W(\mathbf{b}) = \sqrt{\frac{1}{N}\sum_{i=1}^{N} w_i \left(\frac{F_{0,i}^{\mathrm{FE}}(\mathbf{b}) - F_i^{\mathrm{exp}}}{F_i^{\mathrm{exp}}}\right)^2} \tag{5.23}$$

where F_i^{exp} and $F_{0,i}^{\mathrm{FE}}$ are, respectively, the axial load measured in the tensile test, and the corresponding load computed by finite elements (and interpolated for a given experimental point i), $\mathbf{b}$ represents all material parameters being identified, N is the total number of experimental points and w_i is a weight which can be used to favor matching in specific regions of the curves. A measure of the difference between the experimental and numerical values of the final necking radius is defined as

$$T(\mathbf{b}) = \left|\frac{d^{\mathrm{FE}} - d^{\mathrm{exp}}}{d^{\mathrm{exp}}}\right| \tag{5.24}$$

in which d^{FE} and d^{exp} are the final neck diameters computed by finite elements and measured in the tensile test, respectively. In order to make the numerical and the experimental curves match, the following optimization problem is defined:

$$\mathrm{Min}\ \ S(\mathbf{b}) = p_1\frac{W}{W_0} + p_2\frac{T}{T_0} \tag{5.25}$$

where S is the *objective function*, p_1 and p_2 are weight values, and the subscript 0 indicates reference parameters, which are defined according to the optimization approach: $W_0 = T_0 = 1$ for GAs, and W_0 are T_0 the initial values in the case of gradient-based methods. Equation (5.24) helps guiding the optimization procedure to the correct material parameters, as one can find different sets of *design variables* (plastic and damage parameters in the present work) that minimize W, but only one corresponds to the physical reality of the problem.

If this objective function exhibits a strict global minimum, and the optimization scheme is able to determine it, the material parameters of sets S_{d} and S_{ep} would be those that can best represent the experimental behavior. However, as discussed in the next sections, the determination of the global minimum is not a straightforward task. Some tools aiming at finding a minimum using optimization methods are presented in Section 5.4.

It is worth remarking that any finite element simulation of elastic–plastic problems requires a uniaxial true stress–strain function of the material (e.g., the yield stress defined in Eq. (5.22)). Standard tensile tests are frequently used to determine such curves. However, uniaxial stress states in tensile testing can only be assumed up to the onset of microvoid nucleation and necking, thereby limiting its use to relatively small plastic deformations. Metal-forming applications, nonetheless, require material modeling able to handle large plastic strains. The identification strategy discussed herein aims at such problems, in which both the triaxial stress state at necking and microvoid nucleation are accounted for.

5.4 Optimization Methods – Genetic Algorithms and Mathematical Programming

Basically, mathematical programming deals with the minimization of functions in problems that might include constraints. In a general form [51], these problems can be stated as

$$\begin{array}{lll} \text{Minimize} & g_0(\mathbf{b}) & \mathbf{b} \in R^n \\ \text{such that} & h_i(\mathbf{b}) = 0 & i = 1 \ldots l \\ & g_j(\mathbf{b}) \leq 0 & j = 1 \ldots m \\ & \underline{b}_j \leq b_j \leq \overline{b}_j & k = 1 \ldots m \end{array} \quad (5.26)$$

where $\mathbf{b}$ is a point in R^n on which upper bounds ($\overline{b}_k$) and lower bounds ($\underline{b}_k$) are imposed, $g_0(\mathbf{b})$ is the function to be minimized (objective function), $h_i(\mathbf{b}), i = 1, \ldots, l$ are equality constraints and $g_j(\mathbf{b}), j = 1, \ldots, m$, are inequality constraints. The last set of inequalities defines the so-called *side* or *lateral constraints*, since they directly limit the extreme values of the design variables. The objective function and constraints are assumed to be continuous and twice differentiable in R^n. They can be either linear or nonlinear and might be either explicitly or implicitly dependent on $\mathbf{b}$. The region of R^n in which all the constraints are satisfied is called *feasible region of the design space*.

The optimization problem posed by Eq. (5.25) is a particular case of Eq. (5.26), in which only lateral constraints are imposed. This class of problems are nonconvex and nonlinear, demanding an appropriate strategy for a successful solution. As nonconvexity ensures the presence of multiple local minima, a mixed approach combining GAs and gradient-based procedures is proposed. The former aims to skip local minima, whereas the latter intends to refine a coarse solution in the neighborhood of the global minimum. A review of the characteristics of each of these strategies is presented in the next sections.

5.4.1 Genetic Algorithms

Traditional gradient-based optimization algorithms start from a given point in the design space and change this point in order to improve the design. This improvement is based on a descent direction, which is obtained by the derivative information in the vicinity of the point [51]. As the derivatives provide always a measure of the local behavior, this procedure cannot guarantee convergence to a global minimum. Thus, if the objective function is nonconvex (Figure 5.13), gradient-based methods starting from different points could lead to different local minima. For instance, an optimization starting from point ($x = 2.0, y = 1.6$) would converge to the closer minimum, which is located far from the global minimum (Figure 5.14). In order to avoid convergence to local minima and also to overcome the nondifferentiability of some problems, it is advisable to use the so-called evolutionary algorithms (EAs).

EAs are a class of iterative generic population-based stochastic optimization methods that uses concepts and mechanisms based on biological evolution [52]. In the EAs, each candidate solution plays the role of an *individual* (a point in the design space) taken from a *population*, and then some individuals in the population are selected to build next *generations*. *Selection* and *evolution* of the population takes place after recursive application of some operators that mimic biological processes. Such algorithms require neither derivative information nor further knowledge about the functions involved, making them prone to deal with discontinuous and/or nonconvex problems. Additionally, the intrinsic parallel nature of EAs and the possibility to provide a number of potential solutions to a given problem make them well suited to solve multiobjective optimization and scheduling problems. In

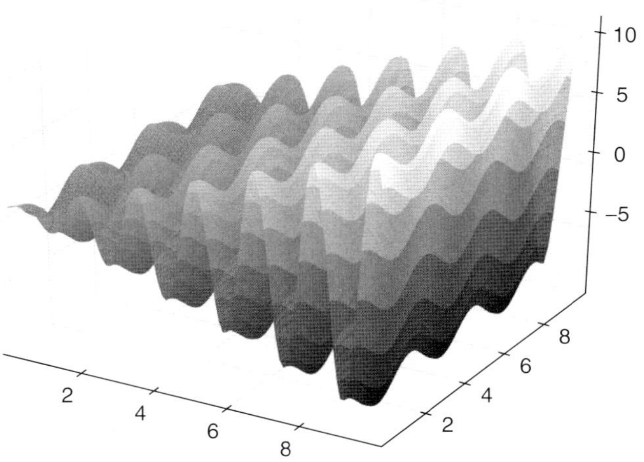

Figure 5.13 Nonconvex function $z\,(x, y) = x\,\sin(4x) + 1.1\,\cos(2y)$.

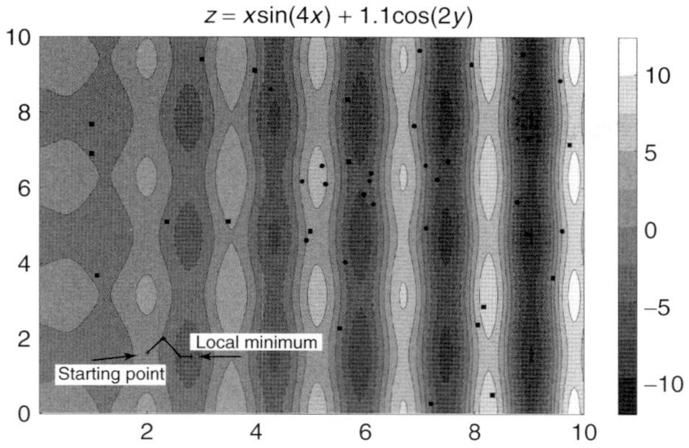

Figure 5.14 Minimization of the function shown in Figure 5.13. Gradient-based method (left, lower corner), starting from ($x = 2.0, y = 1.6$), and genetic algorithms, with the 1st population (■), 5th population (•), and 10th population (∗). The GA sampled a point *close* to the global minimum, whereas the gradient-based algorithm converged to a local minimum.

this section, a special class of EAs known as *genetic algorithms* is detailed. Firstly, a general explanation of the method is given, followed by some details of the implementation considered in this work.

5.4.1.1 **Formulation**

A general optimization problem aims at finding the minimum value of a given objective function $f : S_f \rightarrow R$, where S_f is the set of all feasible points. Each point in the space of feasible solutions is called a *chromosome* or an *individual* and each chromosome is formed by *genes* (design variables). Each gene is usually represented by real (base 10) or binary numbers (base 2), although other representations are also possible. A given collection of individuals is called a *population*, representing a set of possible solutions to the given problem. In this text, it is considered that the population has a fixed size μ_g. Figure 5.15 depicts the tasks employed in the GA.

First, a random set of individuals is generated to account for the first generation ($t = 0$). Considering a given population $\mathbf{P}^t$, some individuals are selected according to a predefined measure of quality (for example, a comparison of their objective function value to other individuals of the same population). The selected individuals (*mates*) are stored in a subset of the current population, $\mathbf{R}^t$ with size $\lambda_g < \mu_g$, also known as the *mating pool* (MP) (Figure 5.16).

Depending on a given probability p_c, pairs of individuals in $\mathbf{R}^t$ are recombined to generate two new individuals (*offspring*), in a process known as *crossover* (Figure 5.17). The new individuals form a new set of candidate solutions $\mathbf{C}^t$, which can be modified by a further operator, known as *mutation* (Figure 5.18).

Mutation changes the value of some genes in the population, generating a new set of individuals $\mathbf{M}^t$ according to a predefined probability. After these procedures, a new generation $\mathbf{P}^{t+1}$ is built by selecting individuals from $\mathbf{P}^t$ and $\mathbf{M}^t$, in a

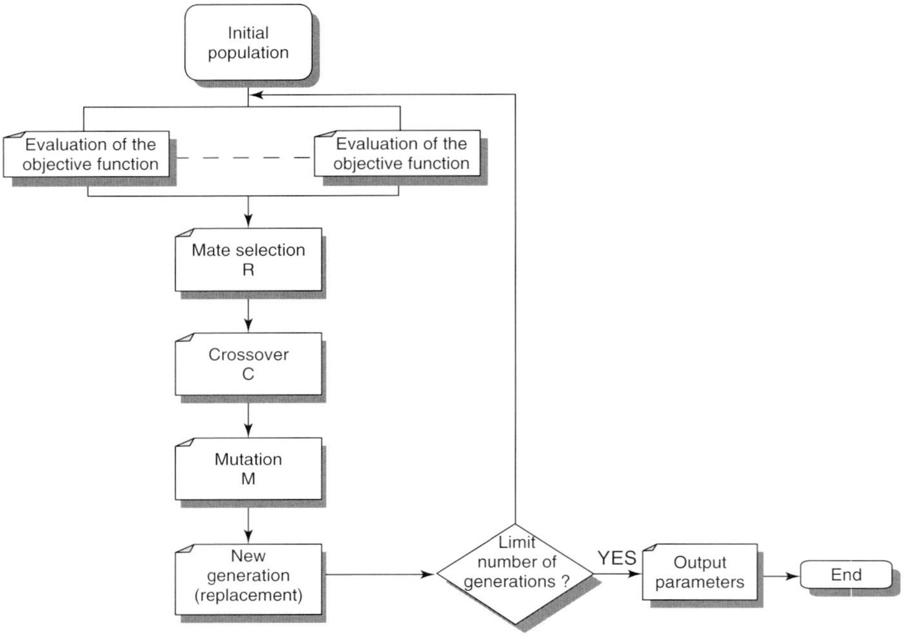

Figure 5.15 Overview of the genetic algorithm.

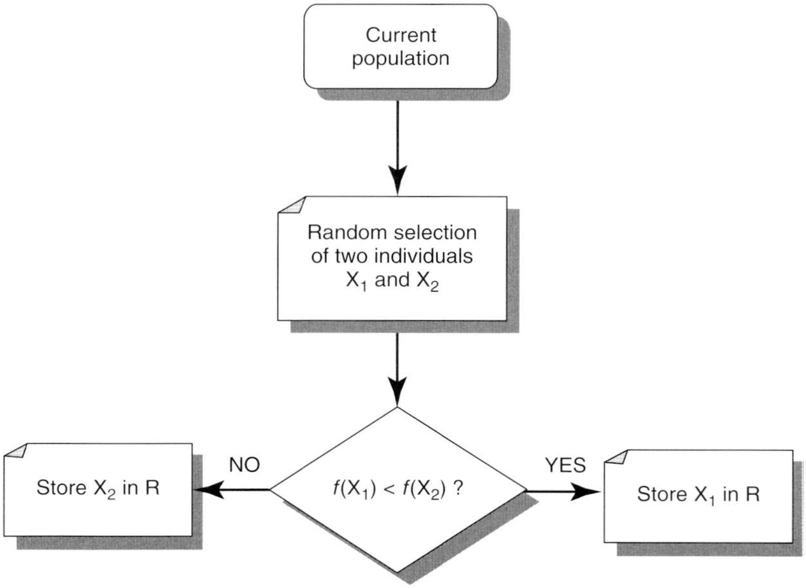

Figure 5.16 Mating pool stage.

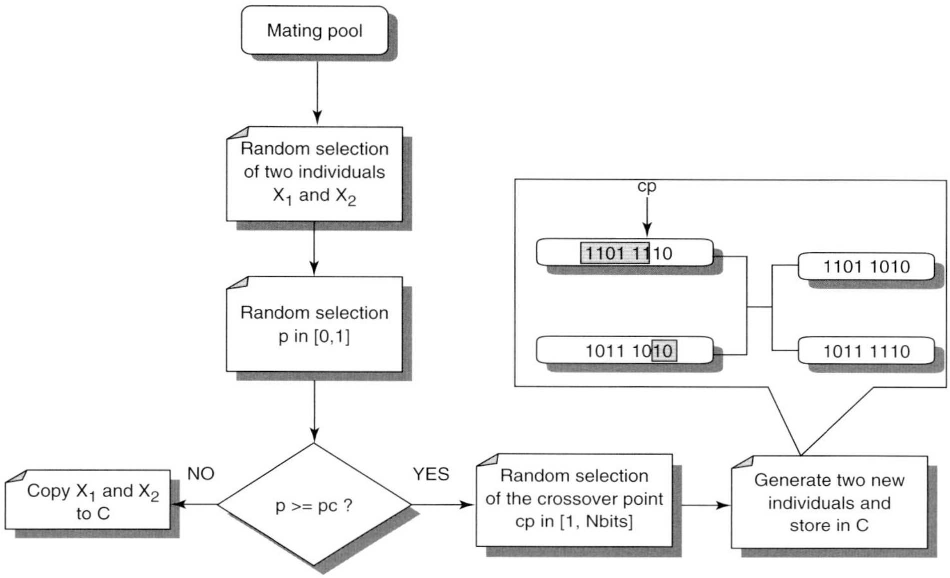

Figure 5.17 Crossover stage.

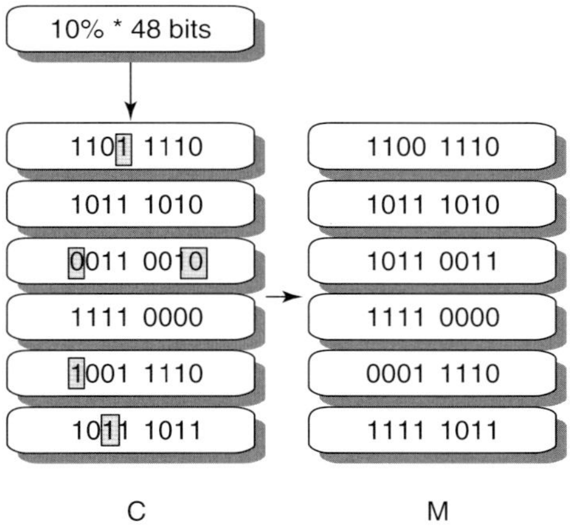

Figure 5.18 Mutation stage.

process known as *replacement*. This process continues until a predefined number of generations is reached or a given criteria is satisfied.

This procedure is known to lead to global optimum in various problems, despite lacking a solid mathematical background. When dealing with GAs, the most accepted explanation for the behavior of the method is known as the *building-block hypothesis* [52]. In this hypothesis, individuals with common parts of their chromosomes, for instance, 1**0** (where the symbol * can be 0 or 1), share a particular schema template. Instead of trying every conceivable combination of bits, the GA combines (by means of crossover and mutation) individuals with better objective functions and common schemas. Thus, simple schemas are tested and recombined at each generation in order to build an optimal (and complete) chromosome. Those simple schemas are called *building blocks*. The main problem with the building-block hypothesis is the absence of mathematical proof of convergence. In fact, it is difficult to ensure the convergence of a GA, unless all possible combinations have been tested. Arabas [53] studied the convergence of floating-point-based EA without crossover. In his work, it is established that a generational (iterative) selection and mutation-based EA will generate at least one point arbitrarily close to the global solution. A more detailed discussion on the implementation of GAs is given by Goldberg [52] and careful study on the mathematical foundations of the GAs is provided by Burjorjee [54].

5.4.1.2 Implementation

In this work, a simple and efficient GA was implemented (Figure 5.15). The initial population is obtained by a homogeneous Gaussian distribution and has a fixed size μ_g. At each generation, the mating pool, **MP**, with $\mu_g/2$ pairs is selected by means of a *binary tournament* and those individuals have a probability p_c of being combined using one-point crossover. Binary tournament means that two individuals from the **MP** are randomly chosen and the fitter of the two is selected as a parent. This procedure is performed twice (in order to select two parents) for each operation of crossover.

The crossover is followed by the mutation of the mating pool according to a given mutation probability, m_p. In sequence, $\mathbf{P}^t$ and $\mathbf{M}^t$ are sorted and a given number of best individuals of both sets are chosen to build the new generation (*elitism*). As some individuals are kept among generations, a database is used to store the chromosome and the *fitness* of the best individuals at each generation. In addition, as the evaluation of the objective function of one individual does not depend on the evaluation of another individual in the same generation, it is recommended to implement this procedure by making use of multiple processors/computers. This particular implementation used the *multiprocessing* python library to achieve this goal, so that the fitness evaluation could be performed by all computers available in the local network. The examples in this text were evaluated using 16 processors, allowing simultaneous computation of 16 individuals.

One key point when using GAs is the termination criterion. In this work, a fixed number of generations (100) was adopted, but when using GA to obtain material parameters, one can also visually compare the experimental and the numerical curves to assess convergence.

5.4.2 Gradient-Based Methods

5.4.2.1 General Procedure

The philosophy underneath most of gradient-based optimization methods is to produce a sequence of improved approximations to the optimum according to the following scheme [51, 55, 56].

1) Start with an initial trial point $\mathbf{b}_1$.
2) Find a suitable direction $\mathbf{Z}_i$ ($i = 1$ in the beginning of the process) which points toward the general direction of the optimum.
3) Find an appropriate step length λ_i^* to move along the direction $\mathbf{Z}_i$.
4) Obtain the new approximation $\mathbf{b}_{i+1}$ as

$$\mathbf{b}_{i+1} = \mathbf{b}_i + \lambda_i^* \mathbf{Z}_i \tag{5.27}$$

5) Test the optimality condition with respect to $\mathbf{b}_{i+1}$. If $\mathbf{b}_{i+1}$ is optimum, stop the procedure. Otherwise, set a new $i = i + 1$ and repeat step (2) onward.

Sometimes the line search defined by steps (3)–(4) is replaced by an alternative update scheme for the design variables. Furthermore, the iterative procedure summarized by Eq. (5.27) is valid for both unconstrained and constrained optimization problems.

In this work, two gradient-based optimization algorithms were employed: sequential linear programming (SLP) and globally convergent method of moving asymptotes (GCMMA). In SLP, there is no line search and Eq. (5.27) is particularized by assuming $\lambda_i^* = 1$. In the GCMMA, different schemes for updating the design variables have been proposed. Although there are versions that use the scheme given by Eq. (5.27), the following sections present a particular implementation using a different approach.

5.4.2.2 Sequential Linear Programming (SLP)

SLP is given by the recursive application of an optimization algorithm in which, at each iteration, both the objective function and the constraints are linearized with respect to the design variables. Thus, in the qth design point, the optimization problem posed in Eq. (5.26) is replaced by

$$\begin{aligned}
&\text{Minimize} && g_0(\mathbf{b}^q) + \left.\frac{\partial g_0(\mathbf{b})}{\partial \mathbf{b}}\right|_q (\mathbf{b} - \mathbf{b}^q) = 0 \quad \mathbf{b} \in R^n \\
&\text{Such that} && h_i(\mathbf{b}^q) + \left.\frac{\partial h_i(\mathbf{b})}{\partial \mathbf{b}}\right|_q (\mathbf{b} - \mathbf{b}^q) = 0 \quad i = 1 \ldots l \\
& && g_j(\mathbf{b}^q) + \left.\frac{\partial g_j(\mathbf{b})}{\partial \mathbf{b}}\right| (\mathbf{b} - \mathbf{b}^q) \leq 0 \quad j = 1 \ldots m \\
& && \underline{b}_k \leq b_k \leq \overline{b}_k \quad k = 1 \ldots n
\end{aligned} \tag{5.28}$$

Care must be taken because the linearization is valid only in a certain neighborhood of the origin. In order to avoid large errors and to improve conservativeness, artificial *lateral* constraints on the design variables are added to the optimization problem. These constraints are called *moving limits*, since they are updated at each optimization iteration step according to the nonlinearity level.

5.4.2.3 Globally Convergent Method of Moving Asymptotes (GCMMA)

The original algorithm of the method of moving asymptote (MMA) Svanberg [57] is based on local function approximations as

$$\tilde{g}(\mathbf{b}) = g(\mathbf{b}^q) + \sum_{+} \frac{\partial g(\mathbf{b}^q)}{\partial z_k}(z_k - z_k^q) + \sum_{-} \frac{\partial g(\mathbf{b}^q)}{\partial y_k}(y_k - y_k^q) \tag{5.29}$$

where z_k and y_k are intermediate variables given by $z_k = 1/(U_k - b_k)$ and $y_k = 1/(b_k - L_k)$, and U_k and L_k are lower and upper asymptotes which define the region of local approximation. After some manipulations, Eq. (5.26) can be rewritten as

$$\tilde{g}(\mathbf{b}) = g(\mathbf{b}^q) + \sum_{+} p_k^q \left(\frac{1}{U_k - b_k} - \frac{1}{U_k^q - b_k^q} \right) + \sum_{-} q_k^q \left(\frac{1}{b_k - L_k} - \frac{1}{b_k^q - L_k^q} \right) \tag{5.30}$$

where

$$p_k^q = \left(U_k^q - b_k^q\right)^2 \frac{\partial g\left(\mathbf{b}^q\right)}{\partial b_k} \quad \text{and} \quad q_k^q = \left(b_k^q - L_k^q\right)^2 \frac{\partial g\left(\mathbf{b}^q\right)}{\partial b_k} \tag{5.31}$$

and the symbols $\sum_{+}$ and $\sum_{-}$ collect the linearization terms with positive and negative derivatives, respectively.

The inequality $L_k^q < b_k^q < U_k^q$ must hold in order to ensure convexity of the approximations and to avoid a null denominator. Since this strategy is not globally convergent, Zillober [58] and Svanberg [59–61] presented modified versions with global convergence characteristics. Zillober [58] introduced a line-search strategy and Svanberg [59] proposed an algorithm consisting of a different manner to calculate coefficients p_k and q_k in Eq. (5.31).

For an approximation around point $\mathbf{b}^q$, one has

$$\text{if} \quad \frac{\partial g(\mathbf{b}^q)}{\partial b_k} > 0, \quad \begin{cases} p_k^q = \left(U_k^q - b_k^q\right)^2 \left(\frac{\partial g(\mathbf{b}^q)}{\partial b_k} + \frac{\rho^q}{2} \left(U_k^q - L_k^q\right) \right) \\ q_k^q = \left(b_k^q - L_k^q\right)^2 \frac{\rho^q}{2} \left(U_k^q - L_k^q\right) \end{cases} \tag{5.32}$$

$$\text{if} \quad \frac{\partial g(\mathbf{b}^q)}{\partial b_k} < 0, \quad \begin{cases} p_k^q = \left(U_k^q - b_k^q\right)^2 \frac{\rho^q}{2} \left(U_k^q - L_k^q\right) \\ q_k^q = \left(b_k^q - L_k^q\right)^2 \left(-\frac{\partial g(\mathbf{b}^q)}{\partial b_k} + \frac{\rho^q}{2} \left(U_k^q - L_k^q\right) \right) \end{cases} \tag{5.33}$$

where ρ^q are strictly positive parameters (in order to ensure convexity of the approximation), which are used together with the asymptotes L_k^q and U_k^q. Note that there is one value of ρ^q for each of the approximated functions (objective or constraint). This modification results in the introduction of a second asymptote, creating an approximation with a global minimum (see Figure 5.19).

After defining the approximations, the key point is a heuristic rule for updating the value of ρ^q. The original scheme [59] gave place to a more efficient strategy proposed by Svanberg [60, 61]. In the latter, the algorithm for solving the optimization subproblem consists of two nested iterative loops. In the external loop, the convex approximation is defined using the values of the objective function, constraints, and their derivatives on the design point. In the internal loop, only the parameter ρ_j is adjusted, making the approximation increasingly convex until $\tilde{g}_j(\mathbf{b}^k) \geq g_j(\mathbf{b}^k)$. The asymptotes L_k^q and U_k^q must be updated at each qth external iteration. Svanberg proposes

$$L_k^q = b_k^q - s_k \left(b_k^{q-1} - L_k^{q-1} \right) \tag{5.34}$$

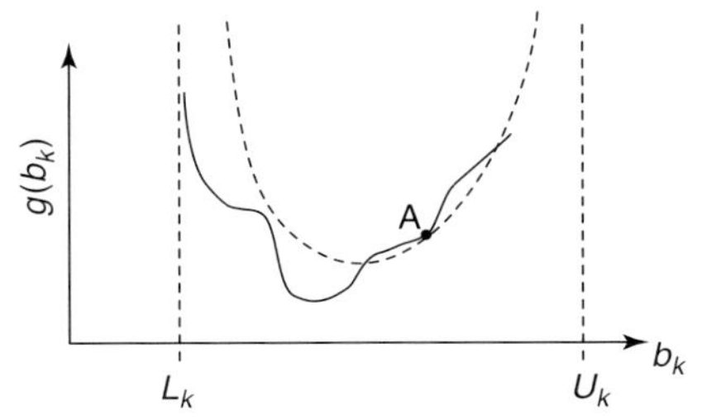

Figure 5.19 Convex approximation using GCMMA.

and

$$U_k^q = b_k^q + s_k \left(U_k^{q-1} - b_k^{q-1} \right) \tag{5.35}$$

where s_k is prescribed heuristically depending on the values of b_k^q, b_k^{q-1}, and b_k^{q-2}, and on the iteration number.

5.5
Sensitivity Analysis

In nonlinear problems such as those involving large strain elastoplasticity with frictional contact and damage, computation of each displacement increment depends on the constitutive model and the stress integration algorithm employed. This dependence directly affects displacements, stresses, and internal variables history. Therefore, in order to perform the displacement sensitivity analysis at the nth load increment, it is necessary to differentiate displacements, stresses, and internal variables at the load increment $n-1$.

There are three main approaches to sensitivity analysis in nonlinear problems. (i) the *analytical method* is efficient and accurate but frequently limited to particular cases. The analytical sensitivity analysis may clearly be difficult when nondifferentiable functions are involved and when complex constitutive laws are considered, especially in large strain situations. The consideration of a damage model falls within the latter case. (ii) The *finite difference method* is very simple and general, but usually requires a much higher computational cost. Moreover, in opposition to analytical methods, the finite difference method does not provide accurate results in large strain situations that require remeshing [62, 63]. (iii) The *semianalytical approach* is computationally cheaper but still suffers the remeshing limitation (Figure 5.20).

This section reviews a modified finite difference sensitivity method presented by Muñoz-Rojas *et al.* [64] for applications to path-dependent nonlinear problems. The method considerably reduces the computational time usually associated to finite differences and successfully handles the limitations regarding remeshing. Moreover, a particularization of the procedure recovers the semianalytical approach developed by Hisada [65].

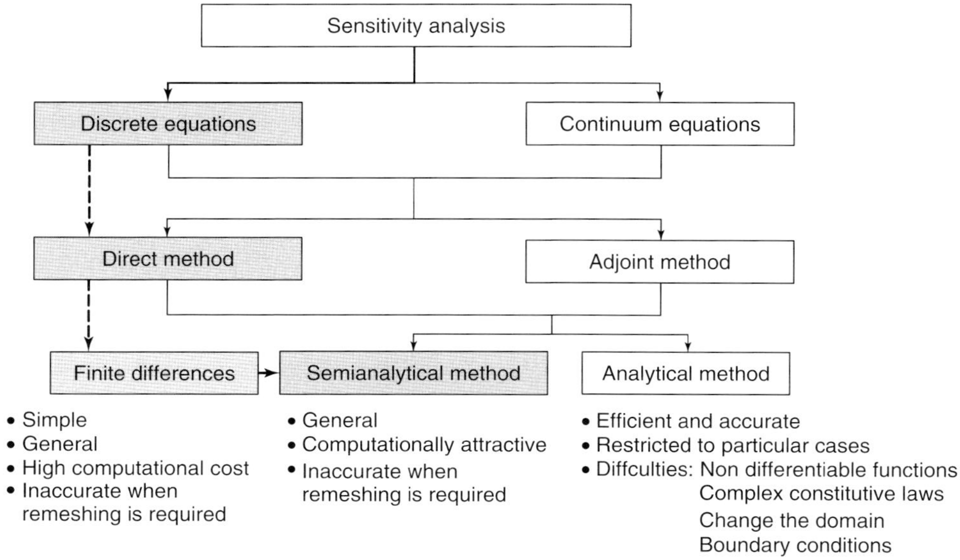

Figure 5.20 Alternatives for sensitivity evaluation.

5.5.1
Modified Finite Differences and the Semianalytical Method

Despite its apparent simplicity, the tensile test involves complex physical phenomena such as large strain, elastoplasticity, and damage, thereby requiring a demanding computational analysis. The numerical simulations uses an incremental hypoelastic finite element code based on a Lagrangian description of motion and additive decomposition of the strain tensor [66]. The stress integration algorithm uses the combination of the method of instantaneous rotation [67] and radial return.

The discretized form of the nonlinear equilibrium equations is given by

$$\mathbf{F}(\mathbf{u}^n, \mathbf{U}^{n-1}, \mathbf{b}) = \mu^n \mathbf{Q}(\mathbf{b}) \tag{5.36}$$

$$\mathbf{U}^n = \sum_{i=1}^{n} \mathbf{u}^i \tag{5.37}$$

where the upper index n stands for the nth load step; $\mathbf{F}$ is the vector of internal forces, which is dependent on the displacement increment vector $\mathbf{u}^n$, on the total displacement vector $\mathbf{U}^{n-1}$, and on the design variables vector $\mathbf{b}$; $\mathbf{Q}$ is the vector of applied loads; and μ^n is an amplitude parameter. By using the Newton–Raphson method, Eq. (5.36) yields

$$\mathbf{F}(\mathbf{u}^n_{\omega+1}, \mathbf{U}^{n-1}, \mathbf{b}) \approx \mathbf{F}(\mathbf{u}^n_{\omega}, \mathbf{U}^{n-1}, \mathbf{b}) + \mathbf{K}_T(\mathbf{u}^n_{\omega+1} - \mathbf{u}^n_{\omega}) = \mu^n \mathbf{Q}(\mathbf{b}) \tag{5.38}$$

$$\mathbf{K}_T = \frac{\partial \mathbf{F}(\mathbf{u}^n_{\omega})}{\partial \mathbf{u}^n} \tag{5.39}$$

where $\mathbf{K}_T$ is the tangent matrix, and, when $\omega = 0$, the term $\mathbf{u}_0^n$ corresponds to the initial estimate.

Equation (5.38) is rearranged so that

$$\mathbf{u}_{\omega+1}^n = \mathbf{u}_\omega^n + \mathbf{K}_T^{-1}[\mu^n \mathbf{Q}(\mathbf{b}) - \mathbf{F}(\mathbf{u}_\omega^n, \mathbf{U}^{n-1}, \mathbf{b})] = \mathbf{u}_\omega^n + \mathbf{K}_T^{-1}\mathbf{R}_\omega^n \tag{5.40}$$

in which $\mathbf{R}_\omega^n$ is the residual of the equilibrium equations in the nth step and ωth iteration.

The problem stated by Eq. (5.36) is slightly modified by a perturbation δb_k on the kth design variable, so that the corresponding perturbed equilibrium equations and their solution are

$$\mathbf{F}(\mathbf{u}_\Delta^n, \mathbf{U}_\Delta^{n-1}, \mathbf{b} + \delta\mathbf{b}_k) = \mu^n \mathbf{Q}(\mathbf{b} + \delta\mathbf{b}_k) \tag{5.41}$$

$$\mathbf{U}_\Delta^n = \sum_{i=1}^{n} \mathbf{u}_\Delta^i \tag{5.42}$$

where $\delta\mathbf{b}_k = \begin{bmatrix} 0 & \ldots & \delta b_k & \cdots & 0 \end{bmatrix}$.

The traditional forward finite differences would estimate the displacement sensitivity by simply performing the operation

$$\frac{\mathrm{d}\mathbf{U}^n}{\mathrm{d}b_k} \approx \frac{\mathbf{U}_\Delta^n - \mathbf{U}^n}{\delta b_k} \tag{5.43}$$

However, this would require the solution of a complete independent analysis for each design variable, which is prohibitively expensive (even worse for central differences). A modified finite difference alternative is detailed in the following paragraphs.

The analysis begins with a separate set of data for the unperturbed and perturbed problems. The load history is divided into *nine* increments, and the procedure starts with the search of equilibrium for the first load (solution) increment of the unperturbed problem. Once this condition is achieved within the tolerance imposed for the iterative solution method, all the nodal and Gauss-point information is stored in a binary file. In sequence, the data of the first perturbed problem is loaded and a new analysis takes place. Noticeably, the converged configuration of the unperturbed problem is taken as the first solution estimate of the perturbed problem. Moreover, the tangent matrix of the unperturbed problem is used in a modified Newton–Raphson scheme, increasing the convergence rate at low cost. The same tolerance is imposed to both unperturbed and perturbed problems. After convergence is reached, the information is stored in a binary file. This procedure is repeated for every design variable. In each load step, after all the perturbed problems have converged, the sensitivity obtained in the increment is determined by finite differences and the total sensitivity up to the step is given by Eqs (5.37), (5.42), and (5.43). At the end of the cycle, the procedure is repeated for each solution increment until the last step.

Hence, the perturbed incremental displacement is given by the iterative solution of

$$\mathbf{u}_{\Delta k,\omega+1}^n = \mathbf{u}_{\Delta k,\omega}^n + \mathbf{K}_T^{-1}\ \mathbf{R}_\omega^n(\mathbf{u}_{\Delta k,\omega}^n;\ \mathbf{U}_{\Delta k}^{n-1};\ \mathbf{b} + \delta\mathbf{b}_k);\ \mathbf{u}_{\Delta k,0}^n = \mathbf{u}^n \tag{5.44}$$

Haftka [68] demonstrated that large errors might be obtained when adopting the solution of the original problem as an initial estimate for the corresponding perturbed problem. This happens because the convergence of the original problem is reached within a certain tolerance, admitting a nonzero residual. Therefore, when $u^n_{\Delta k,0} = u^n$ is used as the initial estimate for the perturbed problem, two situations compete: the effect of the perturbation and the enhancement of the initial estimate (a further decrease of the aforementioned residual). This shortcoming can be avoided by restating Eq. (5.44) as

$$\begin{aligned}\mathbf{u}^n_{\Delta k,\omega+1} = \mathbf{u}^n_{\Delta k,\omega} + \mathbf{K}_T^{-1}[\mathbf{R}^n_\omega(\mathbf{u}^n_{\Delta k,\omega}; \mathbf{U}^{n-1}_{\Delta k}; \mathbf{b} + \delta\mathbf{b}_k) \\ -\mathbf{R}^n(\mathbf{u}^n, \mathbf{U}^{n-1}, \mathbf{b})]; \mathbf{u}^n_{\Delta k,0} = \mathbf{u}^n\end{aligned} \tag{5.45}$$

Convergence of Eqs (5.40) and (5.45) is achieved iteratively so that

$$\begin{aligned}&\frac{\left|\mathbf{R}^n_\omega\left(\mathbf{u}^n_\omega, \mathbf{U}^{n-1}, \mathbf{b}\right)\right|}{\mu^n \mathbf{Q}(\mathbf{b})} \le TOL \quad \text{and} \\ &\frac{\left|\mathbf{R}^n_\omega\left(\mathbf{u}^n_{\Delta k,\omega}; \mathbf{U}^{n-1}_{\Delta k}; \mathbf{b} + \delta\mathbf{b}_k\right) - \mathbf{R}^n\left(\mathbf{u}^n, \mathbf{U}^{n-1}, \mathbf{b}\right)\right|}{\mu^n \mathbf{Q}(\mathbf{b})} \le TOL\end{aligned} \tag{5.46}$$

The total unperturbed and perturbed displacements are then evaluated by Eqs (5.37) and (5.42), and the displacement sensitivity can be approximated by Eq. (5.43).

Sensitivity analysis using central differences is analogous to the procedure described previously. In this case, however, the displacement sensitivity with respect to the parameters is

$$\frac{\Delta \mathbf{U}^n}{\Delta b_k} = \frac{\mathbf{U}^n_{\Delta+} - \mathbf{U}^n_{\Delta-}}{2\delta b_k} \tag{5.47}$$

This method has shown enhanced efficiency when compared to traditional finite differences and admits an extension to handle remeshing situations [64], which, however, will not be detailed in the present contribution. It is worth noting that, if Eq. (5.45) is not iterated to satisfy a prescribed tolerance, but rather computing just *one Newton–Raphson iteration step*, the method approaches Hisada's semianalytical method [65]. On the other hand, if *central differences* are adopted, the additional term involving the residual of the original problem is canceled (Eq. (5.46)). In this case, and if no remeshing is required, Kleinermann's [69] formulation is recovered.

The iterative procedure represented in Eq. (5.45) is interpreted graphically in Figure 5.21. It shows a modified Newton–Raphson method applied to the curve $\mathbf{F}(\mathbf{b} + \delta\mathbf{b})$ using the unperturbed tangent matrix $\mathbf{K}$ and initial estimate $\mathbf{U}^n$, approaching point 2 at each load increment. It can be seen that, if only one iteration is performed, Hisada's semianalytical method is obtained, approximating the solution in point 2 by point 1. Hisada, Chen, and coworkers [70, 71] used this method for a wide range of applications. Provided the perturbation is sufficiently small, the unbalance error was considered acceptable. The present work adopts this strategy.

The sensitivity of the objective function (5.25) is given by

$$\frac{\mathrm{d}S(\mathbf{b})}{\mathrm{d}b_k} = p_1 \frac{1}{W_0}\frac{\mathrm{d}W}{\mathrm{d}b_k} + p_2 \frac{1}{T_0}\frac{\mathrm{d}T}{\mathrm{d}b_k} \tag{5.48}$$

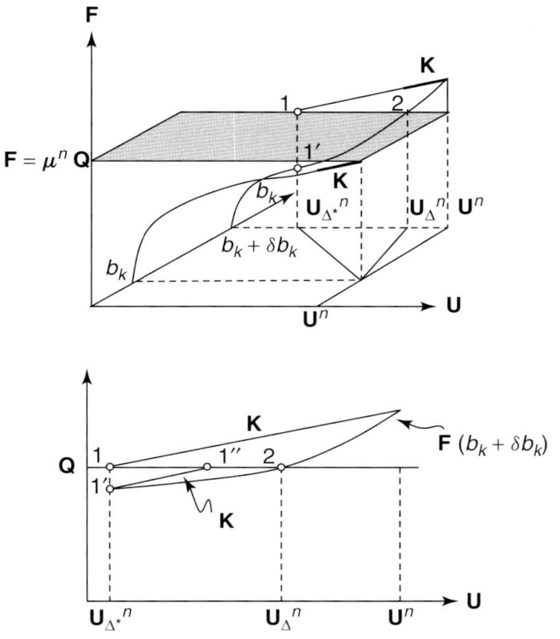

Figure 5.21 Structural response with respect to parameter b_k (a) and iterative refinement using the tangent matrix of the unperturbed problem (b).

where

$$\frac{\mathrm{d}W}{\mathrm{d}b_k} = \frac{1}{2}\left(\frac{1}{N}\sum_{i=1}^{N} w_i\left(\frac{F_{0,i}^{FE}(b_j) - F_i^{\exp}}{F_i^{\exp}}\right)^2\right)^{-1/2}$$
$$\frac{1}{N}\sum_{i=1}^{N} 2w_i\left(\frac{F_{0,i}^{FE}(b_j) - F_i^{\exp}}{F_i^{\exp}}\right)\frac{1}{F_i^{\exp}}\frac{dF_{0,i}^{FE}}{db_k} \tag{5.49}$$

and

$$\frac{\mathrm{d}T}{\mathrm{d}b_k} = \mathrm{sign}\left(\frac{d^{\mathrm{FE}} - d^{\exp}}{d^{\exp}}\right)\left(\frac{1}{d^{\exp}}\frac{\mathrm{d}d^{\mathrm{FE}}}{\mathrm{d}b_k}\right) \tag{5.50}$$

Hence, in the case of the present application, the sensitivity of the reactions and final necking diameter are evaluated, respectively, as

$$\frac{\mathrm{d}\mathbf{F}^n}{\mathrm{d}b_k} \approx \frac{\Delta\mathbf{F}^n}{\Delta b_k} = \frac{\mathbf{F}_\Delta^n - \mathbf{F}^n}{\delta b_k} \tag{5.51}$$

and

$$\frac{\mathrm{d}(d^{FE})^n}{\mathrm{d}b_k} \approx \frac{\Delta(d^{\mathrm{FE}})^n}{\Delta b_k} = \frac{(d^{\mathrm{FE}})_\Delta^n - (d^{\mathrm{FE}})^n}{\delta b_k} \tag{5.52}$$

where $\mathbf{F}_\Delta^n$ and $\mathbf{F}^n$ are the reactions of the perturbed and unperturbed problems at the nth load step, $\left(d^{\mathrm{FE}}\right)_\Delta^n$ and $\left(d^{FE}\right)^n$ are the specimen diameters in the necking section at the nth load step and the upper index FE has been omitted for notation consistency with the equations throughout the work.

5.6
A Mixed Optimization Approach

The key point of the present approach is the combination of a GA with gradient-based procedures. After a given number of generations in the genetic stage, the solution is expected to be in the neighborhood of the global minimum. The result is then introduced as initial estimate in a gradient-based optimization procedure aiming at refining the solution. This strategy is depicted in Figure 5.22.

5.7
Examples of Application

In this section, the mixed optimization procedure described in Sections 5.4–5.6 is applied to identify material parameters of low-carbon steel and aluminum alloys. The procedure is based on tensile tests of cylindrical specimens. Owing to symmetry, the finite element simulations were performed using one-fourth reduced models. Following Cunda [72], linear axisymmetric quadrilateral elements were adopted with one integration point for both deviatoric and volumetric parts. Owing to the low acquisition rate and the global nature of measured parameters, measurement noise was not accounted for. In both examples, the Poisson coefficient was assumed constant and equal to 0.3.

5.7.1
Low Carbon Steel at 25 °C

This example summarizes the results of parameter identification based on a tensile test of low carbon steel at room temperature [73]. The tensile testing specimens were machined according to the standard ABNT NBR ISO 6892 [74] (similar to the ASTM E 8M [75] with 10 mm of nominal diameter at its central section). A universal testing machine EMIC DL30000 coupled to an automatic data-acquisition system was also used. The load–displacement curve was measured using an extensometer with a 50 mm initial gauge length and a crosshead speed of 1.0 mm min^{-1}. Measurements were performed during the whole deformation process up to the final fractured state, as illustrated in Figure 5.23. The final diameter immediately prior to the fracture was also included in the optimization processes in order to overcome the nonconvexity typical of such problems. This effect is accounted for by the second term of Eq. (5.25). In addition, weight parameters of Eq. (5.25) used in the simulations are $p_1 = p_2 = 1$.

The specimen was described using axisymmetry so that one-fourth of the geometry was modeled. The finite element mesh used in the simulations has 200 elements and 231 nodes. The mesh topology and its spatial gradation were adapted from Souza Neto *et al.* [76]. The geometry of the specimen and corresponding mesh are presented in Figures 5.23 and 5.24.

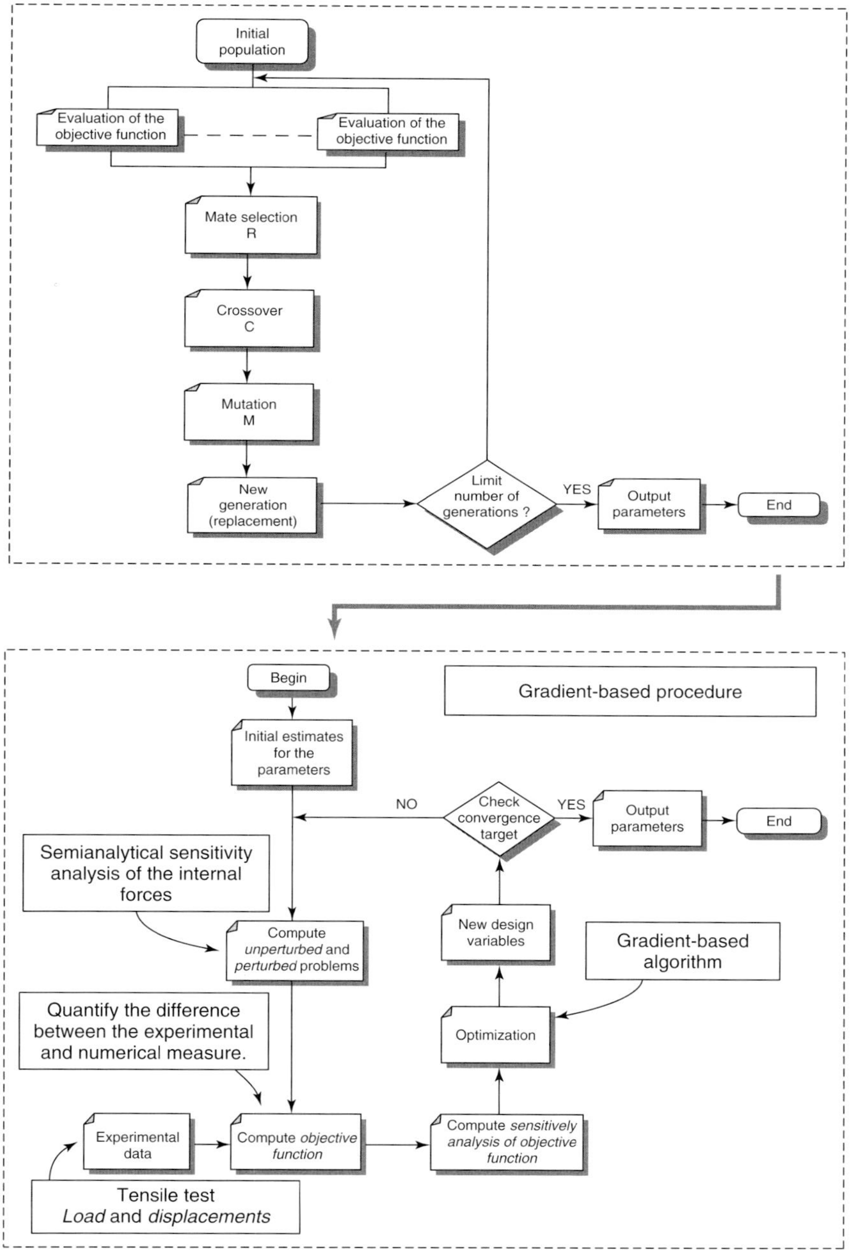

Figure 5.22 Mixed optimization strategy.

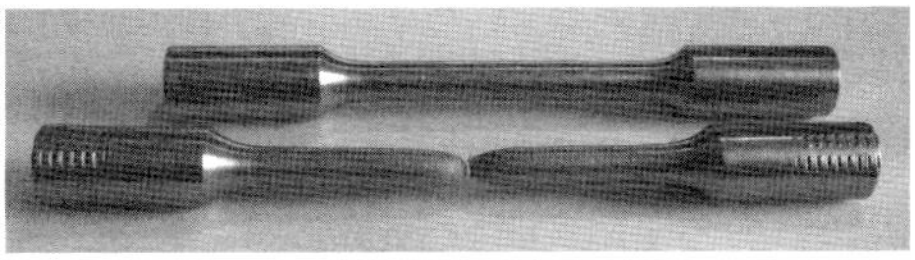

Figure 5.23 Tensile testing specimens.

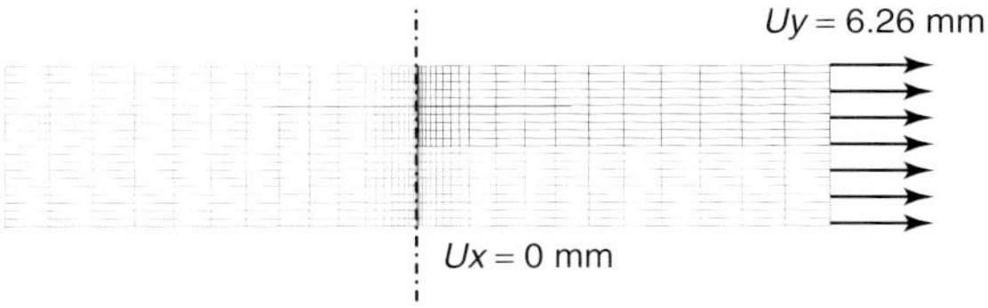

Figure 5.24 Finite element mesh.

In the case of the present low-alloy steel, numerical experiments have shown that some parameters impose a small effect on the global response. Therefore, only the parameters representing the void fraction formation, f_N, standard deviation for the void creation, s_N, the equivalent plastic deformation around which microvoids are nucleated, ε_N, and the critical void fraction for coalescence, f_C, were determined. In addition to the damage parameters, this example uses a yield stress curve given by

$$\sigma_y = \sigma_0 + (\sigma_\infty - \sigma_0)[1 - \exp(-\delta\varepsilon^p)] + \gamma\ \varepsilon^p \tag{5.53}$$

where $\sigma_0, \sigma_\infty, \delta$, and γ are the material parameters. This curve can be obtained from Eq. (5.21) by considering nonzero values for the parameters P_1, P_2, P_3, and P_8. The yield stress and the damage parameters, f_N, s_N, ε_N, and f_C, are adopted as design variables of this problem. The remaining damage parameters f_U, f_{lim}, f_0, and $\Delta\varepsilon$ are assumed to have fixed values of 0.666, 0.99, 0.00, and 0.50, respectively. Therefore, in this case, there are a total of eight design variables.

A problem observed in this simulation is related to the fact that the GA generates some individuals with $f_C < f_N$, which are not physically viable. In this case, a high fitness value is associated with those individuals in order to prevent the appearance of such cases in future generations. The final experimental neck radius was 2.75 mm. The search for the optimized material parameters using the GA approach was applied to 100 generations, leading to a reasonable match to the experimental stress–strain curve, while satisfying the radius constraint (final radius = 2.7423 mm). The deformed mesh superimposed to the deformed specimen is displayed in Figure 5.25. The convergence of the objective function is presented in Figure 5.26 and the material parameters obtained are shown in Tables 5.1 and 5.2. The Young modulus, $E = 210025.30$ MPa, and the parameter $P_1 = 340.0$ MPa, were assumed constant in the simulations.

In this example, the gradient-based GCMMA optimization scheme was used to refine the solution given by the GA. In this stage, the following relaxations were adopted: only the elastic–plastic parameters were allowed to change, the final radius constraint was not introduced, and the stress–strain matching between

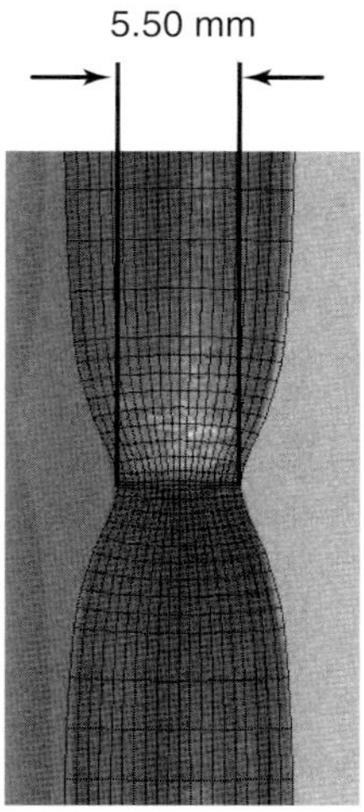

Figure 5.25 Necking and deformed FE mesh.

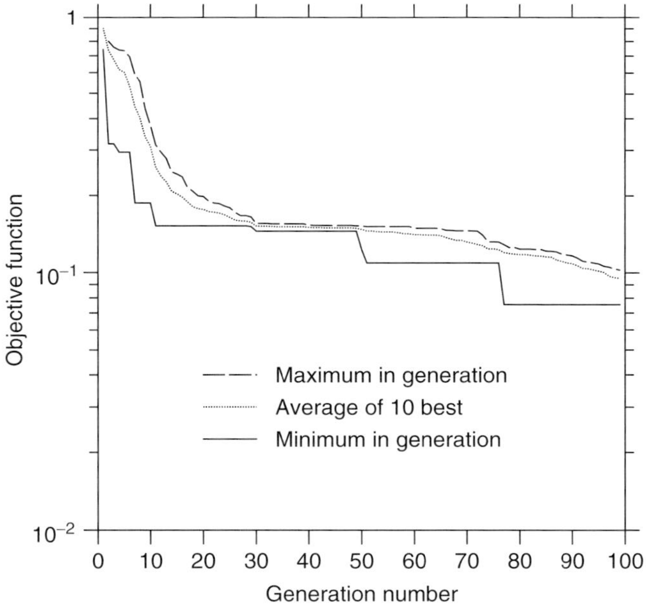

Figure 5.26 Convergence of the genetic algorithm.

Table 5.1 Hardening parameters obtained with genetic algorithms.

P_1	P_2	P_3	P_8
340.00	477.859 24	19.804 497	196.578 69

Table 5.2 Damage parameters obtained with genetic algorithms.

f_N	ε_N	s_N	f_C
2.932 551 3e−3	0.123 167 16	8.266 862 2e−2	0.341 153 47

Table 5.3 Final hardening parameters obtained by using gradient-based optimization.

P_1	P_2	P_3	P_8
337.031 38	470.384 80	17.338 554	207.387 12

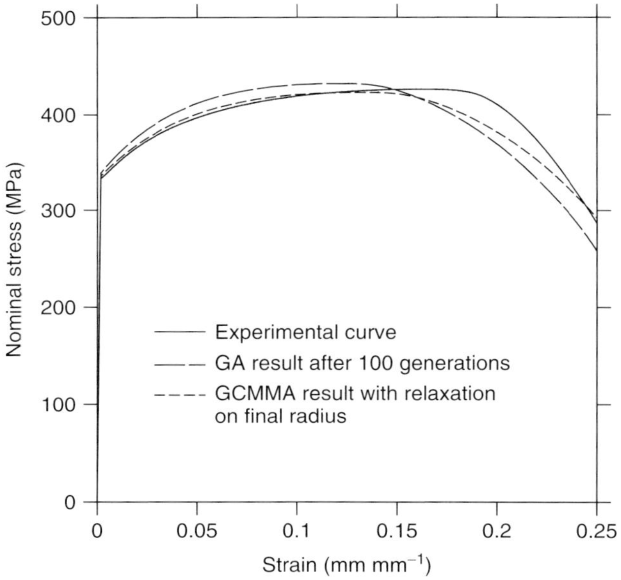

Figure 5.27 Stress–strain curves for the genetic algorithm and after application of the gradient-based optimization procedure.

experimental and numerical curves (Eq. 5.22) was applied only up to the maximum stress point. Such conditions produced a better overall curve matching, but the final neck radius increased to 3.05 mm. In this case, the four elastic–plastic material parameters are given in Table 5.3. The experimental and optimized stress–strain curves are displayed in Figure 5.27.

It is worth remarking that the ratio ε_N/s_N is equal to 1.496, (see Table 5.2) which, as discussed in Section 5.2.1, corresponds to 93% of nucleation level for f_N.

5.7.2
Aluminum Alloy at 400 °C

The second example concerns a tensile test of 7055 aluminum alloy at a constant temperature of 400°C [77]. The dimensions of the specimen are 9.81 mm of nominal diameter at its central section and 75 mm initial gauge length. A crosshead speed of 1.0 mm min^{-1} was applied. The heating was achieved with a heater band that enclosed the specimen and fixture. The device contained two independent heating zones (one at the top, the other at the bottom) to ensure a nearly constant temperature distribution. The thermal insulation to the environment was guaranteed by a thick layer of polymer foam.

The darkened region in Figure 5.28 shows the mesh used in the finite element model, with a total of 500 elements and 561 nodes. The topology of this mesh was proposed by Cunda [72]. The gray area represents the extensometer region. Prescribed displacements are applied incrementally in the right face, from 0 to 5 mm (corresponding to a total of 10 mm for the specimen).

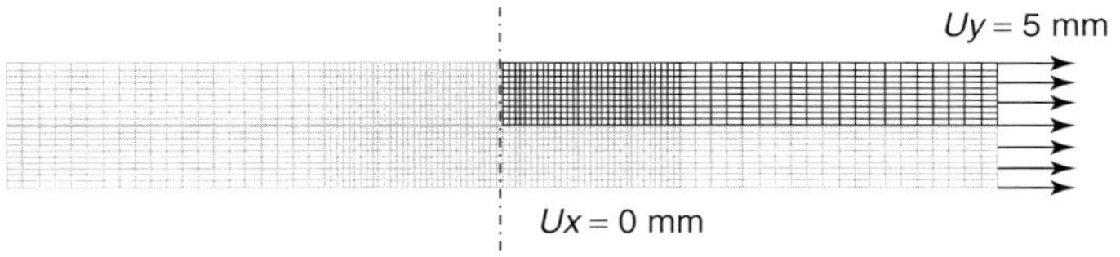

Figure 5.28 Mesh and boundary conditions adopted in the FEM simulation.

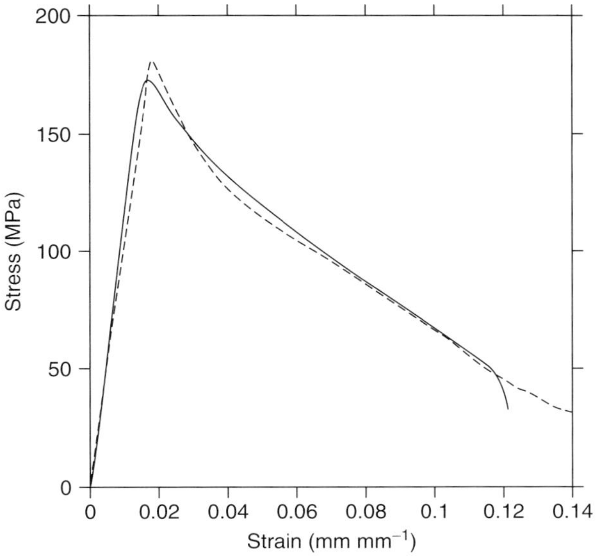

Figure 5.29 Stress–strain curve obtained using genetic algorithms after 100 generations (dotted line).

The design variables considered in this example are the elastic–plastic parameters, $P_1, P_2, P_3, P_6, P_7, P_8$, and the damage parameters, f_N, s_N, ε_N, and f_C. The same fixed damage parameters of the first example are also used in this problem. The Young modulus of the aluminum alloy is $E = 10269.17$ MPa. The objective function shown in Eq. (5.25) is used with $p_1 = 1$ and $p_2 = 0$.

Figure 5.29 presents the results after 100 generations. The thick and the dotted lines correspond to the experimental and numerical curves, respectively. It can be observed in the figure that the specimen exhibited a very ductile behavior. Figure 5.30 shows a detail of the ruptured section. Although the yield stress estimate can be further improved, it is clear that the overall approximation is very good. The parameters obtained with the proposed approach are presented in Tables 5.4 and 5.5. Table 5.5 shows that the ratio ε_N/s_N is equal to 1.452, which corresponds to 93% of the nucleation level for f_N, according to discussion presented in Section 5.2.1. The convergence of the objective function is shown in Figure 5.31.

The material parameters given in Tables 5.4 and 5.5 are used as the starting point for the gradient-based algorithm. In this case, optimization was performed using SLP. Although this class of problems has many local minima, the results obtained using GAs seem to have provided a quasi-global starting point to the mathematical programming approach. The material parameters obtained are given

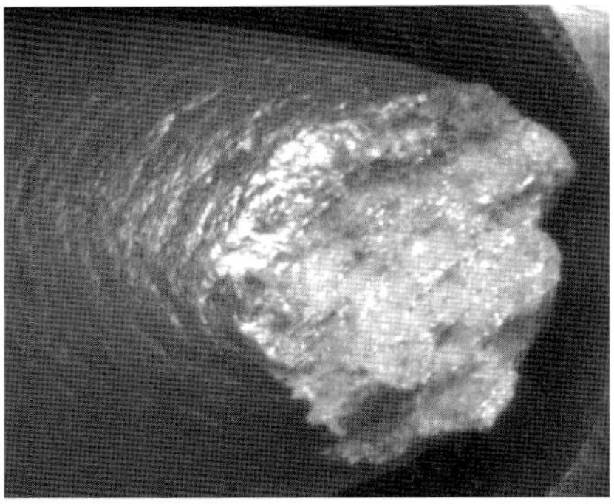

Figure 5.30 Ductile fracture of the aluminum specimen at 400 °C.

Table 5.4 Elastic–plastic parameters obtained with genetic algorithms.

P_1	P_2	P_3	P_6	P_7	P_8
183.643 9	72.502 63	3.349 972	0.4 684 500	0.6 723 966	91.921 26

Table 5.5 Damage parameters obtained with genetic algorithms.

f_N	ε_N	s_N	f_C
1.953 155e−003	3.225 605e−001	2.221 714e−001	9.439 231e−002

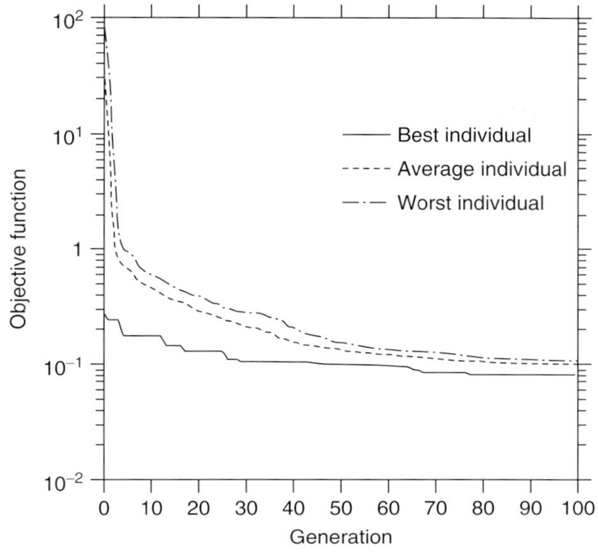

Figure 5.31 Convergence of the genetic algorithm.

Table 5.6 Final elastic–plastic parameters obtained using gradient-based optimization.

P_1	P_2	P_3	P_6	P_7	P_8
177.058 17	82.712 942	3.260 126 2	0.454 193 5	0.615 508 9	91.097 022

Table 5.7 Final damage parameters obtained using gradient-based optimization.

f_N	ε_N	s_N	f_C
2.436 038 4e−03	3.642 685 4e−01	2.057 972 3e−01	1.0 118 938e−01

in Tables 5.6 and 5.7, whereas the final stress–strain curve is depicted in Figure 5.32. In this example, the ratio ε_N/s_N is equal to 1.770 (see Table 5.7), reaching 96% of the nucleation level for f_N.

It is important to note that the formulation adopted does not ensure unicity of the material parameters. This is consistent with results obtained by Springmann and Kuna [78] and means that a different set of plastic and damage parameters could lead to nearly the same dotted curve depicted in Figure 5.28. One way to circumvent this inconvenience is to provide more information to be matched between experiment and simulation [79, 80], as performed in the example discussed in Section 5.7.1.

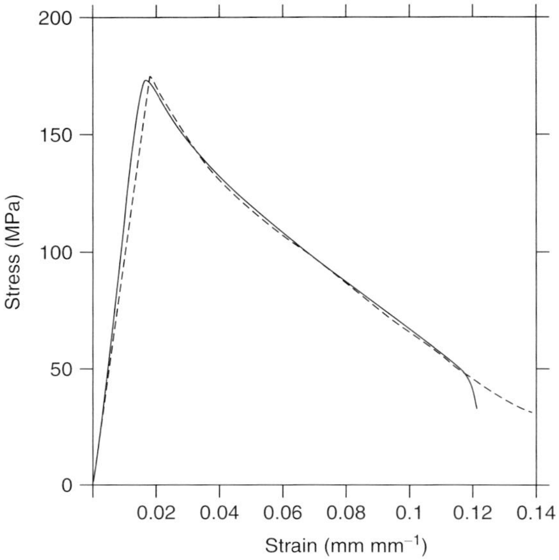

Figure 5.32 Final stress–strain curve obtained with gradient-based optimization (dotted line).

5.8
Concluding Remarks

Application of identification procedures to material modeling has become of paramount importance for the success of simulation of metal-forming processes. Moreover, a combination of damage approaches to the classic elastic–plastic formulations makes it possible, not only to more accurately describe material behavior but also to predict failure onset in complex forming operations. This has prompted the development of new and innovative techniques aiming at improving the quality of the parameters identified and at reducing the overall computation time. The present work is one in which a new mixed approach based on a combination of GA and gradient-based optimization strategies is proposed for the determination of plastic and damage parameters. The method accounts for an elastic–plastic formulation at finite strains, fully coupled to a Gurson-type damage model. The strategy was successfully applied to tensile tests of low carbon steel and aluminum alloy specimens using an objective function combining the force–displacement curve and the final diameter immediately prior to fracture.

Acknowledgments

The GCMMA algorithm used in this work was implemented by Prof. K. Svanberg, who kindly provided the source code to be used in the research applications conducted by Prof. Pablo Muñoz-Rojas. The finite element code METAFOR was used under cooperation agreement between Prof. Guillermo Creus and Prof. Michel Hogge. The use of these codes is gratefully acknowledged.

References

1. Gurson, A.L. (1975) Plastic flow and fracture behavior of ductile materials incorporating void nucleation, growth and coalescence. PhD Dissertation, Brown University.
2. Gurson, A.L. (1977) Continuum theory of ductile rupture by void nucleation and growth: part I-yield criteria and flow rules for porous ductile media. *ASME Journal of Engineering Materials and Technology*, **99**, 2–15.
3. Kachanov, L.M. (1999) Rupture time under creep conditions. *International Journal of Fracture*, **97** (1-4), 11–18.
4. Kachanov, L.M. (1958) On the time to rupture under creep conditions. *Izvestia Akademii Nauk SSSR, Otdelenie Tekhnicheskich Nauk*, **8**, 26–31 (in Russian).
5. Drucker, D.C. and Prager, W. (1952) Soil mechanics and plastic analysis for limit design. *Quarterly of Applied Mathematics*, **10** (2), 157–165.
6. Tvergaard, V. (1981) Influence of voids on shear band instabilities under plane strain conditions. *International Journal of Fracture*, **17**, 389–407.
7. Tvergaard, V. (1982) On localization in ductile materials containing spherical voids. *International Journal of Fracture*, **18**, 237–252.
8. Thomason, P.F. (1985) A three-dimensional model for ductile fracture by the growth and coalescence of microvoids. *Acta Metallurgica*, **33** (6), 1087–1095.
9. Chu, C.C. and Needleman, A. (1980) Void nucleation effects in biaxially stretched sheets. *ASME Journal of Engineering Materials and Technology*, **102**, 249–256.
10. Cunda, L.A.B. and Creus, G.J. (1999) A note on damage analyses in processes with nonmonotonic loading. *Computer Modeling and Simulation in Engineering*, **4** (4), 300–303.
11. Tvergaard, V. (1982) Material failure by void coalescence in localized shear bands. *International Journal of Solids and Structures*, **18** (8), 659–672.
12. Tvergaard, V. and Needleman, A. (1984) Analysis of the cup-cone fracture in a round tensile bar. *Acta Metallurgica*, **32** (1), 157–169.
13. Mori, T. and Tanaka, K. (1973) Average stress in matrix and average elastic energy of materials with misfitting inclusion. *Acta Metallurgica*, **21**, 571–579.
14. Mackenzie, J.K. (1959) The elastic constants of a solid containing spherical holes. *Procedures of Physics Society*, **63B**, 2–11.
15. Johnson, J.N. (1981) Dynamic fracture and spallation in ductile solids. *Journal of Applied Physics*, **52**, 2812–2825.
16. Perzyna, P. (1986) Internal state variable description of dynamic fracture of ductile solids. *International Journal of Solids and Structures*, **22** (7), 797–818.
17. Zhang, Z.L. (1996) A sensitivity analysis of material parameters for the Gurson constitutive model. *Fatigue and Fracture of Engineering Materials and Structures*, **19** (5), 561–570.
18. Tvergaard, V. (1982) Ductile frature by cavity nucleation between larger voids. *Journal of Mechanics and Physics of Solids*, **30** (4), 265–286.
19. Stainier, L. (1996) Modélisation numérique du comportement irréversible des métaux ductiles soumis à grandes déformations avec endommagement. PhD Dissertation, Université de Liège, Belgium (in french).
20. Goya, M., Nagaki, S., and Sowerby, R. (1992) Yield criteria for ductile porous solids. *JSME International Journal*, **35** (3), 310–318.
21. Voyiadjis, G.Z. and Kattan, P.I. (1992) A plasticity-damage theory for large deformation of solids- I. theoretical formulation. *International Journal of Engineering Science*, **30** (9), 1089–1108.

22. Lee, B.J. and Mear, M.E. (1993) An evaluation of Gurson's theory of dilatational plasticity. *ASME Journal of Engineering Materials and Technology*, **115**, 339–344.
23. Klöcker, H. and Tvergaard, V. (2003) Growth and coalescence of non-spherical voids in metals deformed at elevated temperature. *International Journal of Mechanical Sciences*, **45**, 1283–1308.
24. Lee, J.H. and Zhang, Y. (1994) A finite-element work-hardening plasticity model of the uniaxial compression and subsequent failure of porous cylinders including effects of void nucleation and growth – part 1: plastic flow and damage. *ASME Journal of Engineering Materials and Technology*, **116**, 69–79.
25. Besson, J. and Guillemer-Neel, C. (2003) An extension of the Green and Gurson models to kinematic hardening. *Mechanics of Materials*, **35**, 1–18.
26. Cedergren, J., Melin, S., and Lidström, P. (2004) Numerical modeling of P/M steel bars subjected to fatigue loading using an extended Gurson model. *European Journal of Mechanics A/Solids*, **23**, 899–908.
27. Zavaliangos, A. and Anand, A.L. (1992) Thermal aspects of shear localization in microporous viscoplastic solids. *International Journal for Numerical Methods in Engineering*, **33**, 595–634.
28. Thomason, P.F. (1985) Three-dimensional models for the internal neckings at incipient failure of intervoid matrix in ductile porous solids. *Acta Metallurgica*, **33** (6), 1079–1085.
29. Rice, J.R. and Tracey, D.M. (1969) On the ductile enlargement of voids in triaxial stress fields. *Journal of Mechanics and Physics of Solids*, **17**, 201–217.
30. Zhang, Z.L. and Niemi, E. (1995) A new failure criterion for the Gurson-Tvergaard dilatational constitutive model. *International Journal of Fracture*, **70**, 321–334.
31. Wen, J., Huang, Y., Hwang, K.C., Liu, C., and Li, M. (2005) The modified Gurson model accounting for the void size effect. *International Journal of Plasticity*, **21**, 381–395.
32. Aravas, N. and McMeeking, R.M. (1985) Microvoid growth and failure in the ligament between a hole and a blunt crack tip. *International Journal of Fracture*, **29**, 21–38.
33. Kikuchi, M., Miyamoto, H., Otoyo, H., and Kuroda, M. (1991) Ductile fracture of aluminum alloys. *JSME International Journal*, **34** (1), 90–97.
34. Needleman, A. and Tvergaard, V. (1994) Mesh effects in the analysis of dynamic ductile crack growth. *Engineering Fracture Mechanics*, **47** (1), 75–91.
35. Koppenhoefer, K.C. and Dodds, R.H. (1998) Ductile crack growth in pre-cracked CVN specimens: numerical studies. *Nuclear Engineering and Design*, **180**, 221–241.
36. Skallerud, B. and Zhang, Z.L. (2001) On numerical analysis of damage evolution in cyclic elastic-plastic crack growth problems. *Fatigue and Fracture of Engineering Materials and Structures*, **23**, 81–86.
37. Bazant, Z.P., Belytschko, T., and Chang, T.P. (1984) Continuum model for strain softening. *Journal of Engineering Mechanics Division ASCE*, **110**, 1666–1692.
38. Pijaudier-Cabot, G. and Bazant, Z.P. (1987) Nonlocal damage theory. *Journal of Engineering Mechanics Division ASCE*, **113**, 1512–1533.
39. Needleman, A. (1988) A material rate dependence and mesh sensitivity in localization problems. *Computational Methods in Applied Mechanics and Engineering*, **67**, 69–86.
40. Mazars, J. and Bazant, Z.P. (1989) *Cracking and Damage: Strain Localization and Size Effects*, Elsevier, Amsterdam.
41. de Borst, R. and Sluys, L.J. (1991) Localization in a Cosserat continuum under static and loading conditions. *Computational Methods in Applied Mechanics and Engineering*, **90**, 805–827.
42. de Borst, R. and Muhlhaus, H.B. (1992) Gradient-dependent plasticity: formulation and algorithmic aspects. *International Journal for Numerical Methods in Engineering*, **35**, 521–539.
43. Abu Al-Rub, R.K. and Voyiadjis, G.Z. (2005) A direct finite element implementation of the gradient-dependent theory. *International Journal for Numerical Methods in Engineering*, **63** (4), 603–629.

44. Leblond, J.B., Perrin, G., and Devaux, J. (1994) Bifurcation effects in ductile metals with nonlocal damage. *ASME Journal of Applied Mechanics*, **1** (2), 236–242.
45. Tvergaard, V. and Needleman, A. (1995) Effects of nonlocal damage in porous plastic solids. *International Journal of Solids and Structures*, **32** (8/9), 1063–1077.
46. Reusch, F., Svendsen, B., and Klingbeil, D. (2003) A non-local extension of Gurson-based ductile damage modeling. *Computational Materials Science*, **26**, 219–229.
47. Reusch, F., Svendsen, B., and Klingbeil, D. (2003) Local and non-local Gurson-based ductile damage and failure modelling at large deformation. *European Journal of Mechanics A/Solids*, **22**, 779–792.
48. Gologanu, M., Leblond, J.B., and Perrin, G. (1995) A micromechanically based Gurson-type model for ductile porous metals including strain gradient effects. *ASME Net Shape Processing of Powder Materials – Applied Mechanics Division AMD*, **216**, 47–56.
49. Ramaswamy, S. and Aravas, N. (1998) Finite element implementation of gradient plasticity models – part I: gradient-dependent yield functions. *Computer Methods in Applied Mechanics and Engineering*, **163**, 11–32.
50. Ramaswamy, S. and Aravas, N. (1998) Finite element implementation of gradient plasticity models – part II: gradient-dependent evolution equations. *Computer Methods in Applied Mechanics and Engineering*, **163**, 33–53.
51. Arora, J.S. (2007) *Optimization of Structural and Mechanical Systems*, World Scientific Publishing.
52. Goldberg, D.E. (1989) *Genetic Algorithms in Search, Optimization and Machine Learning*, Addison-Wesley.
53. Arabas, J. (2009) Evolutionary method as a random tool for searching in Rn. *Computational Materials Science*, **45**, 21–26.
54. Burjorjee, K.M. (2008) The fundamental problem with the building block hypothesis. *http://arxiv.org/pdf/0810.3356*.
55. Vanderplaats, G.N. (1984) *Numerical Optimization Techniques for Engineering Design*, McGraw-Hill.
56. Rao, S. (1996) *Engineering Optimization: Theory and Practice*, 3rd edn, John Wiley & Sons, Inc.
57. Svanberg, K. (1987) Method of moving asymptotes – a new method for structural optimization. *International Journal for Numerical Methods in Engineering*, **24** (2), 359–373.
58. Zillober, C. (1993) A globally convergent version of the method of moving asymptotes. *Structural Optimization*, **6**, 166–174.
59. Svanberg, K. (1995) A globally convergent version of the MMA without linesearch. First World Congress of Structural and Multidisciplinary Optimization – WCSMO1, Goslar.
60. Svanberg, K. (1999) The MMA for modeling and solving optimization problems. Third World Congress of Structural and Multidisciplinary Optimization – WCSMO3, Niagara Falls/Amherst.
61. Svanberg, K. (1999) A new globally convergent version of the method of moving asymptotes. Internal report: TRITA/MAT-99-OS2, Department of Mathematics, Royal Institute of Technology, Stockholm.
62. Srikanth, A. and Zabaras, N. (2001) An updated Lagrangian finite element sensitivity analysis of large deformations using quadrilateral elements. *International Journal for Numerical Methods in Engineering*, **52**, 1131–1163.
63. Forestier, R., Chastel, Y., and Massoni, E. (2003) 3D inverse analysis using semi-analytical differentiation for mechanical parameter estimation. *Inverse Problems in Science and Engineering*, **11**, 255–271.
64. Muñoz-Rojas, P.A., Fonseca, J.S.O., and Creus, G.J. (2004) A modified finite difference sensitivity analysis method allowing remeshing in large strain path-dependent problems. *International Journal for Numerical Methods in Engineering*, **61**, 1049–1107.
65. Hisada, T. (1988) Sensitivity analysis of nonlinear FEM. Proceedings of the ASCE EMD/GTD/STD Specialty Conference on Probabilistic Methods, pp. 160–163.

66. Ponthot, J.P. and Hogge, M. (1991) The use of the Euler Lagrange finite element method in metal forming including contact and adaptive mesh. Proceedings of the ASME Winter Annual Meeting, Atlanta.
67. Nagtegaal, J.C. (1982) On the implementation of inelastic constitutive equations with special reference to large deformation problems. *Computer Methods in Applied Mechanics and Engineering*, **33**, 469–484.
68. Haftka, R.T. (1985) Sensitivity calculations for iteratively solved problems. *International Journal for Numerical Methods in Engineering*, **21**, 1535–1546.
69. Kleinermann, J.P. (2000) Identification parametrique et optimisation des procedes de mise a forme par problemes inverses. PhD thesis, University of Liège, Belgium.
70. Chen, X. (1994) Nonlinear finite element sensitivity analysis for large deformation elastoplastic and contact problems. PhD thesis, University of Tokyo, Japan.
71. Chen, X., Hisada, T., Nakamura, K., and Mori, M. (1999) Sensitivity analysis of inelastic structures. *Theoretical and Applied Mechanics*, **48**, 39–47.
72. Cunda, L.A.B. (2006) O modelo de Gurson para dano dúctil: estratégia computacional e aplicações, PhD thesis, Universidade Federal do Rio Grande do Sul, Porto Alegre, Brazil.
73. Muñoz-Rojas, P.A., Cardoso, E.L., and Vaz, M. Jr. (2010) Parameter Identification of Damage Models using Genetic Algorithms. *Experimental Mechanics*, **50**, 627–634.
74. ABNT (2002) NBR ISO 6892. *Metallic Materials – Tensile Testing at Ambient Temperature*, ABNT, Rio de Janeiro.
75. ASTM (1987) E 8M-86a. *Tensile Testing of Metallic Materials*, ASTM, Philadelphia.
76. de Souza Neto, E.A., Peric, D., Dutko, M., and Owen, D.R.J. (1996) Design of simple low order finite elements for large strain analysis of nearly incompressible solids. *International Journal of Solids and Structures*, **33**, 3277–3296.
77. Muñoz-Rojas, P.A. and Cardoso, E.L. (2007) Identificação de parâmetros de dano utilizando uma abordagem mista de otimização. 8° Congreso Iberoamericano de Ingenieria Mecânica, Cusco, Peru.
78. Springmann, M. and Kuna, M. (2005) Identification of material parameters of the Gurson–Tvergaard–Needleman model by combined experimental and numerical techniques. *Computational Materials Science*, **32**, 544–552.
79. Cooreman, S., Lecompte, S., Sol, H., Vantomme, J., and Debruyne, D. (2008) Identification of mechanical material behavior through inverse modeling and DIC. *Experimental Mechanics*, **48**, 421–433.
80. Springmann, M. and Kuna, M. (2006) Determination of ductile damage parameters by local deformation fields: measurement and simulation. *Archive of Applied Mechanics*, **75**, 775–797.

6
Semisolid Metallic Alloys Constitutive Modeling for the Simulation of Thixoforming Processes

Roxane Koeune and Jean-Philippe Ponthot

6.1
Introduction

Historically, metal forming has always been a major concern for human beings in their quest to craft different kinds of objects, weapons, tools, and so on. Different techniques for metal forming have been developed, and most of these can be classified into two main categories: casting and forging.

Casting: The alloy is first heated up to its melting point, then the liquid material is poured into a die so that it takes the desired shape. The biggest part of the energy consumed in the process concerns the heating of the material. This kind of process allows to produce a large variety of complex geometries, most notably thin-wall components that allow the manufacturing of lighter parts. However, during solidification, the material tends to shrink, and this inevitably leads to porosities that weaken the mechanical properties of the final product.

Forging: The alloy is kept in the solid state and is deformed into the desired shape. In this case, the consumption of energy is mainly due to the load necessary for producing the prescribed deformation. This kind of process can offer a very good level of mechanical properties but is limited to simpler designs than with casting. Moreover, the waste of material is higher than that in casting.

Semisolid thixoforming: is an intermediate process. It relies on a particular behavior that can be exhibited by semisolid materials. These materials display thixotropy, which is characterized by a solidlike behavior at rest and a liquidlike flow when submitted to shear. This behavior is illustrated in Figure 6.1 where the metal can be cut and spread as easily as butter.

A family of innovative manufacturing methods based on this thixotropic behavior has been developed and has gained interest over the past 30 years. These processes present several advantages, such as energy efficiency, production rates, smooth die filling, low shrinkage porosity, which together lead to near net shape capability and thus to fewer manufacturing steps than with classical methods. Semisolid material processes have already proved to be efficient in several application fields, such as military, aerospace, and most notably automotive industries. Examples of parts that have already been produced by semisolid processing are illustrated

Advanced Computational Materials Modeling: From Classical to Multi-Scale Techniques.
Edited by Miguel Vaz Júnior, Eduardo A. de Souza Neto, and Pablo A. Munoz-Rojas

ISBN: 978-3-527-32479-8

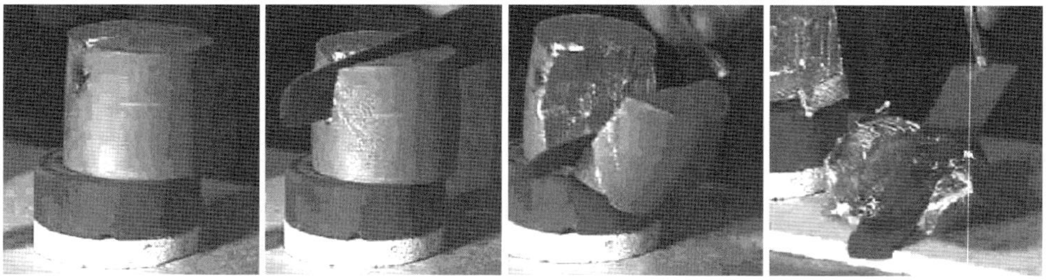

Figure 6.1 Photographic sequence illustrating the thixotropic behavior of a semisolid alloy slug [1].

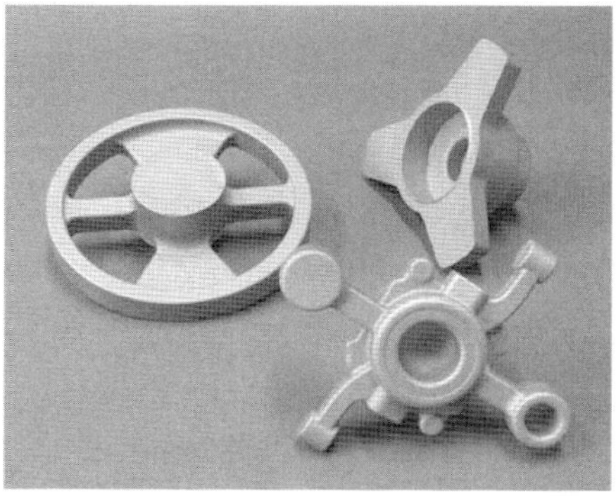

Figure 6.2 Parts that can be produced by semisolid processing.

in Figure 6.2. Thixoforming of aluminum and magnesium alloys is state of the art and a growing number of serial production lines are already operating all over the world. However, so far, there are only few applications of semisolid processing of higher melting point alloys such as steel. This can partly be attributed to the high forming temperature combined with the intense high temperature corrosion that requires new technical solutions. However, the semisolid forming of steels reveals high potential to reduce material as well as energy consumption compared to conventional process technologies. In this way, a lot of effort is currently being put into broadening the range of materials specifically designed for thixoforming.

In this context, simulation techniques exhibit great potential to gain a good understanding of these semisolid manufacturing routes and may be very helpful in the development of the process. Therefore, this chapter focuses on the background rheology and mathematical theories of thixotropy in order to develop constitutive models and to simulate thixoforming processes by means of, for example, the finite elements method.

After a short review on the technical aspects of semisolid metal processing, the specificities of the thixotropic behavior that have been observed experimentally and that need to be inserted in the models are detailed. In essence, thixotropic materials are highly temperature and rate sensitive so that computational modeling must include nonsteady-state and thermomechanical effects.

To set up the theoretical framework in which these simulations are established, continuum mechanics aspects as well as associated computational procedures like kinematics in large deformations or thermomechanical finite deformation constitutive theory are described. Treating semisolid material, containing both liquid and solid in more or less large quantities, can be done using the mathematical framework of both liquid dynamics or solid mechanics. Thus, both formalisms are detailed and compared to each other.

A state of the art in modeling of thixotropy is also gathered. It can be roughly categorized as one-phase or two-phase models. Different existing theories to mathematically reproduce thixotropic behavior are explained.

A thermomechanical and transient one-phase model, using the solid mechanics formalism that is extended here to be able to simulate fluid flows, is then detailed.

Finally, as an illustration of possible applications, the simulation of a few bench tests is conducted. The computed results are described, analyzed, and compared to references.

For the interested reader, there are several books [2–4] devoted to semisolid metal processing.

6.2
Semisolid Metallic Alloys Forming Processes

By definition, a thixotropic material behaves as a solid when allowed to stand still and as a liquid (it flows) when submitted to deformation. The viscosity of a thixotropic material decreases with shearing (it is said that the material *thins*) and increases again at rest (the material *thickens*).

Some examples of thixotropic systems include [5] the following:

- flocculation under interparticle forces: paints, coatings, inks, clay slurries, cosmetics;
- flocculation of droplets: emulsions;
- flocculation of bubbles: foams;
- interlocking of growing crystals: waxes, butter, chocolate;
- agglomeration of macromolecules/entanglement: polymeric melts, sauces; and
- agglomeration of fibrous particles: tomato ketchup, fruit pulps.

Among those examples, a couple of thixotropic materials that can be encountered in everyday life can help to better understand this particular thixotropic behavior: Ketchup flows out of its bottle more easily if shaken previously. Certain kinds of paints stick on paintbrushes and walls but flow and spread when worked with the brush.

Some decades ago, a lot of work was done on the deformation of alloys during solidification in order to better understand some defects inherent to casting. In this environment, at the Massachusetts Institute of Technology (MIT) in the 1970s, Spencer *et al.* [6] discovered that metallic alloys in the semisolid state with some specific microstructure could also display thixotropy.

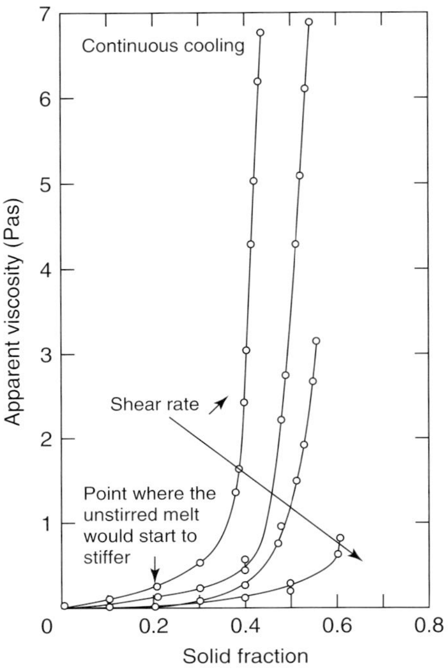

Figure 6.3 Apparent viscosity versus solid fraction for Sn– 15% Pb stirred at various shear rates during continuous cooling [6].

This discovery has given rise to considerable interest and resulted in new metal forming processes – a whole family of semisolid processes.

6.2.1 Thixotropic Semisolid Metallic Alloys

It all started with the very unexpected results of some viscosity measurements of alloys during solidification carried out at the MIT. The solidification of a nonpure material occurs in a finite range of temperatures at which two phases coexist: a liquid and a solid phase. The material is then in the semisolid state. Thus, during a solidification experiment, the percentage of solid contained in the semisolid, called the *solid fraction* – denoted by f_s – increases[1]. In these experiments, the viscosity measured as a function of the solid fraction depending on whether the alloy was continuously stirred or not. In both cases, the apparent viscosity increases with the solid fraction, as the material solidifies. This rise in viscosity speeds up at a certain solid fraction level, as the material stiffens. It is this starting point of sudden stiffening that depends on the conditions of solidification. So, the stirred melt starts to stiffen at a much higher solid fraction than the unstirred one. We can see in Figure 6.3 that this "critical" solid fraction increases with the strain rate.

1) At the same time, the percentage of liquid contained in the semisolid, called the *liquid fraction* – denoted by $f_l = 1 - f_s$ – decreases.

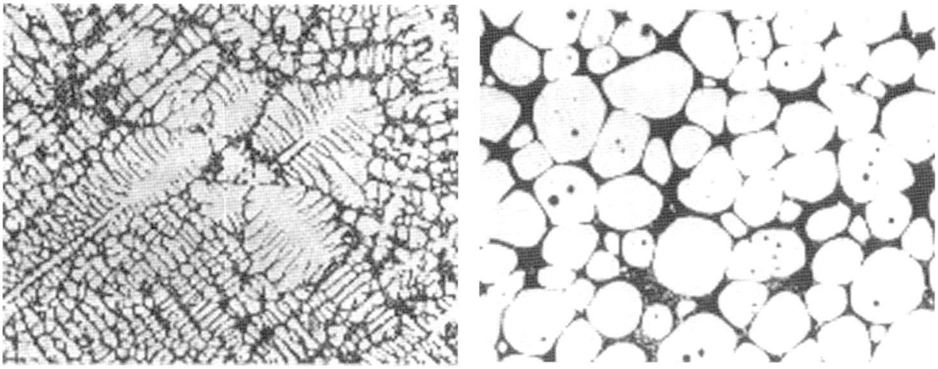

(a) (b)

Figure 6.4 Comparison between dendritic (a) and globular (b) microstructures in a semisolid alloy sample.

These experiments showed that the viscosity of a semisolid metallic alloy was very sensitive to the shear rate. Viscosity is a measure of how easy it is for a material to flow. Thus, at the same solid fraction, the stirred melt would flow more easily than the unstirred one.

Flemings and his MIT group, who discovered this phenomenon, attributed it to the microstructure evolution of the molten alloy. It was well known that a steady solidification led to a dendritic microstructure, with solid dendrites particles lying in the liquid matrix. They proposed that a vigorous agitation during solidification leads to a nondendritic but rather spheroidal or globular microstructure because the shearing would continuously break up the arms of the dendrites. A comparison between dendritic and globular microstructure is shown in Figure 6.4, where the liquid phase is dark and the solid phase is light.

A higher shear rate leads to particles closer to pure spheres and, hence, to a lower viscosity and to a better flow. Thus they found that semisolid metallic alloys with a nondendritic microstructure are thixotropic.

6.2.2
Different Types of Semisolid Processing

As explained in the previous section, the key point of semisolid forming processes is the thixotropy of metallic alloys at the semisolid state. For this, a spheroidal microstructure is required. Thus, semisolid-forming processes take place in several steps:

- production of nondendritic microstructures;
- reheating to the semisolid state; and
- forming.

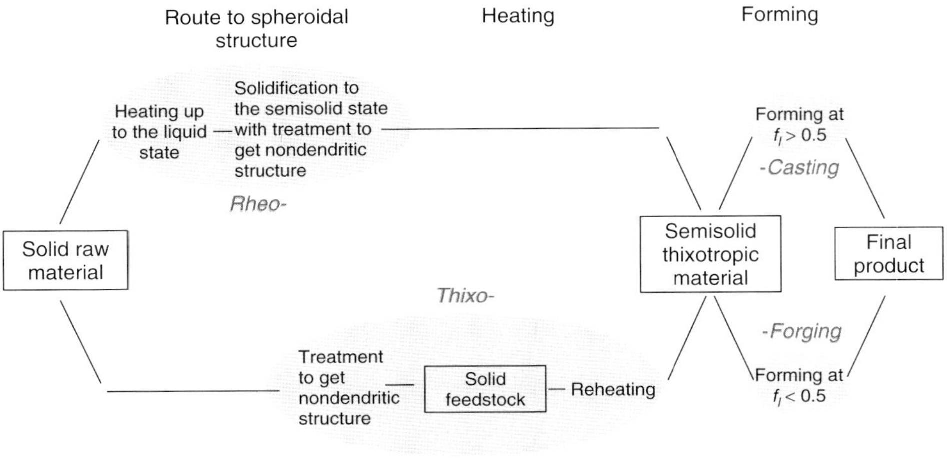

Figure 6.5 Schematic illustration of different routes for semisolid metal processing.

Depending on the type of the process, the reheating step can be avoided. The standard terminology is as follows.

*Rheo*forming refers to a process where the raw material is brought to the total liquid state before being cooled down to the desired semisolid state. In this case, the metal is *liquid* at the start of the forming process and the reheating stage is not needed.

*Thixo*forming refers to a process where an intermediate solidification step occurs. In this case, the metal is *solid* at the start of the forming process.

In addition, semisolid forming processes can be seen as an intermediate family of processes between casting and forging. Therefore, depending on the liquid fraction, the forming stage can be close to casting or to forging. So, in the case where the liquid fraction is relatively high (i.e., above about 50%), the process is closer to *casting*. On the other hand, a process working with lower liquid fraction is closer to *forging*.

The different types of semisolid processes can be categorized as thixocasting, thixoforging, or rheocasting[2], illustrated in Figure 6.5.

6.2.2.1 Production of Spheroidal Microstructure

Conventional partial solidification or remelting leads to solid dendrites in a liquid matrix, which is precisely what is to be avoided in an efficient semisolid process. Therefore, the stage of production of a globular microstructure is critical and requires the development of innovative processes besides regular casting or forging processes. In his book [7], Suéry classifies the processes into two main categories. The first category contains processes where the main microstructure

2) Nowadays, there is no "rheoforging" process.

modification occurs during *solidification*; it does not matter then whether the semisolid thixotropic material is then formed right away (prefix rheo-) or cooled back down to the solid state and reheated before the forming step (prefix thixo-). The other category includes processes where *remelting* plays the major part in structure modification.

Microstructure Modification during Solidification The goal of this kind of processes is to act on the solidification stage in order to avoid the development of dendrites. According to Suéry [7], this can be done in three ways:

- ***Mechanical methods*** impose shearing to the structure under solidification. The shearing can be produced by different kinds of stirring (mechanical, passive, or electromagnetic stirring), by ultrasound or by electric shocks.
- ***Chemical methods*** modify the composition of the alloy to get grain refinement and equiaxed microstructures.
- ***Thermal methods*** use some appropriate cooling conditions to get globular microstructures. For example, maintaining a dendritic structure in the semisolid temperature range for a period of time can produce a relatively coarse globular structure by natural maturing. The hold time needed for maturing depends on the size of the dendrites, which in turn depends on the cooling rate. Thus, the fastest solidification possible is tried for in order to get the finest structure before remelting it.
  Another example is used in the process of new rheocasting (NRC). The alloy is molten at its near-liquidus temperature and then cooled down slowly and homogeneously. The liquid metal that is lightly overheated passes under the liquidus in a rapid and homogeneous way. This produces the nucleation of numerous and thus small grains that remain spheroidal during the slow cooling.

Certain methods can be combined together to get a better efficiency.

Microstructure Modification during Remelting The achievement of the nondendritic microstructure is made during the remelting stage, although the structure of the alloy at the initial solid state is also very important. These processes can be sorted into three kinds [7].

- ***Powder metallurgy***: The partial remelting of a fine powder made of several alloys with different solidus temperatures can lead to a globular microstructure under proper heating conditions.
- ***Deformation processes***: The metallic alloy is submitted to plastic deformation before being remelted. If the deformation is high enough, a fine grain recrystallization occurs during remelting and the liquid penetrates the recrystallized boundaries. There are different processes of this kind depending on the temperature at which the plastic strain is imposed. A route called *strain-induced melt activated* (*SIMA*) involves hot working (above the recrystallization temperature), while the process of *recrystallization and partial remelting* (*RAP*) involves warm working (below the recrystallization temperature).

- ***Thixomolding***™ [8]: This is an exception to the terminology "thixo". ***Thixomolding***™ is a licensed process highly effective for magnesium alloys. Metal pellets are injected into a continuously rotating screw. The shear induced to the material by the screw generates enough energy to heat the pellets into the semisolid state and creates a globular structure. The material is then directly injected into a die.

6.2.2.2 Reheating

The reheating of the material up to the adequate semisolid state is necessary only in thixoforming processes. This stage of the process is a very important one and must be of good quality to meet the desired liquid fraction. Indeed, the liquid fraction is a crucial parameter for the quality of die filling. The liquid fraction and thus the temperature should be as uniform as possible to get a homogeneous behavior.

There exist two main routes for this reheating step: the resistance furnace and the inductive heating.

Heating by Resistance Furnace These furnaces supply the heat by convection and radiation. The heating times are higher than that in induction heating. Above the industrial inconvenience of considerable waste of time, these long heating times lead to a nonhomogeneous microstructure. Indeed, the solid phase maturing depends on the hold time in the semisolid state so that a slow heating can lead to larger solid grain size close to the skin than in the heart of the slug.

It is possible to speed up the resistance heating by improving the heat transfer, thanks to forced air convection or even convection in a liquid, but these means are not very efficient in general.

Inductive Heating Inductive heating, which, unlike resistance furnace, supplies the heat directly inside the billet, and is the most employed heating route in thixoforming. It has the advantage of being much faster than the previous heating by resistance furnace; this results in enhanced productivity and energy efficiency and avoids the problem of grain size mentioned in the case of resistance heating.

A primary circuit transmits electric energy to the metallic billet, which plays the role of secondary circuit. The electric energy is then converted into heat by *Joule effect* losses.

In practice, as illustrated in Figure 6.6, the sample is set inside an inductor supplied with alternative current. This produces an alternative magnetic field that creates induced currents, called *Foucault currents,* inside the billet. These currents dissipate heat inside the sample by the Joule effect.

If no particular care is taken, the induced currents and the resulting heat can be nonhomogeneous inside the billet. By the laws of electromagnetism, also known as *Maxwell's laws,* it can be shown that the intensity of the induced current decreases exponentially from the skin to the heart of the sample. This

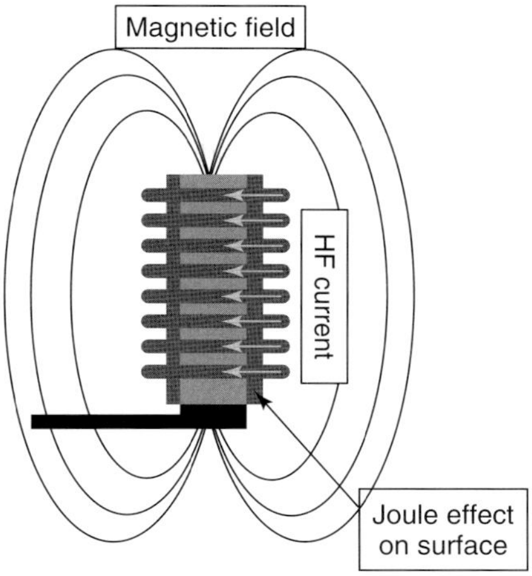

Figure 6.6 Schematic illustration of the inductive heating device.

phenomenon is called the *skin effect*. In fact, the main part of the transferred power is confined close to the surface of the sample. The heart of the billet is heated afterward by thermal conduction from the surface. Therefore, there is a radial inhomogeneity of the temperature (and thus of the liquid fraction). A temperature gradient along the axis can also appear since the length of the inductor is not infinite.

However, unlike resistance heating, these inhomogeneities can be avoided, or at least reduced to a minimum by a careful set up of the inductive heating process parameters. Collot [9] states that thermal and energy efficiency of the induction can be qualified by two main values: the penetration depth and the dissipated power. These quantities depend on various parameters that can be divided into two families depending on whether they characterize the primary (the inductor geometry, the current frequency or intensity) or the secondary circuit (the material properties or the size of the billet). The parameters of the first kind are chosen according to those from the second family.

These parameters are numerous and interconnected and a description of the influence of each is beyond the scope of this chapter. However, a parameter that plays a major role in temperature homogeneity is the evolution of the power supply. In fact, it is important to calibrate the right heating cycle with, for instance, some stages of lower power that leave some time for the conduction to homogenize the temperature. In practice, the determination of such an appropriate heating cycle, which keeps reasonable temperature gradients, is a major concern.

6.2.2.3 **Forming**

The last – but not least – step consists in deforming the slug into the desired shape.

As already mentioned, there exist different kinds of forming processes depending on the liquid fraction obtained after reheating. Among those, it is possible

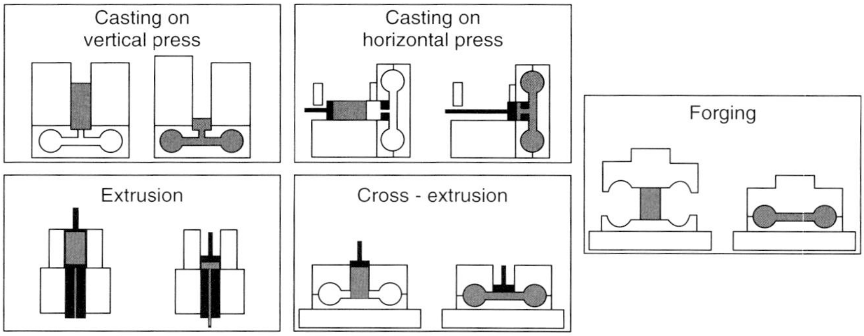

Figure 6.7 Illustration of the main kinds of semisolid forming processes.

to make other differentiations depending on the die device as illustrated in Figure 6.7: extrusion, vertical, or horizontal press, and so on. Nowadays, some processes are already in use in production, while others are still in the development stage. The forming can also be continuous or the billets can be treated one by one.

Semisolid processes are used to treat mainly aluminum, magnesium and steel alloys. Some bronze or copper alloys can also be treated by semisolid processing. The main fields of application are the automotive, aerospace, and military industries. More or less geometrically complicated parts can be produced by semisolid processing, as already illustrated in Figure 6.2. In addition, thixomolding™ mainly produces magnesium alloy components for cellular phones, portable computers, cameras, and so on.

One crucial advantage of thixoforming over solid forging is the reduction in the number of manufacturing steps needed to reach the final step by eliminating some operations (forging steps, clipping, punching, final machining) and incorporating all the remaining operations into a single line [10]. This is illustrated in Figure 6.8 where the component can be obtained in a single pass by thixoforming. However, the transfer from one technology to another is never straightforward and another

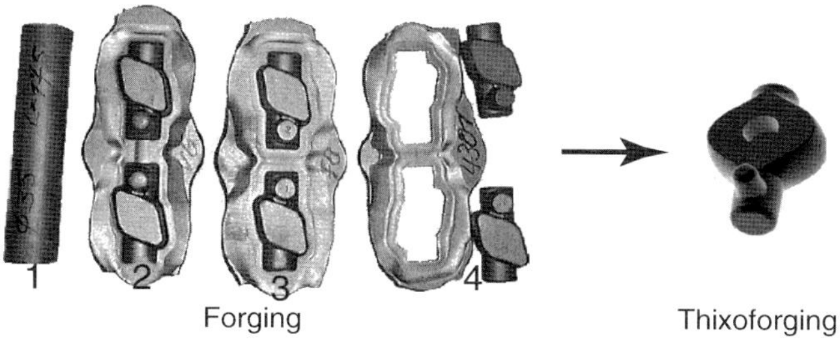

Figure 6.8 From forging to thixoforging of a flange: reduction in the number of manufacturing steps and geometry redesign.

factor that is illustrated by Figure 6.8 is the necessary geometry redesign and resize of the existing forged component for thixoforming. It is, of course, necessary for the redesign to keep connecting and functional dimensions unchanged. In the case of the flange represented in Figure 6.8, the number of sharp edges has been reduced to a minimum. This allows to obtain a laminar flow of the semisolid material and to prevent tool material deterioration.

On the other hand, the redesign can take advantage of the enhanced freedom in part design that is offered by thixoforming.

A point common to all types of processes is that liquid and solid phases can move apart, which causes segregations that can lead to inhomogeneities in the properties of the final product. So, in all cases, the process should be designed to minimize this effect.

Another concern, which is common with classical solid or liquid forming processes, is the lifetime of the die. The manufacturing costs of a die makes it worthwhile to put in significant effort into the design to increase the average life of the die.

6.2.3 Advantages and Disadvantages of Semisolid Processing

As with any manufacturing process, there are certain advantages and disadvantages in semisolid processing.

The main advantages of semisolid processing, relative to die casting, have been gathered by Atkinson [1] as follows:

- Energy efficiency: Metal is not being held in the liquid state over long periods of time.
- Production rates are similar to or better than that for pressure die casting.
- Smooth filling of the die with no air entrapment and low shrinkage porosity gives parts of high integrity (including thin-walled sections) and allows application of the process to higher strength heat-treatable alloys.
- Lower processing temperatures reduce the thermal shock on the die, promoting die life and allowing the use of nontraditional die materials and processing of high melting point alloys such as tool steels that are difficult to form by other means.
- Lower impact on the die also introduces the possibility of rapid prototyping dies.
- Fine, uniform microstructures give enhanced mechanical properties.
- Reduced solidification shrinkage gives dimensions closer to near net shape and justifies the removal of machining steps; the near net shape capability reduces machining costs and material losses.
- Surface quality is suitable for plating.

The main disadvantages are as follows:

- The cost of raw material can be high and the number of suppliers small.
- Process knowledge and experience has to be continually built up in order to facilitate application of the process to new components.

- This leads to potentially higher die development costs.
- Initially at least, personnel require a higher level of training and skills than with more traditional processes.
- Temperature control: Solid fraction and viscosity in the semisolid state are very temperature dependent. Alloys with a narrow temperature range in the semisolid region require accurate control of the temperature.
- Liquid segregation due to nonuniform heating can result in nonuniform composition in the component.

6.3 Rheological Aspects

Modigell and Pape [11] describe rheology as an interdisciplinary science connecting physics, physical chemistry, chemistry, and engineering sciences. The word "rheology" combines the Greek words *rheo*, which means "flow" and *logos*, meaning "science". Rheology deals with simultaneous deformation and flow of materials [12]. It is quantitatively expressed in relations between forces acting on bodies and the resulting deformations.

Probably because of the "flow" connotation, the word rheology is mainly used in fluid mechanics. However, in general, it is concerned with the mechanics of deformable bodies and can be seen as a discipline that aims at modeling material behavior under any of the mathematical frameworks detailed in Section 6.4.2. Two branches of rheology working together have been distinguished by Modigell and Pape [11]. The first one, often called *rheometry*, should give reliable experimental data to set up heuristically defined equations. The other one is theoretical and tries to derive constitutive relations by structural considerations. This section focuses on the second part.

6.3.1 Microscopic Point of View

First, thixotropy in a microscopic point of view is treated, and more specifically, the microscopic origin of thixotropy is explained.

6.3.1.1 Origins of Thixotropy

It has already been mentioned in Section 6.2 that the origin of thixotropy could be found in the globular microstructure of the material. This section focuses on the reasons why such a structure is required and how it leads to thixotropy.

Agglomeration The basic phenomenon of thixotropy is the tendency of the solid particles to agglomerate. Atkinson [1] explains that this agglomeration occurs because particles collide (either because the shear brings them into contact or, if at rest, because of sintering) and, if favorably oriented, form a boundary. "Favorable orientation" means the fact that if the particles are oriented in such a way that a low energy boundary is formed, it will be more energetically favorable for the agglomeration to occur than if a high energy boundary is formed.

Particle Size Once the bonds are formed, the agglomerated particles sinter, with the neck size increasing with time.

Because of shearing, the existing bonds between particles are broken down and the average agglomerate size decreases [1].

Thus, there exists a concurrency between two antagonist phenomena: buildup and breakdown. This will dictate the size of the particles that, in turn, will influence the material behavior.

Physical Basis of Thixotropy When the slurry is at rest, gravity will bring the particles into contact and there is no important shear force to break the bonds. Thus, a 3D network can build up throughout the material and the semisolid will support its own weight and can be handled like a solid.

During a deformation, both buildup and breakdown happen. Actually, shear not only breaks the bonds but also forces particles into contact with each other. Thus, agglomeration still occurs. This process is influenced by the shear rate in two opposing ways. Increasing the rate of shear not only increases the possibility of particle–particle contact but also decreases the time of contact [1]; yet, the formation of a new solid–solid boundary needs time to be accomplished. Overall, the structure is more or less unstructured by the deformation and the material responds by a more or less viscous flow, depending on the shear rate.

Importance of Spheroidal Structure During deformation, bonds are broken and the solid globules are able to roll over each other while the fluid surrounding them acts as a lubricant. The ease with which particles are able to move past each other depends on the liquid fraction, the size of the particles, and the degree of agglomeration. This deformation mechanism is illustrated in Figure 6.9 and explains why a globular structure is required. Indeed, the globular particles move more easily over each other than dendritic phases that tend to interlock during application of an external force.

The viscosity in the steady state depends on the balance between the rate of structure buildup and the rate of breakdown. It also depends on the particle morphology. The closer the shape to that of a pure sphere, the lower the steady-state viscosity [1].

6.3.1.2 Transient Behavior

We have seen that the thixotropic material is highly dependent on the load rate and history. Moreover, during a forming process, the slurry undergoes a sudden increase in shear rate. So, the thixotropic transient behavior is much different than the one under steady-state conditions, and that is the first one that has to be introduced in the simulations.

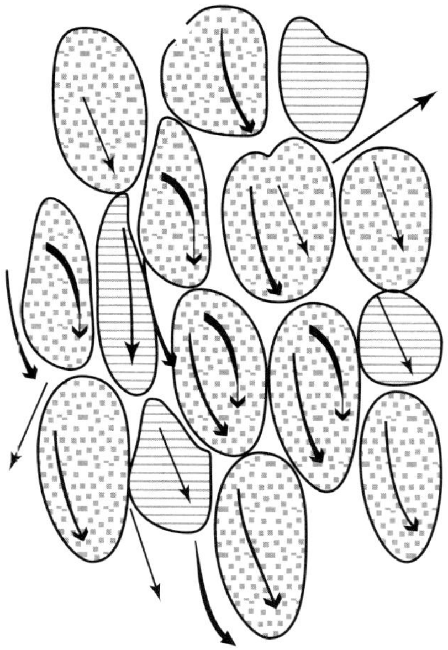

Figure 6.9 Microstructural deformation mechanism in a semisolid metallic alloy with globular microstructure.

Shear-Rate Step-Up and Step-Down Experiments An experiment that is well representative of the thixotropic behavior is the shear rate step up and step down, illustrated in Figure 6.10. This figure represents the results of the numerical simulation of a simple shear test by the finite element method and using the constitutive model described in Section 6.6. It is in qualitative agreement with the experiments. For a thixotropic material at rest, when a step increase in shear rate is imposed, the shear stress will peak and then gradually decrease until it reaches an equilibrium value for the shear rate over time [1]. Similarly, when the shear rate is suddenly decreased, the material responds by an undershoot in shear stress before another stabilization.

Comparison of semisolid Metallic Alloy with Other Thixotropic Systems The explanation of this transient behavior typical of thixotropic systems can be found in a deeper detailed description of the microscopic mechanisms of thixotropy. For this, Atkinson [1] compares semisolid thixotropic metallic alloy with other thixotropic systems. Indeed, the semisolid metallic systems have much in common with flocculated suspensions.

Figure 6.11 represents a classical diagram of the microstructure evolution and the stress response to a change of shear rate during a simple shear experiment on such systems. This behavior also applies for semisolid metallic alloys with a globular microstructure. At equilibrium, the microstructure has got enough time to adapt to a new shear rate level. So, Figure 6.11 distinguishes the equilibrium flow dotted curve from the isostructure evolutions represented by plain lines.

Starting from point "*a*", at which the microstructure consists of large particles agglomerates, the shear rate is first increased from $\dot{\gamma}_1$ to $\dot{\gamma}_2$. In the first step, the flow

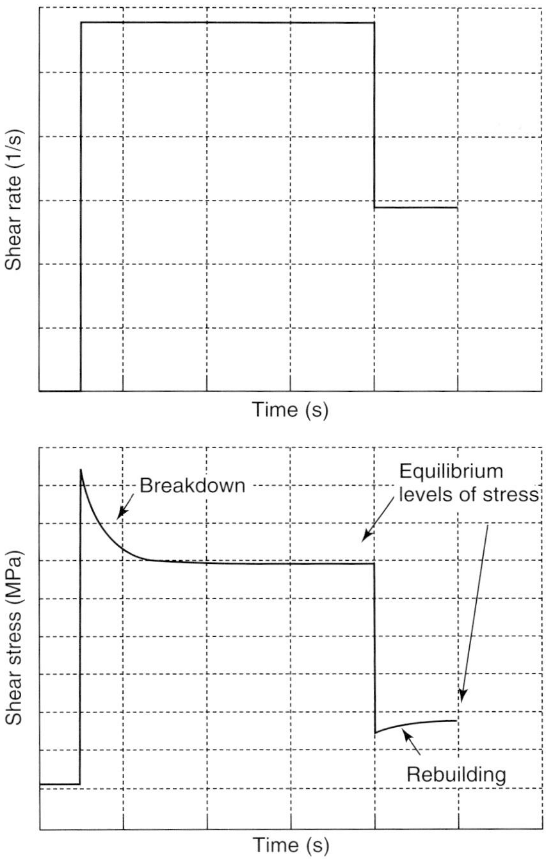

Figure 6.10 Shear rate step-change experiment (numerically generated).

follows the plain line to go toward point "a'" and then gets back on the equilibrium flow curve by a structure break up until the size corresponds to the flow curve that passes through point "b". This point corresponds to a new microstructure made of much smaller agglomerates than the initial one.

Then, the shear rate is reduced back to $\dot{\gamma}_1$ and the flow curve corresponding to the new microstructure is followed up to point "b'". The new equilibrium is then reached by collision and agglomeration of individual particles to get back to point "a" and its corresponding microstructure made of large flocs.

Atkinson [1] studied the similarities and differences between thixotropy in semisolid metallic systems and that in other thixotropic systems. These are associated with the nature of the forces between the particles. In general, the forces between particles include the following: van der Waals attraction, steric repulsion due to adsorbed macromolecules, electrostatic repulsion due to the presence of like charges on the particles and a dielectric medium, and electrostatic

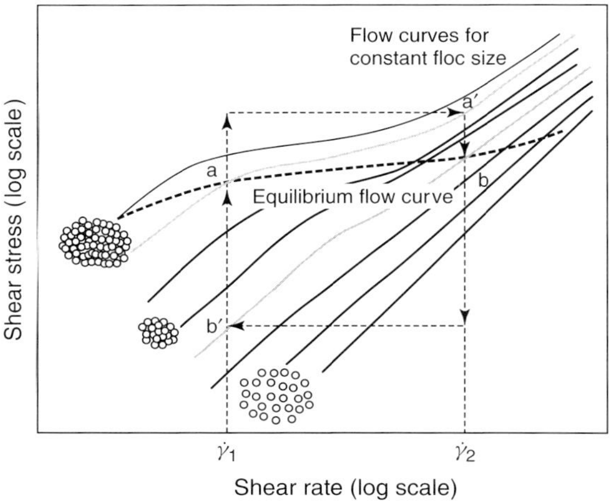

Figure 6.11 Thixotropic behavior in the case of flocculated suspensions [1].

attraction between unlike charges on different parts of the particle (e.g., edge/face attraction between clay particles). In semisolid metallic slurries, none of these forces apply. What must actually be occurring in structural buildup is a process akin to adhesion in wear.

Another difference between thixotropic flocculations and metallic systems is the coarsening of the microstructure over time. In the case of metallic alloys, it has already been mentioned that during an isothermal hold in the semisolid state, the average size of the solid particles gets progressively larger. This has a major influence on the material behavior.

Finally, many thixotropic systems show "reversibility," that is, the slurries have a steady-state viscosity characteristic of a given shear rate at a given solid fraction regardless of past shearing history. However, in semisolid alloy slurry systems, the evolution of particle shape (and size) with time and stirring (Figure 6.13) is irreversible in the sense that a globule cannot convert back to a dendrite. The measured viscosity is then expected to depend on the shearing and thermal history.

Microstructural Explanation of Thixotropic Transient Behavior Quaak [13] proposed the mechanism illustrated in Figure 6.12 as the microstructural basis of thixotropic transient behavior. Right after a change in shear rate, the structure remains unchanged ("isostructure"). During this transient period, structure evolution has no time to occur and the structure still corresponds to that of the previous shear rate. It is not yet very clear whether the isostructural material behaves in a Newtonian way or as a shear thickening flow, but this transient period corresponds to the overshoot (or undershoot in the case of a decrease in shear rate) in the response of the material to the rapid shear rate change observed in Figure 6.10.

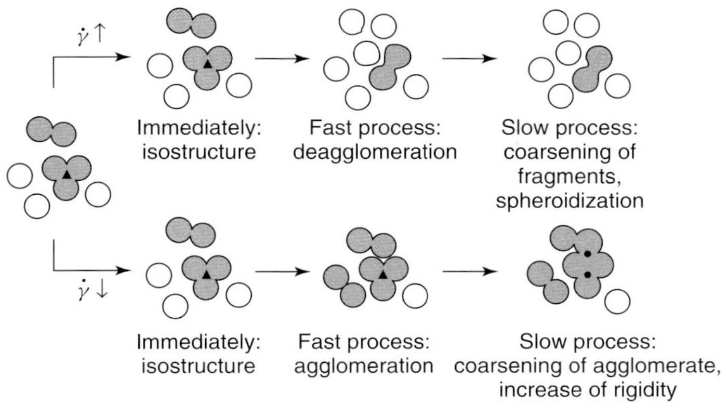

Figure 6.12 Microstructural basis of thixotropic transient behavior [13].

There is then a fast mechanical process of deagglomeration where the bonds are broken, followed by a slower diffusional process where fragments coarsen and progressively get spheroidal. These two processes correspond to the return to equilibrium, with the shear stress evolution going back on the equilibrium shear stress that corresponds to the new shear rate level.

Step-Change Experiment Purpose: Identification of Material Parameters We have seen that the key for the thixotropic behavior is the balance between two antagonist phenomena: buildup and breakdown. Thus, these processes have to be inserted in models of thixotropy and need to be characterized. For this, two relaxation times can be quantified, thanks to shear rate step-change experiments: breakdown time and buildup time. The breakdown time is the characteristic time for the slurry to achieve its steady-state condition after a shear rate change from a lower value to a higher value, while the buildup time is for a change from a higher shear rate to a lower shear rate. The times for breakdown are faster than those for buildup. The breaking up of "bonds" between spheroidal solid particles in agglomerates is likely to be easier than the formation of bonds during shear-rate drops [1].

On the basis of these experiments, some tendencies can be highlighted:

- Regardless of the initial shear rate, both the viscosity and the breakdown time decrease with increasing final shear rate.
- During a change in shear rate, the height of the peak in stress evolution is proportional to the change in shear rate.
- With longer rest times, the peak stress increases and the breakdown time decreases. Atkinson [1] argued that this is consistent with microstructural evidence showing that increasing the rest time increases the solid-particle sizes and the degree of agglomeration. This increase would impede the movement of the particles upon the imposition of the shear stress.
- The overshoot increases with increasing solid fraction.

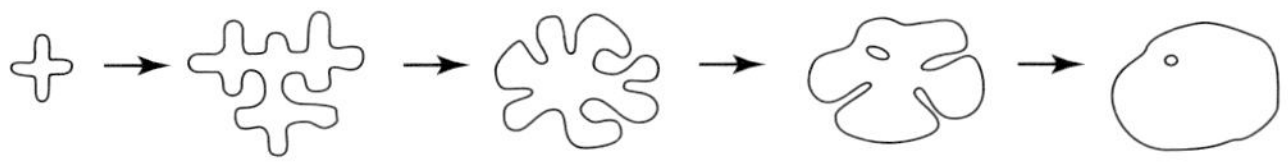

Figure 6.13 Structure evolution during solidification with vigorous agitation [14].

6.3.1.3 Effective Liquid Fraction

During the formation of the globular microstructure, some liquid can get entrapped inside solid grains. For example, during solidification with vigorous agitation, Flemings [14] proposed that the structure evolution was described by Figure 6.13. This figure shows how liquid can get entrapped within solid globules (see also Figure 6.4).

This part of the liquid does not contribute to the flow. Thus, although the liquid fraction may take a certain value governed by the temperature (and indeed kinetics as the thermodynamically predicted liquid fraction is not achieved instantaneously on reheating from the solid state), in practice, the *effective* liquid fraction should be distinguished and is smaller than (or equal to) the liquid fraction. This tends to increase the viscosity in a nonnegligible way since it is very sensitive to the liquid fraction.

During a deformation, solid bonds get broken and some of the entrapped liquid is released, leading to an increase in the effective liquid fraction. This accentuates the ease with which the material under load will flow and causes a decrease in the viscosity.

6.3.2 Macroscopic Point of View

Now that the microscopic basement of thixotropy has been well established, let us focus on some macroscopic facts.

6.3.2.1 Temperature Effects

First of all, temperature affects the microstructure via the liquid fraction, with a significant effect on viscosity.

In addition, viscosity is itself highly dependent on temperature. For a Newtonian fluid (e.g., the liquid matrix in a semisolid slurry), the viscosity decreases with an increase in temperature.

Finally, over time, the microstructure will coarsen by diffusion and this will be accelerated as the temperature increases.

Thus, the thixotropic material behavior depends highly on temperature, its variation as well as its history, and a constitutive model of thixotropy should therefore be thermomechanically coupled.

6.3.2.2 Yield Stress

Nowadays, there is still a dispute over whether thixotropic semisolid alloys display yield and whether they should be modeled as such. Anyway, in terms of modeling semisolid alloy die fill, the use of a yield stress may be appropriate because a vertical billet does not collapse under its own weight unless the liquid fraction is too high [1].

Moreover, Section 6.4.2.8 shows that the introduction of a finite yield stress in the constitutive model is needed for the prediction of the residual stresses. Yet, in forming processes, the residual stresses are an important indicator of the quality of the final product and a good model should be able to take this into account.

6.3.2.3 Macrosegregation

When semisolid metallic slurries are being deformed, a pressure gradient occurs. It induces a relative motion between the liquid and the solid grains. Thus, liquid tends to flow out toward the surface of the component. This phenomenon can be compared to the squeeze of a sponge and is therefore called *sponge effect*.

It creates segregation between the two phases, from which result nonhomogeneous properties, especially in terms of liquid fraction.

Moreover, it may lead to a too high liquid fraction at the boundaries, which causes deficiencies inherent to casting (shrinkage, porosities, etc.).

This phenomenon cannot be totally eradicated, but can be reduced to its minimum by an appropriate choice of process parameters. For this, it must be well understood and its introduction into modeling may help.

6.4 Numerical Background in Large Deformations

Numerical simulations can be of great help in the development of the thixoforming technology. This section gives a brief review of the theoretical background in which such modeling can be implemented.

During a forming process, large deformations occur. Thus, the hypothesis of small strains and of linear elasticity is no longer applicable to simulate such processes, and the numerical problem is thus nonlinear. The aim of this section is to explain and compare different formulations that deal with large deformations, in terms of kinematics and material behavior description.

6.4.1 Kinematics in Large Deformations

In large deformations, the current and initial configurations are significantly different (Figure 6.14). This implies that the expressions of variables, volume integrals, and so on depend on the configuration, which is not the case under the small strains hypothesis. So, the use of a specific formulation is required to deal with large deformations.

6.4.1.1 Lagrangian Versus Eulerian Coordinate Systems

Let $\mathbf{X}$ be the position vector of a material particle in the reference configuration of volume $\mathbb{V}_0$ (Lagrangian coordinates), $\mathbf{x}$ be the position vector of the same material

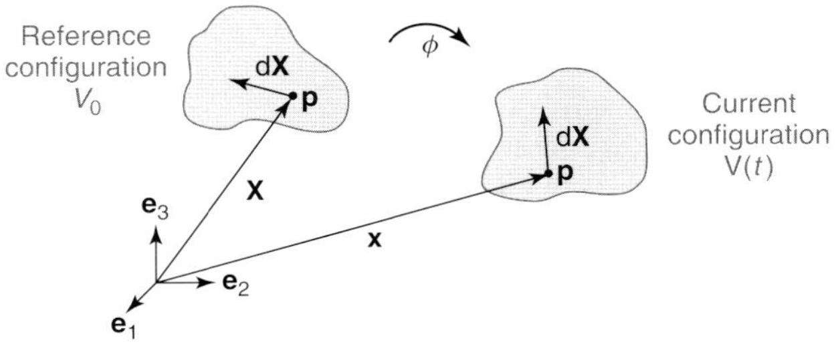

Figure 6.14 Current and reference configurations in large deformations.

particle but in the current configuration of volume $\mathbb{V}(t)$ (Eulerian coordinates), any property can be described as a function of **x** (spatial representation) or **X** (material representation).

In the Lagrangian, also called *material*, coordinate system, the initial position **X** must be known and is used as the independent variable. The displacement is

$$\mathbf{u} = \mathbf{x}(\mathbf{X}, t) - \mathbf{X} \tag{6.1}$$

Thus, the formulation tracks specific identifiable material particles that are carried along with the deformation. In this case, the movement of free boundaries is easily followed and is thus computed automatically. Another advantage of this formulation is that history-dependent materials are relatively easy to handle. However, the knowledge of the history of the particles from the initial conditions is required and is not always easy to get in the case of a fluid flow. In addition, the mesh, which is attached to the material, can be rapidly distorted. For these two reasons, the theory in fluid mechanics is usually developed in the Eulerian representation.

In the Eulerian, also called *spatial*, coordinate system, the initial position **X** is treated as a dependent variable. In the case of fluid mechanics, this initial position **X** is usually unknown. Thus, it is the velocity rather than the displacement that is estimated:

$$\dot{\mathbf{u}} = \dot{\mathbf{x}} = \frac{\mathrm{d}\mathbf{x}}{\mathrm{d}t} \tag{6.2}$$

So, the Eulerian representation observes the velocity at locations that are fixed in space. Since the mesh is attached to spatial coordinates, the material flows through the mesh that remains perfectly undistorted. However, the drawback of such a fixed mesh is that free boundaries are difficult to track.

It is generally assumed that the current position can be written in terms of the reference coordinates by the one-to-one relationship,

$$\mathbf{x} = \phi(\mathbf{X}, t) \tag{6.3}$$

To take advantage of both formulations, while avoiding their drawbacks, mixed Lagrangian–Eulerian approaches have been developed, like the arbitrary Lagrangian–Eulerian (ALE) formulation [15] that is used in the applications shown in Section 6.7.

6.4.1.2 Deformation Gradient and Strain Rate Tensors

This section gives a brief review of some kinematics notations used in large deformations.

The deformation gradient, which describes the motion from the reference configuration to the current one, is a two-point tensor:

$$\mathbf{F} = \frac{\partial \mathbf{x}}{\partial \mathbf{X}} \text{ with } J = \det \mathbf{F} > 0 \tag{6.4}$$

The spatial gradient of velocity is given by

$$\mathbf{L} = \frac{\partial \dot{\mathbf{x}}}{\partial \mathbf{x}} = \dot{\mathbf{F}}\mathbf{F}^{-1} \tag{6.5}$$

and can be split into its symmetric and antisymmetric parts, according to

$$\mathbf{L} = \mathbf{D} + \mathbf{W} \tag{6.6}$$

With

$$\mathbf{D} = \frac{1}{2}(\mathbf{L} + \mathbf{L}^{\mathrm{T}}) \text{ strain rate (symmetric)} \tag{6.7}$$

$$\mathbf{W} = \frac{1}{2}(\mathbf{L} - \mathbf{L}^{\mathrm{T}}) \text{ spin tensor (antisymmetric)} \tag{6.8}$$

By essence, the spin tensor is neglected under the small strains hypothesis and is thus specific to large deformations.

6.4.2 Finite Deformation Constitutive Theory

Behaviors of material submitted to large deformations, as well as metallic alloys at high temperature or, more specifically, thixotropic material, are highly nonlinear. This section thus sets the framework in which a constitutive law of thixotropy can be established.

6.4.2.1 Principle of Objectivity

Constitutive equations must be invariant under changes of frame of reference[3]. Indeed, two different observers of the same phenomenon should observe the same stress tensor in a given body, even if they are in relative motion. Functions and fields are called frame in different or *objective* if they transform according to the rules established for the change of frame. The constitutive equations must be written only in terms of objective quantities. However, the spin tensor **W** and thus the spatial gradient of velocity **L** are not objective; and so is the time derivative of any Eulerian second-order objective tensor (e.g., the Cauchy stress tensor $\boldsymbol{\sigma}$ is objective, and it is easy to demonstrate that $\dot{\boldsymbol{\sigma}}$ is not). If a constitutive law uses time derivative, it should be rewritten into a new form. For example, an objective

3) A frame of reference is a set of points moving in a rigid body motion (linked to an observer).

derivative of the stress tensor can be used, like the Jaumann derivative, which is written as

$$\overset{\nabla J}{\boldsymbol{\sigma}} = \dot{\boldsymbol{\sigma}} - \mathbf{W}\boldsymbol{\sigma} + \boldsymbol{\sigma}\mathbf{W} \tag{6.9}$$

We have seen in Section 6.2 that a thixotropic behavior is mainly characterized by a high dependence on strain rate. So, it is often found in the literature that thixotropic material properties depend on time and that there exist models where the variable time is explicitly introduced into the constitutive law. Yet, time is not an objective quantity and this rate sensitivity (rather than time sensitivity) exhibited by thixotropic material should be taken into account by other means, such as a strain rate dependence since **D** is indeed an objective quantity.

6.4.2.2 **Different Classes of Materials**

Semisolid contains both liquid and solid and behave as a solid or as a fluid depending on local process conditions. Thus, the first distinction to be made is between solid and liquid materials.

It is interesting to note that the designation of solid or liquid is more or less a matter of convenience. Actually, this designation intrinsically depends on the ratio of the material characteristic time τ_{char} with the experiment duration τ_{exp}. This is expressed by the Deborah number [16] : $De = \tau_{\text{char}}/\tau_{\text{exp}}$. So, at room temperature, the characteristic time of glass is of the order of a century while the one of water is of a nanosecond. So, if the experiment duration is some hundreds of years, glass could be considered as a liquid (deterioration of stained glass window in very old churches is an experimental proof that "solids" flow). "It therefore appears that the Deborah number is destined to become the fundamental number of rheology, bringing solids and fluids under a common concept" [16]. The larger the Deborah number, the more solid the material; the smaller the Deborah number, the more fluid it is.

Still, in the present framework, the experiment durations are in the order of the second and the classical concept of liquid and solid remains applicable.

Solids As a solid is deformed beyond a certain level of strain, permanent, nonrecoverable, or plastic deformation can occur. The plastic behavior depends on the loading history. Among solid materials, we can distinguish the different behaviors illustrated in Figure 6.15 for a 1D problem:

- *Elastic*: Reversible and time independent. The mechanical energy is stored into elastic strain. The elastic strain response is instantaneous when the stress is modified. This behavior is generally observed in rubber materials or under the small strains hypothesis.
- *Viscoelastic*: Reversible but time dependent (creep, relaxation). Upon load release, some time is required before total recovery.
- *Elastoplastic*: Irreversible and time independent. Permanent deformations are induced. This behavior is typical of metallic alloys at room temperature.
- *Elastovisco-plastic*: Irreversible and time dependent. This behavior is typical of hot metallic alloys including thixotropic semisolid metallic alloys.

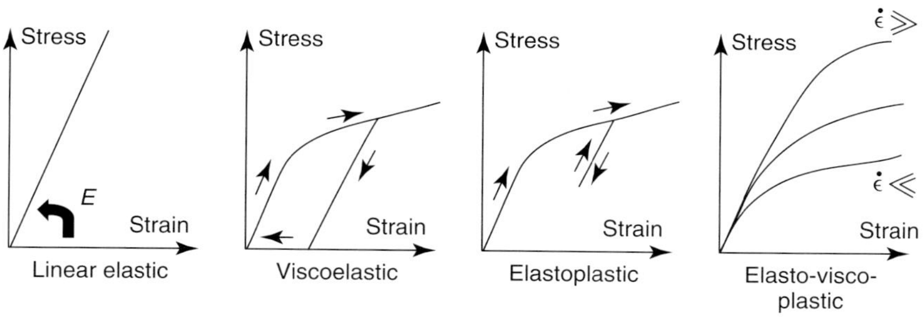

Figure 6.15 Different types of solid material behaviors in a stress–strain diagram ($\dot{\epsilon}$ is the strain rate).

There are two main ways to describe a "solid" constitutive law: hypo- and hyperelastic formulations. In hyperelastic formulations, stresses are derived from a potential while in hypoelastic formulations they are represented by an evolution law of the type

$$\dot{\boldsymbol{\sigma}} = f(\boldsymbol{\sigma}, \mathbf{L}) \tag{6.10}$$

which, in order to fulfill the objectivity principle, has to be rewritten:

$$\overset{\nabla}{\boldsymbol{\sigma}} = f(\boldsymbol{\sigma}, \mathbf{D}) \tag{6.11}$$

In this chapter, only the hypoelastic formulation is treated.

Liquids Liquids have very weak resistance to shear and tensile loading. Liquids cannot prevent themselves from shearing because they cannot develop enough restoring force to balance such a stress (as solids do). This means that any shear stress applied on a liquid leads to steadily increasing strains and, thus, to a flow. Therefore, it does not make sense to relate stress to strain (as it is done for solids), but rather to strain rate. The main material parameter of this relationship is the viscosity η, which is not necessarily constant since internal structural changes induced by the deformation can cause a variation of viscosity. Thus, one can write a canonical constitutive equation of the form:

$$\boldsymbol{\sigma} = g(\mathbf{D}, \eta) \tag{6.12}$$

The deformation is irreversible as the mechanical energy is converted into heat via dissipation through the viscosity.

There are four types of liquid behaviors illustrated in Figure 6.16 for a simple shear test:

- *Newtonian*: Linear viscous behavior, characterized by a constant viscosity.
- *Pseudoplastic*: Also known as *shear-thinning Behavior*, the viscosity is not constant and is called the *apparent viscosity*. It decreases when the shear rate increases.

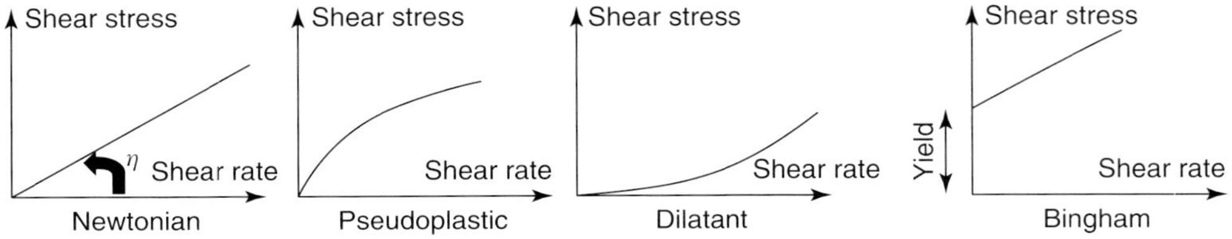

Figure 6.16 Different types of liquid material behaviors.

- *Dilatant*: Also known as *shear-thickening Behavior*, the viscosity increases with the shear rate.
- *Bingham*: The concept of plasticity is introduced in this model by means of a yield stress. This behavior is similar to the rigid[4] viscoplastic solid behavior.

The following sections focus on the mathematical formulation for these different behaviors. First, two ideal behaviors of linear elastic solid and viscous liquid are developed. Then, some more sophisticated models lying between these two "frontiers" are presented and compared.

6.4.2.3 A Corotational Formulation

As explained in Section 6.4.2.1, in order to satisfy the principle of objectivity, some "generalized time derivative," such as the Jaumann derivative had to be introduced. These slightly complicated expressions can be transformed to a much simpler relation provided an appropriate change of frame, such as in a corotational frame, is performed. A corotational frame can be generated as follows.

Given any skew-symmetric tensor $\boldsymbol{\omega}$, it is possible to generate a group of rotations $\boldsymbol{\rho}$, by solving

$$\begin{cases} \dot{\boldsymbol{\rho}} = \boldsymbol{\omega}\boldsymbol{\rho} \\ \boldsymbol{\rho}(t = t*) = \mathbf{I} \end{cases} \tag{6.13}$$

Thanks to this group of rotations, it is possible to generate a change of frame from the fixed Cartesian axes to a frame rotating with a spin $\boldsymbol{\omega}$. This rotating frame is called a *corotational frame*. Field quantities components in the rotating frame can then be expressed in terms of the same quantity in the fixed Cartesian axes as

$$s^{\mathrm{C}} = s \text{ for a scalar} \tag{6.14}$$

$$\mathbf{q}^{\mathrm{C}} = \boldsymbol{\rho}^{\mathrm{T}}\mathbf{q} \text{ for a vector} \tag{6.15}$$

$$\boldsymbol{\sigma}^{\mathrm{C}} = \boldsymbol{\rho}^{\mathrm{T}}\boldsymbol{\sigma}\boldsymbol{\rho} \text{ for a second order tensor} \tag{6.16}$$

$$\mathcal{H}^{\mathrm{C}} = \boldsymbol{\rho}^{\mathrm{T}}[\boldsymbol{\rho}^{\mathrm{T}}\mathcal{H}\boldsymbol{\rho}]\boldsymbol{\rho} \text{ for a fourth order tensor} \tag{6.17}$$

4) In a rigid model, the elastic part of the deformation is negligible. In this case, there is no strain below the yield stress (infinite elastic modulus).

In this corotational frame, the Cauchy stress tensor $\boldsymbol{\sigma}^{\mathrm{C}}$, as well as its time derivative $\dot{\boldsymbol{\sigma}}^{\mathrm{C}}$, can be written, respectively, as

$$\boldsymbol{\sigma}^{\mathrm{C}} = \boldsymbol{\rho}^{\mathrm{T}}\boldsymbol{\sigma}\boldsymbol{\rho} \tag{6.18}$$

$$\dot{\boldsymbol{\sigma}}^{\mathrm{C}} = \boldsymbol{\rho}^{\mathrm{T}}(\dot{\boldsymbol{\sigma}} - \boldsymbol{\omega}\boldsymbol{\sigma} + \boldsymbol{\sigma}\boldsymbol{\omega})\boldsymbol{\rho} \equiv \boldsymbol{\rho}^{\mathrm{T}}\overset{\nabla}{\boldsymbol{\sigma}}\boldsymbol{\rho} \tag{6.19}$$

By comparing the expressions of the objective Jaumann derivative (6.9) with the time derivative of the Cauchy tensor in the corotational frame (6.19), one can notice that an objective derivative, such as the Jaumann rate, transforms into a simple time derivative in the rotating frame whose rotation is such that $\boldsymbol{\omega} = \mathbf{W}$.

For example, relation (6.11) can be rewritten, provided $\boldsymbol{\omega}$ in Eq. (6.13) is identified to the spin tensor $\mathbf{W}$:

$$\dot{\boldsymbol{\sigma}}^{\mathrm{C}} = f^{\mathrm{C}}(\boldsymbol{\sigma}^{\mathrm{C}}, \mathbf{D}^{\mathrm{C}}) \tag{6.20}$$

with

$$\dot{\boldsymbol{\sigma}}^{\mathrm{C}} = \boldsymbol{\rho}^{\mathrm{T}}\overset{\nabla J}{\boldsymbol{\sigma}}\boldsymbol{\rho} \tag{6.21}$$

$$\boldsymbol{\sigma}^{\mathrm{C}} = \boldsymbol{\rho}^{\mathrm{T}}\boldsymbol{\sigma}\boldsymbol{\rho} \tag{6.22}$$

$$\mathbf{D}^{\mathrm{C}} = \boldsymbol{\rho}^{\mathrm{T}}\mathbf{D}\boldsymbol{\rho} \tag{6.23}$$

In the following sections, since all quantities are expressed in the corotational frame, for the sake of simplicity, we omit the exponent "C" for such quantities (thus, in particular, one has $\overset{\nabla J}{\boldsymbol{\sigma}} \rightarrow \dot{\boldsymbol{\sigma}}^{\mathrm{C}}$ written as $\dot{\boldsymbol{\sigma}}$).

However, it may be noted that the corotational frame is only used for solid description, while the fluid description sticks to the fixed Cartesian frame. It should, however, be clear that corotational frame is only a tool to help integrating the constitutive law (6.20) and that once Eq. (6.23) has been integrated, the resulting stress tensor is brought back to fixed cartesian axis by using the inverse of Eq. (6.23); see Ref. [17] for details.

6.4.2.4 Linear Elastic Solid Material Model

In a hypoelastic formulation, Hooke's law of linear elasticity can be expressed as

$$\dot{\boldsymbol{\sigma}} = \mathcal{H} : \mathbf{D} \tag{6.24}$$

where $\mathcal{H}$ is Hooke's tensor (fourth-order constant tensor) and the symbol ":" represents the double contraction of two tensors, that is, $(\mathcal{H} : \mathbf{D})_{ij} = \mathcal{H}_{ijkl} D_{kl}$.

In the case of an isotropic material, the general Hooke's tensor can be reduced to two scalar quantities called the *Lamé coefficients*: λ^{e} and μ^{e}, and Eq. (6.24) can be written as

$$\dot{\boldsymbol{\sigma}} = 2\mu^{\mathrm{e}}\mathbf{D} + \lambda^{\mathrm{e}}\mathrm{tr}(\mathbf{D})\mathbf{I} \tag{6.25}$$

or, in terms of the deviatoric part of the strain rate tensor $\hat{\mathbf{D}} = \mathbf{D} - \frac{1}{3}\mathrm{tr}(\mathbf{D})\mathbf{I}$:

$$\dot{\boldsymbol{\sigma}} = 2G\hat{\mathbf{D}} + K\,\mathrm{tr}(\mathbf{D})\mathbf{I} \tag{6.26}$$

with

- $G = \frac{E}{2(1+\nu)} = \mu^{e}$, the shear modulus
- $K = \frac{E}{3(1-2\nu)} = \lambda^{e} + \frac{2}{3}\mu^{e}$, the bulk modulus
- E, Young's modulus
- ν, Poisson's ratio

The volumetric part of the stress tensor can be seen as the hydrostatic pressure p in the solid, and the constitutive equation is often written as

$$\dot{\sigma} = 2G\hat{\mathbf{D}} + \dot{p}\mathbf{I} \tag{6.27}$$

with $\dot{p} = K\ \mathrm{tr}(\mathbf{D})$.

6.4.2.5 Linear Newtonian Liquid Material Model

For liquids, it is generally assumed that the stress tensor results from two uncoupled phenomena: the hydrostatic pressure p[5] and the flow, which displays viscous resistance. Thus, the stress tensor is split into two parts:

$$\sigma = \mathbf{s}' + p\mathbf{I} \tag{6.28}$$

where $\mathbf{s}'$ is called the *extra stress tensor*. The hydrostatic pressure p does not depend on the strain field and the constitutive law is expressed in terms of the extra stress only. Since liquids are generally isotropic, it can be written by analogy with the elastic constitutive equation:

$$\mathbf{s}' = 2\mu^{\nu}\mathbf{D} + \lambda^{\nu}\ \mathrm{tr}(\mathbf{D})\mathbf{I} \tag{6.29}$$

or in terms of the deviatoric part of the strain rate tensor $\hat{\mathbf{D}}$:

$$\mathbf{s}' = 2\eta\hat{\mathbf{D}} + \left(\lambda^{\nu} + \frac{2}{3}\mu^{\nu}\right)\ \mathrm{tr}(\mathbf{D})\mathbf{I} \tag{6.30}$$

where $\eta = \mu^{\nu}$ is the viscosity of the material.

In many cases, $\lambda^{\nu} + \frac{2}{3}\mu^{\nu}$ is chosen to be equal to zero (Stoke's hypothesis) so that the second term of Eq. (6.30) can be neglected and eventually, we can write

$$\sigma = 2\eta\hat{\mathbf{D}} + p\mathbf{I} \tag{6.31}$$

which has a form similar to relation (6.27).

The schematic 1D representation of both linear solid and liquid models is pictured in Figure 6.17.

5) Usually, fluid and solid formulations use opposite sign conventions for the hydrostatic pressure. As liquids can only sustain compression loads (thus negative 1D stresses in the "solid" convention), the fluid formulations generally consider compression as positive and traction as negative. For a matter of uniformity, the solid convention is adopted all along this chapter.

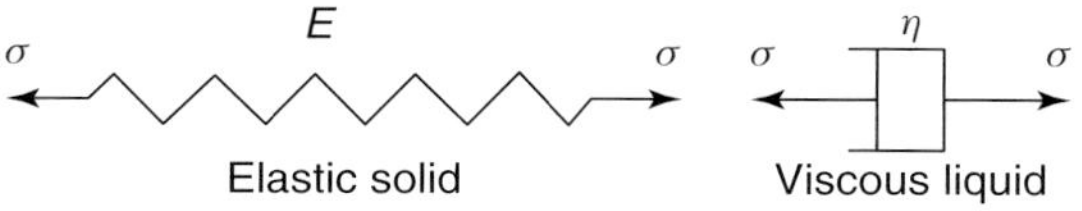

Figure 6.17 Schematic 1D representation of linear solid and liquid behaviors.

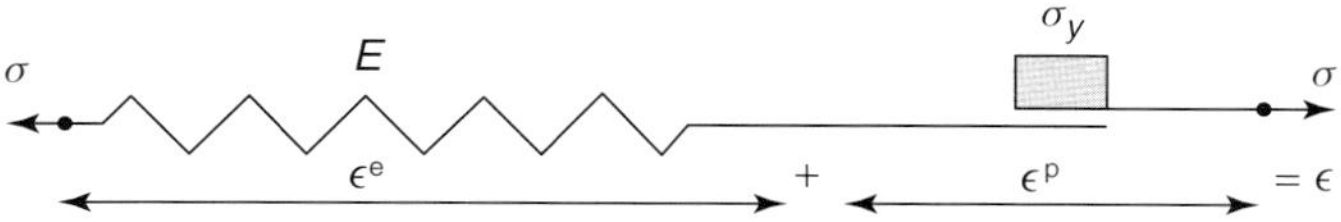

Figure 6.18 Elastoplastic behavior schematic 1D representation.

6.4.2.6 **Hypoelastic Solid Material Models**

Viscous effects are initially neglected and are studied later in this section.

The concept of yield stress σ_y is the key point of the formulation. Below this stress, the material behavior is assumed to be linear elastic and to obey Hooke's law. Beyond it, nonreversible phenomena occur and the material behavior becomes nonlinear.

Basic Hypoelastic Hypotheses The basic assumption for hypoelastic models consists in an additive split of the strain rate into two parts: an elastic and reversible part $\mathbf{D}^e$ and an irreversible plastic part $\mathbf{D}^p$:

$$\mathbf{D} = \mathbf{D}^e + \mathbf{D}^p \tag{6.32}$$

Figure 6.18 illustrates this split in the specific case of a small strains 1D problem, for which the equation $\dot{\epsilon} = \dot{\epsilon^e} + \dot{\epsilon^p}$ implies $\epsilon = \epsilon^e + \epsilon^p$. By obvious inspection of the system in Figure 6.18 and of the subsystem (that must be in equilibrium) in Figure 6.19, one can write in 1D:

$$\sigma = E\epsilon^e = E(\epsilon - \epsilon^p) \tag{6.33}$$

or

$$\dot{\sigma} = E\dot{\epsilon}^e = E(\dot{\epsilon} - \dot{\epsilon}^p) \tag{6.34}$$

which is generalized to 3D as

$$\dot{\boldsymbol{\sigma}} = \mathcal{H} : (\mathbf{D} - \mathbf{D}^p) \tag{6.35}$$

Plastic Part of the Strain Rate Evolution This section focuses on the concepts of plastic strain, of yield stress, and of its evolution.

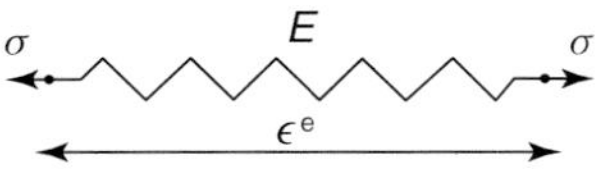

Figure 6.19 Elastic subsystem at equilibrium that leads to the constitutive relation $\sigma = E\epsilon^e$.

- *Yield criterion*: To define the onset of plastic deformation, the stress state is compared to the yield stress. For this quantitative comparison, a scalar quantity is needed. In general, the stress is a tensor and a way to measure its magnitude must be adopted. Thus, the existence of a ***yield surface*** f, defined in the 6D space of stresses, is assumed. This surface detects whether the deformation is elastic or plastic:

$$f(\boldsymbol{\sigma}, T, \mathbf{q}) = 0 \tag{6.36}$$

where T is the temperature and $\mathbf{q}$ is the set of internal parameters, which describes, at the macroscopic level, the microscopic state of the material and which is an image of the past history of the deformation (so far, T and $\mathbf{q}$ are considered as parameters).
This surface divides the space into three regions, as illustrated in Figure 6.20:

$$f < 0: \text{ Elastic deformation domain}$$
$$f = 0: \text{ Plastic deformation domain}$$
$$f > 0: \text{ No-access region}$$

Thus, one of these three areas is characterized by a nonadmissible behavior and a restriction on the behavior evolution to thermodynamically admissible states should be set up by $f(\boldsymbol{\sigma}, T, \mathbf{q}) \leq 0$. This additional condition imposes that, during a plastic deformation, the yield criterion evolves and the undergone thermodynamical states must remain on the evolving function $f = 0$ or, in other words, that the stress state must remain on the subsequent yield surface. The mathematical expression of this condition, which is called the *consistency condition*, is

$$\dot{f} = \frac{\partial f}{\partial \boldsymbol{\sigma}} : \dot{\boldsymbol{\sigma}} + \frac{\partial f}{\partial \mathbf{q}} * \dot{\mathbf{q}} + \frac{\partial f}{\partial T}\dot{T} = 0 \tag{6.37}$$

where the symbol "$*$" represents the product if used with two scalars, the scalar product if used with two vectors of the same dimension $\mathbf{a}$ and $\mathbf{b}$ ($\mathbf{a} * \mathbf{b} = \mathbf{a}^{\mathrm{T}}\mathbf{b} = a_i b_i$), and a double contraction if used with two second-order tensors $\mathbf{A}$ and $\mathbf{B}$ ($\mathbf{A} * \mathbf{B} = \mathbf{A} : \mathbf{B} = A_{ij}B_{ij}$).

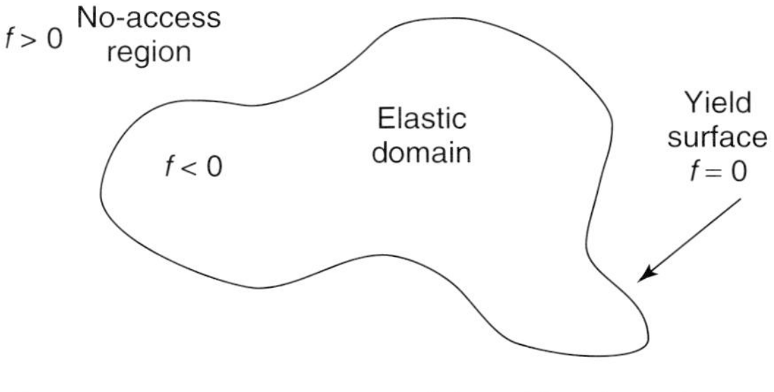

Figure 6.20 2D representation of the yield surface (the actual space is 6D).

The function f can be expressed as a function of the scalar representation of the stress state $\overline{\sigma}$ and of a critical stress σ_{crit}:

$$f = \overline{\sigma} - \sigma_{\text{crit}} \tag{6.38}$$

There exist several expressions for the yield surface function f depending on the criterion used for the equivalent stress $\overline{\sigma}$ and on the critical stress σ_{crit}. In the following sections, the *von Mises criterion* is adopted, which means that the von Mises equivalent stress $\overline{\sigma}^{\text{VM}}$ and the yield stress σ_{y} are used. The equation of this particular yield surface is written as

$$\begin{cases} f(\boldsymbol{\sigma}, T, \mathbf{q}) = \overline{\sigma}^{\text{VM}} - \sigma_{\text{y}}(\mathbf{q}, T) \\ \overline{\sigma}^{\text{VM}} = \left(\dfrac{3}{2}(\mathbf{s} - \mathbf{C}(\mathbf{q}, T)) : (\mathbf{s} - \mathbf{C}(\mathbf{q}, T))\right)^{1/2} \end{cases} \tag{6.39}$$

where $\mathbf{C}$ is the backstress deviatoric tensor relative to kinematic hardening.
In the space of principal stresses (i.e., eigenvalues of the stress tensor), this equation represents a cylinder whose axis is the angle trisector of the principal stresses (hydrostatic state of stresses: $\sigma_1 = \sigma_2 = \sigma_3$), whose radius is equal to $\sqrt{2/3}\sigma_{\text{y}}$, and whose center is at a distance of $\overline{C}$ (called *equivalent backstress*) from the origin, given by

$$\overline{C} = \sqrt{\frac{3}{2}\mathbf{C} : \mathbf{C}} \tag{6.40}$$

- *Flow rule*: It is assumed that the evolution of the plastic part of the strain rate runs along the normal direction to a scalar plastic potential g and also that this plastic potential g is identified as the yield surface f itself (associative plasticity). This gets

$$\mathbf{D}^{\text{p}} = \xi\frac{\partial g}{\partial \boldsymbol{\sigma}} = \xi\frac{\partial f}{\partial \boldsymbol{\sigma}} \tag{6.41}$$

where $\xi(\boldsymbol{\sigma}, T, \mathbf{q})$ is a consistency or flow parameter. It can be calculated using the consistency condition (6.37).
By defining $\Lambda = \xi\left\|\frac{\partial f}{\partial \boldsymbol{\sigma}}\right\|$ and the unit external normal direction $\mathbf{N}$ (such that $\mathbf{N} : \mathbf{N} = 1$) to the yield surface f, the flow rule can be expressed by

$$\mathbf{D}^{\text{p}} = \xi\frac{\partial f}{\partial \boldsymbol{\sigma}} = \Lambda\mathbf{N} \tag{6.42}$$

In the particular case of von Mises criterion and isotropic hardening, the normal direction can be written as

$$\mathbf{N} = \frac{\mathbf{s}}{\sqrt{\mathbf{s} : \mathbf{s}}} \tag{6.43}$$

- *Internal variables evolution laws and hardening*: Internal variables change with irreversible deformations. The parameters of the yield criterion evolve with those internal variables according to hardening laws.
There exist several possibilities of such internal variables, the one picked here being the equivalent plastic strain $\overline{\epsilon}^{\text{p}}$. It is defined through its rate by

$$\dot{\overline{\epsilon}}^{\text{p}} = \sqrt{\frac{2}{3}\mathbf{D}^{\text{p}} : \mathbf{D}^{\text{p}}} \tag{6.44}$$

which, according to Eq. (6.42), leads to

$$\dot{\bar{\epsilon}}^{\mathrm{p}} = \sqrt{\frac{2}{3}}\Lambda \tag{6.45}$$

In the case of the von Mises criterion (Eq.(6.39)), the internal parameters have an impact on the yield stress σ_{y} and on the backstress tensor **C**. So, two types of hardening can be set apart:

- The isotropic hardening refers to the evolution of the yield stress σ_{y}, namely, the radius of the yield surface. It represents an inflation of the yield surface.
- The kinematic hardening refers to the evolution of the backstress tensor **C**, namely, the position of the center of the yield surface in the stress space. It represents a shift of the yield surface.

In the following sections, only the isotropic hardening is considered.
The isotropic hardening law will complete the present constitutive formulation. It describes the evolution of the yield stress during a plastic deformation. For example, a linear hardening is written as [18]

$$\sigma_{\mathrm{y}}(\bar{\epsilon}^{\mathrm{p}}, T) = \sigma_{\mathrm{y}}^{0}\left(1 - \omega_{\sigma_{\mathrm{y}}}(T - T_{\mathrm{ref}})\right) + h(T)\bar{\epsilon}^{\mathrm{p}} \tag{6.46}$$

where $\omega_{\sigma_{\mathrm{y}}}$ is the linear softening coefficient of the initial yield stress σ_{y}^{0}, T_{ref} is the reference temperature at which the yield stress is equal to σ_{y}^{0}, and h is an hardening coefficient (h is constant in the case of linear hardening). Several more or less complex forms can be defined to simulate the yield stress evolution.

Stress Rate As in the elastic case, the stress rate tensor can be split into its volumetric and deviatoric parts:

$$\dot{\boldsymbol{\sigma}} = \dot{\mathbf{s}} + \frac{1}{3}\mathrm{tr}(\dot{\boldsymbol{\sigma}})\mathbf{I} = \dot{\mathbf{s}} + \dot{p}\mathbf{I} \tag{6.47}$$

In many cases, the hydrostatic pressure p does not induce any plastic deformation. Actually, the von Mises criterion, which is written in terms of the deviatoric stress tensor, is independent of the hydrostatic pressure. The plastic strain rate is thus deviatoric (Eqs (6.42) and (6.43)) and the trace of the elastic strain rate tensor is equal to the one of the total strain rate tensor. Considering Eq. (6.26), the pressure variation can be written as

$$\dot{p} = K\,\mathrm{tr}(\mathbf{D}^{\mathrm{e}}) = K\,\mathrm{tr}(\mathbf{D}) \tag{6.48}$$

However, the deviatoric elastic strain rate does depend on the plastic strain rate:

$$\hat{\mathbf{D}}^{\mathrm{e}} = \hat{\mathbf{D}} - \hat{\mathbf{D}}^{\mathrm{p}} = \hat{\mathbf{D}} - \mathbf{D}^{\mathrm{p}} \tag{6.49}$$

Considering Eqs (6.25) and (6.42), the deviatoric stress rate is written as

$$\dot{\mathbf{s}} = 2G\hat{\mathbf{D}}^{\mathrm{e}} = 2G(\hat{\mathbf{D}} - \mathbf{D}^{\mathrm{p}}) = 2G(\hat{\mathbf{D}} - \Lambda\mathbf{N}) \tag{6.50}$$

Viscous Effects To describe a viscoplastic behavior, a von Mises yield criterion can still be used. In this case, the region $f > 0$ is no longer forbidden, but a stress state inside this region means that the behavior is viscoplastic. Thus, the consistency condition (6.37) is no longer applicable and an additional equation is required to describe the evolution of Λ. This evolution law represents the viscous behavior. One largely used example of such law is the Perzyna model:

$$\Lambda = \sqrt{\frac{3}{2}} \left\langle \frac{\overline{\sigma} - \sigma_y}{k(\overline{\epsilon}^{vp})^n} \right\rangle^{1/m} \tag{6.51}$$

where m is a viscosity exponent, n a hardening exponent, k a viscosity parameter, and $\langle\rangle$ the M^{ac} Auley brackets. All these parameters can depend on the temperature and on the internal variables.

Keeping a unified formalism with plasticity, the total strain rate is split here such that $\mathbf{D} = \mathbf{D}^e + \mathbf{D}^{vp}$, where $\mathbf{D}^{vp}$ represents the viscoplastic (irreversible) part of the strain rate. In such a formalism, the flow rule can still be written:

$$\mathbf{D}^{vp} = \Lambda \mathbf{N} = \sqrt{\frac{3}{2}} \left\langle \frac{\overline{\sigma} - \sigma_y}{k(\overline{\epsilon}^{vp})^n} \right\rangle^{1/m} \mathbf{N} \tag{6.52}$$

By combining Eqs (6.45) and (6.51), we get the von Mises criterion generalized to viscoplastic behavior, that is, where σ_{crit} has been generalized to take into account viscous effects:

$$\begin{cases} \overline{f} = \overline{\sigma}^{VM} - \sigma_{crit} = 0 \\ \sigma_{crit} = \sigma_y + k(\overline{\epsilon}^{vp})^n(\dot{\overline{\epsilon}}^{vp})^m \end{cases} \tag{6.53}$$

Figure 6.21 illustrates the elastoviscoplastic model under small strains assumptions and in the case of 1D problem. It is an extension of Figure 6.18 to viscous behavior.

Thermal Strain Rate To take into account the thermal dilatation of the material, an additional contribution to the strain rate is inserted: the thermal strain rate $\mathbf{D}^{th}$. The most common way to express this quantity is as follows [18]:

$$\mathbf{D}^{th} = \alpha \dot{T} \mathbf{I} \tag{6.54}$$

where α is the linear dilatation coefficient.

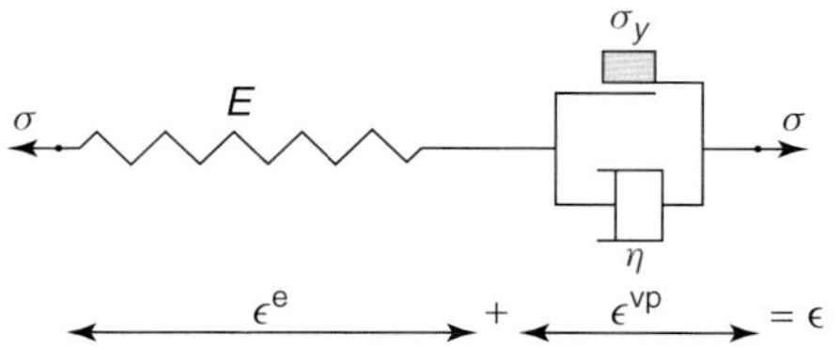

Figure 6.21 Elastoviscoplastic behavior schematic 1D representation.

6.4.2.7 Liquid Material Models

In a liquid formalism, the constitutive law is simply driven by the apparent viscosity evolution. A widely used viscosity law is the Herschell–Bulkley expression:

$$\eta = \frac{\sigma_y^0}{\dot{\bar{\epsilon}}} + k(\dot{\bar{\epsilon}})^{m-1} \tag{6.55}$$

Introducing this expression into the linear liquid formula (6.31) leads to the overall constitutive law:

$$\boldsymbol{\sigma} = 2\left(\frac{\sigma_y^0}{\dot{\bar{\epsilon}}} + k(\dot{\bar{\epsilon}})^{m-1}\right)\hat{\mathbf{D}} + p\mathbf{I} \tag{6.56}$$

6.4.2.8 Comparison of Solid and Liquid Approaches

Behavior of liquids and solids under hydrostatic pressure are comparable, so we focus on the deviatoric stresses to highlight the main differences between both formalisms. First, we look at the linear models. With a fluid formalism, in the incompressible linear (Newtonian) case, we have, according to Eq. (6.31),

$$\text{Liquid approach: } \mathbf{s} = 2\eta\hat{\mathbf{D}} \tag{6.57}$$

With a solid formalism, in the linear (elastic) case, we have, according to Eq. (6.27),

$$\text{Solid approach: } \dot{\mathbf{s}} = 2G\hat{\mathbf{D}} \tag{6.58}$$

We can see that if the deformation is frozen ($\hat{\mathbf{D}} \to 0$, which does not mean that there is no strain left), the deviatoric stresses given by the fluid law (6.57) tends to zero whatever the viscosity.

Let us now compare both formulations when nonlinearities are introduced. In the general viscoplastic model, by rearranging Eqs (6.39), (6.42), (6.43), (6.45), and (6.53), we can show that

$$\mathbf{s} = \frac{2}{3}\left[\sigma_y + k(\bar{\epsilon}^{vp})^n(\dot{\bar{\epsilon}}^{vp})^m\right]\frac{\mathbf{D}^{vp}}{\dot{\bar{\epsilon}}^{vp}} \tag{6.59}$$

which is equivalent to the deviatoric part of the liquid Eq. (6.56) in the particular case of

- No hardening: At high temperatures, Strain hardening is really low and the solid material will deform by viscous flow: $\sigma_y = \sigma_y^0$ and $n = 0$.
- Rigid model: In the case of a liquid, the elastic part of the deformation is very small and can be neglected: $\mathbf{D}^e \simeq 0 \Rightarrow \mathbf{D}^{vp} \simeq \mathbf{D} = \hat{\mathbf{D}}$ and $\dot{\bar{\epsilon}}^{vp} = \dot{\bar{\epsilon}}$.

So, the so-called solid formalism is not restricted to solid materials (at temperature under solidus) and can also represent a fluid flow. In addition, this comparison shows that the implication of the elastic deformation is necessary in case the residual stresses are requested. Indeed, only a nonrigid solid formalism offers the possibility to assess the residual stresses after unloading and cooling down to room temperature. A constitutive law of the thixotropic behavior should be accurate in the whole range of temperatures occurring during the full process, including cooling back to room temperature. A good model of thixotropy should then be able to predict the behavior of a built up semisolid material, of free solid suspensions, or even the elastic behavior (to describe the cooling down to room temperature and thus the residual stresses).

6.5 State-of-the-Art in FE-Modeling of Thixotropy

This section lists the existing models of thixotropic behavior in a nonexhaustive way. They are classified in terms of whether the modeling is one phase or two phase.

The models should introduce the structural considerations discussed in the previous section about rheology and should be thermomechanically coupled.

6.5.1 One-Phase Models

In one-phase models, the material is regarded as a single homogeneous (though mixed) continuous phase and the relative displacement between the phases cannot be taken into account.

This kind of modeling is relatively simple to implement as it concerns only the constitutive law, which is not the case of a two-phase model. We have seen in Section 6.4.2 that, using a liquid formalism, the constitutive law is driven by the apparent viscosity and that a yield stress can be introduced. These two quantities are also fundamental in the solid hypoelastic formulation, as it is driven by viscosity and isotropic hardening[6] laws. Thus, despite the fact that existing one-phase models can differ by the formulation that is chosen (liquid or solid, hypoelastic or hyperelastic, etc.), the models are classified by the choice of viscosity and yield stress evolution form. All the equations that are presented in this section can be implemented in any constitutive formalism.

During the process, the material structure changes with the strain history due to the agglomeration of the particles and the breaking of the grain bonds. Thus, some models use an internal scalar parameter that describes the state of the microstructure. It can be the degree of structural buildup [19], the number of bonds [1], or the average numbers of particles in each agglomerate [12]. Various names, notations, and evolution descriptions can be found in the literature. However, all these formulations can be reduced into a single one common to all models, which

6) No kinematic hardening has been proposed for thixotropy yet. This is probably due to the lack of experimental data. In addition, it is commonly accepted that the thixotropic behavior is mainly driven by viscous effects so that it is not the considering such a complex kinematic hardening.

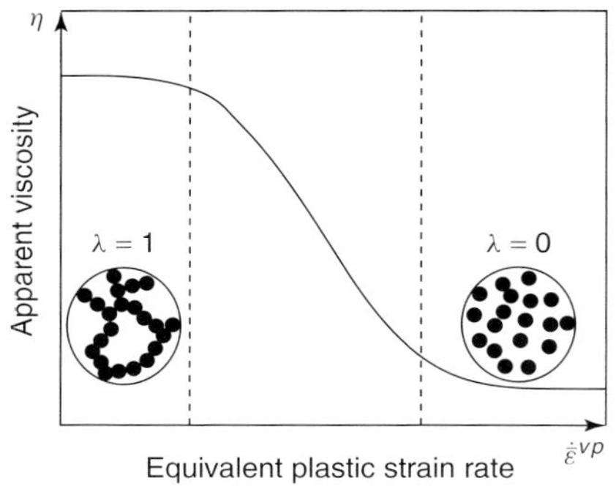

Figure 6.22 Illustration of the cohesion degree.

allows a more direct comparison. This formulation is detailed further (Section 6.6). Here, the structural parameter that has been chosen to present all models is the cohesion degree λ, illustrated in Figure 6.22. It characterizes the degree of buildup in the microstructure. It can take values between 0 and 1: $\lambda = 1$ if the structure is fully built up and $\lambda = 0$ if the structure is fully broken.

Many models also introduce the liquid fraction f_l or the solid fraction f_s as an internal parameter. To ease their comparison, all models are presented here in terms of the liquid fraction. Generally, this only depends on the temperature.

6.5.1.1 Apparent Viscosity Evolution

This section gives a nonexhaustive list of proposed mathematical descriptions of the apparent viscosity in terms of the equivalent viscoplastic strain rate[7] noted $\dot{\bar{\epsilon}}^{vp}$ and, in some cases, of the cohesion degree λ as well as the liquid fraction f_l.

Even though the yield stress is taken as a term of the viscosity in some liquid models, it is not included here since the next section is devoted to the yield stress.

Models Based on the Norton–Hoff Law Many one-phase models [19, 20] are based on the Norton–Hoff law, which in terms of the evolution of viscosity, is also known as the Ostwald–de-Waele [1] relationship:

$$\eta = k(\dot{\bar{\epsilon}}^{vp})^{m-1} \tag{6.60}$$

The complex rheology of thixotropy is taken into account by means of the structural parameter (and sometimes the liquid fraction) via the parameters k and m in various ways. For example, Modigell and Koke [20] suggested

$$\begin{cases} k = \lambda k_1 e^{k_2(1-f_l)} \\ m = m_1 \end{cases} \tag{6.61}$$

7) This is the most general term; it can be reduced as the shear rate $\dot{\gamma}$ in the case of the description of a simple shear test with a liquid formalism.

Or, as another example, Burgos and Alexandrou [19] proposed the following empirical relationships:

$$\begin{cases} k = k_1 e^{k_2 \lambda} \\ m = m_1 + m_2\lambda + m_3\lambda^2 \end{cases} \tag{6.62}$$

where k_1, k_2, m_1, m_2, and m_3 are constant material parameters.

Models of this kind have the advantage of being simple and numerically efficient. The cited models have been implemented within a fluid formalism and validated on shear-rate jump experiments. Thus, relatively simple experiments allow for identification of material parameters.

Micro–Macro Model Favier *et al.* [21] developed a rigid thermoviscoplastic model using micromechanics and homogenization techniques. This original model individualizes the mechanical role of the nonentrapped and entrapped liquid and of the solid bonds and the solid grains in the deformation mechanisms. The microstructure is represented by *coated inclusions* (Figure 6.23); the inclusion is composed of the solid grains and the entrapped liquid, whereas the coating, called the *active zone*, is composed of the solid bonds and the nonentrapped liquid.

To determine the viscosity, a self-consistent approximation is used at two scales. The macroscopic apparent viscosity η is deduced from the microscopic apparent viscosities η_A and η_I of the active zone and of the inclusion, respectively, that are both calculated from the liquid and solid behavior, according to Figure 6.24.

Subscripts "A" and "I" stand for the active zone and the inclusion, respectively, and superscripts "s" and "l" correspond to the solid phase and the liquid phase, respectively.

The variables $f_{A,I}^{s,l}$ are the fraction of the different separated scales. Among them, the fraction of active zone f_A is a material parameter. The solid fraction in the active zone f_A^s is equivalent to the cohesion degree λ. Indeed, if $f_A^s = 0$, there are no solid bonds, which corresponds to $\lambda = 0$. On the other hand, if $f_A^s = 1$, all the liquid is entrapped, which is characteristic of a fully built up structure ($\lambda = 1$). It can be seen in Figure 6.24 that all the remaining "zone fraction" (f_I, f_I^s, f_I^l, and f_A^l) can be deduced from f_A and $f_A^s = \lambda$.

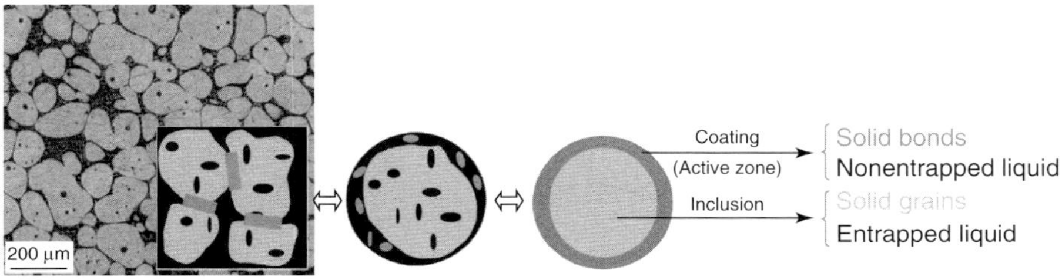

Figure 6.23 Microstructure of semisolid alloy and schematic representation in the micro–macro model [21].

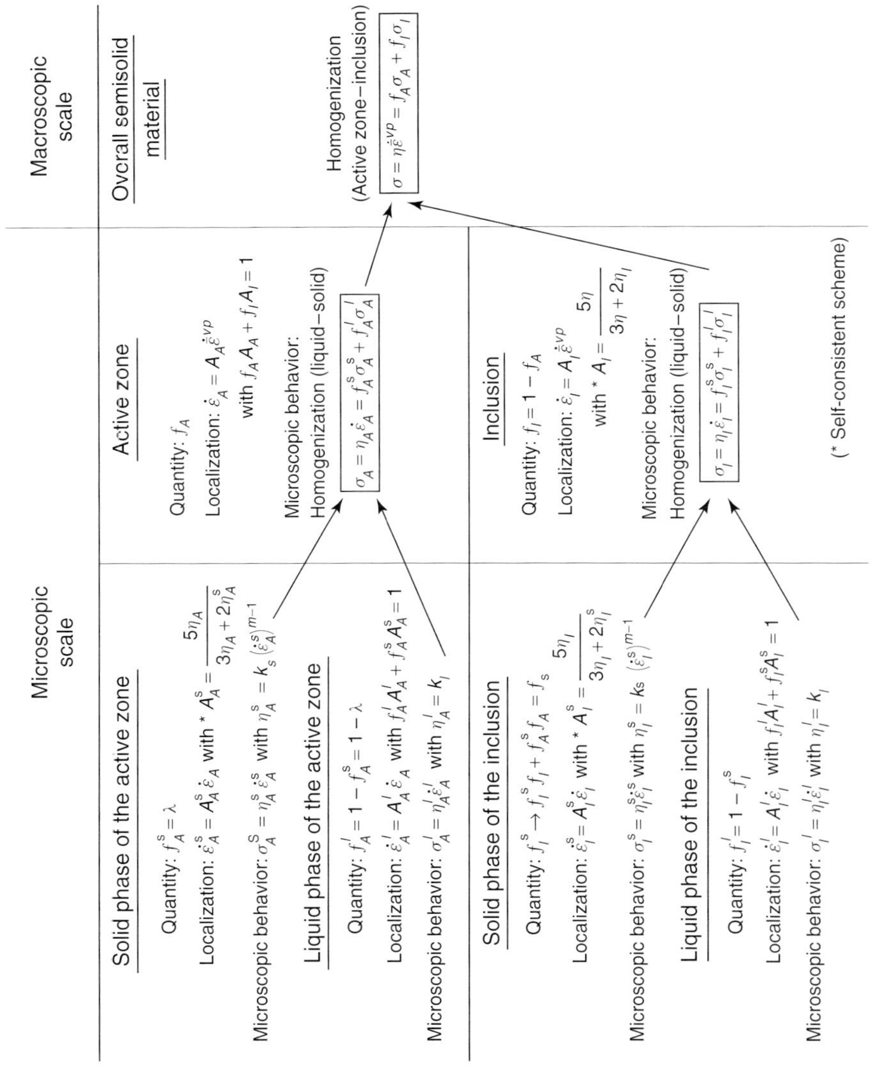

Figure 6.24 Self-consistent approximation at two scales in the micro–macro model.

The variables $A^{s,l}_{A,I}$ are the localization variables of the equivalent plastic strain rate from one scale to a smaller one. For example, A^s_I and A^s_A are the equivalent plastic strain rate localization variables of the solid phase in the inclusion and the active zones, respectively, and A_I is the equivalent plastic strain rate localization variable of the inclusions in the global semisolid material. According to the self-consistent scheme, these variables depend on the viscosities, and the formulation can be reduced to the system of three equations, Eqs (6.63)–(6.65) for three unknowns (η_A, η_I, and η) that is solved numerically by Newton–Raphson iterations:

$$\eta = \eta_A + (1 - f_A)(\eta_I - \eta_A)A_I \tag{6.63}$$

$$\eta_A = k_l + \lambda \left(k_s(\dot{\bar{\epsilon}}^s_A)^{m-1} - k_l\right) A^s_A \quad \text{with} \quad \dot{\bar{\epsilon}}^s_A = A^s_A \frac{1 - A_I(1 - f_A)}{f_A} \dot{\bar{\epsilon}}^{vp} \tag{6.64}$$

$$\eta_I = k_l + \frac{1 - f_l - f_A\lambda}{1 - f_A} \left(k_s(\dot{\bar{\epsilon}}^s_I)^{m-1} - k_l\right) A^s_I \quad \text{with} \quad \dot{\bar{\epsilon}}^s_I = A^s_I A_I \dot{\bar{\epsilon}}^{vp} \tag{6.65}$$

Overall, the predicted viscosity as a function of the strain rate is as represented in Figure 6.25, in the case of tin-15% wt lead alloy. The model predicts a thixotropic behavior characterized by a decreasing viscosity with increasing strain rate, followed by a liquidlike behavior characterized by a very low (but nonzero) constant viscosity. The transition between these "thixotropic" and "liquid" regimes has been observed in real experiments, but is too drastic in the numerical simulations. Real mechanical percolation effect is softer and more complex. This is due to the formulation of the self-consistent model.

This model is more complex than the previous ones and is thus numerically more expensive. However, according to the authors, it has the advantage that it is

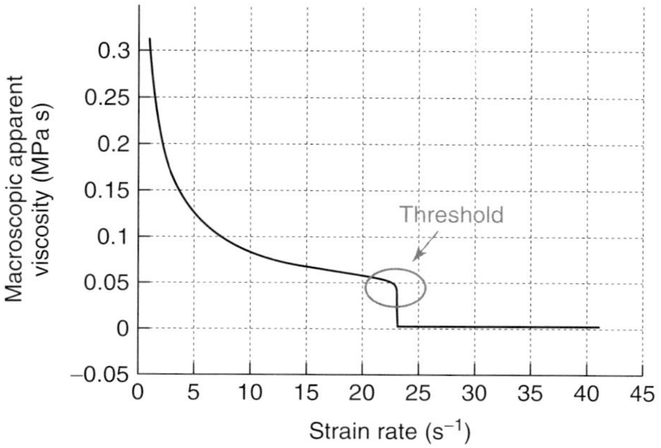

Figure 6.25 Evolution of the macroscopic apparent viscosity as a function of the equivalent plastic strain rate predicted by the micro–macro model in the case of a tin-15% wt lead alloy.

physically and not phenomenologically based, leading to a more straightforward identification of the material parameters. In addition, it gives interesting results in terms of the softening of the material that is experienced beyond a certain level of deformation.

Law Considering the Effective Liquid Fraction Lashkari and Ghomashchi [12] conducted viscometer measurements from which they deduced a viscosity law. This is not strictly modeling since they did not implement it, but the formulation they proposed is original and worth pointing out. Instead of introducing the structural parameter via the power-law parameters, it is the effective liquid fraction that is used. We have seen that some entrapped liquid is released with structure break down; thus, it is clear that the effective liquid fraction depends on both the liquid fraction and the structural parameter.

The authors use a fluid formalism and consider a microstructure made of suspension of clusters. Thus, their results are not applicable if a solid network exists in the material. They proposed the following viscosity law:

$$\eta = \eta_0 (f_l^{\text{eff}})^{-m} \tag{6.66}$$

where η_0 is the viscosity of the liquid matrix and where $f_l^{\text{eff}}(\lambda, f_l)$ is a function of both the liquid fraction and the structural parameter. In their work [12], the authors use another structural parameter, but for a matter of clarity to compare all the different models, their formulation is rewritten here in terms of λ. They proposed the relation

$$f_l^{\text{eff}} = f_l\left[1 - \lambda(1 - f_l)\right] \tag{6.67}$$

This formulation has not been validated yet, but can look attractive because of its simplicity and its need for only two material parameters.

Models with No Structural Parameter Some authors have devised theories that circumvent the use of any structural parameter such as λ, and instead use the viscosity as a direct measure of the structural behavior.

In some cases [1], this is expressed by relating the rate of viscosity to the viscosity difference between the steady state η_e and the current viscosity, which leads to a differential equation of the type

$$\frac{d\eta}{dt} = k(\eta_e(\dot{\overline{\epsilon}}^{\text{vp}}) - \eta)^m \tag{6.68}$$

where η_e can be expressed in different ways, for example, [1]

$$\eta_e = a e^{b(1-f_l)} (\dot{\overline{\epsilon}}^{\text{vp}})^c \tag{6.69}$$

These models have been validated and matched to experiments with reasonable accuracy.

The viscosity can also be expressed only in terms of the strain rate. As illustrated in Figure 6.26, the Cross model [1] assumes that the behavior becomes Newtonian

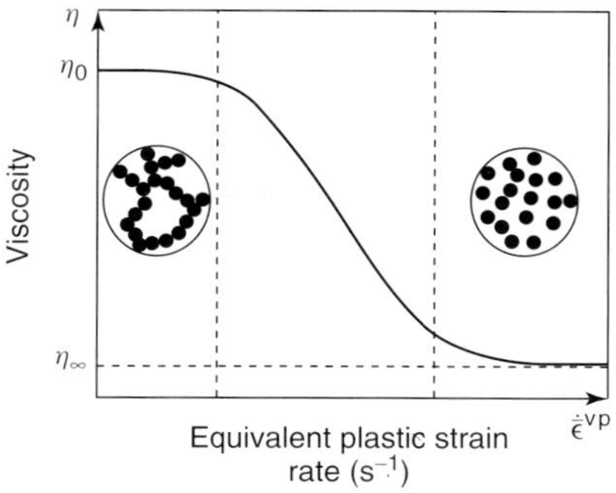

Figure 6.26 Illustration of the "Cross model".

at very low and very high strain rates:

$$\eta = \eta_\infty + \frac{\eta_0 - \eta_\infty}{f(\dot{\bar{\epsilon}}^{\mathrm{vp}})} \tag{6.70}$$

where as $\dot{\bar{\epsilon}}^{\mathrm{vp}} \to 0$, $\eta \to \eta_0$ and as $\dot{\bar{\epsilon}}^{\mathrm{vp}} \to \infty$, $\eta \to \eta_\infty$

and where $f(\dot{\bar{\epsilon}}^{\mathrm{vp}})$ can take different forms as [1]

$$f(\dot{\bar{\epsilon}}^{\mathrm{vp}}) = 1 + k(\dot{\bar{\epsilon}}^{\mathrm{vp}})^m \tag{6.71}$$

or another example is given in Ref. [1]

$$f(\dot{\bar{\epsilon}}^{\mathrm{vp}}) = \left(1 + (k\dot{\bar{\epsilon}}^{\mathrm{vp}})^m\right)^{\frac{n-1}{m}} \tag{6.72}$$

These models are only valid under steady-state conditions and it is discussed in Section 6.3.1.2 that it is the transient behavior that is relevant for the modeling of thixotropic die filling.

6.5.1.2 **Yield Stress Evolution**

At elevated temperatures of the order of the solidus, the yield stress and particularly the hardening are really small. For this reason, many researchers do not include them into their model. Yet, it has been demonstrated that they were important to simulate properly the whole process including the cooling down back to room temperature. In fact, no specific hardening law has been proposed in the frame of the simulation of thixoforming, while only few yield stress descriptions as a function of the liquid fraction and the structural parameters have been developed.

In their fluid dynamics model, Koke and Modigell [20] use the following relationship in addition to the viscosity law (6.61):

$$\sigma_{\mathrm{y}} = \lambda \sigma_{\mathrm{y}}^{0}(f_{\mathrm{l}}) \tag{6.73}$$

where σ_{y}^{0} is the yield stress of a fully built up structure.

Also to complete their viscosity law (6.62), Burgos and Alexandrou [19] conducted shear-rate jump experiments and ended up with the empirical law:

$$\sigma_{\mathrm{y}} = y_{\mathrm{a}}\left[\arctan\frac{y_{\mathrm{b}}y_{\mathrm{c}}}{y_{\mathrm{d}}} + \arctan\left(y_{\mathrm{b}}\frac{\lambda - y_{\mathrm{c}}}{y_{\mathrm{d}} - \lambda}\right)\right] \tag{6.74}$$

where $y_{\mathrm{a,b,c,d}}$ are material parameters.

6.5.2
Two-Phase Models

The main advantage of two-phase models, like those proposed by Kang *et al.* [22], Choi [23], Zavaliangos [24], and Petera [25], is that they predict the relative displacement between the solid and the liquid phase and can thus represent macrosegregation. We have seen that this phenomenon cannot be totally eradicated but that the choice of a judicious set of process parameters can minimize it. Thus, that kind of models can be useful to optimize the process parameters in order to avoid macrosegregation. For example, when working at very low liquid fraction, the phenomenon of macrosegregation is not critical and the use of a two-phase model is not necessary.

In such a two-phase model, a field representing the liquid phase flow is added. During semisolid metal processing, the behavior of the liquid phase, which the solid skeleton is deforming in, can be seen as a fluid flow through a porous medium. This field is generally assumed to be governed by the Darcy law so that the relative movement between the two phases will be linked to the pressure gradient within the liquid. This section presents briefly the main features of this formulation, which is close to the ones that are used in soil mechanics.

6.5.2.1 Two Coupled Fields

Two calculations are done simultaneously: The first one corresponds, as in one-phase models, to the (thermo[8]) mechanical equilibrium and the second one represents the liquid flow.

Mechanical Analysis It ensures the momentum conservation of the whole biphasic system is satisfied, it reads

$$\begin{cases} \nabla\boldsymbol{\sigma} + \rho\mathbf{b} = 0 \text{ on } \mathbb{V}(t) \\ \boldsymbol{\sigma}\mathbf{n} = \mathbf{t} \text{ on } \mathbb{S}(t) \end{cases} \tag{6.75}$$

where $\rho\mathbf{b}$ is the volumetric forces vector (inertial forces included) acting on the current volume $\mathbb{V}(t)$ and $\mathbf{t}$ is the surface force vector acting on the current surface $\mathbb{S}(t)$ whose external unit normal is $\mathbf{n}$.

Liquid-Phase Flow It is governed by the Darcy law, which links the liquid flow through the porous media to the hydrostatic pressure gradient:

$$f_l\dot{\mathbf{u}}_{sl} = \frac{\boldsymbol{\kappa}}{\eta_l}\left(\nabla p - \rho_l\mathbf{g}\right) \tag{6.76}$$

8) As already mentioned, thermal effects should be introduced in the modeling of semisolid processes but, as for one-phase models, the thermal field is not detailed here.

where $\dot{\mathbf{u}}_{sl} = \dot{\mathbf{u}}_s - \dot{\mathbf{u}}_l$ is the relative velocity between the liquid and the solid phases, ρ_l is the liquid phase density, $\mathbf{g}$ is the acceleration of gravity, and $\boldsymbol{\kappa}$ is the permeability tensor.

6.5.2.2 Coupling Sources

The two fields described above are interconnected. Mathematically, the connections can be taken into account by the addition of some terms in one of the governing Eqs. (6.75) and (6.76). However, it also can link variables appearing in these equations. So, this section describes the coupling terms and shows how the unknowns can be reduced to a set of two, which are the solid-phase displacements $\mathbf{u}_s$ (and rates $\dot{\mathbf{u}}_s$) and the hydrostatic pressure p.

Continuity Equation One of the first coupling sources between the liquid flow and the mechanical equilibrium is the continuity equation, which is a consequence of the mass conservations of both phases:

$$\frac{\partial \rho_s}{\partial t} + \boldsymbol{\nabla}(\rho_s \dot{\mathbf{u}}_s) = 0 \qquad (6.77)$$

$$\frac{\partial \rho_l}{\partial t} + \boldsymbol{\nabla}(\rho_l \dot{\mathbf{u}}_l) = 0 \qquad (6.78)$$

with $\rho_s = \rho f_s$ and $\rho_l = \rho f_l$. Thus, we have, by adding Eqs (6.77) and (6.78),

$$\frac{\partial \rho \overbrace{(f_s + f_l)}^{1}}{\partial t} + \boldsymbol{\nabla}(\rho(f_s \dot{\mathbf{u}}_s + f_l \dot{\mathbf{u}}_l)) = 0 \qquad (6.79)$$

In the case of incompressible material $\left(\frac{\partial \rho}{\partial t} = 0\right)$, we have the following continuity equation:

$$\boldsymbol{\nabla}\dot{\mathbf{u}}_s - \boldsymbol{\nabla}\left(f_l \dot{\mathbf{u}}_{sl}\right) = 0 \qquad (6.80)$$

So, the Darcy law (6.76) can be rewritten only in terms of the displacement (rate) of the solid phase and of the pressure p:

$$\boldsymbol{\nabla}\dot{\mathbf{u}}_s = \boldsymbol{\nabla}\left[\frac{\boldsymbol{\kappa}}{\eta_l}\left(\boldsymbol{\nabla} p + (\rho_l \mathbf{g})\right)\right] \qquad (6.81)$$

Interaction Forces In addition, the liquid flow influences the mechanical equilibrium by interaction forces between phases. For the mechanical analysis, the liquid phase is considered as empty and is replaced by its hydrostatic pressure, as illustrated in Figure 6.27. The total internal force $A_T \boldsymbol{\sigma}_T$ acting on the semisolid material is thus made of two contributions: a force $A_s \boldsymbol{\sigma}_s$ acting on the solid region of surface A_s and a force $A_l p\mathbf{I}$ acting on the liquid zone of surface A_l[9] [20]:

$$A_T \boldsymbol{\sigma}_T = A_s \boldsymbol{\sigma}_s + A_l p\mathbf{I} \Rightarrow \boldsymbol{\sigma}_T = \boldsymbol{\sigma}_{\text{eff}} + f_l p\mathbf{I} \qquad (6.82)$$

9) The sign convention of negative compression is maintained though it was not the one adopted by the authors of Ref. [22].

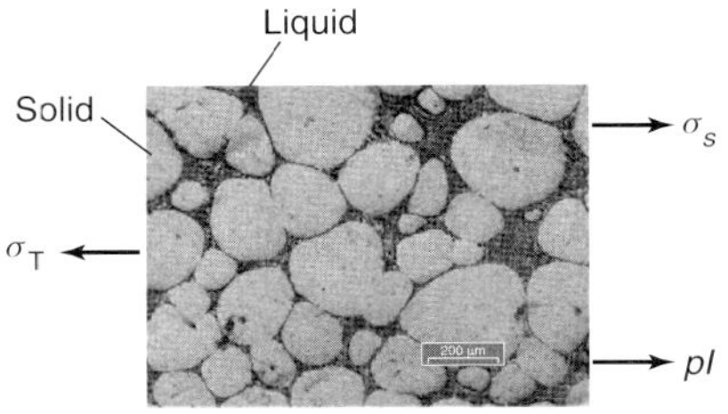

Figure 6.27 Schematic representation of the forces acting inside the semisolid material.

where $\boldsymbol{\sigma}_{\text{eff}} = A_s/A_T\boldsymbol{\sigma}_s$ is the effective stress, meaning the part of the stress tensor that contributes to the constitutive law. Using an appropriate constitutive law that links the stresses to the strains that are deduced from the displacements, the effective stress $\boldsymbol{\sigma}_{\text{eff}}$ is thus related to the displacements of the solid phase $\dot{\mathbf{u}}_s$. Different constitutive laws have been proposed. For example, Kang *et al.* [22] use a viscoplastic model with the porous Shima and Oyane yield criterion [26]. In addition, any of the constitutive laws presented in the previous section about one-phase model can be used.

6.6 A Detailed One-Phase Model

In this section, one particular model [27, 28] is described in detail. The choice of this specific model is motivated by the following arguments.

First, one-phase models are much simpler and thus lighter in terms of implementation and CPU costs. As long as macrosegregation is not a major concern, the consideration of such a complex two-field calculation that is needed by two-phase models is not necessary.

Another important point is the prediction of the residual stresses after unloading and cooling back down to room temperature, which gives a good indication on the quality of the final product and is thus an important feature that a model should include. Along the fact that a billet at rest can sustain its own weight, the use of a finite yield stress σ_y is also necessary for the prediction of the residual stresses. Furthermore, neither liquid formalism nor rigid models can predict such residual stresses.

The model should not only be able to describe the behavior of semi-solid thixotropic structures but also should be able to degenerate to solid, liquid, as well as free suspensions models.

The presented model then uses a solid thermo-elasto-visco-plastic formulation, which has been described in Section 6.4.2.6 and considers the elastic part of the deformation. The von Mises criterion (6.53), with n taken as 0 and with the apparent viscosity $\eta = k(\dot{\overline{\epsilon}}^{\text{vp}})^{m-1}$ defined by analogy with liquid formalism, is used and can be rewritten as

$$\overline{\sigma}^{\text{VM}} - \sigma_y(\dot{\overline{\epsilon}}^{\text{vp}}, \overline{\epsilon}^{\text{vp}}) - \eta(\dot{\overline{\epsilon}}^{\text{vp}}, \overline{\epsilon}^{\text{vp}})\dot{\overline{\epsilon}}^{\text{vp}} = 0 \tag{6.83}$$

Also, the specific behavior of thixotropy, in both steady-state and transient conditions, is integrated by means of two additional nondimensional internal parameters: the liquid fraction f_l and the cohesion degree λ that has been already mentioned in Section 6.5.1.

Overall, the extended consistency equation [17] is written as

$$\overline{\sigma}^{VM} - \sigma_y(f_l, \lambda, \dot{\overline{\epsilon}}^{vp}, \overline{\epsilon}^{vp}) - \eta(f_l, \lambda, \dot{\overline{\epsilon}}^{vp}, \overline{\epsilon}^{vp})\dot{\overline{\epsilon}}^{vp} = 0 \tag{6.84}$$

The following sections detail the expressions of both internal parameters as well as of the isotropic hardening and viscosity laws. The model is then applied to the numerical simulation of several experiments.

6.6.1 Cohesion Degree

The first internal parameter is the cohesion degree λ, discussed in Section 6.5.1.

The evolution of the structural parameter λ is obtained by a differential equation that describes the kinetics between the agglomeration of the solid grains and the destruction of the solid bonds due to shearing. It is assumed that the break down term depends on the amount of existing bounds, while the build-up rate is a function of the fraction of bonds that can still be created. Different expressions of this type have been published [1, 19], whose most general form is given by

$$\dot{\lambda} = \underbrace{a(1-\lambda)}_{\text{build up}} - \underbrace{b\lambda(\dot{\overline{\epsilon}}^{vp})^c e^{d\dot{\overline{\epsilon}}^{vp}}}_{\text{break down}} \tag{6.85}$$

where a, b, c, and d are material parameters. The steady-state (or equilibrium) cohesion degree λ_e can be defined as the one at which

$$\dot{\lambda}(\lambda_e) \rightarrow 0 \Leftrightarrow \lambda_e = \frac{a}{a + b(\dot{\overline{\epsilon}}^{vp})^c e^{d\dot{\overline{\epsilon}}^{vp}}} \tag{6.86}$$

So, another equivalent form of the differential equation (6.85) to evaluate the cohesion degree expresses that the evolution of the cohesion degree is proportional to the distance from equilibrium [20]:

$$\dot{\lambda} = \left(a + b\lambda(\dot{\overline{\epsilon}}^{vp})^c e^{d\dot{\overline{\epsilon}}^{vp}}\right)(\lambda_e - \lambda) \tag{6.87}$$

If Eq. (6.85), or equivalently Eq.(6.87), is solved over a small time step Δt (where it can be assumed that the equivalent plastic strain rate is constant), one gets

$$\lambda(t_{i+1}) = \underbrace{\lambda_e}_{\text{steady-state}} + \underbrace{(\lambda(t_i) - \lambda_e)\, e^{\left(a+b\lambda(\dot{\overline{\epsilon}}^{vp})^c e^{d\dot{\overline{\epsilon}}^{vp}}\right)\Delta t}}_{\text{transient}} \tag{6.88}$$

So far, the cohesion degree does not depend on the liquid fraction, which is not physically based. Actually, the cohesion degree depends on the liquid fraction since it should be zero at liquid state and unity at solid state, whatever the strain rate. There can be no solid bonds at a fully liquid state and a solid structure must be fully built up. Thus, to degenerate properly to a pure solid- or liquid-state behavior, the liquid fraction should be introduced into, for example, Eq. (6.85) as

$$\dot{\lambda} = a(1 - f_l)(1 - \lambda) - bf_l\lambda(\dot{\bar{\epsilon}}^{vp})^c e^{d\dot{\bar{\epsilon}}^{vp}} \tag{6.89}$$

Thus, if $f_l \to 0$, we have

$$\begin{cases} \lambda_e = \dfrac{a(1-f_l)}{a(1-f_l)+bf_l\lambda(\dot{\bar{\epsilon}}^{vp})^c e^{d\dot{\bar{\epsilon}}^{vp}}} \to 1 - f_l \\ \dot{\lambda} \to 0 \\ \lambda \to \lambda_e \end{cases} \tag{6.90}$$

And it is the same if $f_l \to 1$.

6.6.2
Liquid Fraction

The second internal parameter is the volume fraction of liquid f_l present in the mushy state.

In this model, the liquid fraction depends only on the temperature by the steady-state Scheil [12] equation:

$$f_l = \left(\frac{T - T_s}{T_l - T_s}\right)^{\frac{1}{r-1}} \tag{6.91}$$

where r is the equilibrium partition ratio, and T_s and T_l are the solidus and liquidus temperatures, respectively.

An alternative to the liquid fraction is the effective liquid fraction f_l^{eff} described in Section 6.5.1.1. It connects both new internal parameters λ and f_l. Indeed, following Lashkari and Ghomashchi [12], the effective liquid fraction is a combination between both internal parameters f_l and λ. As already discussed, the effective liquid fraction is a corrected liquid fraction. It excludes the liquid that is entrapped inside the solid grains (see micrograph in Figure 6.23) and that does not contribute to the flow. With the breaking of the solid bonds, some of this entrapped liquid is released. In this model, the chosen expression of the effective liquid fraction f_l^{eff} in terms of the liquid fraction f_l and of the cohesion degree λ is simpler than the one in Eq. (6.67) proposed by Lashkari and Ghomashchi [12]:

$$f_l^{eff} = f_l(1 - \lambda) \tag{6.92}$$

6.6.3
Viscosity Law

Viscosity laws based on the Norton–Hoff law $\eta = k(\dot{\bar{\epsilon}}^{vp})^{m-1}$, like those proposed by Burgos [19] or by Modigell [20] (Section 6.5.1.1), are not extensible to a fully broken structure. In this case, $\lambda = 0$ and this parameter cannot act on the apparent viscosity anymore, and this last one keeps increasing with the strain rate [29].

So, in the present model, a combination between the behavior of a built-up structure and the low viscosity of free solid suspensions (whose behavior is

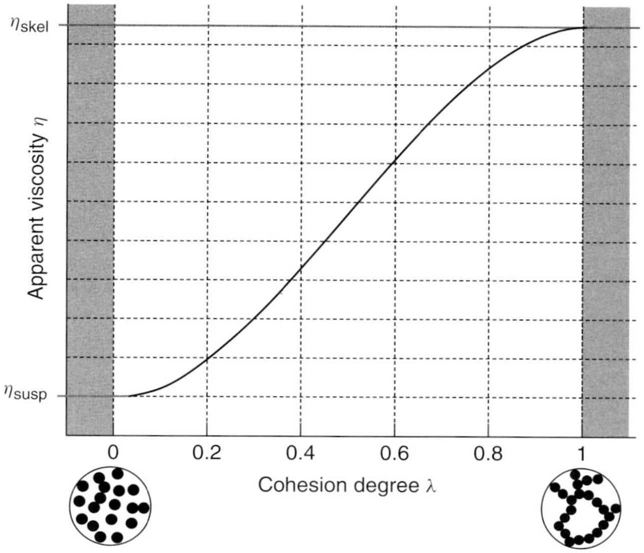

Figure 6.28 Illustration of the viscosity law.

considered as Newtonian[10]) is used. This leads to a smooth cubic interpolation between both behaviors, illustrated in Figure 6.28:

$$\eta = \eta_{susp} + \lambda^2(3 - 2\lambda)(\eta_{skel} - \eta_{susp}) \tag{6.93}$$

where $\eta_{susp} = k_1$ and η_{skel} are the constant viscosity of the free solid suspensions and the viscosity of the solid skeleton, respectively.

The viscosity of the agglomerated clusters η_{skel} is expressed by an extended Norton–Hoff law where both internal parameters are introduced via the viscosity parameters k and m. It is a combination between two viscosity laws proposed in the literature, as it increases with the cohesion degree by an empirical law established by Burgos *et al.* [19] and increases exponentially with the solid fraction [20]. Thus, we have

$$\eta_{skel} = k(\dot{\bar{\epsilon}}^{vp})^{m-1} \tag{6.94}$$

with

$$\begin{cases} k &= k_1 e^{k_2(1-f_l)} e^{k_3\lambda} \\ m &= (m_1 + m_3\lambda^2 + m_4\lambda)\, e^{m_2(1-f_l)} \end{cases} \tag{6.95}$$

where k_1, k_2, k_3, m_1, m_2, m_3, and m_4 are material parameters.

10) An alternative is to consider formulation (6.66) proposed by Lashkari and Ghomashchi [12] to describe the behavior of dispersed suspensions.

6.6.4
Yield Stress and Isotropic Hardening

The isotropic hardening law is based on a simple linear hardening. As a liquid does not display yield, this law also takes into account a decrease in the yield stress with liquid fraction elevation. It is expressed as

$$\sigma_y = (1 - f_l)^{h_2} (\sigma_y^0 + h_1\, \bar{\epsilon}^{vp}) \tag{6.96}$$

where h_1 is the linear hardening coefficient, and it is known that it is almost negligible at high temperatures around the solidus. However, the use of such a term allows to meet a classical linear hardening law at the solid state ($f_l = 0$), particularly when one cools down the component to room temperature.

Then, to introduce the cohesion degree into the isotropic hardening law, the effective liquid fraction f_l^{eff} is used instead of the liquid fraction.

The isotropic hardening law is eventually written as

$$\sigma_y = (1 - f_l^{eff})^{h_2} (\sigma_y^0 + h_1\, \bar{\epsilon}^{vp}) = \left(1 - f_l(1-\lambda)\right)^{h_2} (\sigma_y^0 + h_1\, \bar{\epsilon}^{vp}) \tag{6.97}$$

6.7
Numerical Applications

To illustrate the previously presented model, several numerical simulations have been carried out using the finite elements code METAFOR [30].

To make a first validation of the presented material model, an academic simulation has been conducted. The choice of this test has been motivated by the availability of some results in the literature [22], which offers the possibility to validate the model by comparing its results with the reference. This test will also allow to expose the different features of the model.

6.7.1
Test Description

The test, described in Figure 6.29, is axisymmetric. The initial cylinder is 10 mm high and has a radius of 7.5 mm. One section is discretized using a 10–10 mesh. The die velocity and temperature are 38 mm/s and 150 °C respectively. The Coulomb friction coefficient is 0.3 and the interfacial conductance between the slug and the die is 10 kWm^{-2}K. Room temperature is 25 °C and the heat convection coefficient is equal to 45 W m^{-2}K. The initial temperature is such as the initial liquid fraction will be 37%.

In these simulations, the material under study is a tin–lead alloy (Sn-15% wt Pb). This material is not adapted to industrial applications, but is extensively used in experimental researches on semisolid processing. Indeed, its fusion temperature is low enough to avoid technical problem encountered with steel grades. The material parameters for Sn-15%wt Pb are thus widely available in the literature [19, 21, 22] and are detailed in Table 6.1.

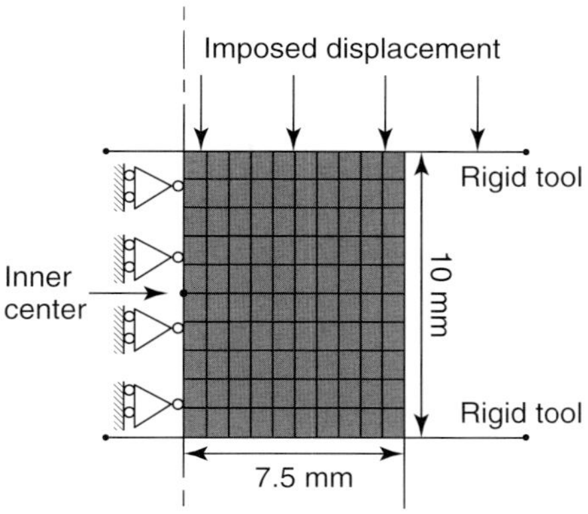

Figure 6.29 Description of the compression test.

Table 6.1 Material parameters for Sn-15% wt Pb alloy.

Mechanical properties:	Young's modulus: $E = 38 - 0.1T$ (GPa)
	Poisson's ratio: $\nu = 0.4$
	Density: $7.7E^3$ (kg m^{-3})
Thermal properties:	Thermal expansion: $23.5E^{-6}$ (1/K)
	Conductivity: 60 W (m °C)$^{-1}$
	Heat Capacity: 220 J (kg °C)$^{-1}$
Cohesion degree:	$a = 0.035$ (1/s)
	$b = 0.15$
	$c = 1.5$ (s)
	$d = 0.001$
	$\lambda_{\text{initial}} = 1$
Liquid fraction:	Equilibrium partition ratio: $r = 1.5$
	Solidus: $T_s = 183$ (°C)
	Liquidus: $T_l = 210$ (°C)
Viscosity law:	$k_1 = 0.45$ (Pa)
	$k_2 = 0.1$
	$k_3 = 12.173$
	$m_1 = 1$
	$m_2 = 0.1$
	$m_3 = m_4 = 0$
Hardening law:	$\sigma_y^0 = 6.86$ (MPa)
	$h_1 = 11.43$ (MPa)
	$h_2 = 2.5$

6.7.2 Results Analysis

6.7.2.1 First Validation of the Model under Isothermal Conditions

In a first step, isothermal simulations have been conducted at a temperature of 199.4 °C, corresponding to a liquid fraction of 37%. Indeed, the chosen reference for the model validation is the simulations of the compression test

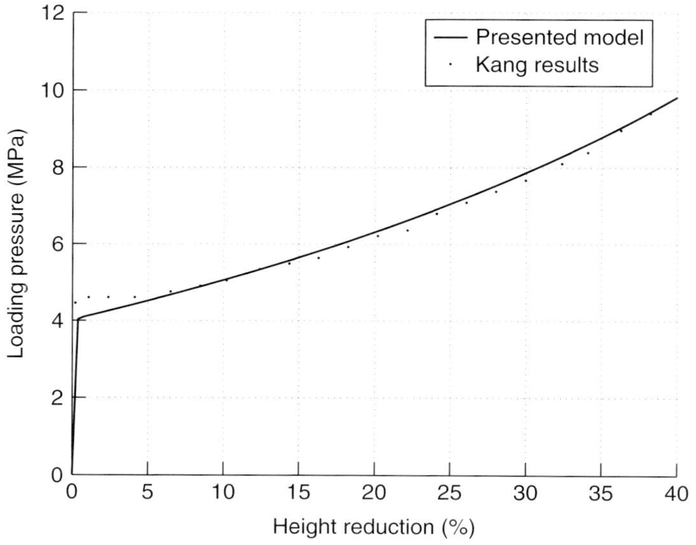

Figure 6.30 Comparison of the loading pressure between isothermal models.

conducted by Kang *et al.* [22] with their isothermal two-phase model. They present results on compression load, so Figure 6.30 shows the reference compression load compared to the one obtained by the presented model. Although, basically, the two models are quite different (as Kang's model is two phase while the presented model is one phase), it reveals a good agreement between them.

6.7.2.2 **Thermomechanical Analysis**

Then, thermomechanical simulations have been carried out. In this case, the contact with the colder die causes some solidification and leads to higher loading pressure (not represented here) than under isothermal conditions (represented in Figure 6.30).

The influence of die velocity and the ability of the model to reproduce the transient thixotropic behavior has been studied by speeding the die velocity up to 500 mm s^{-1} and comparing the viscosity evolutions at the inner center of the cylinder (Figure 6.29) for different die velocities as shown in Figure 6.31.

At the start of the loading, a peak of viscosity can be observed in Figure 6.31. Then, a drop of viscosity occurs with loading. Comparing the behavior under different deformation rates, it can be seen that the initial peak of viscosity is higher for a higher die speed, but the following drop is much more drastic, leading to a very low (but nonzero) viscosity at the end of the test. This is a characteristic of thixotropic behavior.

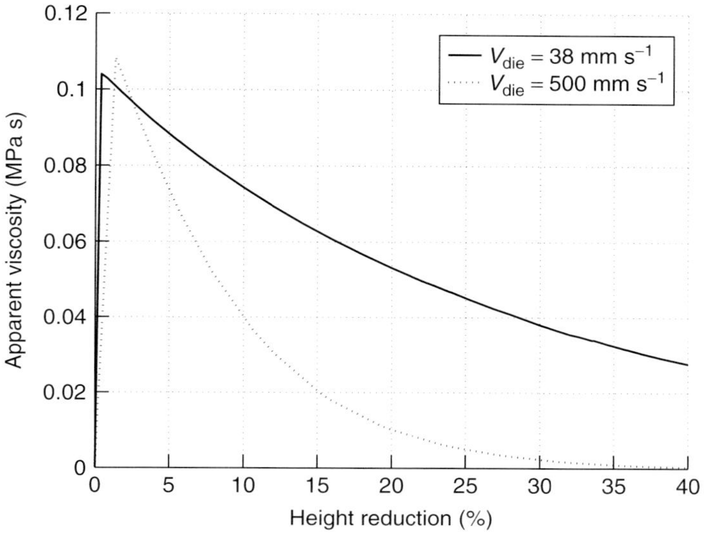

Figure 6.31 Apparent viscosity evolutions under different loading velocities at the inner center of the cylinder.

6.7.2.3 Residual Stresses Analysis

In the last step, the residual stresses have been computed using a mechanical analysis with imposed temperature evolution. This illustrates the aptitude of the presented model to predict such residual stresses, which is one of its main originality. The calculation is then made in two successive steps: the forming stage with a uniform drop of 5 °C, representing the thermal die contact, is followed by the unloading and the cooling down to room temperature.

Figure 6.32 compares the map of stresses at the end of the loading step to the residual stresses. The stress concentration that appears on the upper edge of the flattened cylinder after tool withdrawal is not quantitative because of the quality of the mesh in this area. Nevertheless, the figure gives information about the quality of the component for further use.

6.7.2.4 Internal Variables Analysis

Still in the case of the loading/unloading and cooling down steps, Figure 6.33 shows the evolution of the cohesion degree as well as the liquid fraction compared to the effective liquid fraction at the inner center of the cylinder (Figure 6.29). Before loading, the initial liquid fraction and cohesion degree are set to 37% and 1, respectively. Thus, the effective liquid fraction starts at a value of zero.

During the forming process, some solidification occurs ($f_l \downarrow$) due to the thermo-mechanical contact with the cold die. At the same time, the structure is broken down by shearing, and the cohesion degree decreases. Overall, the effective liquid fraction increases due to the release of some entrapped liquid.

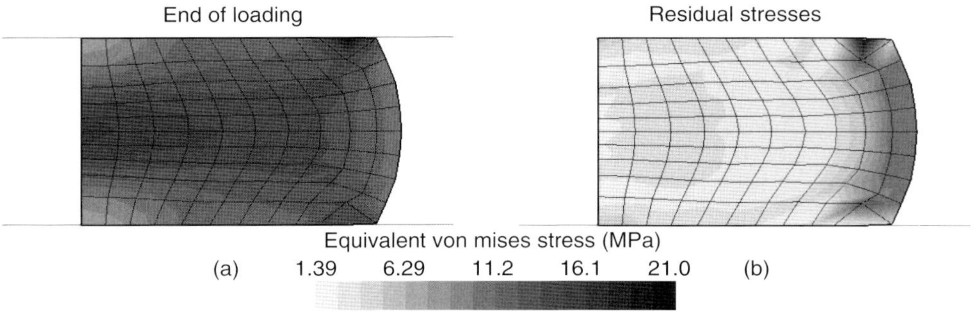

Figure 6.32 Map of stresses at the end of the loading step (a) and map of residual stresses (b).

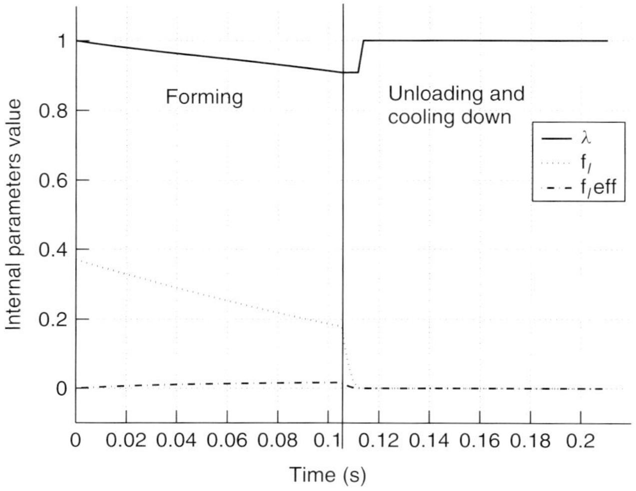

Figure 6.33 Internal parameters evolutions.

Right after unloading, the cohesion degree remains constant for a small period of time. In fact, at this stage, the temperature is still above the solidus and the process of recovery is very slow.

During the cooling down stage, the formulation predicts a fully built-up structure ($\lambda = 1$) when the solidus is reached, which makes physical sense.

6.8
Conclusion

Thixoforming processing of metallic alloys relies on the particular behavior exhibited by semisolid materials with nondendritic microstructure. These materials display thixotropy, which is characterized by a solidlike behavior at rest and a liquidlike flow when submitted to shear.

Processes of this kind have already been commercialized with different alloys such as aluminum or magnesium, but the transfer of the technology to other materials such as steel has not yet been accomplished. However, there is significant interest in commercial use of thixoforming of steels. In this context, accurate modeling can be a great help in die design, predicting appropriate processing conditions, and minimizing defects. Thus, this chapter has focused on such modeling and more particularly on the constitutive formulation of thixotropy.

As background the advantages and the disadvantages of types of semisolid processes have been summarized. Rheological aspects and origins of thixotropy have been laid and mathematical theories of thixotropy introduced. In addition, the numerical framework for large deformation analysis has been introduced particularly the development of constitutive laws.

Then, a review of existing modeling has been presented. Afterward, a specific one phase model has been presented more closely. Some numerical illustrations have also been shown.

What emerges here is that a great deal of models has already been proposed in the literature. They can be categorized into a few types: one or two phase, in the framework of fluid or solid mechanics, using a structural parameter or not. Two-phase models are more complex and should be used only if macrosegregation is a concern. Solid formalism allows predicting residual stresses after unloading and cooling down. The use of a structural parameter offers the possibility to take into account the transient behavior. Finally, the semisolid material is, by essence, very sensitive to temperature, so the consideration of the thermomechanical coupling is really important.

References

1. Atkinson, H.V. (2005) Modelling the semisolid processing of metallic alloys. *Progress in Materials Science*, **50**, 341–412.
2. Hirt, G. and Kopp, R. (eds) (2009) *Thixoforming. Semi-solid Metal Processing*, John Wiley and Sons, Ltd, Chichester.
3. Atkinson, H.V. (eds) (2008) *Modelling of Semi-Solid Processing*, Shaker, Ltd, Aachen.
4. Suery, M. (eds) (2002) *Mise en forme des Alliages Métalliques à l'état Semi-solide.*, Lavoisier, Ltd, Paris.
5. Atkinson, H.V. (ed.) (2008) Introduction to modelling semi-solid processing, in *Modelling of Semi-Solid Processing*, Shaker, Ltd, Aachen, pp. 1–19.
6. Spencer, D.B., Mehrabian, R., and Flemings, M.C. (1972) Rheological behavior of Sn-15 pct Pb in the crystallization range. *Metallurgical and Materials Transactions*, **3**, 1925–1932.
7. Suéry, M. (ed.) (2002) Obtention du matériau de base, in *Mise en forme des Alliages métalliques à l'état Semi-solide*, Lavoisier, Ltd, Paris, pp. 21–36.
8. Decker, R.F. (1990) Solid base for new technology. *Foundry Trade Journal*, **7**, 634–635.
9. Collot, M. (2002) Réchauffage, in *Mise en forme des Alliages métalliques à l'état Semi-solide*, (ed. M. Suéry), Lavoisier, Ltd, Paris, pp. 37–84.
10. Rassili, A., Robelet, M., Bigot, R., and Fischer, D. (2007) Thixoforming of steels and industrial applications, Proceedings of the 10th Esaform Conference on Material Forming, April

18-21, Zaragoza, Spain, E. Cueto, F. Chinesta.

11. Modigell, M. and Pape, L. (2008) Fundamentals of rheology, in *Modelling of Semi-Solid Processing*, (ed. H.V. Atkinson), Shaker, Ltd, Aachen, pp. 20–49.
12. Lashkari, O. and Ghomashchi, R. (2007) The implication of rheology in semi-solid metal processes: An overview. *Journal of Materials Processing Technology*, **182**, 229–240.
13. Quaak, C.J. (1996) Rheology of Partially Solidified Aluminium Alloys and Composites. PhD Thesis, Technische Universiteit, Delft, The Netherlands.
14. Flemings, M.C. (1991) Behavior of metal alloys in the semisolid state. *Metallurgical and Materials Transactions*, **22A**, 269–293.
15. Donea, J., Antonio, H., Ponthot, J.-Ph., and Rodríguez-Ferran, A. (2004) Arbitrary Lagrangian-Eulerian methods, in *Encyclopedia of Computational Mechanics - Volume 1: Fundamentals* (eds E. Stein, R. Borst, and Hughes, T.J.R.), John Wiley & Sons.
16. Reiner, M. (1964) The Deborah Number. Talk presented at the Fourth International Congress on Rheology, August, Providence, USA, Physics Today.
17. Ponthot, J.P. (2002) Unified stress update algorithms for the numerical simulation of large deformation elasto-plastic and elasto-viscoplastic processes. *International Journal of Plasticity*, **18**, 91–126.
18. Adam, L. and Ponthot, J.P. (2005) Thermomechanical modeling of metals at finite strains: First and mixed order finite elements. *International Journal of Solids and Structures*, **42**, (21-22), 5615–5655.
19. Burgos, G.R., Alexandrou, A.N., and Entov, V.M. (2001) Thixotropic rheology of semisolid metal suspensions. *Journal of Materials Processing Technology*, **110**, 164–176.
20. Modigell, M. and Koke, J. (2001) Rheological modelling on semi-solid metal alloys and simulation of thixocasting processes. *Journal of Material Processing Technology*, **111**, 53–58.
21. Cézard, P., Favier, V., Bigot, R., Balan, T., and Berveiller, M. (2005) Simulation of semi-solid thixoforging using a micro-macro constitutive equation. *Computational Materials Science*, **32**, 323–328.
22. Kang, C.G. and Yoon, J.H. (1997) A finite-element analysis on the upsetting process of semi-solid aluminum material. *Journal of Materials Processing Technology*, **66**, 76–84.
23. Ko, D.C., Min, G.S., Kim, B.M., and Choi, J.C. (2000) Finite element analysis for the semi-solid state forming of aluminium alloy considering induction heating. *Journal of Material Processing Technology*, **100**, 95–104.
24. Zavaliangos, A. and Lawley, A. (1995) Numerical simulation of thixoforming. *Journal of Materials Engineering and Performance*, **4**, 40–47.
25. Petera, J., Modigell, M., and Hufschmidt, M. (2004) Special issue modelling and optimisation: physical simulation and material testing. *International Journal of Forming Processes*, **7**, 123–140.
26. Shima, S. and Oyane, M. (1976) Plasticity theory for porous metals. *International Journal of Mechanical Sciences*, **18**, 285–291.
27. Koeune, R. and Ponthot, J.P. (2007) A one phase thermomechanical model for semi-SOLID thixoforming. *Solid State Phenomena*, **141-143**, 629–635.
28. Koeune, R. and Ponthot, J.P. An improved constitutive model for the numerical simulation of semi-solid thixoforming. (2009) *Journal of Computational and Applied Mathematics*, DOI: 10.1016/j.cam.2009.08.085.
29. Koeune, R. and Ponthot, J.P. (2007) Thermomechanical one-phase modeling of semi-solid thixoforming. Proceedings of the International Conference on Computational Methods for Coupled Problems in Science and Engineering, May 21-23, Ibiza, Spain, E. Oñate, M. Papadrakakis, B. Schrefler.
30. Metafor, University of Liège, Aerospace and Mechanical Engineering department, Computational Mechanics Unit (2009) *http://www.ltas-mnl.ulg.ac.be* (accessed 27 october 2009).

7
Modeling of Powder Forming Processes; Application of a Three-invariant Cap Plasticity and an Enriched Arbitrary Lagrangian–Eulerian FE Method

Amir R. Khoei

7.1
Introduction

Powder metallurgy is a highly developed method of manufacturing reliable ferrous and nonferrous parts. The powder metallurgy process is cost-effective, because it minimizes machining, produces good surface finish, and maintains close dimensional tolerances. The method is a material-processing technique utilized to achieve a coherent near-to-net shape industrial component. The often extremely high tolerance requirements of the parts and the cost for hard machining of a sintered component are a challenge for die pressing. One of the main difficulties that exists in the compaction-forming process of powders includes a nonhomogeneous density distribution, which has wide ranging effects on the final performance of the compacted part. The variation of density results in cracks and also in localized deformation in the compact, producing regions of high density surrounded by lower density material, leading to compact failure. The lack of homogeneity is primarily caused by friction, due to interparticle movement, as well as relative slip between powder particles and the die wall. The die geometry and the sequence of movement result in a lack of homogeneity of density distribution in a compact. Thus, the success of compaction forming depends on the ability of the process in imparting a uniform density distribution in the engineered part. In order to perform such analysis, the complex mechanisms of compaction process must be drawn into a mathematical formulation with the knowledge of material behavior.

A number of constitutive models have been developed for the compaction of powders over the last three decades, including micromechanical models [1–3], flow formulations [4], and solid mechanics models [5–11]. The porous material model, generally known as a *modified von Mises criterion* [12], has been used for the simulation of powder-forming processes. This model includes the influence of the hydrostatic stress component, and satisfies the symmetry and convexity conditions required for the development of a plasticity theory. The yielding of porous materials is more complicated than that of fully dense materials, because the onset of yielding is influenced not only by the deviatoric stress components

Advanced Computational Materials Modeling: From Classical to Multi-Scale Techniques.
Edited by Miguel Vaz Júnior, Eduardo A. de Souza Neto, and Pablo A. Munoz-Rojas

ISBN: 978-3-527-32479-8

but also by the hydrostatic stress. It has been shown that the yield functions proposed by various researchers satisfy the required conditions and reduce to the von Mises yield function for fully dense materials. However, this model neglects the hardening factor associated with the densification process, and does not predict the dependence of compressive yield stress on relative density, as described by the large discrepancies between experiment and theory.

The granular material model, which has been used for the modeling of frictional materials such as soil or rock, is then adopted to describe the behavior of powder and granular materials. This model reflects the yielding, frictional, and densification characteristics of powder along with strain and geometrical hardening, which occurs during the compaction process. The experimental results of Watson and Wert [13] and Brown and Abou-Chedid [14] demonstrated that the constitutive modeling of geological and frictional materials can be utilized to construct suitable phenomenological constitutive models, which capture the major features of the response of initially loose powders to the complex deformation processing histories encountered in the manufacture of engineering components by powder metallurgy techniques. In particular, they suggested that a two-mechanism model, such as Drucker–Prager or Mohr–Coulomb and elliptical cap models, which are widely used for geological materials and exhibit pressure-dependent behavior can be applied for modeling the behavior of powder materials. These models consist of two yield surfaces: a "distortion surface," which controls the ultimate shear strength of materials and a "consolidation" or "cap" surface, which has an elliptical shape and captures the hardening behavior of materials under compression.

The cone-cap model based on a density-dependent Drucker–Prager yield surface and a noncentered ellipse was developed by Haggblad and Oldenburg [15], Aydin *et al.* [16], Lewis and Khoei [17], Brandt and Nilsson [18], and Gu *et al.* [19]. A double-surface cap plasticity model was proposed by Khoei and Azizi [20] for the nonlinear behavior of powder materials. This model is based on the combination of a convex yield surface consisting of an exponential failure envelope and an elliptical hardening cap. The model comprises two surfaces, one to reflect shear failure and the other to capture densification. A density-dependent endochronic theory was developed by Khoei *et al.* [21, 22] and Khoei and Bakhshiani [23] based on coupling between the deviatoric and hydrostatic behavior to simulate the compaction process of powder material. A generalized single-cap plasticity with an isotropic hardening rule was developed by Khoei and Azami [24, 25], which generate the elliptical yield surface and double-surface cap plasticity as special cases. Recently, a three-invariant plasticity model was developed by Khoei and DorMohammadi [26] based on the isotropic–kinematic hardening rule for powder-forming process. An overview of various plasticity models was addressed by Khoei [27] for the behavior of powder die-pressing.

A feature of powder compaction simulation common with that of solid mechanics constitutive models is the use of a Lagrangian kinematics formulation. This approach has shown to be adequate for problems that do not exhibit large mass fluxes among different parts of the sample. But in practical problems, as those that appear in realistic design processes, the Lagrangian approach leads to highly distorted and usually useless meshes [17]. This difficulty can be particularly observed in higher order elements. Because of severe distortion of elements, the determinant of Jacobian matrix may become negative at quadrature points, aborting the calculations or causing numerical errors. In order to solve these problems, the mesh-adaptive strategies were implemented by Khoei and Lewis [9]. However, it is computationally expensive and information must be interpolated from the old mesh to new mesh. Thus, an arbitrary Lagrangian–Eulerian (ALE) technique is employed by Khoei *et al.* [28] to powder-forming process, which can alleviate many of the drawbacks of the traditional Lagrangian formulation. ALE formulation is able to overcome the finite element distortion while representing the boundaries correctly. In the ALE technique, the computational grid need not adhere to the material nor be fixed in space but can be moved arbitrarily. In this method, the reference configuration is applied to describe the motion instead of material configuration in Lagrangian and spatial configuration in Eulerian formulation. In Lagrangian phase, the mesh and material movements are identical. In the Eulerian phase, the mesh is allowed to have an arbitrary motion, independent of material motion, keeping the mesh regular. Since the mesh and material movements are uncoupled, the convective term appears in the balance of momentum equation. In the ALE description, the choice of the material, spatial, or any arbitrary configuration yields to a Lagrangian, Eulerian, or ALE description, respectively.

The extended ALE finite element technique was recently developed by Khoei *et al.* [29] to model the large plastic deformation with moving boundaries. In this approach, an ALE analysis is implemented into the large deformation extended finite element method (X-FEM) based on an operator splitting technique. In X-FEM, the discontinuities are taken into account by adding appropriate functions into the standard approximation through a partition of unity method [30]. The discontinuity in the displacement field is modeled by additional degrees of freedom at nodal points of elements, which have been cut by discontinuity. This description allows releasing the computational grids from conforming to the shape of the discontinuity. In order to perform an extended ALE finite element technique, a typical X-FEM analysis is first carried out with updated Lagrangian approach. The Eulerian phase is then applied to update the mesh, while the material interface is independent of the FE mesh. Special care has to be taken with respect to the integration of constitutive equations, often denoted as the stress update, since the stress field is usually discontinuous across the elements due to the fact that stress values are only evaluated at discrete integration points. To handle this, an approach called the *Godunov scheme* is used here for the stress update. This uncoupled approach makes easy the extension of a pure extended Lagrangian FE code to the ALE technique and allows the use of original updated Lagrangian program to solve the relevant ALE equations.

The plan of this chapter is as follows: in Section 7.2, the three-invariant cap plasticity model with isotropic–kinematic hardening rule is developed for pressure-sensitive materials. The constitutive elastoplastic matrix and its components are also extracted in this section. In Section 7.3, the ALE is presented in large deformation problems. The algorithm of uncoupled ALE solution together with the mesh motion strategy and stress update procedure based on Godunov method for transferring of variables from Lagrangian mesh to relocated mesh is demonstrated in this section. In Section 7.4, the X-ALE-FEM technique is described on the basis of the combined ALE and X-FEM methods to reduce the mesh distortion that occurred in conventional large X-FEM deformation. Finally, some concluding remarks are given in Section 7.5.

7.2
Three-Invariant Cap Plasticity

The mechanical behavior of powders involves several interacting micromechanical processes. First, at low pressure, particle sliding occurs leading to particle rearrangement. The second stage involves both elastic and plastic deformation of the particles via their contact areas leading to geometric hardening (i.e., plastic deformation and void closure). Lastly, at very high pressure, the flow resistance of the material increases rapidly due to material strain hardening. Therefore, it is necessary that the constitutive model of powder captures the various behaviors of compaction process. On the basis of these requirements, the constitutive model is developed here for both isotropic and kinematic hardening behaviors based on three invariants of stress states, J_1, J_{2D}, and J_{3D}. As the first step in derivation of plasticity model, a shear-failure function is defined as

$$F_{\mathrm{f}} = f_{\mathrm{d}}^2 - \left(\frac{f_{\mathrm{d}}}{f_{\mathrm{h}}}\right)^2 J_1^2 \tag{7.1}$$

where J_1 is the first invariant of stress tensor, and f_{h} and f_{d} are the hydrostatic and deviatoric material functions, respectively, defined as functions of hardening parameters. The function F_{f} captures the pressure-dependence of the shear strength of material, which increases with more compressive mean stresses, as shown in Figure 7.1a. It must be noted that different combinations of f_{h} and f_{d} lead to different shapes of the yield surface. The initial yield surface f_0 is offset from the pressure-dependence function F_{f}, as shown in Figure 7.1b.

Considering the shear-failure function F_{f}, the yield function can be written in (J_1, J_{2D}) stress space as

$$F_1 = \frac{2}{3} J_{2D} - F_{\mathrm{f}} = 0 \tag{7.2}$$

or

$$F_1 = \frac{2}{3} J_{2D} + \left(\frac{f_{\mathrm{d}}}{f_{\mathrm{h}}}\right)^2 J_1^2 - f_{\mathrm{d}}^2 = 0 \tag{7.3}$$

where J_{2D} is the second invariant of deviatoric stress tensor. In the above relation, we define the coefficients ϕ_{h} and ϕ_{d} in order to indicate the effect of material functions f_{h} and f_{d} in the yield function. The yield function (7.3) can be therefore rewritten as

$$F_2 = \frac{2}{3} J_{2D} + \left(\frac{\phi_{\mathrm{d}} f_{\mathrm{d}}}{\phi_{\mathrm{h}} f_{\mathrm{h}}}\right)^2 J_1^2 - (\phi_{\mathrm{d}} f_{\mathrm{d}})^2 = 0 \tag{7.4}$$

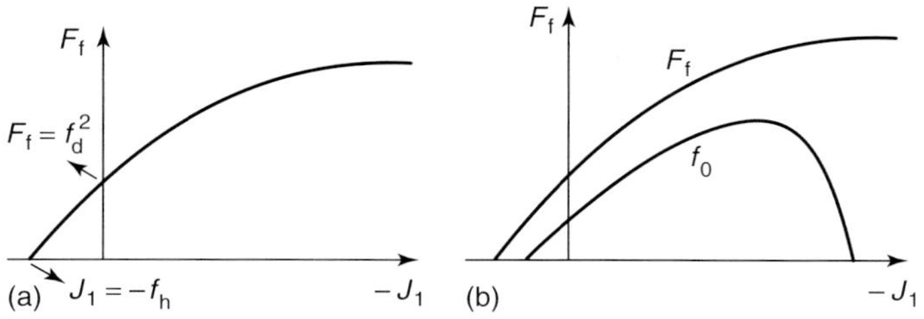

Figure 7.1 (a) The shear-failure function F_f. (b) The shear-failure function F_f and the initial yield surface f_0.

Finally, applying the effect of third invariant of deviatoric stress tensor J_{3D} into Eq. (7.4), that is, the triangularity of deviatoric trace along the hydrostatic axis, the yield function can be written in its final representation as

$$F(\boldsymbol{\sigma}, \boldsymbol{\alpha}, \kappa) = \psi J_{3D}^{2/3} + \frac{2}{3} J_{2D} + \left(\frac{\phi_d f_d}{\phi_h f_h}\right)^2 J_1^2 - (\phi_d f_d)^2 = 0 \tag{7.5}$$

where ψ is a constant, and $\boldsymbol{\alpha}$ and κ are the kinematic and isotropic hardening parameters, respectively. It must be mentioned that the material functions f_h and f_d can be decomposed into two parts, the isotropic and kinematic parts, which control the shape of yield surface (7.5). Figure 7.2 presents the 3D representation of yield surface (7.5) for the isotropic and kinematic hardening behavior of material.

The isotropic and kinematic hardening parameters κ and $\boldsymbol{\alpha}$ evolve with plastic deformation. The evolution of κ is related to the mean stress, and more directly to the volumetric plastic strain ε_v^p, that is, $\kappa = \varepsilon_v^p$, while the evolution of $\boldsymbol{\alpha}$ is related to the deviatoric plastic strain $\mathbf{e}^p$. As can be expected, the kinematic hardening parameter $\boldsymbol{\alpha} = \{\alpha_1\ \alpha_2\ \alpha_3\}^T$ can be decomposed in two directions J_1 and J_{2D} in meridian plane, which contains two parts as follows:

$$\boldsymbol{\alpha} = a_1 \exp\left(a_2 \left((\mathbf{e}^p)^T : \mathbf{e}^p\right)^{a_3}\right) \mathbf{m} + \left(a_4 \left((\mathbf{e}^p)^T : \mathbf{e}^p\right)^{a_3}\right) \mathbf{e}^p \tag{7.6}$$

where the first term controls the movement of yield surface in $\tilde{J}_1$ axis and the second term controls the movement of yield surface in perpendicular direction to $\tilde{J}_1$ axis.

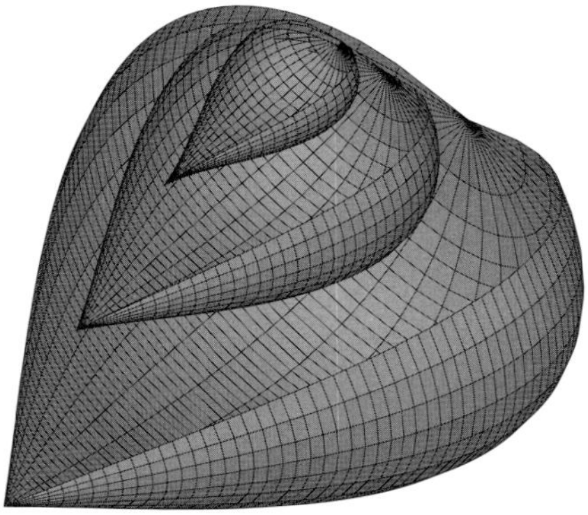

Figure 7.2 Trace of 3D representation of yield surface (7.5) in principal stress space for the isotropic and kinematic hardening behavior of material.

In relation (7.6), $\mathbf{m}$ is the unit vector, defined as $\mathbf{m} = \{1\,1\,1\}^{\mathrm{T}}$, and a_1, a_2, a_3, and a_4 are the material parameters. The three components of the kinematic hardening parameter $\boldsymbol{\alpha}$, namely, α_1, α_2, and α_3, determine the values of the yield surface movements in the directions of principal stresses σ_1, σ_2, and σ_3, respectively.

7.2.1
Isotropic and Kinematic Material Functions

The material functions f_{h} and f_{d} control the size and movement of the yield surface and are functions of hardening parameters. These functions can be decomposed into two parts, the isotropic and kinematic parts, as

$$f_{\mathrm{h}} = f_{\mathrm{h}_{\mathrm{isotropic}}} + f_{\mathrm{h}_{\mathrm{kinematic}}} \tag{7.7}$$

$$f_{\mathrm{d}} = f_{\mathrm{d}_{\mathrm{isotropic}}} + f_{\mathrm{d}_{\mathrm{kinematic}}} \tag{7.8}$$

The isotropic part of material functions f_{h} and f_{d} are the exponential increasing functions of the isotropic hardening parameter $\kappa = \varepsilon_{\mathrm{v}}^{\mathrm{p}}$, defined as

$$f_{\mathrm{h}_{\mathrm{isotropic}}} = (b_1 + b_2 \exp(b_3 \varepsilon_{\mathrm{v}}^{\mathrm{p}}))\delta(\varepsilon_{\mathrm{v}}^{\mathrm{p}}) \tag{7.9}$$

$$f_{\mathrm{d}_{\mathrm{isotropic}}} = (c_1 + c_2 \exp(c_3 \varepsilon_{\mathrm{v}}^{\mathrm{p}}))\delta(\varepsilon_{\mathrm{v}}^{\mathrm{p}}) \tag{7.10}$$

where b_1, b_2, b_3, c_1, c_2, and c_3 are the material parameters and $\delta(\varepsilon_{\mathrm{v}}^{\mathrm{p}})$ is defined as

$$\delta(\varepsilon_{\mathrm{v}}^{\mathrm{p}}) = \begin{cases} 1 & \text{if } \varepsilon_{\mathrm{v}}^{\mathrm{p}} \neq 0 \\ 0 & \text{if } \varepsilon_{\mathrm{v}}^{\mathrm{p}} = 0 \end{cases} \tag{7.11}$$

In order to determine the kinematic parts of material functions (7.7) and (7.8), consider two different stress spaces σ_i and Ξ_i; the former is located in the center of yield surface before kinematic hardening and the latter is placed in the center of yield surface after kinematic hardening (Figure 7.3). The distance of centers of two coordinate systems can be defined by α_i, in which the relationship between the principal stresses in two stress spaces is defined as

$$\Xi_i = \sigma_i + \alpha_i \tag{7.12}$$

The definitions of the parameters $J_{1\alpha}$ and $J_{2D\alpha}$ are similar those of the invariants of stress and deviatoric stress tensors:

$$J_{1\alpha} = \alpha_1 + \alpha_2 + \alpha_3 \tag{7.13}$$

$$J_{2D\alpha} = \frac{1}{6}[(\alpha_1 - \alpha_2)^2 + (\alpha_2 - \alpha_3)^2 + (\alpha_3 - \alpha_1)^2] \tag{7.14}$$

Considering the definition of the principal stresses in two stress spaces, defined by Eq. (7.12), the three-invariant cap plasticity (7.5) can be written in the new stress space as

$$\psi J_{IIID}^{2/3} + \frac{2}{3} J_{IID} + \left(\frac{\phi_{\mathrm{d}} f_{\mathrm{d}}}{\phi_{\mathrm{h}} f_{\mathrm{h}}}\right)^2 J_I^2 - \phi_{\mathrm{d}}^2 f_{\mathrm{d}}^2 = 0 \tag{7.15}$$

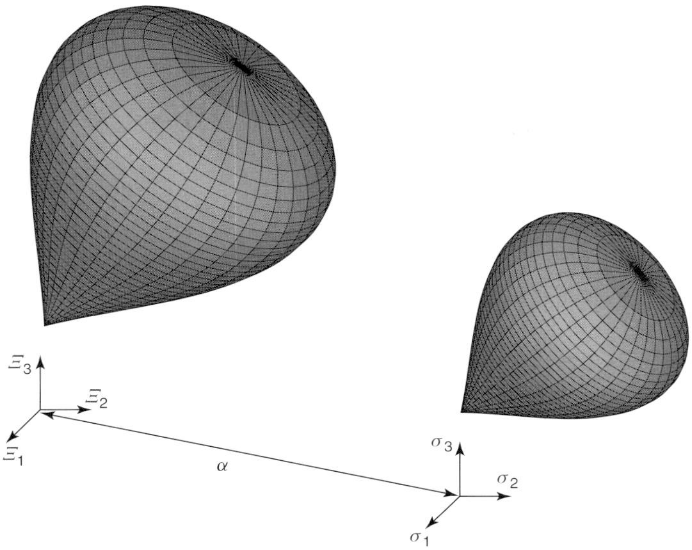

Figure 7.3 The stress spaces before kinematic hardening σ_i and after kinematic hardening Ξ_i.

where J_I is the first invariant of stress tensor and J_{IID} and J_{IIID} are the second and third invariants of deviatoric stress tensors in the second stress space, defined as

$$J_I = J_1 + J_{1\alpha} \tag{7.16}$$

$$J_{IID} = J_{2D} + J_{2D\alpha} + J_{\sigma\alpha} \tag{7.17}$$

where

$$J_{\sigma\alpha} = \frac{2}{\sqrt{3}}\sqrt{J_{2D}}\left(\cos\omega\left(\alpha_1 - \frac{1}{2}\alpha_2 - \frac{1}{2}\alpha_3\right) + \frac{\sqrt{3}}{2}\sin\omega(\alpha_2 - \alpha_3)\right) \tag{7.18}$$

where

$$\omega = \frac{1}{3}\cos^{-1}\left(\frac{3\sqrt{3}}{2}\frac{J_{3D}}{J_{2D}^{3/2}}\right) \qquad 0^\circ \le \omega \le 60^\circ \tag{7.19}$$

Substituting relations (7.16) and (7.17) into the yield surface (7.15) in the absence of third invariants of deviatoric stress with respect to Eq. (7.5), results in

$$\left(\frac{\phi_\text{d} f_\text{d}}{\phi_\text{h} f_\text{h}}\right)^2 = -\frac{2}{3}\frac{(J_{2D\alpha} + J_{\sigma\alpha})}{(2J_{1\alpha}J_1 + J_{1\alpha}^2)} \tag{7.20}$$

According to Eq. (7.20), the kinematic parts of material functions f_h and f_d can be therefore written as

$$f_{\text{h}\,\text{kinematic}} = \frac{1}{\phi_\text{h}}(2\,J_{1\alpha}J_1 + J_{1\alpha}^2)^{1/2} \tag{7.21}$$

$$f_{\text{d}\,\text{kinematic}} = \frac{1}{\phi_\text{d}}\left(-\frac{2}{3}(J_{2D\alpha} + J_{\sigma\alpha})\right)^{1/2} \tag{7.22}$$

Finally, the material functions f_h and f_d can be defined by substituting Eqs (7.9), (7.10), (7.21), and (7.22) into Eqs (7.7) and (7.8) as

$$f_h = \left(b_1 + b_2 \exp\left(b_3 \varepsilon_v^p\right)\right) \delta\left(\varepsilon_v^p\right) + \frac{1}{\phi_h}\left(2 J_{1\alpha} J_1 + J_{1\alpha}^2\right)^{1/2} \tag{7.23}$$

$$f_d = \left(c_1 + c_2 \exp\left(c_3 \varepsilon_v^p\right)\right) \delta\left(\varepsilon_v^p\right) + \frac{1}{\phi_d}\left(-\frac{2}{3}\left(J_{2D\alpha} + J_{\sigma\alpha}\right)\right)^{1/2} \tag{7.24}$$

7.2.2
Computation of Powder Property Matrix

The object of the mathematical theory of plasticity is to provide a theoretical description of the relationship between stress and strain, or more commonly, between increments of stress and increments of strain using the assumption that the material behaves plastically only after a certain limiting value has been exceeded. The elastoplastic constitutive relation in its incremental form can be presented by $d\boldsymbol{\sigma} = \mathbf{D}_{ep} d\boldsymbol{\varepsilon}$, with $d\boldsymbol{\sigma}$ denoting the incremental stress vector, $d\varepsilon$ the incremental strain vector, and $\mathbf{D}_{ep}$ the constitutive elastoplastic matrix. The yield surface $F(\boldsymbol{\sigma}, \boldsymbol{\alpha}, \kappa) = 0$ determines the stress level at which the plastic deformation begins. The material property matrix $\mathbf{D}_{ep}$ is defined as

$$\mathbf{D}_{ep} = \mathbf{D}_e - \frac{(\partial F/\partial \boldsymbol{\sigma})^T \mathbf{D}_e^T \mathbf{D}_e (\partial Q/\partial \boldsymbol{\sigma})}{H + (\partial F/\partial \boldsymbol{\sigma})^T \mathbf{D}_e (\partial Q/\partial \sigma)} \tag{7.25}$$

where $\mathbf{n} = \partial F/\partial \boldsymbol{\sigma}$ and $\mathbf{n}_g = \partial Q/\partial \boldsymbol{\sigma}$ are the normal vectors to the yield and potential plastic surfaces, respectively, and H is the hardening plastic modulus defined as

$$H = -\frac{\partial F}{\partial \mu}\frac{d\mu}{d\lambda} \tag{7.26}$$

where $d\lambda$ is the plastic multiplier and $\mu = (\mathbf{e}_p, \varepsilon_v^p)$.

In order to derive the constitutive elastoplastic matrix and its components, we need to calculate $\mathbf{D}_e$, $\mathbf{n}$, $\mathbf{n}_g$, and H in Eq. (7.25). The normal vector to the yield surface is determined by

$$\mathbf{n} = \frac{\partial F}{\partial \boldsymbol{\sigma}} = \frac{\partial F}{\partial J_1}\frac{\partial J_1}{\partial \boldsymbol{\sigma}} + \frac{\partial F}{\partial J_{2D}}\frac{\partial J_{2D}}{\partial \boldsymbol{\sigma}} + \frac{\partial F}{\partial J_{3D}^{2/3}}\frac{\partial J_{3D}^{2/3}}{\partial \boldsymbol{\sigma}} \tag{7.27}$$

where

$$\frac{\partial F}{\partial J_1} = 2 J_1 \left(\frac{\phi_d f_d}{\phi_h f_h}\right)^2 + \frac{\phi_d^2 f_d^2}{\phi_h^3 f_h^3}\frac{2 J_1^2 J_{1\alpha}}{\sqrt{-(2/3)(2 J_{1\alpha} J_1 + J_{1\alpha}^2)}} \tag{7.28}$$

$$\frac{\partial F}{\partial J_{2D}} = \frac{2}{3} + 2 J_1^2 \frac{f_d \phi_d^2}{\phi_h^2 f_h^2}\left(\frac{\partial f_d}{\partial J_{2D}}\right) - 2\phi_d^2 f_d \left(\frac{\partial f_d}{\partial J_{2D}}\right) \tag{7.29}$$

$$\frac{\partial F}{\partial J_{3D}^{2/3}} = \psi + 2 J_1^2 \frac{f_d \phi_d^2}{\phi_h^2 f_h^2}\left(\frac{\partial f_d}{\partial J_{3D}^{2/3}}\right) - 2\phi_d^2 f_d \left(\frac{\partial f_d}{\partial J_{3D}^{2/3}}\right) \tag{7.30}$$

The hardening plastic modulus H can be determined by substituting the yield surface (7.5) into definition (7.26) as

$$H = -\left(\left(\frac{\partial F}{\partial f_{\rm h}}\frac{\partial f_{\rm h}}{\partial \mathbf{e}_{\rm p}}\frac{d\mathbf{e}_{\rm p}}{d\lambda} + \frac{\partial F}{\partial f_{\rm d}}\frac{\partial f_{\rm d}}{\partial \mathbf{e}_{\rm p}}\frac{d\mathbf{e}_{\rm p}}{d\lambda}\right) + \left(\frac{\partial F}{\partial f_{\rm h}}\frac{\partial f_{\rm h}}{\partial \varepsilon_{\rm v}^{\rm p}}\frac{d\varepsilon_{\rm v}^{\rm p}}{d\lambda} + \frac{\partial F}{\partial f_{\rm d}}\frac{\partial f_{\rm d}}{\partial \varepsilon_{\rm v}^{\rm p}}\frac{d\varepsilon_{\rm v}^{\rm p}}{d\lambda}\right)\right) \tag{7.31}$$

where

$$\frac{\partial F}{\partial f_{\rm h}} = -2\,J_1^2\left(\frac{\phi_{\rm d}^2 f_{\rm d}^2}{\phi_{\rm h}^2 f_{\rm h}^3}\right) \tag{7.32}$$

$$\frac{\partial F}{\partial f_{\rm d}} = 2\,J_1^2\left(\frac{\phi_{\rm d}^2 f_{\rm d}}{\phi_{\rm h}^2 f_{\rm h}^2}\right) - 2\phi_{\rm d}^2 f_{\rm d} \tag{7.33}$$

$$\frac{\partial f_{\rm h}}{\partial \varepsilon_{\rm v}^{\rm p}} = \left(b_2 b_3 \exp(b_3 \varepsilon_{\rm v}^{\rm p})\right)\delta(\varepsilon_{\rm v}^{\rm p}) \tag{7.34}$$

$$\frac{\partial f_{\rm d}}{\partial \varepsilon_{\rm v}^{\rm p}} = \left(c_2\, c_3 \exp(c_3 \varepsilon_{\rm v}^{\rm p})\right)\delta(\varepsilon_{\rm v}^{\rm p}) \tag{7.35}$$

$$\frac{\partial f_{\rm h}}{\partial \mathbf{e}_{\rm p}} = -\frac{1}{\phi_{\rm h}}\frac{(J_1 + J_{1\alpha})}{\sqrt{-(2\,J_{1\alpha}J_1 + J_{1\alpha}^2)}}\left(2\frac{\partial \alpha_1}{\partial \mathbf{e}_{\rm p}} + \frac{\partial \alpha_2}{\partial \mathbf{e}_{\rm p}}\right) \tag{7.36}$$

$$\frac{\partial f_{\rm d}}{\partial \mathbf{e}_{\rm p}} = \frac{(\alpha_2 - \alpha_1)}{3\phi_{\rm d}\sqrt{J_{2D\alpha} + J_{\sigma\alpha}}}\left(\frac{\partial \alpha_2}{\partial \mathbf{e}_{\rm p}} - \frac{\partial \alpha_1}{\partial \mathbf{e}_{\rm p}}\right) + \frac{\sqrt{J_{2D}}(\cos\omega - \sqrt{3}\sin\omega)}{2\phi_{\rm d}\sqrt{3(J_{2D\alpha} + J_{\sigma\alpha})}}\left(\frac{\partial \alpha_1}{\partial \mathbf{e}_{\rm p}} - \frac{\partial \alpha_2}{\partial \mathbf{e}_{\rm p}}\right) \tag{7.37}$$

7.2.3
Model Assessment and Parameter Determination

7.2.3.1 Model Assessment

In order to demonstrate the performance of the proposed plasticity model in the prediction of powder material behavior, experimental tests must be performed to determine and calibrate the parameters of the material functions $f_{\rm h}$ and $f_{\rm d}$, defined by Eqs (7.23) and (7.24), in the yield surface (7.5). These two material functions control the size and movement of the yield surface, and are decomposed into the isotropic and kinematic parts, given by Eqs (7.7) and (7.8), as functions of the hardening parameters κ and $\boldsymbol{\alpha}$, or directly the plastic volumetric strain $\varepsilon_{\rm v}^{\rm p}$ and the deviatoric plastic strain $\mathrm{e}_{\rm p}$. It must be noted that the kinematic hardening parameter $\boldsymbol{\alpha}$ indicates the movement of the yield surface in the direction of the J_1 axis and the direction perpendicular to it.

It is worth mentioning that different values of f_h and f_d result in different aspects of the yield surface (7.5). Considering that the first two terms of Eq. (7.5) are zero, it can be written as

$$\left(\frac{\phi_d f_d}{\phi_h f_h}\right)^2 J_1^2 - \phi_d^2 f_d^2 = 0 \tag{7.38}$$

The above equation generally yields three roots, the points of intersection of the yield surface with the J_1-axis, that is, $J_1 = \pm\phi_h f_h$ and one more from $f_d = 0.0$, which has been defined in Eq. (7.24). If $f_d = 0.0$ does not lead to any value for J_1, the yield surface of Eq. (7.5) yields to two roots for J_1, $J_1 = \pm\phi_h f_h$, which result in an elliptical shape in the meridian plane, as shown in Figure 7.4a. In this case, the yield surface grows with densification, eventually becoming independent of the hydrostatic stress J_1 at full dense material, where the von Mises yield surface is generated. This yield surface was developed by the authors for porous metal and sintered powder based on an extension of the von Mises concept [12].

If $f_d = 0.0$ leads to the value of J_1 between $-\phi_h f_h$ and $+\phi_h f_h$, the cone-cap yield surface can be produced from Eq. (7.5) based on the intersection points of $J_1 = -\phi_h f_h$ and the value obtained from $f_d = 0.0$ for different values of isotropic hardening. In this case, the yield surface grows with densification and reduces to the Drucker–Prager yield function for full dense bodies, as shown in Figure 7.4b. This yield surface is very similar to the double-surface cap models, that is, a combination of the Mohr–Coulomb or Drucker–Prager and elliptical surfaces, which has been extensively used by authors to demonstrate the behavior of powder and granular materials [15–19]. It is worth mentioning that as the double-surface plasticity consists of two different yield functions, special treatment is necessary in order to avoid numerical difficulties in the intersection of these two surfaces [24]; however, the single-yield surface (7.5) does not have such a drawback. Furthermore, the parameter ψ in the yield surface (7.5) causes the triangularity of deviatoric trace along the hydrostatic axis. In this case, the yield surface is similar to the irregular hexagonal pyramid of the Mohr–Coulomb and cone-cap yield surface employed by researchers for the description of soil and geomaterial behavior.

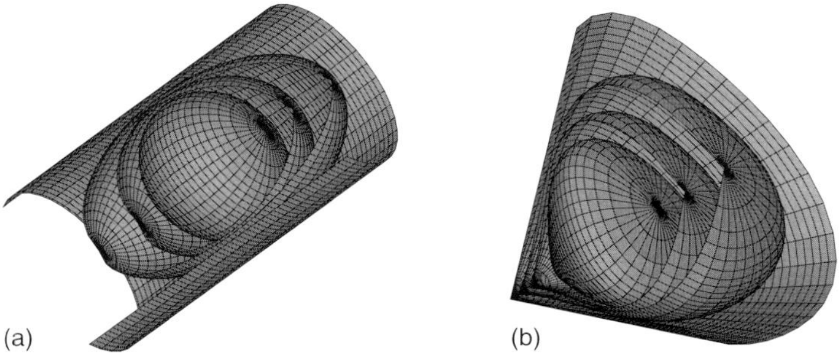

Figure 7.4 Trace of the 3D yield function in the principal stress space for the isotropic hardening behavior. (a) The elliptical yield surface. (b) The cone-cap yield surface.

7.2.3.2 Parameter Determination

The important issue in the prediction of powder material behavior is the identification of the parameters of the proposed plasticity model. The calibration procedure for three-invariant isotropic–kinematic cap plasticity is carried out, based on a series of isostatic and triaxial tests. An organized approach to determine model parameters is to utilize an optimization routine. Mathematically, an objective function and a search strategy are necessary for the optimization. The objective function, which represents the constitutive model, captures the material behavior and can be used in a simultaneous optimization against a series of experimental data. The simplest search strategy is based on the direct search approach that is proved to be reliable, and its relative simplicity makes it quite easy to program into the code.

For the proposed constitutive model, a total of 10 material constants need to be determined for the material functions f_h and f_d. The parameters of isotropic parts of f_h and f_d (i.e., b_1, b_2, b_3, and c_1, c_2, c_3) are first evaluated using the confining pressure test, where the values of J_{2D} and J_{3D} in the yield surface (7.5) are zero. The parameters of the kinematic parts of f_h and f_d (i.e., a_1, a_2, a_3, and a_4) are then estimated performing the least squares method (LSM) method on the data obtained by a series of triaxial tests. These parameters control various aspects of predicted stress–strain curves obtained numerically by fitting the stress path to the triaxial and confining pressure tests. The procedure of parameter determination is performed as follows:

- **Step 1**: On the basis of the results obtained from the confining pressure test, the values of J_1 are evaluated using the yield surface (7.5) where the values of J_{2D} and J_{3D} are zero. From the values of volumetric strain ε_v, the elastic and plastic volumetric strains, ε_v^e and ε_v^p, are estimated. The parameters b_1, b_2, and b_3 in the isotropic part of f_h are computed. The parameters c_1, c_2, and c_3 in the isotropic part of f_d are then calculated by an LSM on the data obtained from the confining pressure test.
- **Step 2**: Applying the results of the triaxial tests and the isotropic parameters of f_h and f_d obtained from step 1, the kinematic parameters of f_h and f_d, that is, a_1, a_2, and a_3, in the first term of relation (7.6) are first estimated. The parameter a_4 in the second term of relation (7.6) is then obtained by performing the LSM on the data obtained from the triaxial tests.

In order to determine and calibrate the powder parameters of the proposed plasticity model for an iron-based powder (95% by weight), a set of compaction experiments performed by Doremus *et al.* [31] is proposed here. Both isostatic and triaxial tests are driven to determine and calibrate the parameters of the model. The raw material is composed of iron, copper, wax zinc stearate, in which the last two components are admixed as internal lubricants. The density of the solid phase is about 7.54 g cm^{-3} and the tap powder density is about 3.67 g cm^{-3}. The particles have irregular shapes and their sizes are between 10 and 100 μm. The compacted specimen has an initial height H_0 of 42 cm and diameter D_0 of 20 cm. The triaxial tests consist of an initial isostatic compaction step up to a pressure

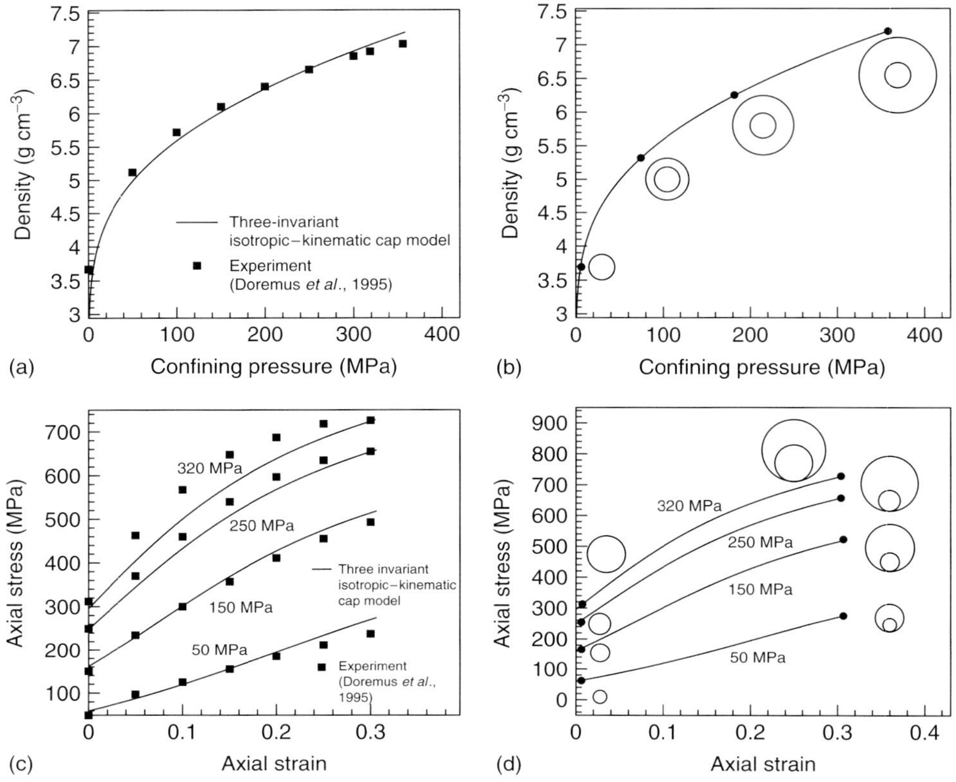

Figure 7.5 Iron powder component. (a, b) The variation of density with hydrostatic pressure in the isostatic compression test. (c, d) The axial stress versus axial strain in triaxial tests.

value of 400 MPa, followed by a subsequent uniaxial compaction step. This step is carried out by keeping pressure constant and increasing the axial stress up to the maximum value of 1250 MPa. The material parameters calibrated for the proposed yield surface are as follows:

$$\begin{array}{llll} a_1 = 2.0e-17 & a_2 = 35.16 & a_3 = 0.08 & a_4 = -0.05 \\ b_1 = 260.2 & b_2 = 8.586e-2 & b_3 = -9.2 & \\ c_1 = 610.5 & c_2 = 1.75 & c_3 = -10.85 & \end{array}$$

The variation of density with hydrostatic pressure is presented in Figure 7.5a for the isostatic test. This evolution is the characteristic of powders. It shows a good agreement between experimental and numerical results for the isostatic compression step. Figure 7.5b illustrates the isotropic hardening behavior of powder due to expansion of the yield surface on increasing the hydrostatic pressure. Clearly, the effect of confining pressure on isotropic hardening of powder can be observed in this figure. Figure 7.5c,d correspond to the complete triaxial compression tests. The variations of the axial stress with axial strain are shown in Figure 7.5c at different values of hydrostatic pressure. Figure 7.5d presents the relevant yield surfaces corresponding to the isotropic and kinematic hardening behavior in complete triaxial tests. As can be observed, the proposed model captures the behavior of powder in complete triaxial experiment. This representation clearly shows how the yield surface grows isotropically and moves kinematically on increasing the axial strain.

7.3 Arbitrary Lagrangian–Eulerian Formulation

The ALE technique was developed to overcome the Lagrangian finite element distortion by taking the advantages of both Lagrangian and Eulerian methods. The technique has been proposed in various solid and fluid mechanics, including incompressible hyperelasticity [32], nonlinear dynamic analysis [33], strain localization [34], nonlinear solid mechanics [35, 36], metal forming [37–39], hyperelastoplasticity [40, 41], and finite strain plasticity [42]. A fully coupled implicit ALE formulation was presented by Bayoumi and Gadala [43] for large deformation dynamic problems. The ALE technique was proposed by Khoei *et al.* [44] in large plastic deformation of pressure-sensitive materials. A key issue in ALE formulation is an efficient mesh motion technique in order to achieve the satisfactory results. There are various mesh relocation techniques, based on a uniform distribution of the equivalent plastic strain indicator [37], the transfinite mapping algorithm using the nodal relocation [38], the ALE split operator [40], and so on. These techniques are able to retain the shape of the elements by equalizing their size and by avoiding shape distortion without changing the mesh topology.

The main difficulty in extending the ALE formulation from fluid to solid mechanics is the path-dependent behavior of material models used in solid mechanics. The constitutive equation of ALE nonlinear mechanics contains a convective term which reflects the relative motion between the physical motion and the mesh motion. In fluid mechanics, the convective effects only occur in the mass balance, momentum balance, and energy balance, but not in the constitutive equation. The correct treatment of this convective term in the constitutive equation is the key point in ALE nonlinear solid mechanics. The most popular approach to deal with the convective term is the use of a split, or fractional-step method. Each time step is first divided into a Lagrangian phase and an Eulerian phase. Convection is neglected in the material phase, which is thus identical to a time step in a standard Lagrangian analysis. The stress and plastic internal variables are then transferred from Lagrangian mesh to the relocated mesh in order to evaluate the relative mesh–material motion in the convection phase. To handle this, an approach called the *Godunov scheme* is used for the stress update [44].

7.3.1 ALE Governing Equations

In the ALE description, three different configurations are considered: the material domain Ω_0, spatial domain Ω, and reference domain $\hat{\Omega}$, which is called *ALE domain*. The material motion is defined by $x_i^{\mathrm{m}} = f_i(X_j,t)$, with X_j denoting the material point coordinates and $f_i(X_j,t)$ a function which maps the body from the initial or material configuration Ω_0 to the current or spatial configuration Ω. The initial position of material points is denoted by x_i^{g} called the *reference or ALE coordinate* in which $x_i^{\mathrm{g}} = f_i(X_j,0)$. The reference domain $\hat{\Omega}$ is defined to describe the mesh motion and is coincident with mesh points, so that it can be denoted by computational domain. The mesh motion is defined by $x_i^{\mathrm{m}} = \hat{f}_i(x_j^{\mathrm{g}},t)$. The material coordinate can be then related to ALE coordinate by $x_i^{\mathrm{g}} = \hat{f}_i^{-1}(x_j^{\mathrm{m}},t)$. The mesh displacement can be defined by

$$u_i^{\mathrm{g}}(x_j^{\mathrm{g}},t) = x_i^{\mathrm{m}} - x_i^{\mathrm{g}} = \hat{f}_i(x_j^{\mathrm{g}},t) - x_i^{\mathrm{g}} \tag{7.39}$$

It must be noted that the mesh motion can be simply obtained from material motion replacing the material coordinate by ALE coordinate. The mesh velocity can be defined as

$$v_i^{\mathrm{g}}(x_j^{\mathrm{g}},t) = \frac{\partial \hat{f}_i(x_j^{\mathrm{g}},t)}{\partial t} = \left.\frac{\partial x_i^{\mathrm{m}}}{\partial t}\right|_{x_j^{\mathrm{g}}} \tag{7.40}$$

in which the ALE coordinate x_j^{g} and material coordinate X_j in material velocity are fixed. In ALE formulation, the convective velocity c_i is defined using the difference between the material and mesh velocities as

$$c_i = v_i^{\mathrm{m}} - v_i^{\mathrm{g}} = \left.\frac{\partial x_i^{\mathrm{m}}}{\partial x_j^{\mathrm{g}}}\frac{\partial x_j^{\mathrm{g}}}{\partial t}\right|_{X_k} = \frac{\partial x_i^{\mathrm{m}}}{\partial x_j^{\mathrm{g}}} w_j \tag{7.41}$$

where the material velocity $v_i^{\mathrm{m}} = (\partial x_i^{\mathrm{m}}/\partial t)_{X_j}$ can be obtained using the chain rule expression with respect to the ALE coordinate x_j^{g} and time t. In Eq. (7.41), the referential velocity w_i is defined by $w_{\mathrm{i}} = (\partial x_i^{\mathrm{g}}/\partial t)_{X_j}$. The above relationship between the convective velocity c_i, material velocity v_i^{m}, mesh velocity v_i^{g}, and referential velocity w_i is frequently used in ALE formulation.

Now, the general relationship between material time derivatives and referential time derivatives of any scalar function f_i can be written as

$$\left.\frac{\partial f_i}{\partial t}\right|_{X_j} = \left.\frac{\partial f_i}{\partial t}\right|_{x_j^{\mathrm{g}}} + \left.\frac{\partial f_i}{\partial x_i^{\mathrm{g}}}\frac{\partial f_i}{\partial t}\right|_{X_j} = \left.\frac{\partial f_i}{\partial t}\right|_{x_j^{\mathrm{g}}} + \frac{\partial f_i}{\partial x_j^{\mathrm{m}}} c_j \tag{7.42}$$

The above equation can be used to deduce the fundamental conservation laws of continuum mechanics, that is, the momentum, mass, and constitutive equations, in nonlinear ALE description.

In the ALE technique, the governing equations can be derived by substituting the relationship between the material time derivatives and referential time derivatives, that is, Eq. (7.42), into the continuum mechanics governing equations. This substitution gives rise to convective terms in the ALE equations, which account for the transport of material through the grid. Thus, the momentum equation in ALE formulation can be written in a manner similar to the updated Lagrangian description by consideration of the material time derivative terms as

$$\rho \dot{v}_i^{\mathrm{m}} = \sigma_{ji,j} + \rho b_i \tag{7.43}$$

where ρ is the density, σ is the Cauchy stress, and b_i is the body force. In the above equation, the material time derivative of velocity $\dot{v}_i^{\mathrm{m}}$ can be obtained by specializing the general relationship (7.42) to $\dot{v}_i^{\mathrm{m}}$ as

$$\dot{v}_i^{\mathrm{m}} = \left.\frac{\partial v_i^{\mathrm{m}}}{\partial t}\right|_{x_j^{\mathrm{g}}} + \frac{\partial v_i^{\mathrm{m}}}{\partial x_j^{\mathrm{m}}} c_j \tag{7.44}$$

Substituting Eq. (7.44) into Eq. (7.43), the momentum equation can be then written as

$$\rho \left(\left.\frac{\partial v_i^{\mathrm{m}}}{\partial t}\right|_{x_j^{\mathrm{g}}} + \frac{\partial v_i^{\mathrm{m}}}{\partial x_j^{\mathrm{m}}} c_j \right) = \frac{\partial \sigma_{ij}}{\partial x_j^{\mathrm{m}}} + \rho b_i \tag{7.45}$$

The mass balance in ALE formulation can be similarly derived by specializing the general relationship (7.42) to the density ρ as

$$\left.\frac{\partial \rho}{\partial t}\right|_{x_j^{\mathrm{g}}} + \frac{\partial \rho}{\partial x_j^{\mathrm{m}}} c_j = -\rho \left.\frac{\partial v_j^{\mathrm{m}}}{\partial x_j^{\mathrm{m}}}\right|_{X_j} \tag{7.46}$$

Finally, in order to describe the constitutive equation for nonlinear ALE formulation, the general relationship (7.42) is specialized to the stress tensor, thus

$$\left.\frac{\partial \boldsymbol{\sigma}}{\partial t}\right|_{x_j^{\mathrm{g}}} + \frac{\partial \boldsymbol{\sigma}}{\partial x_j^{\mathrm{m}}} c_j = \mathbf{q} \tag{7.47}$$

where $\mathbf{q}$ accounts for both the pure straining of the material and the rotational terms that counteract the nonobjectivity of the material stress rate [40].

The basis of any mechanical initial boundary value problem in the framework of the material description is the balance of momentum equation. In the framework of the referential configuration, we have also considered the mass balance and the constitutive equations, which are defined as partial differential equations in the case of the referential description. In the quasi-static problems, the inertia force $\rho\mathbf{a}$ is negligible with respect to other forces of momentum equation; hence, the equilibrium equations in ALE and Lagrangian descriptions are exactly identical. In addition, considering the constant value of density ρ, the balance of mass equation results in $\partial v_j^{\mathrm{m}}/\partial x_j^{\mathrm{m}}|_{X_j} = 0$, which is already satisfied. Thus, the governing equations in ALE formulation for quasi-static problems can be summarized into Eqs (7.45) and (7.47).

7.3.2
Weak Form of ALE Equations

In order to present the weak form of initial boundary value problems in ALE description, the mass balance and the balance of linear momentum can be written in the integral form over the spatial domain Ω, multiplied by the test functions $\delta\rho$, $\delta\mathbf{v}$, and $\delta\boldsymbol{\sigma}$. Clearly, there must be a relationship between the strong form and weak form of governing equations, in which these two forms are identical. The weak form of momentum equation is obtained by multiplying the strong form of Eq. (7.45) by the test function $\delta\mathbf{v} \in \mathrm{U}^0$, where $\mathrm{U}^0 = \{\delta\mathbf{v}|\delta\mathbf{v} \in \mathrm{C}^0, \delta\mathbf{v} = \mathbf{0} \text{ on } \Gamma_\mathrm{v}\}$ and Γ_v indicates the part of the boundary in which the velocities are prescribed. Consider $\mathbf{v} \in \mathrm{U}$ is the trial solution with $\mathrm{U} = \{\mathbf{v}|\mathbf{v} \in \mathrm{C}^0, \mathbf{v} = \hat{\boldsymbol{v}} \text{ on } \Gamma_\mathrm{v}\}$ and $\hat{\boldsymbol{v}}$ is the prescribed velocities in Γ_v, the integration over the spatial domain results in

$$\int_\Omega \delta\mathbf{v}\rho\left(\left.\frac{\partial\mathbf{v}^\mathrm{m}}{\partial t}\right|_{\mathbf{x}^\mathrm{g}} + \frac{\partial\mathbf{v}^\mathrm{m}}{\partial\mathbf{x}^\mathrm{m}}\mathbf{c}\right)\mathrm{d}v = \int_\Omega \delta\mathbf{v}(\mathrm{div}_{x^\mathrm{m}}\,\boldsymbol{\sigma} + \rho\mathbf{b})\,\mathrm{d}v \tag{7.48}$$

or

$$\int_\Omega \delta\mathbf{v}\rho\left.\frac{\partial\mathbf{v}^\mathrm{m}}{\partial t}\right|_{\mathbf{x}^\mathrm{g}}\mathrm{d}v + \int_\Omega \delta\mathbf{v}\rho\frac{\partial\mathbf{v}^\mathrm{m}}{\partial\mathbf{x}^\mathrm{m}}\mathbf{c}\,\mathrm{d}v = \int_\Omega \delta\mathbf{v}\;\mathrm{div}_{x^\mathrm{m}}\,\boldsymbol{\sigma}\,\mathrm{d}v + \int_\Omega \delta\mathbf{v}\rho\mathbf{b}\,\mathrm{d}v \tag{7.49}$$

To eliminate the stress derivatives, the first term on the right-hand side of the above equation is rewritten using the integration part by part as

$$\begin{aligned}&\int_\Omega \delta\mathbf{v}\rho\left.\frac{\partial\mathbf{v}^\mathrm{m}}{\partial t}\right|_{\mathbf{x}^\mathrm{g}}\mathrm{d}v + \int_\Omega \delta\mathbf{v}\rho\frac{\partial\mathbf{v}^\mathrm{m}}{\partial\mathbf{x}^\mathrm{m}}\mathbf{c}\,\mathrm{d}v\\ &= -\int_\Omega \mathrm{div}_{x^\mathrm{m}}\delta\mathbf{v}\,\boldsymbol{\sigma}\,\mathrm{d}v + \int_\Omega \delta\mathbf{v}\rho\mathbf{b}\,\mathrm{d}v + \int_{\Gamma_t} \delta\mathbf{v}\hat{\mathbf{t}}\,\mathrm{d}\Gamma\end{aligned} \tag{7.50}$$

where Γ_t refers to the part of boundary in which the traction vector $\hat{\mathbf{t}}$ is prescribed.

Since the mass balance is enforced in the referential description as a partial differential equation, a weak form must be developed. Considering the trial solution as $\rho \in \mathrm{C}^0$, the weak form of the balance of mass can be obtained by integration of the strong form of mass balance, given in Eq. (7.46) over the spatial domain Ω, which has been multiplied by a test function $\delta\rho \in \mathrm{C}^0$ as

$$\int_\Omega \delta\rho\left(\left.\frac{\partial\rho}{\partial t}\right|_{\mathbf{x}^\mathrm{g}} + \frac{\partial\rho}{\partial\mathbf{x}^\mathrm{m}}\mathbf{c} + \rho\left.\frac{\partial\mathbf{v}^\mathrm{m}}{\partial\mathbf{x}^\mathrm{m}}\right|_\mathrm{X}\right)\mathrm{d}v = 0 \tag{7.51}$$

in which only the first derivatives appear with respect to the mass density ρ and velocity $\mathbf{v}$.

Similar to the momentum equation and mass balance, the weak form of constitutive equation for nonlinear ALE formulation can be obtained by multiplying the strong form of Eq. (7.47) with a test function $\boldsymbol{\delta\sigma}$ and integrating over the spatial domain as

$$\int_\Omega \delta\boldsymbol{\sigma}\left(\left.\frac{\partial\boldsymbol{\sigma}}{\partial t}\right|_{\mathbf{x}^\mathrm{g}} + \frac{\partial\boldsymbol{\sigma}}{\partial\mathbf{x}^\mathrm{m}}\mathbf{c}\right)\mathrm{d}v = \int_\Omega \delta\boldsymbol{\sigma}\mathbf{q}\,\mathrm{d}v \tag{7.52}$$

or

$$\int_\Omega \delta\boldsymbol{\sigma}\left.\frac{\partial\boldsymbol{\sigma}}{\partial t}\right|_{\mathbf{x}^\mathrm{g}}\mathrm{d}v + \int_\Omega \delta\boldsymbol{\sigma}\frac{\partial\boldsymbol{\sigma}}{\partial\mathbf{x}^\mathrm{m}}\mathbf{c}\,\mathrm{d}v = \int_\Omega \delta\boldsymbol{\sigma}\mathbf{q}\,\mathrm{d}v \tag{7.53}$$

7.3.3 ALE Finite Element Discretization

In the finite element method (FEM), the reference domain $\hat{\Omega}$ is subdivided into a number of elements, in which for each element e the ALE coordinates $\mathbf{x}^{\mathrm{g}}$ is defined as

$$\mathbf{x}^{\mathrm{g}}(\xi) = \sum_{I=1}^{N^e} N_I(\xi)\mathbf{x}_I^{\mathrm{g}} \tag{7.54}$$

where ξ denotes the parent element coordinates, $N_I(\xi)$ is the interpolation shape function, $\mathbf{x}_I^{\mathrm{g}}$ stands for the ALE coordinates of node I, and N^e is the number of nodes of element e. The mesh displacement field can be then written following Eq. (7.39) as

$$\boldsymbol{u}^{\mathrm{g}}(\xi) = \mathbf{x}^{\mathrm{m}}(\xi) - \mathbf{x}^{\mathrm{g}}(\xi) = \sum_{I=1}^{N^{\mathrm{e}}} N_I(\xi)\boldsymbol{u}_I^{\mathrm{g}} \tag{7.55}$$

Thus, the mesh, material, and convective velocities can be defined as

$$\begin{aligned}
\mathbf{v}^{\mathrm{g}}(\xi) &= \left.\frac{\partial \mathbf{x}^{\mathrm{m}}}{\partial t}\right|_{x^{\mathrm{g}}} = \sum_{I=1}^{N^{\mathrm{e}}} N_I(\xi)\mathbf{v}_I^{\mathrm{g}} \\
\mathbf{v}^{\mathrm{m}}(\xi) &= \left.\frac{\partial \mathbf{x}^{\mathrm{m}}}{\partial t}\right|_{X} = \sum_{I=1}^{N^{\mathrm{e}}} N_I(\xi)\mathbf{v}_I^{\mathrm{m}} \\
\mathbf{c}(\xi) &= \mathbf{v}^{\mathrm{m}} - \mathbf{v}^{\mathrm{g}} = \sum_{I=1}^{N^{\mathrm{e}}} N_I(\xi)(\mathbf{v}_I^{\mathrm{m}} - \mathbf{v}_I^{\mathrm{g}}) = \sum_{I=1}^{N^{\mathrm{e}}} N_I(\xi^{\mathrm{e}})\mathbf{c}_I(t)
\end{aligned} \tag{7.56}$$

Furthermore, the internal variables, such as density and stresses, can be approximated in the same manner as

$$\begin{aligned}
\rho(\xi, t) &= \sum_{I=1}^{N^{\mathrm{e}}} N_I^{\rho}(\xi^{\mathrm{e}})\rho^I(t) \\
\sigma(\xi, t) &= \sum_{I=1}^{N^{\mathrm{e}}} N_I^{\sigma}(\xi^{\mathrm{e}})\sigma^I(t)
\end{aligned} \tag{7.57}$$

where N_I^{ρ} and N_I^{σ} stand for the shape functions of density and stress, respectively. These shape functions may differ from those used to approximate the displacement field N_I. It must be noted that because the convective terms appear in governing equations, the implementation of standard Galerkin finite element formulation may result in numerical instabilities. This point is especially true for severe dynamic systems. One way to alleviate these difficulties is to employ the Petrov–Galerkin formulation. In this approach, different sets of shape functions are used to interpolate the trial and test functions for displacement, stress, and density.

Substituting the material and the convective velocity ($\boldsymbol{v}^{\mathrm{m}}$ and $\boldsymbol{c}$), given in Eq. (7.56), and the density and the stresses (ρ and $\boldsymbol{\sigma}$), given in Eq. (7.57), into the weak form of the balance of linear momentum, given in Eq. (7.50), yields to

$$\mathbf{M}\frac{d\mathbf{v}^{m}}{dt} + \mathbf{L}\mathbf{v}^{m} + \mathbf{f}^{\,int} = \mathbf{f}^{\,ext} \tag{7.58}$$

where

$$\mathbf{M} = \int_{\Omega} \rho \mathbf{N}^{T}\mathbf{N}\,dv, \quad \mathbf{L} = \int_{\Omega} \rho \mathbf{N}^{T}\mathbf{c}\frac{d\mathbf{N}}{d\mathbf{x}^{m}}\,dv \tag{7.59}$$

and

$$\mathbf{f}^{\,int} = \int_{\Omega} \frac{d\mathbf{N}^{T}}{d\mathbf{x}^{m}}\boldsymbol{\sigma}\,dv, \quad \mathbf{f}^{\,ext} = \int_{\Omega} \rho \mathbf{N}^{T}\boldsymbol{b}\,dv + \int_{\Gamma} \mathbf{N}^{T}\hat{\mathbf{t}}\,d\Gamma \tag{7.60}$$

As mentioned earlier, the term of inertia forces can be neglected in the quasi-static problems. Thus, the momentum Eq. (7.58) can be simplified to $\mathbf{f}^{int} = \mathbf{f}^{\,ext}$.

In a similar manner, the FE formulation for the mass balance can be obtained by substituting the material velocity from Eq. (7.56) and the density from Eq. (7.57) into Eq. (7.51) as

$$\mathbf{M}^{\rho}\frac{d\rho}{dt} + \mathbf{L}^{\rho}\rho + \mathbf{K}^{\rho}\rho = \mathbf{0} \tag{7.61}$$

where

$$\begin{aligned}\mathbf{M}^{\rho} &= \int_{\Omega} \mathbf{N}^{\rho^{T}}\mathbf{N}^{\rho}\,dv, \quad \mathbf{L}^{\rho} = \int_{\Omega} \mathbf{N}^{\rho^{T}}\mathbf{c}\frac{d\mathbf{N}^{\rho}}{d\mathbf{x}^{m}}\,dv,\\ \mathbf{K}^{\rho} &= \int_{\Omega} \mathbf{N}^{\rho^{T}}\,\mathrm{div}_{x^{m}}\,\boldsymbol{v}^{m}\mathbf{N}^{\rho}\,dv\end{aligned} \tag{7.62}$$

Finally, the FE formulation for the constitutive equation can be obtained by replacing the convective velocity from Eq. (7.56) and the stresses from Eq. (7.57) into Eq. (7.53) as

$$\mathbf{M}^{\sigma}\frac{d\boldsymbol{\sigma}}{dt} + \mathbf{L}^{T}\boldsymbol{\sigma} = \mathbf{q} \tag{7.63}$$

where

$$\mathbf{M}^{\sigma} = \int_{\Omega} \mathbf{N}^{\sigma^{T}}\mathbf{N}^{\sigma}\,dv, \quad \mathbf{L}^{\sigma} = \int_{\Omega} \mathbf{N}^{\sigma^{T}}\mathbf{c}\frac{d\mathbf{N}^{\sigma}}{d\mathbf{x}^{m}}\,dv \tag{7.64}$$

7.3.4 Uncoupled ALE Solution

There are basically two methods of solutions for the governing equations of ALE formulation (Eqs (7.58), (7.61), and (7.63)): the fully coupled solution and the uncoupled solution [45]. In the fully coupled solution method, no further simplifications can be considered and the various terms must be calculated simultaneously. This approach was used in the works of Yamada and Kikuchi [32] and Bayoumi and Gadala [43]. In the uncoupled solution technique, we do not consider the fully coupled equations and the whole process can be decoupled into a Lagrangian phase and an Eulerian phase by employing a splitting operator. Such

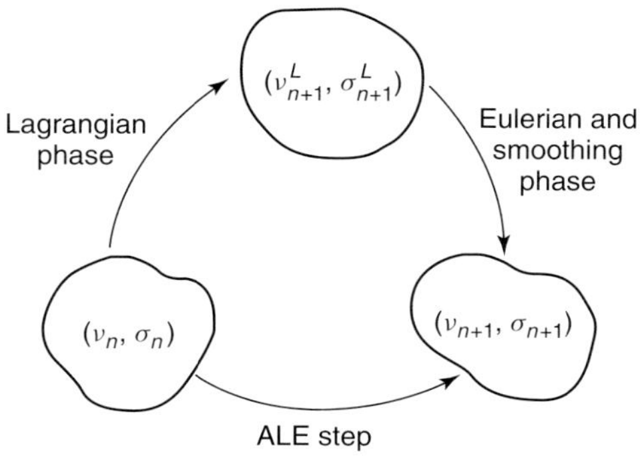

Figure 7.6 Decomposition of the ALE step into a Lagrangian phase and an Eulerian phase combined with a smoothing phase.

a technique has been employed by Rodriguez-Ferran *et al.* [40] and Khoei *et al.* [44]. In the present study, the uncoupled ALE solution is applied, since it makes it possible to upgrade a standard Lagrangian finite element program to the ALE case with as little expenditure as possible. In this case, the Lagrangian computation can be included as a substep of the new ALE computation.

The basis of splitting operator in uncoupled solution is to separate the material (Lagrangian) phase from convective (Eulerian) phase, which is combined with a smoothing phase (Figure 7.6). In the Lagrangian phase, the convective effects are neglected, so the material body deforms from its material configuration to its spatial one. In this framework, the nodal and quadrature points may lead eventually to a high distortion of the spatial discretization after the Lagrangian step. In order to reduce this distortion a smoothing phase is then applied, which leads to the final spatial discretization. This allows the computation of mesh velocity, which leads to the convective velocity, or the Eulerian phase. The advantage of ALE splitting operator is that the calculations are performed in the Lagrangian step with no convective terms to achieve the equilibrium. When the equilibrium is achieved in the Lagrangian step, the Eulerian step is performed by transferring the internal variables from the Lagrangian mesh to the relocated mesh. Special care must be taken with respect to data transferring, since the stress field is evaluated at discrete integration points and is usually discontinuous across the elements. To handle this difficulty, the superconvergent patch recovery (SPR) technique is implemented here together with the Godunov scheme for the stress update.

7.3.4.1 Material (Lagrangian) Phase

In material phase, the convective terms are neglected, so the momentum equation is identical to a time step in a standard Lagrangian analysis. Thus, the FE formulation of the momentum balance (7.58) in quasi-static analysis becomes

$$\int_{\Omega} \frac{d\mathbf{N}^{\mathrm{T}}}{d\mathbf{x}^{\mathrm{m}}} \boldsymbol{\sigma} \, dv = \int_{\Omega} \rho \mathbf{N}^{\mathrm{T}} \mathbf{b} \, dv + \int_{\Gamma} \mathbf{N}^{\mathrm{T}} \hat{\mathbf{t}} \, d\Gamma \tag{7.65}$$

which is a static equilibrium equation with no time, velocity, and convective terms in ALE momentum balance.

The ALE constitutive Eq. (7.63) in material phase can be simplified as

$$\mathbf{M}^{\sigma} \frac{d\boldsymbol{\sigma}}{dt} = \mathbf{q} \tag{7.66}$$

which needs to be integrated at each time step to update the stress from $\boldsymbol{\sigma}_n$ at time t_n to $\boldsymbol{\sigma}_{n+1}^{\mathrm{L}}$ after the Lagrangian phase. It means that in the absence of convective terms, the grid points move attached to material particles. Thus, the Lagrangian phase can be performed with the same stress update algorithm used in Lagrangian simulation, which handles the constitutive equation at the Gauss point level.

7.3.4.2 **Smoothing Phase**

In order to reduce the mesh distortion in spatial configuration, a remeshing procedure must be applied between the Lagrangian and the Eulerian phases. The algorithm can produce smoother meshes without redefining the element connectivity. Since the mesh moves independently from the material, we obtain the mesh velocity, which can be used to compute the convective velocity. Various remeshing strategies have been proposed by Ghosh and Raju [37], Gadala and Wang [38], and Rodriguez-Ferran *et al.* [40]. In this study, simple methods based on the "Laplacian approach" and the "mid-area averaging technique" are used. In these approaches, the mesh distortion is controlled by moving the inner nodes in an appropriate way. In addition, the boundary nodes remain on the boundary by allowing only a tangential movement to those nodes.

The Laplacian approach is a popular and simple smoothing strategy, which has been used by researchers to produce smoother meshes. In this technique, the spatial position of the smoothed node $\mathbf{x}_i$ can be computed using the spatial position after the Lagrangian phase $\mathbf{x}_i^{\mathrm{L}}$ as

$$\mathbf{x}_i = \frac{1}{(2-w)N} \sum_{e=1}^{N} (\mathbf{x}_{e1} + \mathbf{x}_{e2} - w\mathbf{x}_{e3}) \tag{7.67}$$

where $\mathbf{x}_i$ presents the spatial position of node i, and N is the number of four-node elements connecting to node i (typically $N = 4$). For each element $1 \leq e \leq N$, $\mathbf{x}_{e1}$, and $\mathbf{x}_{e2}$ are the coordinates of the nodes of element e connected to $\mathbf{x}_i$ by an edge, and $\mathbf{x}_{e3}$ is the coordinate of the node of element e at the opposite corner of $\mathbf{x}_i$, as shown in Figure 7.7. In the above relation, w is the weighting factor $0 \leq w \leq 1$, in which $w = 0$ yields the commonly used Laplacian scheme.

The mid-area averaging technique is a modification of the Laplacian approach. In this method, the considered node is in the centroid of all connected elements, and the area of different elements is taken into account. Thus, the spatial position of the smoothed node $\mathbf{x}_i$ can be computed as

$$\mathbf{x}_i = \frac{\sum_{e=1}^{N} A_e \mathbf{x}_{Se}}{\sum_{e=1}^{N} A_e} \tag{7.68}$$

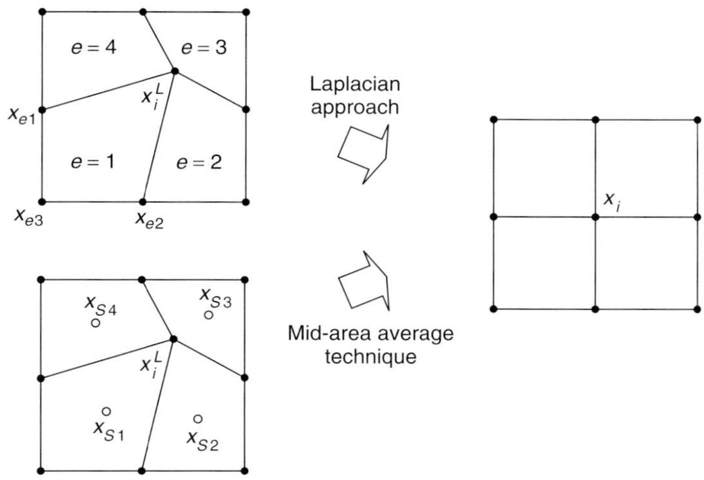

Figure 7.7 Remeshing procedure in the smoothing phase using the Laplacian approach and mid-area averaging technique.

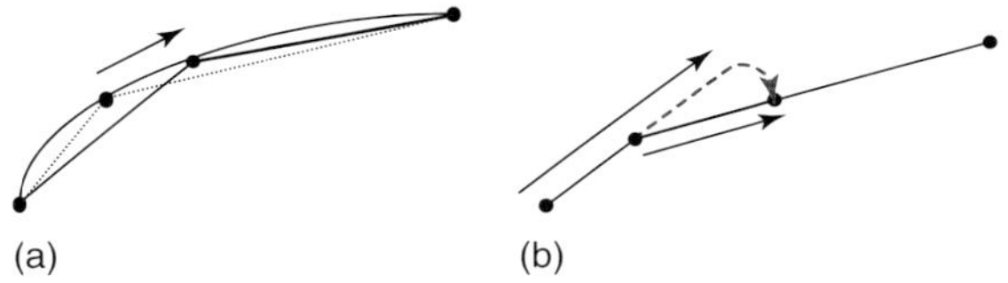

Figure 7.8 Illustration of boundary motion. (a) Curve treatment. (b) Extension.

where $\mathbf{x}_{Se}$ is the location of the centroid of element e (Figure 7.7), A_e is the area of element e and N indicates the number of four-node elements connecting to node i. After smoothing the position of all nodal points $\mathbf{x}_{n+1}$ at time t_n, the convective term $\mathbf{c}_{n+1}$ can be computed using the material (Lagrangian) displacement $\mathbf{u}_{n+1}^{\mathrm{L}}$ and mesh displacement $\mathbf{u}_{n+1}^{g}$ for quasi-static problems as $\mathbf{c}_{n+1} = \mathbf{u}_{n+1}^{\mathrm{L}} - \mathbf{u}_{n+1}^{g}$.

In order to perform the smoothing procedure for boundary nodes, it is assumed that the nodal points remain on the boundary by allowing only a tangential movement to these nodes. In this case, the boundary nodes are allowed to move in the normal direction following the motion of material points. For boundaries with preknown deformed shape, it is sufficient to assign a motion in the normal direction equal to the known material motion. To assign a desired tangential motion for nodal points, the following algorithm is performed here. First, a polynomial of second order is constructed using the considered node and two connected nodes on the boundary. The position of mid-node is then corrected according to the position of two connected nodes, as shown in Figure 7.8a. This procedure will not work properly when the determinant of matrix of its solution becomes zero. To avoid this problem, the approach is modified in the manner so that the extension of mid-node lies on the next connecting line, as shown in Figure 7.8b.

7.3.4.3 Convection (Eulerian) Phase

The final part of the operator splitting technique includes data transferring of solution obtained by the Lagrangian phase onto the new relocated mesh, developed through the mesh smoothing algorithm. In the Eulerian (or convection) phase, the convective terms which were neglected during the Lagrangian phase are taken into account. Since we are dealing with history-dependent materials and due to the fact that different material integration points have different histories, these quantities must be updated in order to compute the history-dependent variables in the next time step. These variables are computed at discrete integration points, which normally lie inside the element. This yields discontinuous fields and is troublesome, since the spatial gradients of these variables are required. In order to overcome these difficulties, a smooth gradient field is obtained on the basis of the Godunov technique that circumvents the computation of history variable gradients.

The FE formulation of constitutive Eq. (7.63) in the convection phase can be written as

$$\mathbf{M}^{\sigma}\frac{\mathrm{d}\boldsymbol{\sigma}}{\mathrm{d}t}+\mathbf{L}^{\mathrm{T}}\boldsymbol{\sigma}=0 \tag{7.69}$$

which needs to be integrated at each time step to update the stress from $\boldsymbol{\sigma}^{\mathrm{L}}_{n+1}$ to $\boldsymbol{\sigma}_{n+1}$ at time t_{n+1}. As noted above, the main difficulty in Eq. (7.69) is the stress gradient, which cannot be properly computed at the element level. In order to avoid computing gradients of the discontinuous fields, the Godunov technique is implemented here to transfer the internal variables ρ^{L}_{n+1}, $\bar{\varepsilon}^{\mathrm{p}^{\mathrm{L}}}_{n+1}$, and $\boldsymbol{\sigma}^{\mathrm{L}}_{n+1}$ from the Lagrangian mesh to the relocated mesh.

The Godunov method assumes a piecewise constant field of the solution of internal variables after the Lagrangian phase. In the finite element framework, this is the situation if one-point quadratures are employed; however, to allow for a subsequent generalization to multiple-point quadratures, the finite element can be subdivided into various subelements, each corresponding to the influence domain of a Gauss point [40]. Considering the scalar quantity ψ, be any components of the stress $\boldsymbol{\sigma}$, density ρ, or the effective plastic strain $\bar{\varepsilon}^{\mathrm{p}}$, the value of internal variable ψ_{n+1} at time t_{n+1} can be obtained from the Lagrangian solution ψ^{L}_{n+1} as

$$\psi_{n+1}=\psi^{\mathrm{L}}_{n+1}-\frac{\Delta t}{2A}\sum_{s=1}^{N_S}\left\{f_s(\psi^{\mathrm{Lc}}_{n+1}-\psi^{\mathrm{L}}_{n+1})[1-\alpha_0\,\mathrm{sign}\,(f_s)]\right\} \tag{7.70}$$

where A is the area of subelement, N_s is the number of edges of subelement, and ψ^{Lc}_{n+1} is the value of ψ^{L}_{n+1} in the contiguous subelement across edge s (Figure 7.9). The upwind parameter α_0 is in the range of $0\le\alpha_0\le 1$, where $\alpha_0=1$ corresponds to a full-donor approximation and $\alpha_0=0$ is a centered approximation. In the above relation, f_s is the flux of convective velocity c across edge s defined as

$$f_s=\int_s \mathbf{c}\cdot\mathbf{n}\,\mathrm{d}s \tag{7.71}$$

On the basis of this approach, if, for instance, quadrilaterals with 2×2 integration points are employed, each element is divided into four subelements, as shown in

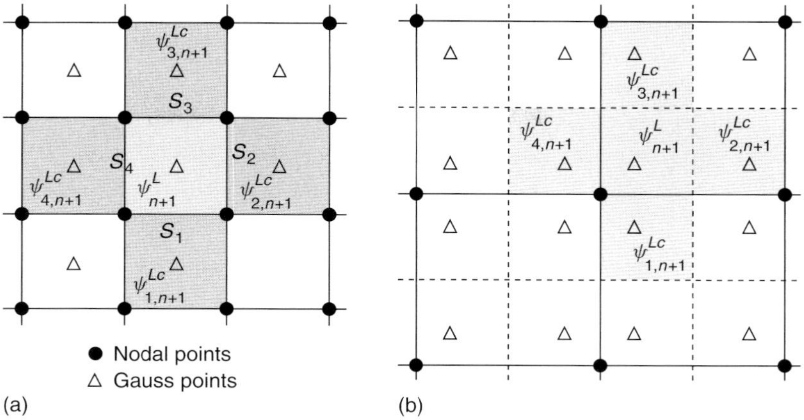

Figure 7.9 Illustration of the Godunov technique. (a) One-point quadrature. (b) Multiple-point quadrature.

Figure 7.9. In each subelement, ψ is assumed to be constant, and represented by the Gauss point value. Thus, ψ is a piecewise constant field with respect to the mesh of subelements, and relation (7.70) can be employed to update the value of ψ for each subelement.

7.3.5 Numerical Modeling of an Automotive Component

In order to illustrate the capability of ALE computational algorithm, the compaction of an automotive component is analyzed numerically. The constitutive relations of single plasticity model, developed in Section 7.2, along with the ALE finite element formulation presented above have been implemented in a nonlinear finite element code to evaluate the capability of the model in simulating the large plastic deformation of pressure-sensitive material. The problem has been solved with displacement control by increasing the punch movements and predicting the compaction forces at different displacements. In the FE simulations, the tools are modeled as rigid bodies, because the elastic deformation of the tools has only an insignificant influence on the density distribution of green component. An axisymmetric automotive part is compacted from iron powder with a mechanical press and a multiplaten die set, as shown in Figure 7.10a. A Lagrangian finite element modeling was carried out for this component by Azami and Khoei [25]. Also, the measured density distribution was obtained for this example by Shen *et al.* [46].

The automotive part is simulated numerically using the single plasticity model in the formwork of Lagrangian and ALE methods. The simulation of component is performed using an axisymmetric finite element analysis, including 376 quadrilateral elements, as shown in Figure 7.10b. The compaction is employed by means of the action of two bottom punches; a lower inner punch, and a lower outer punch

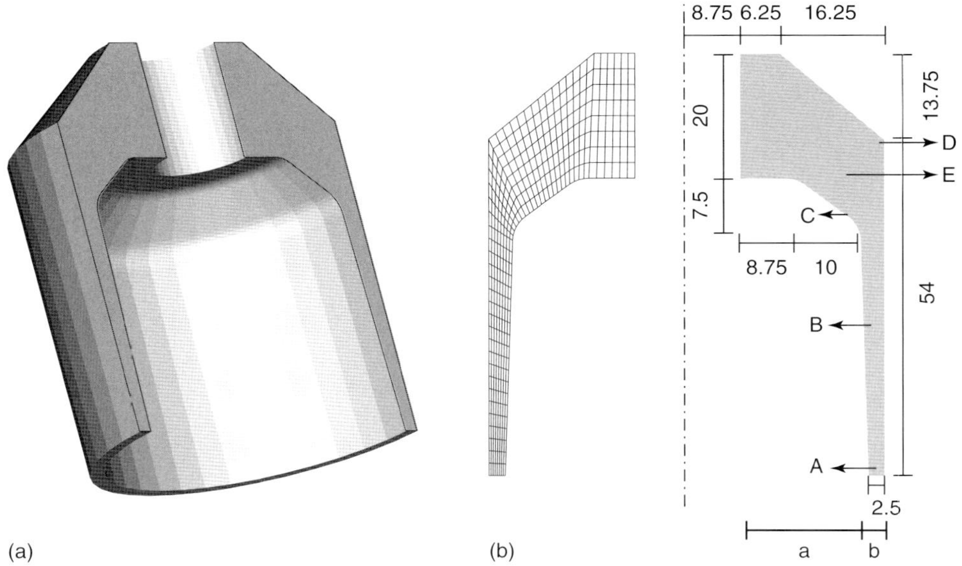

Figure 7.10 An automotive component. (a) A schematic of mechanical press and a multiplaten die set. (b) The problem geometry and computational mesh.

labeled "a" and "b," respectively, in Figure 7.10. The geometry of uncompacted powder in its position before compaction is presented in this figure. The simulation has been performed using the remaining pressing distance of lower inner punch of 15.31 mm and lower outer punch of 48.84 mm. The convergence tolerance is set to 5×10^{-5} and the analysis is performed using 10, 50, and 400 time increments. The analysis is carried out with the iron powder material parameters obtained in Section 7.2.3.2.

In the simulation presented here, the friction effects are neglected. Therefore, the results are a qualitative approximation to the real compaction process, where the friction effects have to be taken into account [47–50]. Owing to high distortion of finite element mesh, particularly around the corners, the Lagrangian simulation proceeds only up to 82% of compaction. In Figure 7.11, the deformed Lagrangian mesh is presented together with the deformed ALE mesh at the compaction of 82%. The mesh distortion can be efficiently reduced in the ALE analysis by means of a very simple ALE remeshing strategy using equal length of elements between the top and bottom boundaries. In Figure 7.12, the distributions of predicted relative density contours are shown at different bottom punch movements for the ALE analysis. These results can be compared with those obtained by Azami and Khoei [25] using a Lagrangian FE modeling. The applicability of the cap plasticity model in the formwork of ALE approach to handle the complete compaction simulation of powder is clearly evident in this figure. The evolution of top punch vertical reaction force with its vertical displacement is depicted in Figure 7.13a,

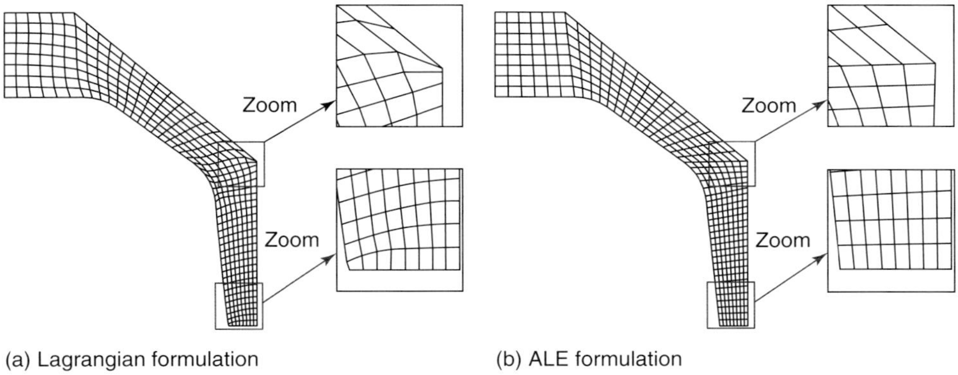

Figure 7.11 An automotive component. The mesh configuration at the compaction of 82% using the Lagrangian and ALE formulations.

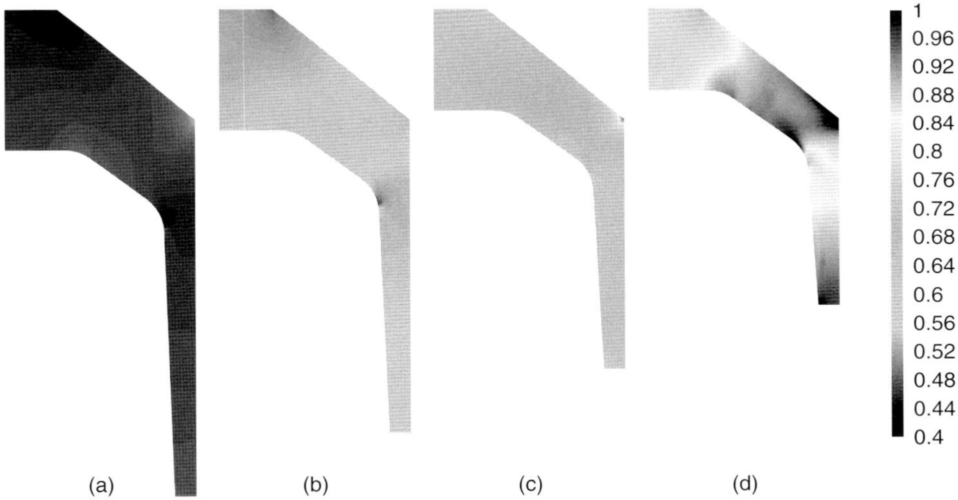

Figure 7.12 The distribution of relative density contours for an automotive component using the ALE formulation at different punch movements, $d = 7.7$, 15.5, 23.3, and 31.1mm.

which has been compared with those reported in Ref. [25]. Good agreements can be observed between the ALE and Lagrangian results. In order to illustrate the accuracy of proposed ALE algorithm, the numerical simulation has been carried out for different number of time increments, that is, 10, 50, and 400 time steps. Figure 7.13b presents a comparison between various time steps of the top punch reaction force with its vertical displacement. This example adequately presents the capability of ALE technique in modeling powder-forming processes using a cap plasticity model.

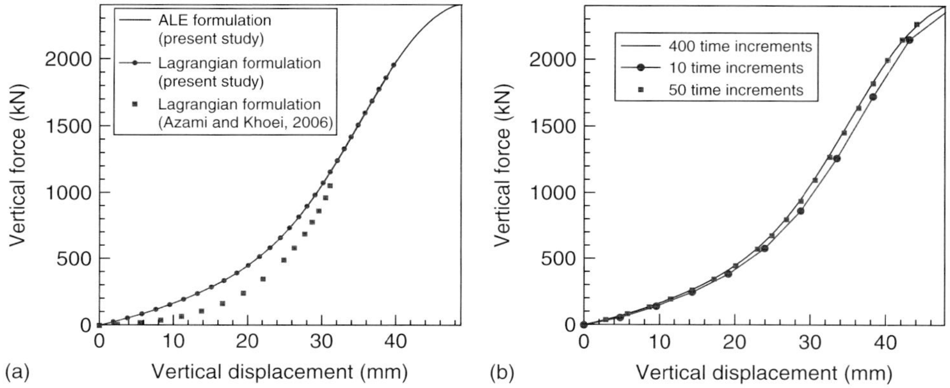

Figure 7.13 The evolution of top punch force with displacement for an automotive component. (a) A comparison between Lagrangian and ALE models. (b) A comparison for various time increments.

7.4
Enriched ALE Finite Element Method

In large deformation problems with discontinuities and material interfaces, the implementation of adaptive mesh refinement in different stages of analysis is of great importance. The requirement of mesh refinement may involve condiderable cost in terms of capacity and time. The arbitrary Lagrangian–Eulerian technique, introduced in the preceding section, alleviates the drawbacks of traditional Lagrangian formulation; however, it is necessary to perform an innovative procedure by allowing the discontinuities to be mesh independent. The X-FEM has been successfully applied to problems exhibiting discontinuities and inhomogeneities, such as cracks, holes, or material interfaces. The technique is a powerful and accurate approach to model discontinuities, which are independent of the FE mesh topology. In this method, the discontinuities, or interfaces, are not considered in the mesh generation operation and special functions, which depend on the nature of discontinuity, are included into the finite element approximation through a partition of unity method [30].

The technique was first introduced by Dolbow [51], Belytschko and Black [52], and Moës *et al.* [53] to model cracks and inhomogeneities. The method allows modeling the entire crack geometry independently of the mesh, and completely avoids the need to remesh as the crack grows. The X-FEM was used to model crack growth and arbitrary discontinuities by enriching the discontinuous approximation in terms of a signed distance and level sets functions [54–57]. The most up-to-date computational aspects of X-FEM have been presented in linear crack problems, including the crack growth with frictional contact [58], cohesive crack propagation [59], quasi-static crack growth [60], fatigue crack propagation [61], stationary and growing cracks [62], and three-dimensional crack propagation [63]. An overview of the technique was addressed by Bordas *et al.* [64] in the framework of an object-oriented-enriched finite element programming. However, less X-FEM modeling has been reported in elastoplasticity. Maximum implementation of X-FEM in plasticity problems are found in the following: plasticity forming of powder compaction [65], plasticity of frictional contact on arbitrary interfaces [66, 67], large X-FEM deformation [68, 69], ALE-X-FEM model for large plastic deformation [29, 70], localization phenomenon in higher order Cosserat theory [71], crack propagation in plastic fracture mechanics [72, 73], and elastoplastic fatigue crack growth [74].

In this section, an enriched FE model is incorporated with the ALE technique in large plastic deformation based on an operator splitting technique. The constitutive equation of the ALE technique contains the convective term, which reflects the relative motion between the physical motion and the mesh motion. The evaluation of the convective term in the constitutive equation is performed using the split, or fractional-step method. In order to perform an enriched arbitrary Lagrangian–Eulerian FE analysis, a typical X-FEM analysis is first carried out in the Lagrangian phase using the updated Lagrangian approach. The Eulerian phase is then applied to update the mesh, while the material interface is independent of the FE mesh.

7.4.1 The Extended-FEM Formulation

In order to introduce the concept of X-FEM method, consider an interface, Γ_c, in domain Ω, as shown in Figure 7.14a. We are interested in the construction of a finite element approximation to the field $\mathbf{u} \in \Omega$, which can be discontinuous along the interface surface Γ_c. The traditional approach is to generate the mesh to conform to the line of discontinuity surface as shown in Figure 7.14b, in which the element edges align with Γ_c. While this strategy certainly creates a discontinuity in the approximation, it is cumbersome if the line Γ_c evolves in time, or if several different configurations for Γ_c are to be considered. Here, we intend to model the discontinuity along interface Γ_c with extrinsic enrichment, in which the uniform mesh of Figure 7.14c is capable of modeling the discontinuity in $\mathbf{u}$, when the circled nodes are enriched with enrichment functions.

In X-FEM, the enrichment functions are associated with new degrees of freedom and the approximation of the displacement field as

$$\mathbf{u}(\mathbf{x}) = \sum_I N_I(\mathbf{x})\bar{\mathbf{u}}_I + \sum_J N_J(\mathbf{x})\psi(\mathbf{x})\mathbf{a}_J \tag{7.72}$$

where $n_I \in \mathbf{n}_T$ and $n_J \in \mathbf{n}_e$, in which $\mathbf{n}_T$ denotes the set of all nodes of the domain, and $\mathbf{n}_e$ is the set of nodes of elements split by the interface, as indicated in Figure 7.14c. The first term of the above equation denotes the standard FEM approximation and the second term indicates the enrichment function considered in X-FEM. In this equation, $\bar{\mathbf{u}}_I$ is the classical nodal displacement, $\mathbf{a}_J$ is the nodal degrees of freedom corresponding to the enrichment function $\psi(\mathbf{x})$, and $N(\mathbf{x})$ is the standard shape function.

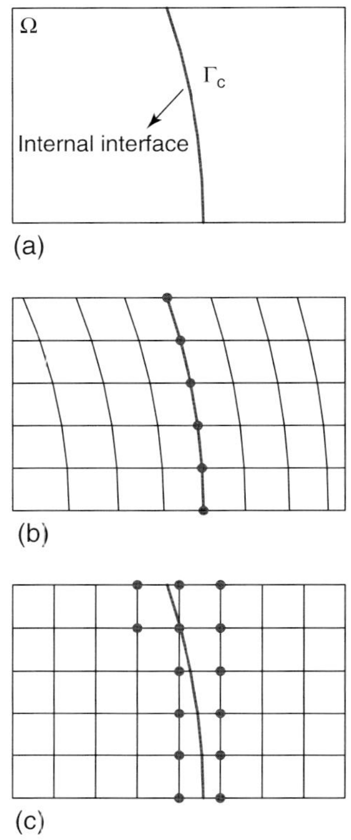

Figure 7.14 Modeling of internal interfaces. (a) Problem definition. (b) The FE mesh which conforms to the geometry of discontinuity. (c) A uniform mesh in which the circled nodes have additional degrees of freedom and enrichment functions.

The choice of enrichment function in displacement approximation is dependent on the conditions of problem. If the discontinuity is a result of different types of material properties, then the level set function is proposed as an enrichment function; however, if the discontinuity is due to different displacement fields on either side of the discontinuity, such as contact surface, then the Heaviside function is appropriate. The *level set method* is a numerical scheme for tracking the motion of interfaces. This method, which is used to predict the geometry of boundaries, is very suitable for bimaterial problems in which the displacement field is continuous but there is a jump in the strain field. In this technique, the interface is implicitly represented by assigning to node I of the mesh, located at the distance φ_{I} from the interface. The sign of its value is negative on one side and positive on the other. The level set function can be then obtained by interpolating the nodal values using the standard FE shape functions as

$$\varphi(\mathbf{x}) = \sum_{I} \varphi_I N_I(\mathbf{x}) \tag{7.73}$$

where the above statement indicates the summation over the nodes that belong to elements cut by the interface. A discontinuity is represented by the zero value of level set φ. The new degrees of freedom $\mathbf{a}_J$ in Eq. (7.72) are considered corresponding to the level set enrichment function, in order to attribute to the nodes that belong to the set of $\mathbf{n}_e$. In order to improve the numerical computation in X-FEM, it is preferable to have a uniform distribution of shape functions around the boundary of discontinuity by employing the nodes that belong to the elements in the neighbor of discontinuity elements. Generally, two types of enrichment function have been implemented to model the discontinuity as a result of different types of material properties. The first function is based on the absolute value of the level set function, which indeed has a discontinuous first derivative on the interface as

$$\psi^1(\mathbf{x}) = \left|\sum_I \varphi_I N_I(\mathbf{x})\right| \tag{7.74}$$

An extension of the above function that improves the previous enrichment strategy has a ridge centered on the interface and zero value on the elements that are not crossed by the interface. It was shown by Anahid and Khoei [69] that the convergence rate of the enrichment function ψ^2 is very close to the optimal FE convergence. The modified level set function ψ^2 is defined as

$$\psi^2(\mathbf{x}) = \sum_I |\varphi_I| N_I(\mathbf{x}) - \left|\sum_I \varphi_I N_I(\mathbf{x})\right| \tag{7.75}$$

Considering the enriched approximation of displacement field defined by Eq. (7.72), the value of $\mathbf{u}(\mathbf{x})$ on an enriched node K in set $\mathbf{n}_e$ can be written as

$$\mathbf{u}(\mathbf{x}_K) = \bar{\mathbf{u}}_K + \psi(\mathbf{x}_K)\mathbf{a}_K \tag{7.76}$$

Since $\psi(\mathbf{x}_K)$ is not necessarily zero, the above expression is not equal to the real nodal value $\bar{\mathbf{u}}_K$. Thus, the enriched displacement field (Eq. (7.72)) can be corrected as

$$\mathbf{u}(\mathbf{x}) = \sum_I N_I(\mathbf{x})\bar{\mathbf{u}}_I + \sum_J N_J(\mathbf{x})(\psi(\mathbf{x}) - \psi(\mathbf{x}_J))\mathbf{a}_J \quad \text{for} \quad n_I \in \mathbf{n}_T \text{ and } n_J \in \mathbf{n}_e \tag{7.77}$$

On the basis of the new definition in the last term of relation (7.77), the expected property can be obtained as $\mathbf{u}(\mathbf{x}_K) = \bar{\mathbf{u}}_K$. Using the enrichment function $\psi(\mathbf{x})$ based on its nodal values as $\psi(\mathbf{x}) = \sum_I N_I(\mathbf{x})\psi_I$, Eq. (7.77) can be rewritten as

$$\mathbf{u}(\mathbf{x}) = \sum_I N_I(\mathbf{x})\bar{\mathbf{u}}_I + \sum_J \left(N_J(\mathbf{x}) \left(\sum_K N_K(\mathbf{x})\psi_K - \psi_J \right) \mathbf{a}_J \right) \tag{7.78}$$

It must be noted that the numerical integration of the weak form with the standard Gauss quadrature points for elements cut by the interface must be improved because of the existence of discontinuous displacement gradient through the interface.

7.4.2
An Enriched ALE Finite Element Method

In X-ALE-FEM analysis, the X-FEM method is performed together with an operator splitting technique, in which each time step consists of two stages: Lagrangian (material) and Eulerian (smoothing) phases. In material phase, the X-FEM analysis is carried out on the basis of an updated Lagrangian approach. It means that the convective terms are neglected and only material effects are considered. The time step is then followed by an Eulerian phase in which the convective term is considered into account. In this step, the nodal points move arbitrarily in the space so that the computational mesh has regular shape and the mesh distortion can be prevented; however, the material interface is independent of the FE mesh.

Figure 7.15 presents the mesh configuration including the material interface before and after the Eulerian phase. In this figure, the position of the interface has been shown in two different cases. In Figure 7.15a, the interface does not move from one element to another during smoothing phase. On the other hand, the number of elements which have been cut by the interface does not change during smoothing phase, while in Figure 7.15b, the material interface may move from one element to another. As can be seen, elements 1, 3, and 4 have been cut by the interface before mesh motion procedure; however, only element 3 is cut by the interface after Eulerian phase. Thus, the number of enriched nodes may be different during the X-ALE-FEM analysis, which results in different number of degrees of freedom in two successive steps. There are two main requirements, which need to be considered in the smoothing phase:

1) Owing to the movement of nodal points in the mesh motion process, a procedure must be applied to determine the new nodal values of level set enrichment function.
2) In the extended finite element analysis, the number of Gauss quadrature points for numerical integration of elements cut by the interface can be determined using the subquadrilaterals obtained by partitioning procedure. However, in the case that the material interface leaves from one element to another during

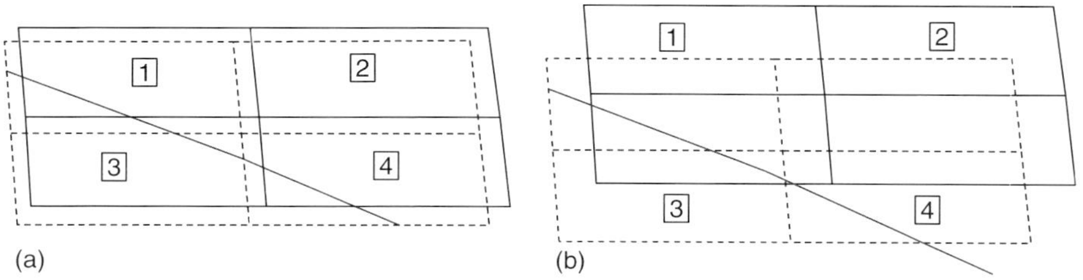

Figure 7.15 Mesh configuration before and after the Eulerian phase together with the material interface in X-ALE-FEM analysis. The dashed and solid elements correspond to mesh configuration before and after mesh motion, respectively.

the mesh update procedure, the number of Gauss quadrature points of an element may differ before and after mesh motion. Hence, an accurate and efficient technique must be applied into the Godunov scheme to update the stress values.

7.4.2.1 **Level Set Update**

During smoothing phase, the nodal points are relocated in order to keep the computational mesh in regular shape; however, the material interface is independent of the FE mesh, as shown in Figure 7.16. In this figure, the dashed lines illustrate the old mesh and the relocated elements are depicted by solid lines. The aim is to obtain the level set value for node I after Eulerian phase. For this purpose, we must determine the distance of relocated node I from the interface. The intersection of interface with the edges of old elements can be calculated at the end of Lagrangian step, labeled *points A, B, C, and D* in Figure 7.16a.

The procedure used here to update the nodal values of level set is as follows. The arc ABCD is first approximated by several straight lines that connect the intersection of the interface with the edges of old elements, that is, lines AB, BC, and CD in Figure 7.16b. The support domain of node I has been indicated by elements 1, 2, 3, and 4 in Figure 7.16a. Among the elements of the support domain, we define the extended support domain (ESD) of node I that contains those elements cut by the interface at the end of Lagrangian phase. As can be seen from Figure 7.16b, the ESD domain of node I includes elements 2, 3, and 4. For each element of ESD, the distance of the relocated node I is calculated from the element interface. The exact value of the level set function for the relocated node I is the minimum value of distances d_1, d_2, and d_3, as shown in Figure 7.16b. In order to determine the sign of the level set value for the relocated node I, the above procedure is performed once again to evaluate the value of the level set for the old node I. If the sign of these two level set values are similar, it means that the old and relocated nodes lie to the same side of interface. Thus, the sign of level set value for the relocated node I must be identical with its sign for the old node I.

However, the proposed algorithm is not appropriate for the determination of the level set value of nodal points, which are not enriched during the Lagrangian phase, and not included in an element split by the interface after mesh motion

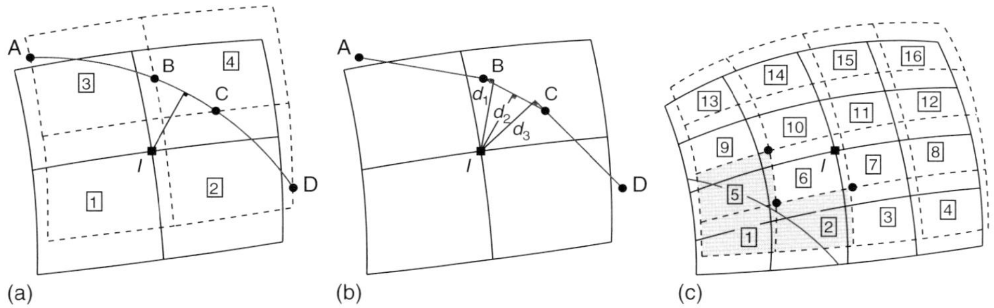

Figure 7.16 The procedure for determination of level set nodal values after mesh motion.

(e.g., node I in Figure 7.16c). In this case, the ESD domain of node I in Figure 7.16c contains no elements, and the definition of ESD must be modified. In order to modify the domain of ESD for these nodal points, we first indicate the support domain of node I (e.g., elements 6, 7, 10, and 11 in Figure 7.16c). All nodal points of this domain that were enriched during the Lagrangian phase are selected (circled nodes in Figure 7.16c). Those elements in the union of the circled nodes' support domains, which are cut by the interface, can be considered as the modified ESD domain of node I. The procedure will be then followed by the determination of the level set value for node I, as demonstrated above. By substituting the nodal values of the level set in Eq. (7.73), the level set function can be finally expressed. The iso-zero of the level set function determines the location of the interface.

7.4.2.2 Stress Update and Numerical Integration

A key point of the Godunov-like stress update procedure is that the number of Gauss points before and after mesh motion must be equal. In addition, the natural coordinates of Gauss quadrature points must remain constant during the smoothing phase. However, these conditions may not be necessarily satisfied in the Eulerian phase of an X-ALE-FEM analysis, since the elements cut by the interface are divided into subpolygons whose Gauss points are used for numerical integration. It must be noted that for the elements cut by the interface boundary, the standard Gauss quadrature points are insufficient for numerical integration, and may not adequately integrate the interface boundary. Thus, it is necessary to modify the element quadrature points to accurately evaluate the contribution to the weak form for both sides of the interface. In the standard FE method, the numerical integration can be performed by discretizing the domain as $\Omega = \bigcup_{e=1}^{m} \Omega_e$, in which m is the number of elements and Ω_e is the element subdomain. In X-FEM, the elements located on interface boundary can be partitioned by subpolygons Ω_s, with the boundaries aligned with the material interface, that is, $\Omega_e = \bigcup_{s=1}^{m_s} \Omega_s$, in which m_s denotes the number of subpolygons of the element. It is important that the Gauss points of subpolygons are only used for numerical integration of the elements cut by the interface and no new degrees of freedom are added to the system. Different algorithms may be applied to generate these subpolygons based on subtriangles and subquadratics; however, for the following two reasons, the subtriangles are not suitable in an X-ALE-FEM analysis:

1) Owing to the relative motion between the interface and elements, the natural coordinates of Gauss quadrature points may differ during smoothing phase, even though the interface does not leave from one element to another in this phase.
2) During the evolution of relative motion between the interface and nodal points, it is possible that a nodal point lies close to the interface. In this case, partitioning the bounded part between the node and interface by subtriangles results in serious numerical errors.

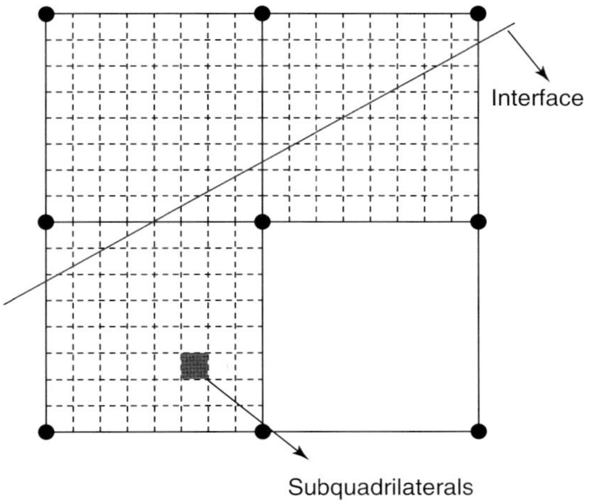

Figure 7.17 The subquadrilateral partitioning for numerical integration of elements cut by the interface.

Thus, the numerical integration of elements cut by the interface is performed here on the basis of subquadrilaterals, as shown in Figure 7.17. In this technique, it is not necessary to conform the subquadrilaterals to the geometry of the interface; however, it is needed to have enough subdivisions to reduce the error of numerical integration. Considering that each split element contains 8×8 subdivision, the total number of Gauss points is 64 for each split element, while a set of 2×2 Gauss points is used for numerical integration of standard elements. On the basis of the subquadrilaterals integration scheme, the natural coordinates of Gauss quadrature points are independent of the interface position and do not change during the smoothing phase. Furthermore, the most important feature is that the error of numerical integration can be significantly reduced in this technique. However, the key requirement of the Godunov scheme is the equivalent of integration points before and after the mesh update process, which is not satisfied in the Eulerian phase of an X-ALE-FEM analysis, when the interface leaves one element to another, or moves into the relocated element during the smoothing procedure. In this circumstance, four different cases may occur regarding the numbers of Gauss quadrature points:

1) The element is not cut by the interface either before or after mesh motion (e.g., elements 1, 2, 3, and 7 in Figure 7.18a,b). In this case, the FE standard Gauss points are used for numerical integration before and after smoothing phase. Thus, the number of Gauss points and their natural coordinates remain constant, and the Godunov scheme can be used without any modification.
2) The element, which has been cut by the interface before smoothing phase, is still included by the interface after the mesh updating procedure (e.g., elements 4, 5, and 9 in Figure 7.18). On the basis of the new integration scheme, both the old and relocated elements contain 64 Gauss quadrature points and the natural

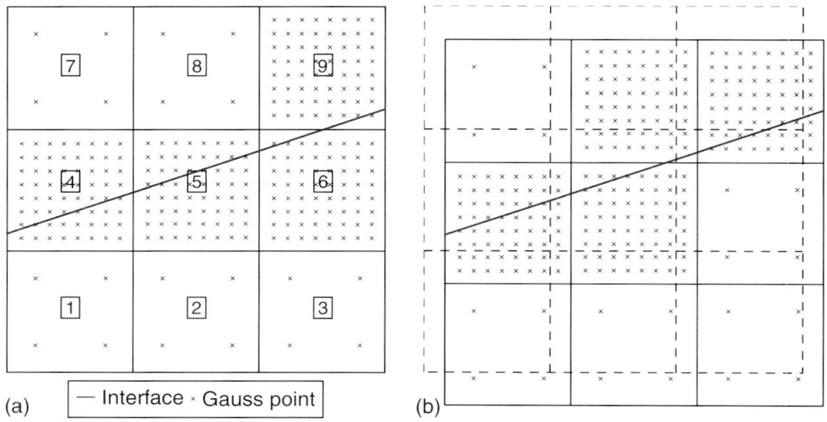

Figure 7.18 The distribution of Gauss integration points in the X-ALE-FEM analysis. (a) Old mesh. (b) Relocated mesh.

coordinates of Gauss points do not change. Hence, the Godunov scheme can be applied without any modification.

3) The interface leaves from one element to another during the smoothing phase. In this case, the element, which was split by the interface before the mesh updating procedure, does not contain the interface after mesh motion (e.g., element 6 in Figure 7.18). It therefore results in different numbers of Gauss points in the old and relocated elements. As mentioned earlier, the old and relocated elements contain 64 and 4 Gauss quadrature points, respectively. Thus, an efficient algorithm must be applied into the Godunov technique to update the stress from $\sigma_{n+1}^{\mathrm{L}}$ obtained from the Lagrangian phase to σ_{n+1} after the Eulerian phase. First, the internal variables such as stresses are updated from the 8×8 Gauss points in the old element to a virtual set of 8×8 FE standard Gauss points in the relocated mesh via a Godunov update algorithm. The stress values must be then transferred from the virtual Gauss points to relocated 2×2 Gauss points in the updated element, as shown in Figure 7.19. According to this figure, for each FE Gauss point q, there is a support domain that consists of three nearest virtual Gauss points. The stress value at q can be interpolated from the stress values of these three nearest Gauss points.
4) The material interface moves into the relocated element during the Eulerian phase. In this case, the relocated element is cut by the interface while it was not cut before the mesh motion (e.g., element 8 in Figure 7.18). In this case, the stress values have been calculated at the 2×2 standard FE Gauss points of the old element at the end of Lagrangian phase. These values must be updated using the Godunov algorithm to obtain the stress values at the virtual 2×2 Gauss points in the relocated element. It must be noted that the stress values are required at the relocated 8×8 Gauss quadrature points. For this to happen, the stress values are first computed at the nodal points of the relocated element by using an averaging technique from the related values at the nearest

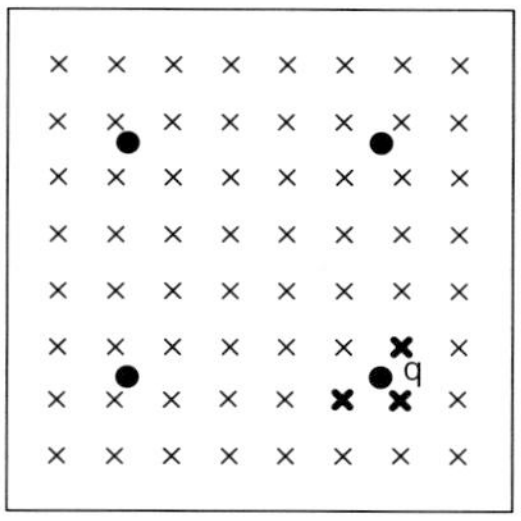

Figure 7.19 The support domain of Gauss integration point q. • The 2×2 Gauss points corresponding to relocated mesh. × The virtual Gauss points in relocated mesh.

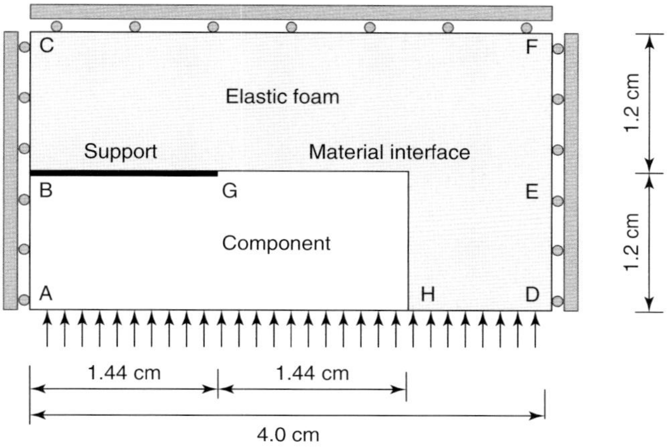

Figure 7.20 Coining test with horizontal and vertical interfaces. Problem definition.

Gauss points. The required stress values at the relocated 8×8 Gauss points can be then obtained using the standard FE shape functions of the relocated nodal values.

7.4.3
Numerical Modeling of the Coining Test

In order to illustrate the accuracy of the X-ALE-FEM technique in large plasticity deformation, the coining problem with horizontal and vertical moving boundaries is analyzed numerically. The example is solved using FEM and X-ALE-FEM techniques and the results are compared. The initial mesh used for X-ALE-FEM method is independent of discontinuity interface, while the FE mesh needs to be conformed to the geometry of discontinuity in FEM analysis. In order to perform a real comparison, similar number of elements is used in the FEM and X-ALE-FEM analyses. The example is simulated by a plain strain representation and the convergence tolerance is set to 5×10^{-14}. The geometry and boundary conditions of the coining test are shown together with the problem definition in Figure 7.20.

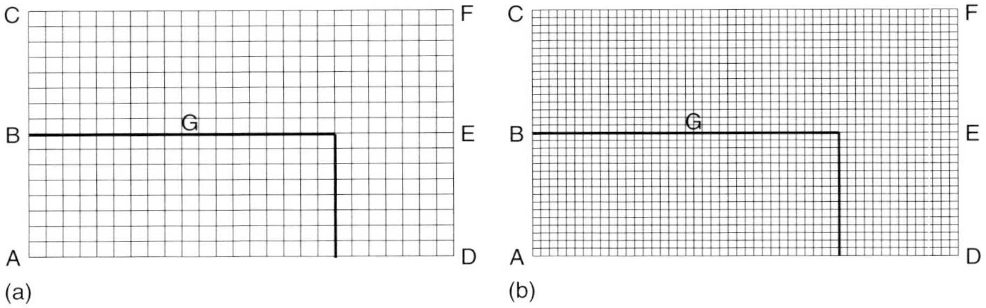

Figure 7.21 Coining test with horizontal and vertical interfaces. The FEM and X-ALE-FEM meshes of (a) 400 elements and (b) 1600 elements.

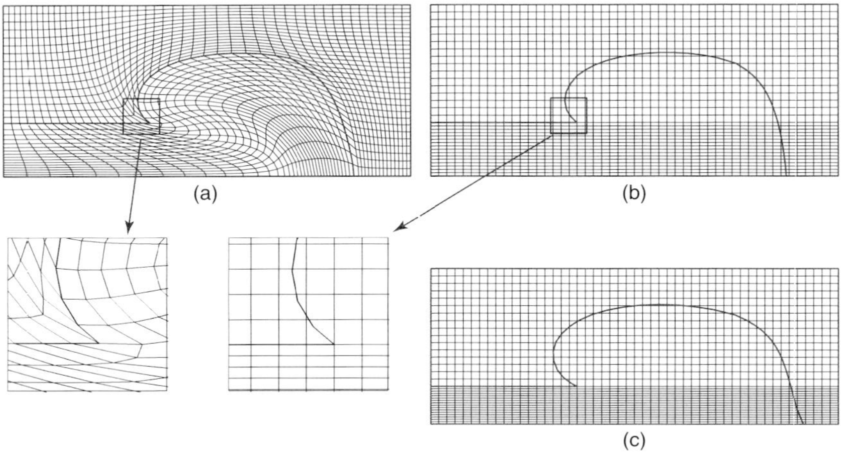

Figure 7.22 Coining test with horizontal and vertical interfaces. The deformed configuration using fine meshes. (a) The FEM at $d = 0.66$ cm. (b) The X-ALE-FEM at $d = 0.66$ cm. (c) The X-ALE-FEM at $d = 0.816$ cm.

The movement of the component is constrained along the line BG, and the contact surfaces AC, CF, and DF are assumed to be frictionless. An elastoplastic metallic component of von Mises behavior with the Young modulus of 2.1×10^6 Kg/cm^2, Poisson ratio of 0.35, yield stress of 2400 Kg/cm^2, and a hardening parameter of 3.0×10^3 Kg/cm^2 is pressed from the bottom edge into the elastic foam with the Young modulus of 2.1×10^5 Kg/cm^2 and Poisson ratio of 0.35.

In Figure 7.21, the computational meshes corresponding to FEM and X-ALE-FEM analyses are presented together with the internal material interface at the initial stage of pressing. In order to implement the prescribed displacement condition along the edge BG, the internal interface for both simulations coincides with the edge of elements at the initial configuration. However, the movement of horizontal and vertical interfaces is independent of the grid in X-ALE-FEM, as

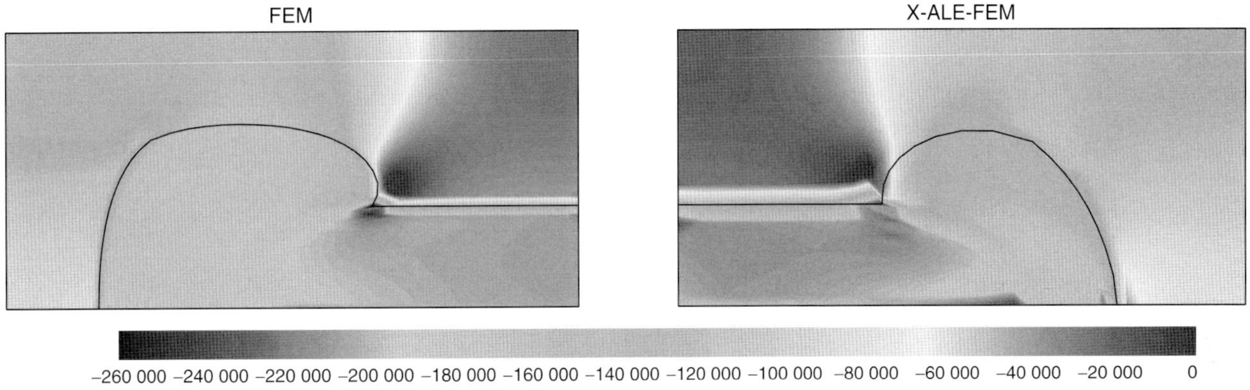

Figure 7.23 Coining test with horizontal and vertical interfaces. The normal stress σ_y contours of the whole domain at $d = 0.5$ cm. A comparison between FEM and X-ALE-FEM analyses.

the compaction proceeds. It must be noted that the X-FEM analysis of this problem leads to similar results obtained by FEM modeling, as the interface passes through the nodal points. The simulation of coining test poses a major difficulty using the Lagrangian formulation due to the highly distorted mesh particularly around the corner, aborting the calculations and causing numerical errors. Owing to the high distortion and elongation of elements around the corner, the FE analysis aborts at 0.66 cm height reduction. However, the X-ALE-FEM analysis proceeds pressing until 0.816 cm by using a mesh smoothing strategy based on a simple ALE remeshing strategy with the equal height and width of elements prescribed in the regions of ABED and BCFE.

In Figure 7.22a,b, the deformed FEM and X-ALE-FEM configurations are shown for the fine mesh at die-pressing of 0.66 cm. Also depicted in Figure 7.22c is the final deformed configuration of the X-ALE-FEM mesh at 0.816 cm height reduction. In order to perform a comparison between the FEM and X-ALE-FEM results, the normal stress contours are illustrated in Figure 7.23 for both techniques at the pressing of 0.5 cm. In Figure 7.24, the variations of the reaction force and the horizontal displacement of the point H are shown with vertical displacement. A good agreement can be observed between the FEM and X-ALE-FEM techniques. This example clearly shows that the proposed enriched-ALE-FEM method can be efficiently used to model large elastoplastic deformations in solid mechanics problems.

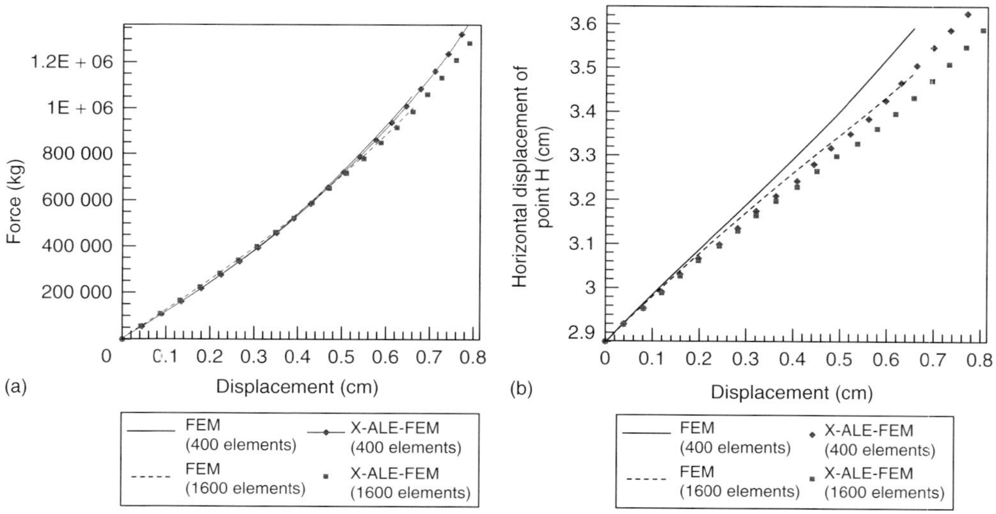

Figure 7.24 A comparison between FEM and X-ALE-FEM analyses for the coining test with horizontal and vertical interfaces. (a) The variation of reaction force with vertical displacement. (b) The variation of horizontal displacement of point H with vertical displacement.

7.5 Conclusion

In this chapter, a new computational model was presented for numerical simulation of large plastic deformations of pressure-sensitive materials with special reference to powder compaction processes. A three-invariant cap plasticity model was developed on the basis of an isotropic–kinematic hardening rule for powder materials, which can be compared with some common double-surface plasticity models proposed for powders in the literature. The material functions were introduced on the basis of the isotropic and kinematic hardening rules to control the size and movement of the yield surface in the J_1-axis and the perpendicular direction to the J_1-axis. The constitutive elastoplastic matrix and its components were derived by using the definition of yield surface, material functions, and nonlinear elastic behavior, as a function of the hardening parameters. The calibration procedure for three-invariant isotropic–kinematic cap plasticity was demonstrated on the basis of a series of isostatic, triaxial, and uniaxial compression tests on an iron-based powder component.

The three-invariant cap plasticity was performed within the framework of an enriched ALE formulation. The enriched FEM technique was employed by implementation of the enrichment functions to approximate the displacement fields of elements located on discontinuities due to different material properties. The X-FEM method was applied by performing a splitting operator to separate the material (Lagrangian) phase from the convective (Eulerian) phase. The Lagrangian phase was carried out by partitioning the domain with subquadrilaterals whose Gauss points were used for integration of the domain of the elements. The ALE governing equation was derived by substituting the relationship between the material time derivative and grid time derivative into the continuum mechanics governing equations. The weak form of initial boundary problems and its FE discretization were presented in the ALE technique. The algorithm of the uncoupled ALE solution was demonstrated together with the mesh motion strategy. The analysis was carried out according to the Lagrangian phase at each time step until the required convergence was attained. The Eulerian phase was then applied to keep the mesh configuration regular. In the Eulerian phase, the nodal points were relocated arbitrarily, and the material interface was independent of the FE mesh. A technique was proposed to update the nodal values of a level set and the natural coordinates of the Gauss quadrature points, which could be different before and after the mesh updating procedure. Furthermore, a technique was applied to update the stress values from the old Gauss points to the new ones based on the Godunov stress updating scheme.

Acknowledgments

The author thanks his former graduate students and his research assistants in the Department of Civil Engineering, Sharif University of Technology, who have contributed to the advances in the application of computational plasticity to powder forming; particularly, M. Anahid, A. R. Azami, S. O. R. Biabanaki, H. DorMohammadi, S. H. Keshavaz, and K. Shahim.

References

1. Fleck, N.A. (1995) On the cold compaction of powders. *Journal of Mechanics and Physics of Solids*, **43**, 1409–1431.
2. Antony, S.J., Momoh, R.O., and Kuhn, M.R. (2004) Micromechanical modeling of oval particulates subjected to bi-axial compression. *Computational Materials Science*, **29**, 494–498.
3. Skrinjar, O. and Larsson, P.L. (2004) On discrete element modeling of compaction of powders with size ratio. *Computational Materials Science*, **31**, 131–146.
4. Lewis, R.W., Jinka, A.G.K., and Gethin, D.T. (1993) Computer aided simulation of metal powder die compaction process. *Powder Metallurgy International*, **25**, 287–293.
5. Aydin, I., Briscoe, B.J., and Sanliturk, K.Y. (1994) Density distributions during the compaction of alumina powders: A comparison of a computational prediction with experiment. *Computational Materials Science*, **3**, 55–68.
6. Akisanya, A.R., Cocks, A.C.F., and Fleck, N.A. (1997) The yield behavior of metal powders. *International Journal of Mechanical Sciences*, **39**, 1315–1324.
7. Khoei, A.R. and Lewis, R.W. (1998) Finite element simulation for dynamic large elasto-plastic deformation in metal powder forming. *Finite Elements in Analysis and Design*, **30**, 335–352.
8. Khoei, A.R. and Lewis, R.W. (1999) Adaptive finite element remeshing in a large deformation analysis of metal powder forming. *International Journal for Numerical Methods in Engineering*, **45**, 801–820.
9. Gasik, M. and Zhang, B. (2000) A constitutive model and FE simulation for the sintering process of powder compacts. *Computational Materials Science*, **18**, 93–101.
10. Lewis, R.W. and Khoei, A.R. (2001) A plasticity model for metal powder forming processes. *International Journal of Plasticity*, **17**, 1652–1692.
11. Cedergren, J., Sorensen, N.J., and Bergmark, A. (2002) Three-dimensional analysis of compaction of metal powder. *Mechanics of Materials*, **34**, 43–59.
12. Oliver, J., Oller, S., and Cante, J.C. (1996) A plasticity model for simulation of industrial powder compaction processes. *International Journal of Solids and Structures*, **33**, 3161–3178.
13. Watson, T.J. and Wert, J.A. (1993) On the development of constitutive relations for metallic powders. *Metallurgical Transactions A*, **24**, 1993–2071.
14. Brown, S.B. and Abou-Chedid, G. (1994) Yield behavior of metal powder assemblages. *Journal of Mechanics and Physics of Solids*, **42**, 383–398.
15. Haggblad, H.A. and Oldenburg, M. (1994) Modeling and simulation of metal powder die pressing with use of explicit time integration. *Modeling and Simulation in Materials Science and Engineering*, **2**, 893–911.
16. Aydin, I., Briscoe, B.J., and Sanliturk, K.Y. (1996) The internal form of compacted ceramic components. A comparison of a finite element modeling with experiment. *Powder Technology*, **89**, 239–254.
17. Lewis, R.W. and Khoei, A.R. (1998) Numerical modeling of large deformation in metal powder forming. *Computer Methods in Applied Mechanics and Engineering*, **159**, 291–328.
18. Brandt, J. and Nilsson, L. (1999) A constitutive model for compaction of granular media, with account for deformation induced anisotropy. *Mechanics of Cohesive Frictional Materials*, **4**, 391–418.

19. Gu, C., Kim, M., and Anand, L. (2001) Constitutive equations for metal powders: applications to powder forming processes. *International Journal of Plasticity*, **17**, 147–209.
20. Khoei, A.R. and Azizi, S. (2004) Numerical simulation of 3D powder compaction processes using cone-cap plasticity theory. *Materials and Design*, **26**, 137–147.
21. Khoei, A.R., Bakhshiani, A., and Mofid, M. (2003) An endochronic plasticity model for finite strain deformation of powder forming processes. *Finite Elements in Analysis and Design*, **40**, 187–211.
22. Khoei, A.R., Bakhshiani, A., and Mofid, M. (2003) An endochronic plasticity model for numerical simulation of industrial powder compaction processes. *Communication in Numerical Methods in Engineering*, **19**, 521–534.
23. Khoei, A.R. and Bakhshiani, A. (2004) A hypoelasto-plastic finite strain simulation of powder compaction processes with density dependent endochronic model. *International Journal of Solids and Structures*, **41**, 6081–6110.
24. Khoei, A.R. and Azami, A.R. (2005) A single cone-cap plasticity with an isotropic hardening rule for powder materials. *International Journal of Mechanical Sciences*, **47**, 94–109.
25. Azami, A.R. and Khoei, A.R. (2006) 3D computational modeling of powder compaction processes using a three-invariant hardening cap plasticity model. *Finite Elements in Analysis and Design*, **42**, 792–807.
26. Khoei, A.R. and DorMohammadi, H. (2007) A three-invariant cap plasticity with isotropic-kinematic hardening rule for powder materials: Model assessment and parameter calibration. *Computational Materials Science*, **41**, 1–12.
27. Khoei, A.R. (2005) *Computational Plasticity in Powder Forming Process*, Elsevier.
28. Khoei, A.R., Azami, A.R., Anahid, M., and Lewis, R.W. (2006) A three-invariant hardening plasticity for numerical simulation of powder forming processes via the arbitrary Lagrangian-Eulerian FE model. *International Journal for Numerical Methods in Engineering*, **66**, 843–877.
29. Khoei, A.R., Anahid, M., and Shahim, K. (2008) An extended arbitrary Lagrangian – Eulerian finite element method for large deformation of solid mechanics. *Finite Elements in Analysis and Design*, **44**, 401–416.
30. Melenk, J.M. and Babuska, I. (1996) The partition of unity finite element method: basic theory and applications. *Computer Methods in Applied Mechanics and Engineering*, **139**, 289–314.
31. Doremus, P., Geindreau, C., Marttin, A., Debove, L., Lecot, R., and Dao, M. (1995) High pressure triaxial cell for metal powder. *Powder Metallurgy*, **38**, 284–287.
32. Yamada, T. and Kikuchi, F. (1993) An arbitrary Lagrangian-Eulerian finite element method for incompressible hyperelasticity. *Computer Methods in Applied Mechanics and Engineering*, **102**, 149–177.
33. Huerta, A. and Casadei, F. (1994) New ALE applications in non-linear fast-transient solid dynamics. *Engineering Computation*, **11**, 317–345.
34. Pijaudier-Cabot, G. and Bodé, L. (1995) Arbitrary Lagrangian– Eulerian finite element analysis of strain localization in transient problems. *International Journal for Numerical Methods in Engineering*, **38**, 4171–4191.
35. Wang, J. and Gadala, M.S. (1997) Formulation and survey of ALE method in nonlinear solid mechanics. *Finite Elements in Analysis and Design*, **24**, 253–269.
36. Gadala, M.S. and Wang, J. (1998) ALE formulation and its application in solid mechanics. *Computer Methods in Applied Mechanics and Engineering*, **167**, 33–55.
37. Ghosh, S. and Raju, S. (1996) R-S adapted arbitrary Lagrangian– Eulerian finite element method of metal-forming with strain localization. *International Journal for Numerical Methods in Engineering*, **39**, 3247–3272.
38. Gadala, M.S. and Wang, J. (1999) Simulation of metal forming processes with finite element method. *International Journal for Numerical Methods in Engineering*, **44**, 1397–1428.

39. Davey, K. and Ward, M.J. (2002) A practical method for finite element ring rolling simulation using the ALE flow formulation. *International Journal of Mechanical Sciences*, **44**, 165–195.
40. Rodriguez-Ferran, A., Casadei, F., and Huerta, A. (1998) ALE stress update for transient and quasistatic processes. *International Journal for Numerical Methods in Engineering*, **43**, 241–262.
41. Rodriguez-Ferran, A., Perez-Foguet, A., and Huerta, A. (2002) Arbitrary Lagrangian– Eulerian (ALE) formulation for hyperelastoplasticity. *International Journal for Numerical Methods in Engineering*, **53**, 1831–1851.
42. Armero, F. and Love, E. (2003) An arbitrary Lagrangian – Eulerian finite element method for finite strain plasticity. *International Journal for Numerical Methods in Engineering*, **57**, 471–508.
43. Bayoumi, H.N. and Gadala, M.S. (2004) A complete finite element treatment for the fully coupled implicit ALE formulation. *Computational Mechanics*, **33**, 435–452.
44. Khoei, A.R., Anahid, M., Shahim, K., and DorMohammadi, H. (2008) Arbitrary Lagrangian-Eulerian method in plasticity of pressure-sensitive material with reference to powder forming process. *Computational Mechanics*, **42**, 13–38.
45. Belytschko, T., Liu, W.K., and Moran, B. (2000) *Nonlinear Finite Elements for Continua and Structures*, John Wiley & Sons, Inc.
46. Shen, W.M., Kimura, T., Takita, K., and Hosono, K. (2001) in *Simulation of Materials Processing: Theory, Methods and Applications* (ed. K. Mori), A.A. Balkema, pp. 1027–1032.
47. Keshavaz, S.H., Khoei, A.R., and Khaloo, A.R. (2008) Contact friction simulation in powder compaction process based on the penalty approach. *Material and Design*, **29**, 1199–1211.
48. Khoei, A.R., Keshavaz, S.H., and Khaloo, A.R. (2008) Modeling of large deformation frictional contact in powder compaction processes. *Applied Mathematical Modeling*, **32**, 775–801.
49. Khoei, A.R., Biabanaki, S.O.R., Vafa, A.R., Yadegaran, I., and Keshavaz, S.H. (2009) A new computational algorithm for contact friction modeling of large plastic deformation in powder compaction processes. *International Journal of Solids and Structures*, **46**, 287–310.
50. Khoei, A.R., Biabanaki, S.O.R., Vafa, A.R., and Taheri-Mousavi, S.M. (2009) A new computational algorithm for 3D contact modeling of large plastic deformation in powder forming processes. *Computational Materials Science*, **46**, 203–220.
51. Dolbow, J.E. (1999) An extended finite element method with discontinuous enrichment for applied mechanics. PhD thesis, Northwestern University.
52. Belytschko, T. and Black, T. (1999) Elastic crack growth in finite elements with minimal remeshing. *International Journal for Numerical Methods in Engineering*, **45**, 601–620.
53. Moës, N., Dolbow, J.E., and Belytschko, T. (1999) A finite element method for crack growth without remeshing. *International Journal for Numerical Methods in Engineering*, **46**, 131–150.
54. Daux, C., Moës, N., Dolbow, J., Sukumar, N., and Belytschko, T. (2000) Arbitrary branched and intersecting cracks with the extended finite element method. *International Journal for Numerical Methods in Engineering*, **48**, 1741–1760.
55. Sukumar, N., Chopp, D.L., Moës, N., and Belytschko, T. (2001) Modeling holes and inclusions by level sets in the extended finite-element method. *Computer Methods in Applied Mechanics and Engineering*, **190**, 6183–6200.
56. Belytschko, T., Moës, N., Usui, S., and Parimi, C. (2001) Arbitrary discontinuities in finite elements. *International Journal for Numerical Methods in Engineering*, **50**, 993–1013.
57. Stolarska, M., Chopp, D.L., Moës, N., and Belytschko, T. (2001) Modeling crack growth by level sets in the extended finite element method. *International Journal for Numerical Methods in Engineering*, **51**, 943–960.
58. Dolbow, J., Moës, N., and Belytschko, T. (2001) An extended finite element method for modeling crack growth with frictional contact. *Computer Methods in*

Applied Mechanics and Engineering, **190**, 6825–6846.

59. Moës, N. and Belytschko, T. (2002) Extended finite element method for cohesive crack growth. *Engineering Fracture Mechanics*, **69**, 813–833.
60. Sukumar, N. and Prévost, J.H. (2003) Modeling quasi-static crack growth with the extended finite element method. Part I: Computer implementation. *International Journal of Solids and Structures*, **40**, 7513–7537.
61. Chopp, D.L. and Sukumar, N. (2003) Fatigue crack propagation of multiple coplanar cracks with the coupled extended finite element/fast marching method. *International Journal of Engineering Science*, **41**, 845–869.
62. Ventura, G., Budyn, E., and Belytschko, T. (2003) Vector level sets for description of propagating cracks in finite elements. *International Journal for Numerical Methods in Engineering*, **58**, 1571–1592.
63. Areias, P.M.A. and Belytschko, T. (2005) Analysis of three-dimensional crack initiation and propagation using the extended finite element method. *International Journal for Numerical Methods in Engineering*, **63**, 760–788.
64. Bordas, S., Nguyen, P.V., Dunant, C., Guidoum, A., and Nguyen-Dang, H. (2007) An extended finite element library. *International Journal for Numerical Methods in Engineering*, **71**, 703–732.
65. Khoei, A.R., Shamloo, A., and Azami, A.R. (2006) Extended finite element method in plasticity forming of powder compaction with contact friction. *International Journal of Solids and Structures*, **43**, 5421–5448.
66. Khoei, A.R. and Nikbakht, M. (2007) An enriched finite element algorithm for numerical computation of contact friction problems. *International Journal of Mechanical Sciences*, **49**, 183–199.
67. Kim, T.Y., Dolbow, J., and Laursen, T. (2007) A mortared finite element method for frictional contact on arbitrary interfaces. *Computational Mechanics*, **39**, 223–236.
68. Khoei, A.R., Biabanaki, S.O.R., and Anahid, M. (2008) Extended finite element method for three-dimensional large plasticity deformations. *Computer Methods in Applied Mechanics and Engineering*, **197**, 1100–1114.
69. Anahid, M. and Khoei, A.R. (2008) New development in extended finite element modeling of large elasto-plastic deformations. *International Journal for Numerical Methods in Engineering*, **75**, 1133–1171.
70. Anahid, M. and Khoei, A.R. (2010) Modeling of moving boundaries in large plasticity deformations via an enriched arbitrary Lagrangian–Eulerian FE method. *Scientia Iranica, Transaction A. Journal of Civil Engineering*, 17, 141–160.
71. Khoei, A.R. and Karimi, K. (2008) An enriched-FEM model for simulation of localization phenomenon in Cosserat continuum theory. *Computational Materials Science*, **44**, 733–749.
72. Prabel, B., Combescure, A., Gravouil, A., and Marie, S. (2006) Level set X-FEM non-matching meshes: application to dynamic crack propagation in elasto-plastic media. *International Journal for Numerical Methods in Engineering*, **69**, 1553–1569.
73. Elguedj, T., Gravouil, A., and Combescure, A. (2006) Appropriate extended functions for X-FEM simulation of plastic fracture mechanics. *Computer Methods in Applied Mechanics and Engineering*, **195**, 501–515.
74. Elguedj, T., Gravouil, A., and Combescure, A. (2007) A mixed augmented Lagrangian-extended finite element method for modeling elastic-plastic fatigue crack growth with unilateral contact. *International Journal for Numerical Methods in Engineering*, **71**, 1569–1597.

8
Functionally Graded Piezoelectric Material Systems – A Multiphysics Perspective

Wilfredo Montealegre Rubio, Sandro Luis Vatanabe, Gláucio Hermogenes Paulino, and Emílio Carlos Nelli Silva

8.1
Introduction

Functionally graded materials (FGMs) possess continuously graded properties and are characterized by spatially varying microstructures created by nonuniform distributions of the reinforcement phase as well as by interchanging the role of reinforcement and matrix (base) materials in a continuous manner. The smooth variation of properties may offer advantages such as local reduction of stress concentration and increased bonding strength [1–3].

Standard composites result from the combination of two or more materials, usually resulting in materials that offer advantages over conventional materials. At the macroscale observation, traditional composites (e.g., laminated) exhibit a sharp interface among the constituent phases that may cause problems such as stress concentration and scattering (if a wave is propagating inside the material), among others. However, a material made using the FGM concept would maintain some of the advantages of traditional composites and alleviate problems related to the presence of sharp interfaces at the macroscale. The design of the composite material itself is a difficult task, and the design of a composite where the properties of its constituent materials change gradually in the unit cell domain is even more complex.

There are two ways to design FGM composites: macro- and microscale approach. In macroscale approach, the conventional piezoelectric active element is replaced by a functionally graded piezoelectric material. Therefore, all or some of the properties (piezoelectric, dielectric, or elastic properties) vary along a specific direction, usually along its thickness, based on a specific gradation function [4–8]. In microscale approach, composites can be modeled by a unit cell with infinitesimal dimensions, which is the smallest structure that is periodic in the composite. By changing the volume fraction of the constituents, the shape of the inclusions, or even the topology of the unit cell, we can obtain different effective properties for the composite material [9]. The calculation of effective properties is necessary to obtain its performance. This can be achieved by applying the homogenization method that plays an important role in the design method.

Advanced Computational Materials Modeling: From Classical to Multi-Scale Techniques.
Edited by Miguel Vaz Júnior, Eduardo A. de Souza Neto, and Pablo A. Munoz-Rojas

ISBN: 978-3-527-32479-8

This chapter is organized as follows. In Section 8.2, a brief introduction about piezoelectricity is presented. In Section 8.3, the concept of FGM is introduced. In Section 8.4, the formulation of the finite element method (FEM) for graded piezoelectric structures is presented, and in Section 8.5 the influence of property scale in piezotransducer performance is described, aiming at ultrasonic applications. The influence of microscale is discussed in Section 8.6, including a brief description of homogenization method. Finally, in Section 8.7, concluding remarks are provided.

8.2 Piezoelectricity

The piezoelectric effect, according to the original definition of the phenomenon discovered by Jacques and Pierre Curie brothers in 1880 [10], is the ability of certain crystalline materials to develop an electric charge that is proportional to a mechanical stress. Thus, piezoelectricity is an interaction between electrical and mechanical systems. The direct piezoelectric effect, the development of an electric charge upon the application of a mechanical stress, is described as [11]

$$P_i = d_{ijk} T_{jk} \tag{8.1}$$

where P_i is a component of the electric polarization (charge per unit area), d_{ijk} are the components of piezoelectric coupling coefficient, and T_{jk} are components of the applied mechanical stress. The converse effect, the development of a mechanical strain upon the application of an electric field to the piezoelectric, is described by Nye [11]:

$$S_{ij} = d_{ijk} E_k \tag{8.2}$$

where S_{ij} is the strain produced and E_k is the applied electric field. In both cases, the piezoelectric coefficients d_{ijk} are numerically identical.

The piezoelectric effect is strongly linked to the crystal symmetry. Piezoelectricity is limited to 20 of the 32 crystal classes. The crystals that exhibit piezoelectricity have one common feature: the absence of a center of symmetry within the crystal. This absence of symmetry leads to polarity, the one-way direction of the charge vector. Most of the important piezoelectric materials are also ferroelectric [10]. The piezoelectric effect occurs naturally in several materials (quartz, tourmaline, and Rochelle salt), and can be induced in other polycrystalline materials; for example, the barium titanate ($BaTiO_3$), polyvinylidene fluoride (PVDF), and the lead zirconate titanate (PZT). Nevertheless, in these nonnatural materials, the piezoelectric effect must be induced through a process of electric polarization [12]. Basically, in the polarization process, the piezoelectric material is heated to an elevated temperature while in the presence of a strong DC field (usually higher than 2000 V mm^{-1}). This polarizes the ceramic (aligning the molecular dipoles of the ceramic in the direction of the applied field) and provides it with piezoelectric properties.

The piezoelectric effect can be described using a set of basic equations. The constitutive relations for piezoelectric media give the coupling between the mechanical and the electrical parts of a piezoelectric system. Thus, the equations of linear piezoelectricity, as given in the IEEE Standard on Piezoelectricity [13], and by using Einstein's summation convention (see also Ref. [14]), can be written as

$$\begin{aligned} T_{ij} &= C_{ijkl}^{E}\, S_{kl} - e_{kij} E_k \\ D_i &= e_{ikl}\, S_{kl} + \varepsilon_{ik}^{S}\, E_k \end{aligned} \tag{8.3}$$

where $i, j, k, l = 1, 2, 3$.

Terms T_{ij}, S_{kl}, D_i, and E_k are, respectively, components of the mechanical stress tensor (Newton per square meter), components of the mechanical strain tensor, components of the electric flux density (coulomb per square meter), and components of the electric field vector (volts per meter). The term C_{ijkl}^{E} represents the components of the elastic stiffness constant tensor, which are evaluated at constant electric field (in newton per square meter). Terms e_{ikl} and ε_{ik}^{S} are, respectively, components of the piezoelectric constant tensor (in coulomb per square meter), and components of the dielectric constant tensor evaluated at constant strain (in fermi per meter).

The components of the strain tensor S_{kl} are defined by

$$S_{kl} = \frac{1}{2}(u_{k,l} + u_{l,k}) \tag{8.4}$$

where u_l is the component no. l of the displacement vector (meters), and where $u_{k,l} = \partial u_k / \partial x_l$.

The electric and magnetic fields inside of a medium are described by Maxwell's equations, which relate the fields to the microscopic average properties of the material. When the quasistatic approximation is introduced [14], the electric field is derivable from a scalar electric potential:

$$E_i = -\phi_{,i} \tag{8.5}$$

where ϕ_i is the electric potential (volts). The following Maxwell's equation is also needed for describing a piezoelectric medium:

$$D_{i,i} = 0 \tag{8.6}$$

Finally, the equation of motion for a piezoelectric medium, not subjected to body forces, may be written:

$$T_{ij,j} = \rho \ddot{u}_i \tag{8.7}$$

where ρ represents the density of the material (in kilograms per cubic meter), being $\ddot{u}_l = \partial^2 u_l / \partial t^2$, and t the time (in seconds). The mechanical stress tensor T_{ij} is symmetric [13].

However, Eq. (8.3) can be expressed as a tensorial equation, which is a more compact expression for constitutive equations of piezoelectric materials. Thus, the constitutive piezoelectric model is formulated as [14]

$$\mathbf{T} = \mathbf{c}^E\mathbf{S} - \mathbf{e}^T\mathbf{E}$$
$$\mathbf{D} = \mathbf{e}\mathbf{S} + \boldsymbol{\varepsilon}^S\mathbf{E} \tag{8.8}$$

where $\mathbf{c}^E$ is the fourth-order elastic tensor, where the components are evaluated by constant electrical field. The term $\boldsymbol{\varepsilon}^S$ is the second-order tensor of dielectric constants, whose components are calculated by constant strain. The third-order tensor of piezoelectric coefficients is expressed by the term $\mathbf{e}$. Finally, terms $\mathbf{T}$ and $\mathbf{D}$ are the second-order stress tensor and the electric displacement vector, respectively. The symbol T indicates transpose.

8.3
Functionally Graded Piezoelectric Materials

8.3.1
Functionally Graded Materials (FGMs)

FGMs are composite materials whose properties vary gradually and continuously along a specific direction within the domain of the material. The property variation is generally achieved through the continuous change of the material microstructure [1] (Figure 8.1); in other words, FGMs are characterized by spatially varying microstructures created by nonuniform distributions of the constituent phases. This variation can be accomplish by using reinforcements with different properties, sizes, and shapes, as well as by interchanging the role of reinforcement and matrix (base) material in a continuous manner. In this last case, the volume fraction of

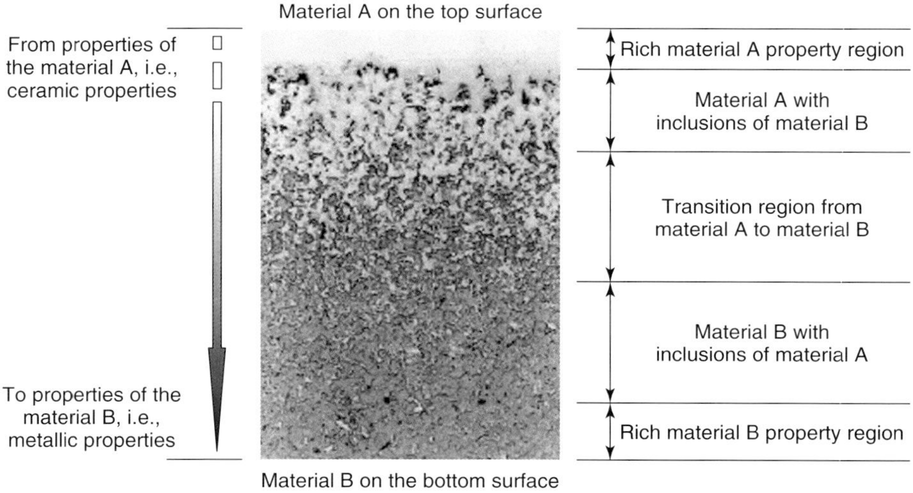

Figure 8.1 Microstructure of an FGM that is graded from material A to material B [15].

material phases continuously varies from 0 to 100% between two points of the structure; for instance, the material A of Figure 8.1 is gradually replaced by material B, smoothly varying through a transition zone [15, 16].

The main advantage of FGMs is their characteristic of "combining" advantage of desirable features of their constituent phases. For example, if a metal and a ceramic are used as a material base, FGMs could take advantage of heat and corrosion resistance typical of ceramics, and mechanical strength and toughness typical of metals; accordingly, a part of Figure 8.1 can represent a thermal barrier on the top surface (material A), by tacking advantage of the thermal properties of ceramics (low thermal conductivity and high melting point), while another part represents a material with high tensile strength and high resilience (metallic material B), on the bottom side [17], without any material interface. The absence of interfaces produces other interesting features such as reduction of thermal and mechanical stress concentration [18] and increased bonding strength and fatigue-lifetime [19].

The concept of FGMs is bioinspired (biomimic), as these materials occur in nature; for instance, in bones and teeth of animals [20–22], and trees such as the bamboo [23]. A dental crown is an excellent example of the FGM concept in natural structures: teeth require high resistance to friction and impact on the external area (enamel), and a flexible internal structure for reasons of fatigue and toughness [21, 22]. Other interesting example is the bamboo. Bamboo stalks are optimized composite materials that naturally exploit the concept of FGMs, as shown in Figure 8.2. The bamboo culm is an approximately cylindrical shell that is divided periodically by transversal diaphragms at nodes. Between 20 and 30% of the cross-sectional area of the culm is made of longitudinal fibers that are nonuniformly distributed through the wall thickness, the concentration being most dense near the exterior (Figure 8.2). The orientation of these fibers makes bamboo an orthotropic material with high strength along to fibers and low strength along its transverse direction [23].

In engineering applications, the FGM concept was originally proposed around 1984–1985 [17], when Japanese scientists researched advanced materials for aerospace industry; specifically, materials that bear the temperature gradient generated when a spacecraft returns to earth. In this case, the temperature gradient is approximately 1000 °C from the outside to the inside of the aircraft. They designed an FGM structure with ceramic properties on the outer surface, exposed to high temperature, and with properties of a thermally conductive material on the inner

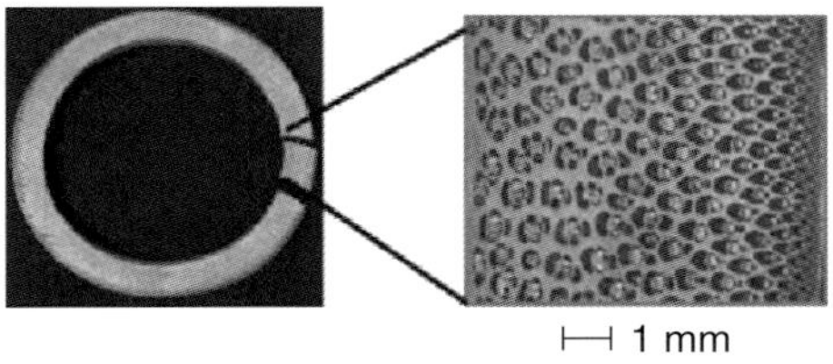

Figure 8.2 Cross section of bamboo culm showing radial nonuniform distribution of fibers through the thickness [23].

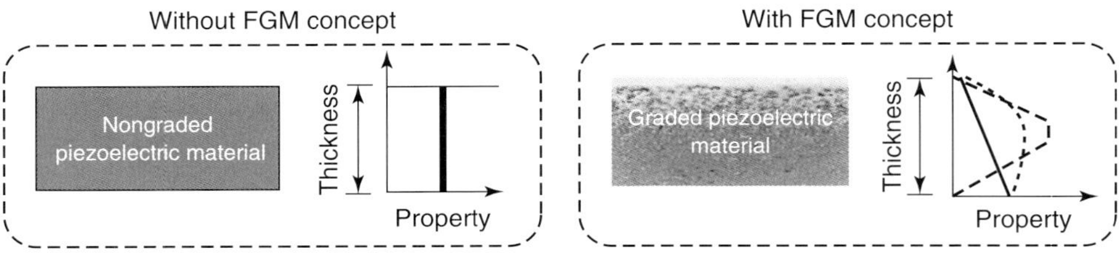

Figure 8.3 Sketch of a graded piezoelectric material: (a) nongraded piezoelectric material and (b) graded piezoelectric material based on FGM concept and considering several laws of gradation.

surface. Since then, the development of manufacture methods, design, and modeling techniques of FGMs has been the focus of several research groups worldwide. These researches focus mainly on thermal applications [18, 24]; bioengineering [20–22, 25]; industrial applications such as the design of watches, baseball cleats, razor blades, among others;[1)] and since the late 1990s, the design and fabrication of piezoelectric structures [4–8, 19, 26–30].

8.3.2 FGM Concept Applied to Piezoelectric Materials

As previously mentioned, piezoelectric materials have the property to convert electrical energy into mechanical energy and vice versa. Their main applications are in sensors and electromechanical actuators, as resonators in electronic equipment, and acoustic applications, as ultrasound transducers, naval hydrophones, and sonars. A recent trend has been the design and manufacture of piezoelectric transducers based on the FGM concept [19, 26–29]; in this case, the conventional single piezoelectric material is replaced by a graded piezoelectric material (Figures 8.3–8.5) and, consequently, some or all properties (elastic, piezoelectric, and/or dielectric properties) may change along one specific direction in which several functions or laws of gradation can be used (Figure 8.3).

Several authors have highlighted the advantages of the FGM concept when applied to piezoelectric structures [4–8, 19, 26, 28, 30]. In static or dynamic applications, the main advantages are reduction of the mechanical stress [26, 31], improved stress redistribution [26], maximization of output displacement, and increased bonding strength (and fatigue-life). For exemplifying the FGM concept when applied to piezoelectric structures, the case of a bimorph actuator can be considered. This kind of actuators is traditionally composed of two piezoelectric materials with opposite direction of polarization, and both are mechanically coupled by a metallic phase (electrode) (Figure 8.4). By applying an electric field along the thickness, the transducer will bend due to the fact the piezoelectric material will deform

1) More details are available in http://fgmdb.kakuda.jaxa.jp/others/e_product.html.

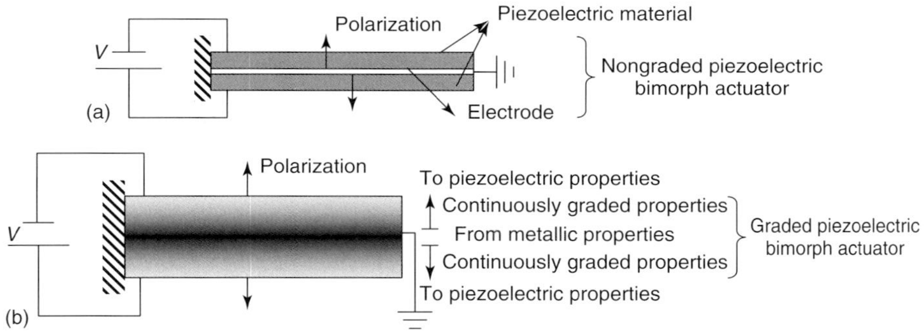

Figure 8.4 FGM concept applying to bimorph transducers: (a) nongraded piezoelectric transducer and (b) graded transducer from electrode (metallic) properties in the middle to piezoelectric properties on the top and bottom surfaces.

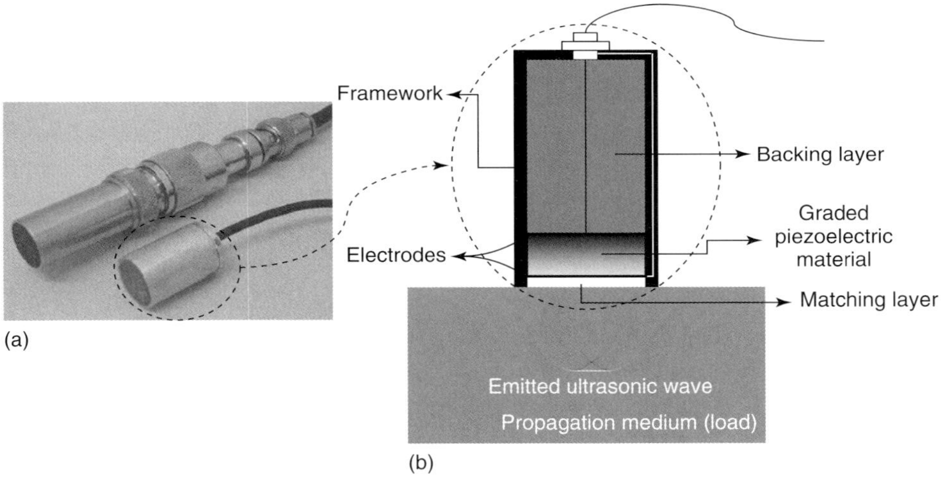

Figure 8.5 FGM concept applied to piezoelectric ultrasonic transducers: (a) photograph of a typical assembly of an ultrasonic transducer and (b) sketch of the same transducer considering a graded piezoelectric material.

in opposite directions. In dynamic applications, a critical issue in these bimorph actuators is the fatigue phenomena, which depends on the stress distribution into the actuator. It is clear that the stress distribution is not uniform due to the presence of material interfaces; specifically, between electrode/piezoelectric-material. The material interface will cause mechanical stress concentration, reduction of the fatigue limit, and accordingly, reduction of the transducer lifetime. However, by applying the FGM concept to a bimorph actuator, the material interface can be reduced or eliminated, as electrode properties can be smoothly varied from center to piezoelectric properties on the top and bottom surfaces (Figure 8.4).

In relation to durability, Qiu *et al.* [19] study the durability of graded piezoelectric bimorph actuators considering durability at low frequency (quasistatic operation) and durability at resonant frequency. In the former case, the graded and nongraded piezoelectric actuators are excited by 500 and 160 V (peak to peak), respectively. In a durability test, at 20 Hz (much lower than the resonance frequency of the first mode), the results show that the graded piezoelectric actuator did not break down after 300 h (2.16×10^7 cycles) of actuation, while the nongraded bimorph break down due to a crack after 138 h (about 10^7 cycles). In the second case (durability test at first resonance frequency), the graded and nongraded piezoelectric actuators are excited to 40 and 28 V (peak to peak), respectively. The average lifetime of a nongraded bimorph actuator is 24 min, and considering the FGM concept, graded actuators fail in the test period of 240 min.

In other application of piezoelectric materials, the FGM concept allows reducing reflection waves inside piezoelectric ultrasonic transducers [5] and obtaining acoustic responses with smaller time waveform (larger bandwidth) [4, 7, 8, 30] than nongraded piezoelectric transducers. These advantages are desirable for improving the axial resolution in medical imaging and nondestructive testing applications [32]. Particularly, considering that the piezoelectric material of an ultrasonic transducer is graded from a nonpiezoelectric to a piezoelectric material (or from piezoelectric property $e_{33} = 0$ on the top electrode to $e_{33} \neq 0$ on the bottom electrode), as shown in Figure 8.5, the larger bandwidth of the graded piezoelectric transducer is produced because the acoustic pulse is generated mainly from the surface with higher piezoelectric properties. In other words, as explained by Yamada *et al.* [4], the induced piezoelectric stress $T_3 = -e_{33}\ E_3$ is higher on the surface with $e_{33} \neq 0$ than on the surface with $e_{33} = 0$. Hence, the volumetric force $F_v = \partial T_3/\partial z$ (or spatial derivative of the induced piezoelectric stress), which is responsible for the generation of acoustic waves, is equal to zero on the surface without piezoelectric properties, and it has its maximum value on the opposite surface [4]. Accordingly, a single ultrasonic pulse is generated by an impulse excitation.

In addition, in piezoelectric ultrasonic transducers, the FGM concept could be used to reduce the material interfaces between piezoelectric-material/backing and piezoelectric-material/matching[2] (Figure 8.5).

2) Generally, ultrasound transducers are composed of an active element (piezoelectric material) bonded to two passive and non-piezoelectric elements (backing and matching layers); thus, when the piezoelectric material is electrically excited, the transducer produces an ultrasonic wave in a specific medium (load), as shown in Figure 8.5. The backing layer damps the back wave, and the matching layer matches the piezoelectric material and load acoustic impedances. The active element of most acoustic transducers used today is a piezoelectric ceramic.

8.4 Finite Element Method for Piezoelectric Structures

8.4.1 The Variational Formulation for Piezoelectric Problems

The dynamic equations of a piezoelectric continuum can be derived from the Hamilton principle, in which the Lagrangian and the virtual work are properly adapted to include the electrical contributions as well as the mechanical ones. According to this principle, the displacements and electrical potentials are those that satisfy the following equation [33]:

$$\delta \int_{t_1}^{t_2} L\,\mathrm{d}t + \int_{t_1}^{t_2} \delta W\,\mathrm{d}t = 0 \tag{8.9}$$

where L is the Lagrangian term, W is the external work done by mechanical and electrical forces, and the term δ represents the variational. The Lagrangian is given by Tiersten [33]:

$$L = \int_{\Omega} \left(\frac{1}{2}\rho\, \dot{\mathbf{u}}^T \dot{\mathbf{u}} - H \right) \mathrm{d}\Omega \tag{8.10}$$

where the terms H and $\mathbf{u}$ are the electrical enthalpy and the displacement vector, respectively. The integration of Eq. (8.10) is performed over a piezoelectric body of volume Ω. Considering that the surface S of the body Ω is subjected to prescribed surface tractions ($\mathbf{F}$) and surface charge per unit area (Q), the virtual work δW is given by the following expression [33]:

$$\delta W = \int_S (\delta \mathbf{u}^T \mathbf{F} - \delta\varphi\, Q)\,\mathrm{d}S \tag{8.11}$$

where φ is the electrical potential. On the other hand, the enthalpy is expressed as

$$H = P - \mathbf{E}^T \mathbf{D} \tag{8.12}$$

with $\mathbf{E}$ being the electrical field vector, $\mathbf{D}$ the electrical displacement vector, and P the potential energy, which is given as

$$P = \frac{1}{2}\mathbf{S}^T \mathbf{T} + \frac{1}{2}\mathbf{E}^T \mathbf{D} \tag{8.13}$$

where $\mathbf{S}$ and $\mathbf{T}$ are the second-order strain and stress tensors, respectively. By replacing Eqs (8.13) in (8.12), and using this result in Eq. (8.10), we obtain

$$L = \int_{\Omega} \left(\frac{1}{2}\rho\, \dot{\mathbf{u}}^T \dot{\mathbf{u}} - \frac{1}{2}\mathbf{S}^T \mathbf{T} + \frac{1}{2}\mathbf{E}^T \mathbf{D} \right) \mathrm{d}\Omega \tag{8.14}$$

On the other hand, by substituting in Eq. (8.14) the constitutive equations of a piezoelectric material, which are expressed in Eq. (8.8), the Lagrangian is expressed as

$$L = \int_{\Omega} \frac{1}{2} \left(\rho\, \dot{\mathbf{u}}^T \dot{\mathbf{u}} - \mathbf{S}^T \mathbf{c}^E \mathbf{S} + \mathbf{S}^T \mathbf{e}^T \mathbf{E} + \mathbf{E}^T \mathbf{e}\, \mathbf{S} + \mathbf{E}^T \boldsymbol{\varepsilon}^S \mathbf{E} \right) \mathrm{d}\Omega \tag{8.15}$$

and by replacing both Eqs (8.11) and (8.15) in Hamilton's formula (8.9), we deduce

$$\int_{t_1}^{t_2}\left(\int_{\Omega}(\rho\,\delta\dot{\mathbf{u}}^T\dot{\mathbf{u}} - \delta\mathbf{S}^T\mathbf{c}^E\mathbf{S} + \delta\mathbf{S}^T\mathbf{e}^T\mathbf{E} + \delta\mathbf{E}^T\mathbf{e}\,\mathbf{S} + \delta\mathbf{E}^T\boldsymbol{\varepsilon}^S\mathbf{E})\,\mathrm{d}\Omega + \int_S(\delta\mathbf{u}^T\mathbf{F} - \delta\varphi\,Q)\,\mathrm{d}S\right)\mathrm{d}t = 0 \tag{8.16}$$

To complete the variational formulation for a piezoelectric medium, the first term of Eq. (8.16) is integrated by parts with relation to time:

$$\int_{t_1}^{t_2}\rho\delta\dot{\mathbf{u}}^T\dot{\mathbf{u}}\,\mathrm{d}t = \rho\delta\mathbf{u}^T\dot{\mathbf{u}}\Big|_{t_1}^{t_2} - \int_{t_1}^{t_2}\rho\delta\mathbf{u}^T\ddot{\mathbf{u}}\,\mathrm{d}t = -\int_{t_1}^{t_2}\rho\delta\mathbf{u}^T\ddot{\mathbf{u}}\,\mathrm{d}t \tag{8.17}$$

and by substituting Eq. (8.17) in Eq. (8.16), the final expression of the variational piezoelectric problem is found:

$$\int_{\Omega}(-\rho\,\delta\mathbf{u}^T\ddot{\mathbf{u}} - \delta\mathbf{S}^T\mathbf{c}^E\mathbf{S} + \delta\mathbf{S}^T\mathbf{e}^T\mathbf{E} + \delta\mathbf{E}^T\mathbf{e}\,\mathbf{S} + \delta\mathbf{E}^T\boldsymbol{\varepsilon}^S\mathbf{E})\,\mathrm{d}\Omega + \int_S(\delta\mathbf{u}^T\mathbf{F} - \delta\varphi\,Q)\,\mathrm{d}S = 0 \tag{8.18}$$

8.4.2
The Finite Element Formulation for Piezoelectric Problems

The FEM is an approximation technique for finding the solution of complex constitutive relations, as expressed in Section 8.1. The method consists of dividing the continuum domain Ω, into subdomains V^e, named *finite elements*. These elements are interconnected at a finite number of points, or nodes, where unknowns are defined. In piezoelectric domains, unknowns usually are displacements and electrical potentials. Within each finite element, unknowns are uniquely defined by the values they assume at the element nodes, by using interpolation functions, usually named *shape functions* [34].

For piezoelectric problems, the FEM considers that the displacement field $\mathbf{u}$ and electrical potential φ for each finite element e, are, respectively, approximated by nodal displacements $\mathbf{u}_e$, nodal electrical potentials $\boldsymbol{\varphi}_e$, and shape functions; thus [34]

$$\mathbf{u}^e = \mathbf{N}_u\mathbf{u}_e \text{ and } \varphi^e = \mathbf{N}_\varphi\boldsymbol{\varphi}_e \tag{8.19}$$

where $\mathbf{N}_u$ and $\mathbf{N}_\varphi$ are shape functions for mechanical and electrical problems, respectively. By deriving Eq. (8.19), the strain tensor $\mathbf{S}$ and electric field $\mathbf{E}$ can be written in the following form:

$$\mathbf{S}^e = \mathbf{B}_u\mathbf{u}_e \text{ and } \mathbf{E}^e = -\mathbf{B}_\varphi\,\boldsymbol{\varphi}_e \tag{8.20}$$

where $\mathbf{B}_\mathbf{u}$ and $\mathbf{B}_\varphi$ are the strain-displacement and voltage-gradient matrices, respectively, which can be expressed as

$$\mathbf{B}_\mathbf{u} = \begin{bmatrix} \frac{\partial}{\partial x} & 0 & 0 & 0 & \frac{\partial}{\partial z} & \frac{\partial}{\partial y} \\ 0 & \frac{\partial}{\partial y} & 0 & \frac{\partial}{\partial z} & 0 & \frac{\partial}{\partial x} \\ 0 & 0 & \frac{\partial}{\partial z} & \frac{\partial}{\partial y} & \frac{\partial}{\partial x} & 0 \end{bmatrix}^T \mathbf{N}_\mathbf{u} \quad \text{and} \quad \mathbf{B}_\mathbf{u} = \begin{Bmatrix} \frac{\partial}{\partial x} \\ \frac{\partial}{\partial y} \\ \frac{\partial}{\partial z} \end{Bmatrix} \mathbf{N}_\mathbf{u} \tag{8.21}$$

The terms x, y, and z are the Cartesian coordinates. By substituting Eqs (8.19) and (8.20) in Eq. (8.18), the variational piezoelectric problem becomes (where subscript e indicates finite element) for each finite element domain:

$$\begin{aligned} &(\delta \mathbf{u}^e)^T \left\{ -\left(\int_{V^e} \rho \mathbf{N}_\mathbf{u}^T \mathbf{N}_\mathbf{u} \, \mathrm{d}V^e \right) \ddot{\mathbf{u}}^e - \left(\int_{V^e} \mathbf{B}_\mathbf{u}^T \mathbf{c}^E \mathbf{B}_\mathbf{u} \, \mathrm{d}V^e \right) \mathbf{u}^e - \left(\int_{V^e} \mathbf{B}_\mathbf{u}^T \mathbf{e}^T \mathbf{B}_\varphi \, \mathrm{d}V^e \right) \varphi^e \right. \\ &\quad \left. + \int_{S^e} \mathbf{N}_\mathbf{u}^T \mathbf{F} \, \mathrm{d}S^e \right\} + \ldots \\ &(\delta \boldsymbol{\varphi}^e)^T \left\{ -\left(\int_{V^e} \mathbf{B}_\varphi^T \mathbf{e} \mathbf{B}_\mathbf{u} \, \mathrm{d}V^e \right) \mathbf{u}^e + \left(\int_{V^e} \mathbf{B}_\varphi^T \boldsymbol{\varepsilon}^S \mathbf{B}_\varphi \, \mathrm{d}V^e \right) \boldsymbol{\varphi}^e - \int_{S^e} \mathbf{N}_\varphi^T Q \, \mathrm{d}S^e \right\} = 0 \end{aligned} \tag{8.22}$$

By grouping the terms that multiply $\delta \mathbf{u}_e^T$ and $\delta \boldsymbol{\varphi}_e^T$ in Eq. (8.22), two sets of matrix equations are obtained, yielding for each finite element, the following piezoelectric finite element formulation:

$$\begin{bmatrix} \mathbf{M}_{\mathbf{uu}}^e & 0 \\ 0 & 0 \end{bmatrix} \begin{Bmatrix} \ddot{\mathbf{u}}^e \\ \ddot{\boldsymbol{\varphi}}^e \end{Bmatrix} + \begin{bmatrix} \mathbf{K}_{\mathbf{uu}}^e & \mathbf{K}_{\mathbf{u}\varphi}^e \\ \mathbf{K}_{\mathbf{u}\varphi}^e & -\mathbf{K}_{\varphi\varphi}^e \end{bmatrix} \begin{Bmatrix} \mathbf{u}^e \\ \boldsymbol{\varphi}^e \end{Bmatrix} = \begin{Bmatrix} \mathbf{F}_p^e \\ \mathbf{Q}_p^e \end{Bmatrix} \tag{8.23}$$

where

$$\begin{aligned} \mathbf{M}_{\mathbf{uu}}^e &= \sum_{N_{\text{ele}}} \iiint \mathbf{N}_\mathbf{u}^T \, \rho\,(x, y, z)\, \mathbf{N}_\mathbf{u} \, \mathrm{d}x \, \mathrm{d}y \, \mathrm{d}z; \\ \mathbf{K}_{\mathbf{uu}}^e &= \sum_{N_{\text{ele}}} \iiint \mathbf{B}_\mathbf{u}^T \, \mathbf{c}^E (x, y, z)\, \mathbf{B}_\mathbf{u} \, \mathrm{d}x \, \mathrm{d}y \, \mathrm{d}z; \\ \mathbf{K}_{\mathbf{u}\varphi}^e &= \sum_{N_{\text{ele}}} \iiint \mathbf{B}_\mathbf{u}^T \, \mathbf{e}^T (x, y, z)\, \mathbf{B}_\varphi \, \mathrm{d}x \, \mathrm{d}y \, \mathrm{d}z; \\ \mathbf{K}_{\varphi\varphi}^e &= \sum_{N_{\text{ele}}} \iiint \mathbf{B}_\varphi^T \, \boldsymbol{\varepsilon}^S (x, y, z)\, \mathbf{B}_\varphi \, \mathrm{d}x \, \mathrm{d}y \, \mathrm{d}z \end{aligned} \tag{8.24}$$

$$\mathbf{F}_\mathrm{p}^e = \int_S \mathbf{N}_\mathbf{u}^T \mathbf{F} \, \mathrm{d}S^e \quad \mathbf{Q}_\mathrm{p}^e = -\int_S \mathbf{N}_\varphi^T \mathbf{Q} \, dS^e \tag{8.25}$$

The terms $\mathbf{M}_{\mathbf{uu}}^e$, $\mathbf{K}_{\mathbf{uu}}^e$, $\mathbf{K}_{\mathbf{u}\varphi}^e$, and $\mathbf{K}_{\varphi\varphi}^e$ are, respectively, the mass, elastic, piezoelectric, and dielectric matrices; the terms x, y, and z explicitly represent the dependency of material properties with position, in graded piezoelectric systems.

According to FEM theory, matrices and vectors in Eq. (8.23) must be rearranged for the whole domain Ω by a process called *assembly*. Thus, global matrices and vectors of piezoelectric constitutive equations result from assembling the vectors and matrices of single elements [34]. In Eq. (8.24), assembly is represented by the summation symbol, and the term N_{ele} represents the total number of finite elements.

In addition, the matrix equation (8.23) may be adapted for a variety of different analyses, such as static, modal, harmonic, transient types [34], which are given (the global piezoelectric system is considered):

- **Static analysis**:

$$\begin{bmatrix} \mathbf{K}_{\mathrm{uu}} & \mathbf{K}_{\mathrm{u}\varphi} \\ \mathbf{K}_{\mathrm{u}\varphi}^T & -\mathbf{K}_{\varphi\varphi} \end{bmatrix} \begin{Bmatrix} \mathbf{u} \\ \boldsymbol{\varphi} \end{Bmatrix} = \begin{Bmatrix} \mathbf{F}_p \\ \mathbf{Q}_p \end{Bmatrix} \tag{8.26}$$

- **Modal analysis**:

$$\begin{aligned} & -\lambda \begin{bmatrix} \mathbf{M}_{\mathrm{uu}} & 0 \\ 0 & 0 \end{bmatrix} \begin{Bmatrix} \boldsymbol{\Psi}_{\mathrm{u}} \\ \boldsymbol{\Psi}_{\varphi} \end{Bmatrix} \\ & + \begin{bmatrix} \mathbf{K}_{\mathrm{uu}} & \mathbf{K}_{\mathrm{u}\varphi} \\ \mathbf{K}_{\mathrm{u}\varphi}^T & -\mathbf{K}_{\varphi\varphi} \end{bmatrix} \begin{Bmatrix} \boldsymbol{\Psi}_{\mathrm{u}} \\ \boldsymbol{\Psi}_{\varphi} \end{Bmatrix} = \begin{Bmatrix} 0 \\ 0 \end{Bmatrix} \text{ with } \lambda = \omega^2 \end{aligned} \tag{8.27}$$

- **Harmonic analysis**:

$$\left(-\Omega_c^2 \begin{bmatrix} \mathbf{M}_{\mathrm{uu}} & 0 \\ 0 & 0 \end{bmatrix} + \begin{bmatrix} \mathbf{K}_{\mathrm{uu}} & \mathbf{K}_{\mathrm{u}\varphi} \\ \mathbf{K}_{\mathrm{u}\varphi}^T & -\mathbf{K}_{\varphi\varphi} \end{bmatrix} \right) \begin{Bmatrix} \mathbf{u}_0 \\ \boldsymbol{\varphi}_0 \end{Bmatrix} = \begin{Bmatrix} \mathbf{F}_0 \\ \mathbf{Q}_0 \end{Bmatrix} \tag{8.28}$$

- **Transient analysis**:

$$\begin{aligned} & \begin{bmatrix} \mathbf{M}_{\mathrm{uu}} & 0 \\ 0 & 0 \end{bmatrix} \begin{Bmatrix} \ddot{\mathbf{u}} \\ \ddot{\boldsymbol{\varphi}} \end{Bmatrix} + \frac{1}{\omega_0} \begin{bmatrix} \mathbf{K}'_{\mathrm{uu}} & \mathbf{K}'_{\mathrm{u}\varphi} \\ \mathbf{K}'^T_{\mathrm{u}\varphi} & -\mathbf{K}'_{\varphi\varphi} \end{bmatrix} \begin{Bmatrix} \dot{\mathbf{u}} \\ \dot{\boldsymbol{\varphi}} \end{Bmatrix} \\ & + \begin{bmatrix} \mathbf{K}_{\mathrm{uu}} & \mathbf{K}_{\mathrm{u}\varphi} \\ \mathbf{K}_{\mathrm{u}\varphi}^T & -\mathbf{K}_{\varphi\varphi} \end{bmatrix} \begin{Bmatrix} \mathbf{u} \\ \boldsymbol{\varphi} \end{Bmatrix} = \begin{Bmatrix} \mathbf{F}_p \\ \mathbf{Q}_p \end{Bmatrix} \end{aligned} \tag{8.29}$$

where λ and ω represent eigenvalue and natural frequency, respectively. The term $\boldsymbol{\psi}$ represents eigenvectors, and the term Ω_c is the circular frequency of a harmonic input excitation. Each of these analysis cases requires specific conditioning and computation techniques. A full description of theses techniques is a topic that is beyond the scope of this chapter.

8.4.3
Modeling Graded Piezoelectric Structures by Using the FEM

When graded piezoelectric structures are simulated, properties must change continuously inside piezoelectric domain, which means that matrices of Eq. (8.23) must be described by a continuous function that depends on Cartesian position (x, y, z). In literature, there are several material models applied to estimate the

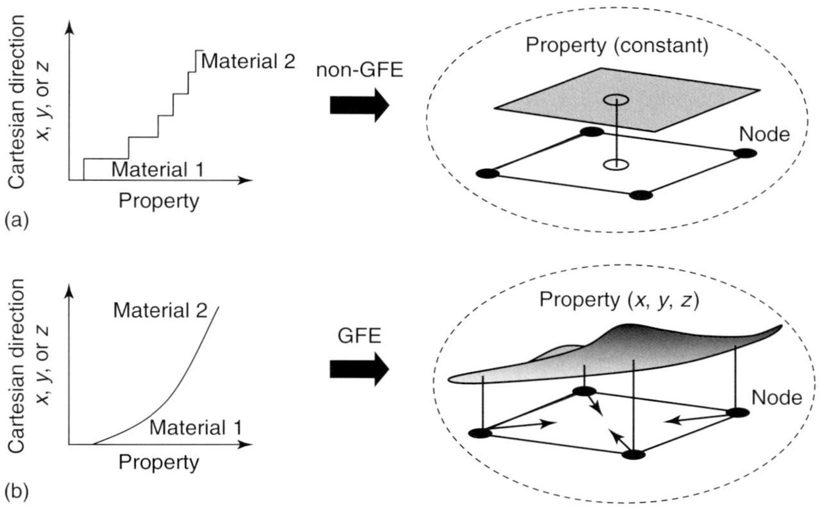

Figure 8.6 Finite element modeling of FGMs: (a) homogeneous finite element and (b) graded finite element.

effective properties of composite materials, which could be used in FGM, such as the Mori–Tanaka and the self-consistent models [35]. Other works treat the nonhomogeneity of the material, inherent in the problem, by using homogeneous finite elements with constant material properties at the element level; in this case, properties are evaluated at the centroid of each element (Figure 8.6a). This element-wise approach is close to the multilayer one, which depends on the number of layers utilized; accordingly, a layer convergence analysis must be performed in addition to finite element convergence, as shown in Ref. [30], for reducing the difference between the multilayer approach and the continuous material distribution. On the other hand, the multilayer approach leads to undesirable discontinuities of the stress [26] and strain fields [36] and discontinuous material gradation [30].

A more natural way of representing the material distribution in a graded material is based on graded finite elements (GFEs), which incorporate the material property gradient at the size scale of the element (Figure 8.6b) and reduce the discontinuity of the material distribution. Specifically, for static problems, Kim and Paulino [36] and Silva *et al.* [37] demonstrate that, by using the generalized isoparametric formulation (GIF), smoother and more accurate results are obtained in relation to element-wise approach. Essentially, the GIF leads to GFEs, where the material property gradient is continuously interpolated inside each finite element based on property values at each finite element node (Figure 8.6b). By employing the GIF, the same interpolation functions are used to interpolate displacements and electric potential, spatial coordinates (x, y, z), and material properties inside each finite element. Specifically, finite element shape functions N are used as interpolation functions.

For material properties, and by using the GIF, the density, ρ, and the elastic, C^E_{ijkl}, piezoelectric, e_{ikl}, and dielectric, ε^S_{ik}, material properties of a piezoelectric finite element can be written as

$$\rho\left(x,y,z\right)=\sum_{n=1}^{n_d}N_n\left(x,y,z\right)\ \rho_n,\quad C^E_{ijkl}\left(x,y,z\right)=\sum_{n=1}^{n_d}N_n\left(x,y,z\right)\ \left(C^E_{ijkl}\right)_n,$$

$$e_{ikl}\left(x,y,z\right)=\sum_{n=1}^{n_d}N_n\left(x,y,z\right)\ \left(e_{ikl}\right)_n,$$

$$\varepsilon^S_{ik}\left(x,y,z\right)=\sum_{n=1}^{n_d}N_n\left(x,y,z\right)\left(\varepsilon^S_{ik}\right)_n\quad \text{for } i,j,k,l=1,2,3 \tag{8.30}$$

where n_d is the number of nodes per finite element. When the GFE is implemented, the material properties must remain inside the integrals in Eq. (8.24), and they must be properly integrated. On the contrary, in homogeneous finite elements, these properties usually are constants.

8.5 Influence of Property Scale in Piezotransducer Performance

8.5.1 Graded Piezotransducers in Ultrasonic Applications

A very interesting application of piezoelectric materials is in ultrasonics. Piezoelectric ultrasonic transducers are mainly applied to nondestructive tests and medical images. In the last case, ultrasonic imaging has quickly replaced conventional X-rays in many clinical applications because of its image quality, safety, and low cost. Usually, a piezoelectric ultrasonic transducer is composed of a backing and matching layers and a piezoelectric disk, as shown in Figure 8.5. The piezoelectric disk is capable of transmitting and receiving pressure waves directed into a propagation medium such as the human body. Such transducers normally comprise single or multistacking element piezoelectric disks, which vibrate in response to an applied voltage for radiating a front-side wave in a specific medium (solid, liquid, or air), or produce an electrical potential when a pressure wave is received.

To obtain high-quality images, the ultrasonic transducer must be constructed so as to produce specified frequencies of pressure waves. Generally speaking, low-frequency pressure waves provide deep penetration into the medium (e.g., the human body); however, they produce poor-resolution images due to the length of the transmitted wavelengths. On the other hand, high-frequency pressure waves provide high resolution, however, with poor penetration. Accordingly, the selection of a transmitting frequency involves balancing resolution and penetration concerns. Unfortunately, there is a trade-off between resolution and deeper penetration. Traditionally, the frequency selection problem has been addressed by selecting the highest imaging frequency that offers adequate penetration for a given application. For example, in adults' cardiac imaging, frequencies in the 2–3 MHz range are typically selected in order to penetrate the human chest wall [38].

Recently, a new method has been studied in an effort to obtain both high resolution and deep penetration: treating piezoelectric ultrasonic transducers as graded structures (based on FGM concept). Hence, by focusing on the piezoelectric material, the piezoelectric disk of Figure 8.5 is assumed as a graded piezoelectric disk [30].

For studying this kind of graded piezoelectric transducers, which are based on the acoustic transmission line theory, and referred to as *functionally graded piezoelectric ultrasonic transducers* – FGPUTs here, a simple and nonexpensive computational cost approach is used [39]. According to acoustic transmission line theory, an FGPUT can be represented as a three-port system (Figure 8.7): one electric and two mechanical ports. The acoustic interaction of the graded piezoelectric disk with the propagating medium and the backing layer (both media are considered to be semi-infinite) is represented by mechanical ports. The electric port represents the electric interaction between the graded piezoelectric disk and the electric excitation circuit. This electrical circuit has a signal generator (E_g) and an internal resistance (R). Nevertheless, to complete the analytical model, an intermediate system must be assumed: the matching layer. The iteration between matching layer and graded piezoelectric disk is modeled as a system with two mechanical ports. In acoustic transmission line theory, the thickness of electrodes is supposed to be sufficiently small compared to acoustic wavelength involved, so

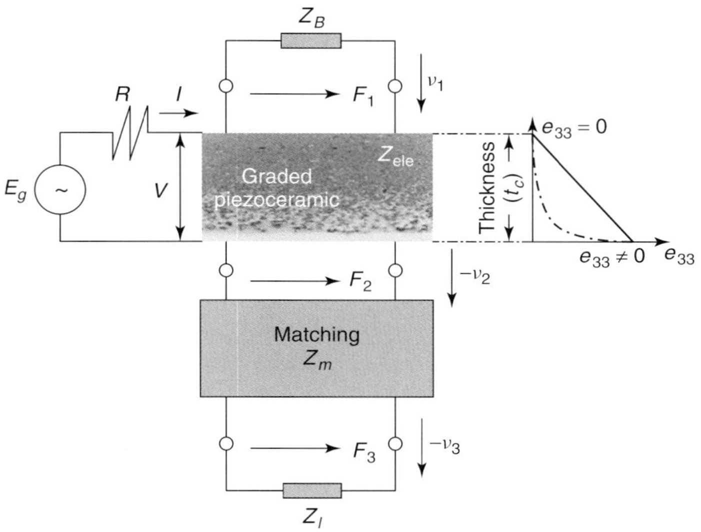

Figure 8.7 Sketch of an FGPUT, where Z_B, Z_{ele}, Z_m, and Z_l represent the electrical impedance of backing layer, graded piezoceramic, matching layer, and load, respectively.

the perturbation of the pressure distribution on load-medium, caused by these electrodes, can be neglected.

The goal of the FGPUT modeling, based on the acoustic transmission line theory, is to find a relationship between the electric current (I) and the electric potential (V) in the electric circuit with the force (F_3) and the particle velocity (v_3) radiated into the propagation media (Figure 8.7). This relationship is expressed in matrix form as [30]

$$\begin{pmatrix} V \\ I \end{pmatrix} = \begin{pmatrix} T_{11} & T_{12} \\ T_{21} & T_{22} \end{pmatrix} \begin{pmatrix} F_3 \\ v_3 \end{pmatrix} = [A][M] \begin{pmatrix} F_3 \\ v_3 \end{pmatrix} \tag{8.31}$$

where

$$\begin{pmatrix} V \\ I \end{pmatrix} = \begin{pmatrix} A_{11} & A_{12} \\ A_{21} & A_{22} \end{pmatrix} \begin{pmatrix} F_2 \\ v_2 \end{pmatrix} = [A] \begin{pmatrix} F_2 \\ v_2 \end{pmatrix}$$

$$\text{and} \quad \begin{pmatrix} F_2 \\ v_2 \end{pmatrix} = \begin{pmatrix} M_{11} & M_{12} \\ M_{21} & M_{22} \end{pmatrix} \begin{pmatrix} F_3 \\ v_3 \end{pmatrix} = [M] \begin{pmatrix} F_3 \\ v_3 \end{pmatrix} \tag{8.32}$$

Matrices **A** and **M**, in Eqs (8.32) and (8.33), essentially depend on gradation function assumed in the graded piezoelectric disk. Details of mathematical procedure for finding matrices **A** and **B** can be found in Ref. [30]. Two gradation functions are studied by Rubio *et al.* [30]: linear and exponential functions, where the gradation is considered along the thickness direction for the piezoelectric property e_{33} (Figure 8.7). In both cases, piezoelectric properties are graded from a nonpiezoelectric material, on the top surface, to a piezoelectric one, on the bottom surface (or from $e_{33} = 0$ to $e_{33} \neq 0$).

To complete the FGPUT modeling and based on Eqs (8.32) and (8.33), one can find the expressions, in the frequency domain, of the transmission transfer function (TTF) and the input electrical impedance of the graded piezoelectric disk (Z_{ele}). The TTF is a relationship between the mechanical force on load-medium (F_3) and the electric potential of the voltage generator (E_g), and Z_{ele} is a relationship between the electric current (I) and the electric potential (V) in the electric circuit. Hence, terms TTF and Z_{ele} can be written as follows:

$$\text{TTF} = \frac{F_3}{E_g} = \frac{Z_l}{T_{11}Z_l + T_{12} + R\,(T_{21}Z_l + T_{22})} \tag{8.33a}$$

$$Z_{\text{ele}} = \frac{V}{I} = \frac{A_{12}}{A_{22}} \tag{8.33b}$$

Overall, on the basis of the acoustic transmission line theory and considering FGPUTs, one can explore the FGM concept in medical imaging applications. Thus, it is assumed that an FGPUT, with a thickness of the piezoelectric disk equal to 1 mm, as shown in Figure 8.5, radiates a ultrasonic wave inside a water medium[3)] when the piezoelectric disk is excited with a half-sine electrical wave, whose fundamental frequency is $f_0 = 2.3$ MHz.

3) Water has acoustics impedance close to human tissue one [40].

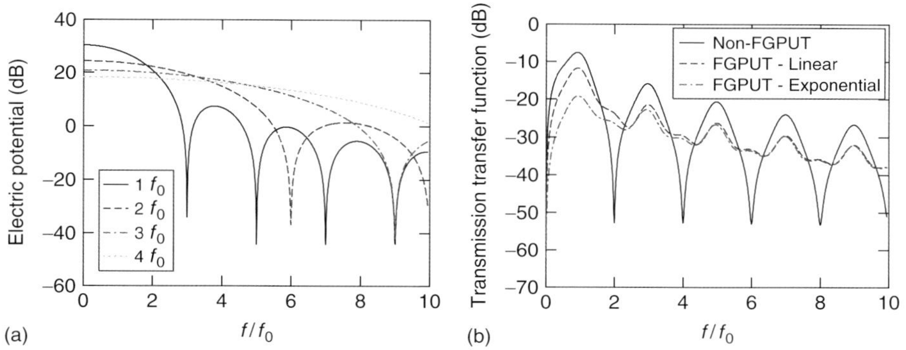

Figure 8.8 (a) Spectrum of several electrical input excitations (each with different fundamental frequency) and (b) normalized-frequency transmission transfer function (TTF) for both non-FGPUT and FGPUT with linear and exponential gradation functions [30].

Figure 8.8a shows the frequency spectra of an input signal at $f_0 = 2.3$ MHz; additionally, Figure 8.8a also shows frequency spectra of other input signals, however, with fundamental frequencies equal to $2f_0$, $3f_0$, and $4f_0$. From Figure 8.8a, it is clear that when the fundamental frequency is increased, its spectrum exhibits higher broadband; however, with less amplitude. On the other hand, Figure 8.8b presents the frequency spectra of the TTF, which is calculated by using Eq. (8.33a). For the non-FGPUT, it is observed that its frequency spectrum "falls to zero" in even-order modes ($f/f_0 = 2, 4, 6, \ldots$), while for the FGPUT (considering linear and exponential gradations), its frequency spectra do not fall to zero either for even or for odd order modes. For this reason, the bandwidth of an FGPUT is only limited by the bandwidth of the input excitation; on the other hand, the bandwidth of a non-FGPUT is limited by both input excitation and TTF bandwidths. However, the FGPUT represents a system with less gain in relation to non-FGPUT one. As a result, the FGPUT is a transducer with less output power (or less power delivered to fluid), because gradation functions depict an FGPUT with less "regions" of high piezoelectric properties than the non-FGPUT. On the other hand, the output signal (Figure 8.9), which is the dot product between TTF and input signal spectra, clearly highlights the incremented broadband, which is achieved by using the FGM concept, with both linear and exponential gradation functions.

The larger bandwidth of the FGPUT is produced because the acoustic pulse is generated mainly from the surface with high piezoelectric properties, while the opposite surface generates small vibration. In other words, as explained in Ref. [4], the induced piezoelectric stress $T_3 = -e_{33}E_3$ is higher on the surface with $e_{33} \neq 0$ than on the surface with $e_{33} = 0$. Thus, the volumetric force $F_\nu = \partial T_3/\partial z$ (spatial derivative of the induced stress), which is responsible for acoustic wave generation, is equal to zero on the surface without piezoelectric properties, and it has its maximum value on the opposite surface. For this reason, a single ultrasonic pulse is generated by an impulse excitation, which exhibits higher bandwidth.

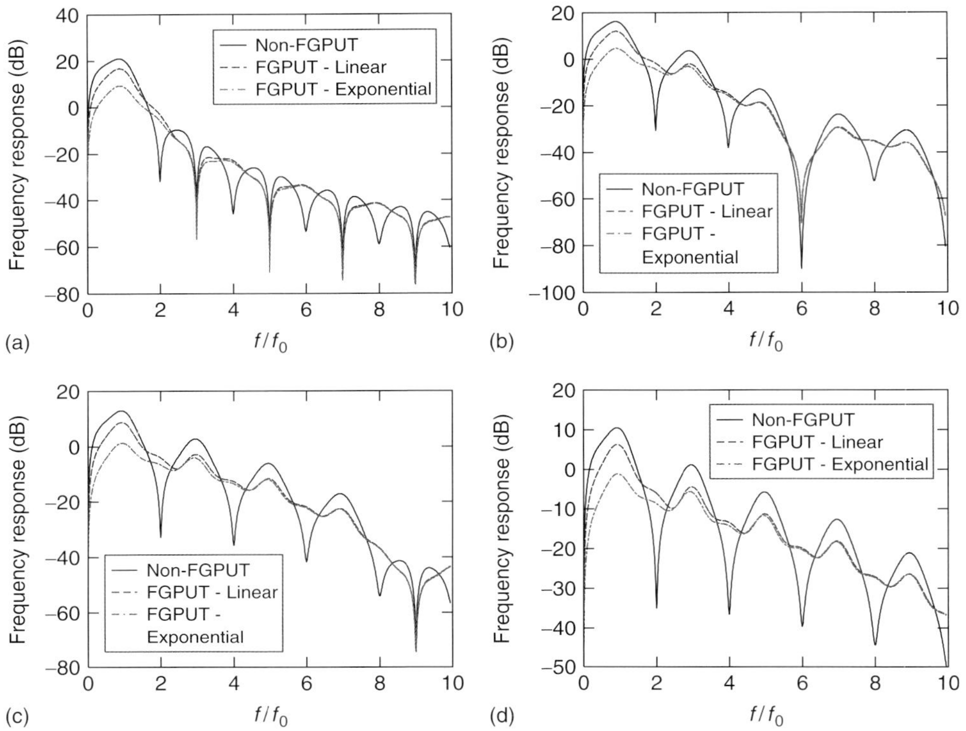

Figure 8.9 Spectrum of output signals for non-FGPUT and FGPUT systems considering several input excitations: (a) at $1f_0$; (b) at $2f_0$; (c) at $3f_0$; and (d) at $4f_0$.

Another interesting aspect is the analysis of the electrical impedance of an FGPUT with linear and exponential gradation functions, which can be computed by using Eq. (8.33b) [30]. Figure 8.10 shows normalized-frequency electrical impedance curves (focusing on thickness vibration modes), for both linear and exponential gradation functions of the piezoelectric property e_{33} (Figure 8.7). Specifically, Figure 8.10 shows the electrical impedance of only the graded piezoelectric disk; in other words, the FGPUT is simulated without backing and matching layers. It is observed that in FGPUT, electrical impedance curves appear with even and odd vibration modes; by contrast, in the non-FGPUT, curves appear with only odd vibration modes. This result indicates that it is possible to achieve, by using the FGM concept, more or less resonance modes in selective frequencies according to the gradation function used.

From Figures 8.8–8.10, it is observed that FGPUTs arise as a new and versatile alternative for applications in medical and nondestructive images. Specifically, in medical images, FGPUTs can obtain high resolution and deep penetration, when operated by using the harmonic imaging technique because they exhibit large bandwidth (Figures 8.8b and 8.9). Hence, it is possible to excite an object

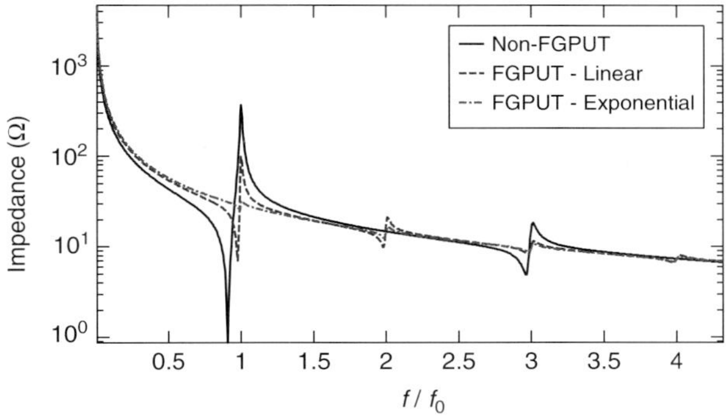

Figure 8.10 Electric impedance calculated by using Eq. (8.33b): for a non-FGPUT and an FGPUT considering linear and exponential gradations.

to be imaged, such as human tissues, by transmitting at a low (and therefore deeply penetrating) fundamental frequency (f_0) and receiving at a harmonic wave having a higher frequency (e.g., human tissues develop and return their own non-fundamental frequencies, for instance, $2f_0$), which can be used to form a high-resolution image of the object. In fact, in medical imagining application and by using FGPUTs, a wave having a frequency less than 2 MHz can be transmitted into the human body (e.g., human chest for cardiac imaging) and one or more harmonic waves having frequencies equal, and/or greater than 3 MHz can be received to form the image. By imaging in this manner (using FGPUTs in conjunction with the harmonic imaging technique), deep penetration can be achieved without a concomitant loss of image resolution.

8.5.2 Further Consideration of the Influence of Property Scale: Optimal Material Gradation Functions

As observed in the above section, the gradation function can influence the performance of graded piezoelectric transducers. This fact has been confirmed by several authors, for example, Almajid *et al.* [26], Taya *et al.* [28], and Rubio *et al.* [30]. This suggests using an optimization method for finding the optimal gradation function along a specific direction. Among the optimization methods, the topology optimization method (TOM) has shown to be a successfully technique for determining the best-property gradation function for a specific static or dynamic application [41–43].[4)]

For exemplifying the above idea, one can design an FGPUT, finding the optimal gradation function, in order to maximize a specific objective function. Thus, for

4) Details about topology optimization method can be found in the work by Bendsøe and Sigmund [44].

FGPUT design, the topology optimization problem can be formulated for finding the optimal gradation law that allows achieving multimodal or unimodal frequency response. These kinds of response define the type of generated acoustic wave, either short pulse (unimodal response) or continuous wave (multimodal response). Furthermore, the transducer is required to oscillate in a thickness extensional mode (pistonlike mode), aiming at acoustic wave generation applications.

For unimodal transducers, the electromechanical coupling of a desirable mode k must be maximized, and the electromechanical coupling of the adjacent modes (mode number $k + a_1$ with $a_1 = 1, 2, \ldots, A_1$, and/or $k - a_2$ with $a_2 = 1, 2, \ldots, A_2$) must be minimized. Additionally, the resonance frequencies of the modes $k_1 = k + a_1$ with $a_1 = 1, 2, \ldots, A_1$ must be maximized, and the resonance frequencies of the modes $k_2 = k - a_2$ with $a_2 = 1, 2, \ldots, A_2$ must be minimized. For multimodal transducers, electromechanical couplings of a mode set must be maximized, and their resonance frequencies must be approximated. Accordingly, for unimodal (F_1) and multimodal (F_2) transducers, the objective functions can be formulated as follows [45]:

$$F_1 = w_k\left(A_{r_k}\right) - \left[\frac{1}{\alpha_1}\left(\sum_{k_1=1}^{A_1} w_{1k_1}\left(A_{r_{k_1}}\right)^{n_1}\right)\right]^{1/n_1} - \left[\frac{1}{\alpha_2}\left(\sum_{k_2=1}^{A_2} w_{2k_2}\left(A_{r_{k_2}}\right)^{n_2}\right)\right]^{1/n_2} + \left[\frac{1}{\alpha_3}\left(\sum_{k_1=1}^{A_1} w_{3k_1}\left(\lambda_{r_{k_1}}\right)^{n_3}\right)\right]^{1/n_3} - \left[\frac{1}{\alpha_4}\left(\sum_{k_2=1}^{A_2} w_{4k_2}\left(\lambda_{r_{k_2}}\right)^{n_4}\right)\right]^{1/n_4} \tag{8.34}$$

with

$$\alpha_1 = \sum_{k_1=1}^{A_1} w_{1k_1};\ \alpha_2 = \sum_{k_2=1}^{A_2} w_{2k_2};\ \alpha_3 = \sum_{k_1=1}^{A_1} w_{3k_1};$$

$$\alpha_4 = \sum_{k_2=1}^{A_2} w_{4k_2};\ n_m = -1, -3, -5, -7\ldots;\ m = 1, 2, 3, 4$$

$$F_2 = \left[\frac{1}{\alpha_1}\left(\sum_{k=1}^{m} w_k\left(A_{r_k}\right)^{n_1}\right)\right]^{1/n_1} - \left[\frac{1}{\alpha_2}\sum_{k=1}^{m}\frac{1}{\lambda_{0_k}^2}\left(\lambda_{r_k}^2 - \lambda_{0_k}^2\right)\right]^{1/n_2} \tag{8.35}$$

with

$$\alpha_1 = \sum_{k=1}^{m} w_k;\quad \alpha_2 = \sum_{k=1}^{m}\frac{1}{\lambda_{0_k}^2};\quad \lambda_{r_k} = \omega_{r_k}^2;$$
$$n_1 = -1, -3, -5, -7\ldots;\quad n_2 = \pm 2, \pm 4, \pm 6, \pm 8\ldots$$

where, for unimodal transducers (Eq. (8.34)), terms A_{r_k}, $A_{r_{k_1}}$, and $A_{r_{k_2}}$ represent the electromechanical coupling (measured by the piezoelectric modal constant – PMC [46]) of the desirable mode, and left and right adjacent modes, respectively. Terms

w_k, $w_{i_{k_1}}$ $(i = 1, 3)$, and $w_{j_{k_2}}$ $(j = 2, 4)$ are the weight coefficients for each mode considered in the objective function F_1. Finally, terms $\lambda_{r_{k_1}}$ and $\lambda_{r_{k_2}}$ represent eigenvalues of the left and the right modes with relation to the desirable one (mode number k), and the term n is a given power. For multimodal transducers (Eq. (8.35)), the constant m is the number of modes considered; the terms λ_{r_k} and λ_{0_k} are the current and desirable (or user-defined) eigenvalues for mode k $(k = 1, 2, \ldots, m)$, respectively; and ω_{r_k} are the resonance frequencies for mode k $(k = 1, 2, \ldots, m)$.

The optimization problem is formulated as finding the material gradation of FGPUT, which maximizes the multiobjective function F_1 or F_2 subjected to a piezoelectric volume constraint. This constraint is implemented to control the amount of piezoelectric material into the two-dimensional design domain, Ω. The optimization problem is expressed as

$$\begin{aligned} &\underset{\rho_{\mathrm{TOM}}(x,y)}{\text{maximize}} && F_i \; i = 1 \text{ or } 2 \\ &\text{subjected to} && \int_{\Omega} \rho_{\mathrm{TOM}}(x, y)\mathrm{d}\Omega - \Omega_s \leq 0; \\ & && 0 \leq \rho_{\mathrm{TOM}}(x, y) \leq 1 \\ & && \text{equilibrium and constitutive equations} \end{aligned} \tag{8.36}$$

where $\rho_{\mathrm{TOM}}(x, y)$ is the design variable (pseudodensity) at Cartesian coordinates x and y. The term Ω_s describes the amount of piezoelectric material in the two-dimensional domain Ω.

The last requirement (mode shape tracking) is achieved by using the modal assurance criterion (MAC) [47], which is used to compare eigenmodes and to track the desirable eigenvalue and/or eigenvector along the iterative process of the TOM. Besides, to treat the gradation in FGPUT, material properties are continuously interpolated inside each finite element based on property values at each finite element node, as explained in Section 8.4.3. On the other hand, the continuum approximation of material distribution (CAMD) concept [48] is used to continuously represent the pseudodensity distribution. The CAMD considers that design variables inside each finite element are interpolated by using, for instance, the finite element shape functions N. Thus, the pseudodensity ρ^e_{TOM} at each GFE e can be expressed as

$$\rho^e_{\mathrm{TOM}}(x, y) = \sum_{i=1}^{n_d} \rho^n_{\mathrm{TOM}_i} N_i(x, y) \tag{8.37}$$

where $\rho^n_{\mathrm{TOM}_i}$ and N_i are, respectively, the nodal design variable and shape function for node $i (i = 1, \ldots, n_d)$, and n_d is the number of nodes at each finite element. This formulation allows a continuous distribution of material along the design domain instead of the traditional piecewise material distribution.

Additionally, to achieve an explicit gradient control, a projection technique can be implemented as explained in Ref. [49].

To illustrate the design of an FGPUT based on TOM, one can consider the design domain shown in Figure 8.11, for designing a unimodal and a multimodal FGPUT considering plane strain assumption. The design domain is specified as a

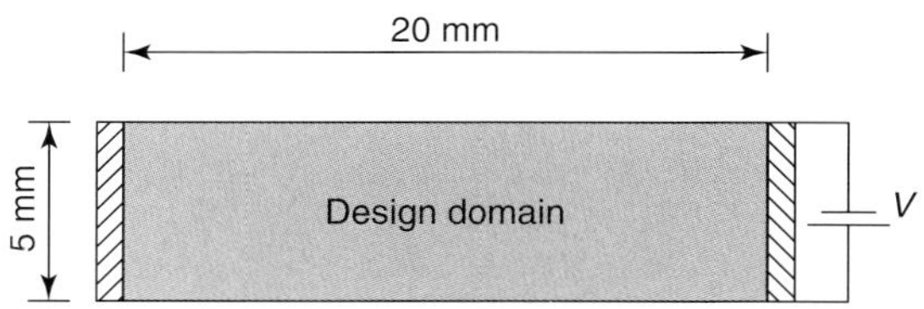

Figure 8.11 Design domain for FGPUT design.

20 mm × 5 mm rectangle with two fixed supports at the end of the left- and right-hand sides. The idea is to simultaneously distribute two types of materials into the design domain. The material *type 1* is represented by a PZT-5A piezoelectric ceramic and the material *type 2* is a PZT-5H. Initially, the design domain contains only PZT-5A material and a material gradation along thickness direction is assumed. In addition, a mesh of 50 × 30 finite elements is adopted.

Figure 8.12 shows the result when a unimodal FGPUTs is designed. It is observed that for the unimodal transducer, the optimal material gradation depicts an FGPUT with rich region of piezoelectric material PZT-5A in the middle and rich region of piezoelectric material PZT-5H on the top and bottom surfaces (Figure 8.12a).[5] The material gradation is found to allow the electromechanical coupling value (measured by the PMC [46]) of the pistonlike mode to increase by 59% while the PMC values of adjacent modes decrease approximately by 75% (Figure 8.12b).

The designed multimodal transducer is shown in Figure 8.13. The final material gradation represents an FGPUT with regions rich in piezoelectric properties PZT-5A around layers 10 and 23 and PZT-5H on the top and bottom surfaces. The optimal material gradation increases the PMC value of the pistonlike mode by 15%, while the PMC values of the left and right adjacent modes are increased by 15 and 181%, respectively. In both uni- and multimodal designs, the pistonlike mode is retained, which represents the mode with highest electromechanical coupling.

8.6
Influence of Microscale

The combination of a piezoelectric material (polymer or ceramic) with other materials (including air-filled voids) usually results in new composites, called *piezocomposites*, that offer substantial advantages over conventional piezoelectric

5) Observe that the optimization problem is treated as layerlike optimization problem; in other words, the design variables are assumed to be equal at each interface between finite elements. This approach allows manufacturing FGPUTs by sintering a layer-structured ceramic green body without using adhesive material.

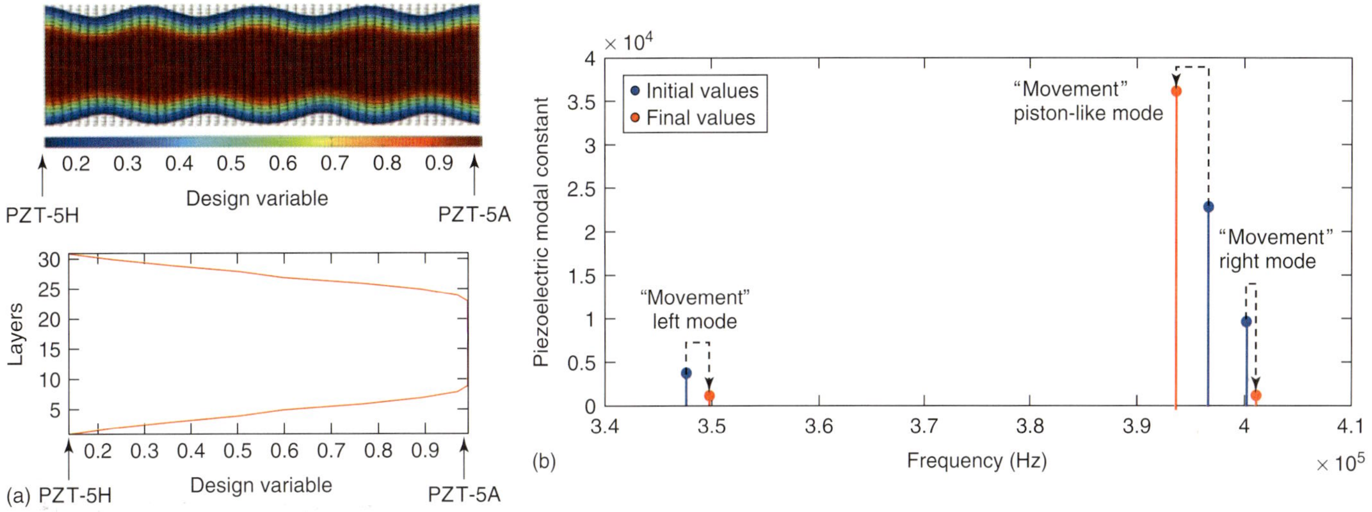

Figure 8.12 Design of a unimodal FGPUT: (a) pistonlike mode (dashed and solid lines, respectively, depict nondeformed and deformed structures) and final material distribution and (b) initial and final frequency response [45].

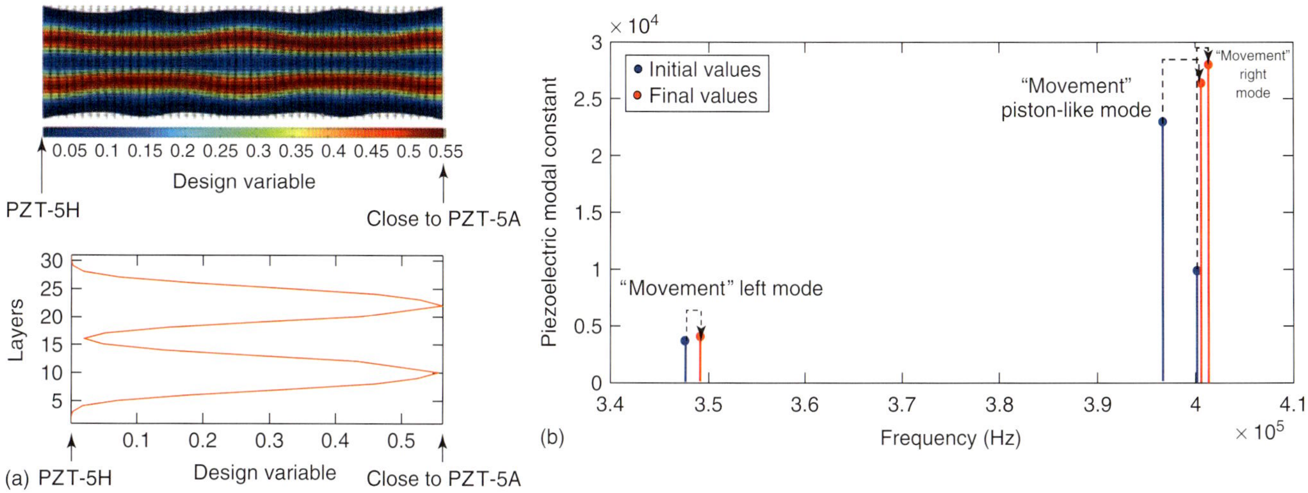

Figure 8.13 Design of a multimodal FGPUT: (a) pistonlike mode (dashed and solid lines, respectively, depict nondeformed and deformed structures) and final material distribution and (b) initial and final frequency response [45].

materials. The advantages are high electromechanical coupling, which measures energy conversion in the piezocomposite and therefore its sensitivity, and low acoustic impedance, which helps to transmit acoustic waver to media such as human body tissue or water [50–52].

A composite can be modeled by considering its unit cell with infinitesimal dimensions, which is the smallest structure that is periodic in the composite. By changing the volume fraction of the constituents, the shape of the inclusions, or even the topology of the unit cell, we can obtain different effective properties for the composite material [9].

In composite applications, we assume that the excitation wavelengths are so large that the detailed structure of the unit cell does not matter, and the material may be considered as a new homogeneous medium with new effective material properties. Then, the excitation (acoustic, for example) will average out over the fine-scale variations of the composite medium, in the same way as averaging occurs in the micron-sized grain structure in a conventional pure ceramic. When wavelengths are not large enough relative to the size of the unit cell, the composite will present a dispersive behavior with scattering occurring inside the unit cells, making its behavior extremely difficult to model. Homogeneous behavior can be assured by reducing the size of the unit cell relative to the excitation wavelength. However, it is not always possible to guarantee that the size of the microstructure (or unit cell) is smaller than the wavelength.

The homogenization method allows the calculation of effective properties of a complex periodic composite material from its unit cell or microstructure topology, that is, types of constituents and their distribution in the unit cell [53, 54]. This is a general method for calculating effective properties and has no limitations regarding volume fraction or shape of the composite constituents. The main assumptions are that the unit cell is periodic and that the scale of the composite part is much larger than the microstructure dimensions [55–57]. There are other methods that allow calculation of effective properties of a composite material. However, the main advantage of the homogenization technique is that it needs only the information about the unit cell that can have any complex shape. A brief introduction to this method is given in Section 8.6.2.

Assuming that the composite is a homogeneous medium, its behavior can be characterized by Eq. (8.3), by substituting all properties by the effective properties of the composite (or homogenized properties) into these equations [58]. These effective properties can be obtained using the homogenization method presented in Section 8.6.2. Therefore, the constitutive equations of the composite material considering homogenized properties become

$$\begin{cases} \langle \mathbf{T} \rangle = \mathbf{c}_H^E \langle \mathbf{S} \rangle - \mathbf{e}_H \langle \mathbf{E} \rangle \\ \langle \mathbf{D} \rangle = \mathbf{e}_H^t \langle \mathbf{S} \rangle + \boldsymbol{\varepsilon}_H^S \langle \mathbf{E} \rangle \end{cases} \tag{8.38}$$

or

$$\begin{cases} \langle \mathbf{S} \rangle = \mathbf{s}_H^E \langle \mathbf{T} \rangle + \mathbf{d}_H \langle \mathbf{E} \rangle \\ \langle \mathbf{D} \rangle = \mathbf{d}_H^t \langle \mathbf{T} \rangle + \boldsymbol{\varepsilon}_H^T \langle \mathbf{E} \rangle \end{cases} \tag{8.39}$$

where

$$\langle \cdots \rangle = \frac{1}{|V|} \int_V dV \tag{8.40}$$

and the subscript "H" refers to the homogenized properties. $\mathbf{s}_H^E$ is the homogenized compliance tensor under short-circuit conditions, $\boldsymbol{\varepsilon}_H^T$ is the homogenized clamped body dielectric tensor, and $\mathbf{d}_H$ is the homogenized piezoelectric stress tensor. The relations among the properties in Eqs (8.40) and (8.41) are [13]

$$\mathbf{s}^E = \left(\mathbf{c}^E\right)^{-1} \qquad \boldsymbol{\varepsilon}^T = \boldsymbol{\varepsilon}^S + \mathbf{d}^t \left(\mathbf{s}^E\right)^{-1} \mathbf{d} \qquad \mathbf{d} = \left(\mathbf{s}^E\right) \mathbf{e} \tag{8.41}$$

In the following sections, the subscript "H" is omitted for the homogenized properties for the sake of brevity. As a convention, the polarization axis of the piezoelectric material is considered in the third (or z) direction.

8.6.1
Performance Characteristics of Piezocomposite Materials

The main applications of piezocomposites are the generation and detection of acoustic waves. It can be classified as low frequency (hydrostatic operation mode, such as some hydrophones and naval sonars) and high frequency (ultrasonic transducers for imaging). In low-frequency applications, the operation of the device is quasistatic since the operational frequency of the device is generally smaller than the first resonance frequency of the device.

In piezocomposite design, there are several important parameters that directly influence its performance. An ultrasonic transducer, for example, requires a combination of properties such as large piezoelectric coefficient (d_h or g_h, explained below), low density, and mechanical flexibility [59]. However, these properties usually lead to *trade-offs*. To make a flexible ultrasonic transducer, it would be desirable to use the large piezoelectric effects in a poled piezoelectric ceramic; however, ceramics are brittle and stiff, lacking the required flexibility, while polymers having the desired mechanical properties are not piezoelectrics. This problem can be simplified dealing with *figure of merit*, which combines the most sensitive parameters in a form allowing simple comparison of property coefficients. So, the main problem in piezocomposite design is to combine the components in such a manner as to achieve the desired features of each component and also try to maximize the figure of merit [59]. Besides, Newmham *et al.* [59] also studied the connectivity of the individual phases.

The figures of merit that describe the performance of the piezocomposites explained below are obtained by considering only the constitutive properties (neglecting the effects of inertia) as described in Refs [58, 60].

8.6.1.1 Low-Frequency Applications

Consider an orthotropic composite under hydrostatic pressure P as shown in Figure 8.14.

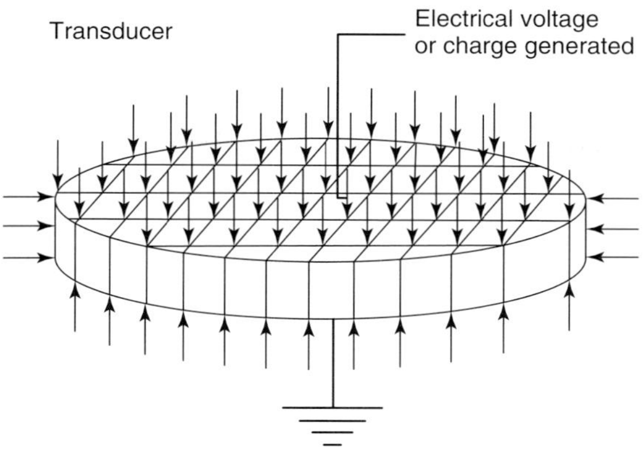

Figure 8.14 Piezocomposite transducer under hydrostatic pressure.

The composite response can be measured by three different quantities [61]:

- **Hydrostatic coupling coefficient ($\boldsymbol{d}_h$):**

$$d_h = \frac{\langle \mathbf{D}_3 \rangle}{P} = d_{33} + d_{23} + d_{13} \tag{8.42}$$

- **Figure of merit ($\boldsymbol{d}_h \boldsymbol{g}_h$):**

$$g_h = \frac{\langle E_3 \rangle}{P} = \frac{d_h}{\varepsilon_{33}^T} \Rightarrow d_h g_h = \frac{d_h^2}{\varepsilon_{33}^T} \tag{8.43}$$

- **Hydrostatic electromechanical coupling factor ($\boldsymbol{k}_h$):**

$$k_h = \sqrt{\frac{d_h^2}{\varepsilon_{33}^T s_h^E}} \tag{8.44}$$

where $s_h = (\langle \varepsilon_1 \rangle + \langle \varepsilon_2 \rangle + \langle \varepsilon_3 \rangle)/P$ is the dilatational compliance. For an orthotropic material, $s_h^E = s_{11}^E + s_{22}^E + s_{33}^E + s_{12}^E + s_{13}^E + s_{23}^E$, and the coefficients s_{kl}^E are those defined in Eq. (8.39).

The quantities d_h and g_h measure the response of the material in terms of electrical charge and electrical voltage generated, respectively, when subjected to a hydrostatic pressure field considering a null electric field ($\langle E_3 \rangle = 0$, for short-circuit conditions) and null electric displacement ($\langle D_3 \rangle = 0$, for open circuit conditions), respectively. $d_h g_h$ is the product of d_h and g_h. The coefficient k_h measures the overall acoustic/electric power conversion. The expressions for d_h and s_h^E can be obtained by substituting the hydrostatic pressure into Eq. (8.39) and considering a null electric field ($\langle E_3 \rangle = 0$). The expression for g_h can be obtained in the same way, but considering null electric displacement ($\langle D_3 \rangle = 0$) in Eq. (8.39).

These quantities can be written in terms of the properties described in Eq. (8.38) by using Eq. (8.41). The definitions presented above consider an orthotropic piezocomposite material. If a transversely isotropic composite in the 12 (or xy) plane is considered, then $s_{13}^E = s_{23}^E$, $s_{11}^E = s_{22}^E$, and $d_{13} = d_{23}$.

8.6.1.2 High-Frequency Applications

In ultrasonic applications, thin plates of the piezocomposite are excited near their thickness-mode resonance. In this case, the quantity that describes the performance of the ultrasonic transducer is given by Smith and Auld [60]:

- **Electromechanical coupling factor (k_t)**

$$k_t = \sqrt{\frac{e_{33}^2}{c_{33}^D \varepsilon_{33}^S}} \tag{8.45}$$

where the properties are the same as defined in Eq. (8.38) and $c_{33}^D = c_{33}^E + (e_{33})^2/\varepsilon_{33}^S$.

8.6.2 Homogenization Method

The homogenization method was initially developed to solve partial differential equations whose parameters vary rapidly in space. In engineering field, this method has been used to obtain effective properties of composite materials [62], allowing us to save the effort. For example, imagine a perforated beam as illustrated in Figure 8.15a. If we were to build an FEM model of the beam by considering all the holes, it would be very difficult to model and the computational cost would be prohibitive. However, it can be understood as a continuous beam (no holes) made of a material with properties equal to the effective properties of a "composite" material whose unit cell consists of a square with a circular hole inside, that is, a homogenized material. Therefore, if we have the effective properties of this composite material, the beam can be modeled as a homogeneous medium by building a simple FEM model with corresponding homogenized properties.

The same concept can be applied to model a wall made of bricks, for example, as illustrated in Figure 8.15b; however, it is important to mention that the

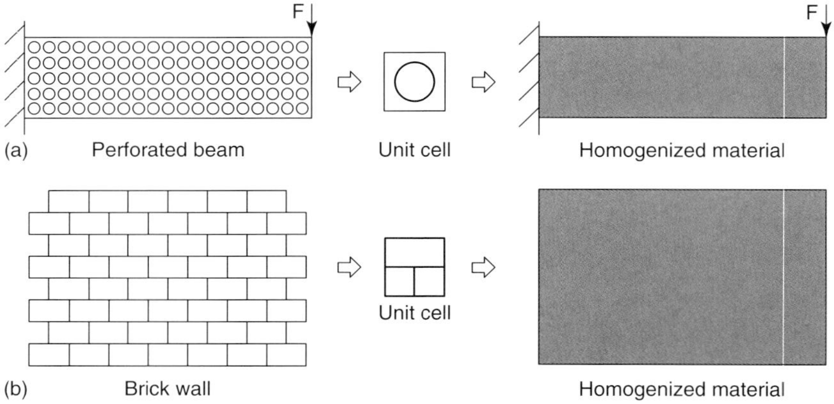

Figure 8.15 Homogenization concept: (a) perforated beam and (b) brick wall.

size of phenomenon we are interested in analyzing will determine whether homogenization concept can be applied. The wall can be understood as a composite material, whose unit cell is described in Figure 8.15b. In another case, suppose that we want to model a bullet hitting the wall. If the bullet size is much larger than the unit cell size, homogenization can be applied and the entire wall can be modeled as a homogeneous material with corresponding effective properties. However, if the bullet size is of the same order of wall unit cell size (i.e., the brick), then homogenization cannot be applied and a detailed FEM model of the wall must be built taking into account the unit cell details.

Homogenization equations are obtained by first expanding displacement $\mathbf{u}$ in zeroth-order and first-order terms. The zeroth-order terms represent the "average" values of displacement over the piezocomposite domain scale $(\mathbf{x})$, while the first-order terms represent the variation of displacement in the unit cell domain scale $(\mathbf{y})$, that is,

$$\mathbf{u}^{\varepsilon} = \mathbf{u}_0(\mathbf{x}) + \varepsilon \mathbf{u}_1(\mathbf{x}, \mathbf{y}) \tag{8.46}$$

where ε is a small number. This expression is substituted in the energy formulation for the medium and, by using variational calculus, the so-called homogenization equations related only to the $\mathbf{y}$ scale are extracted. Essentially, the meaning of homogenization equations consists in applying different load cases to the unit cell to calculate its response to these load cases. On the basis of these responses, the composite effective properties are obtained. Since the unit cell may have a complex shape, these equations are solved by using FEM.

The basic homogenization equations applied to calculate the effective properties of elastic materials are presented below:

$$\frac{1}{|Y|}\int_Y \left[c_{ijkl}\left(\mathbf{x}, \mathbf{y}\right) \left(\delta_{im}\delta_{jn} + \frac{\partial \chi_i^{(mn)}}{\partial y_j} \right) \right] S_{kl}\left(\mathbf{v}\right) dY = 0 \qquad \forall \mathbf{v} \in H_{\text{per}}\left(Y, R^3\right) \tag{8.47}$$

$$H_{\text{per}}\left(Y, R^3\right) = \{\mathbf{v} = (v_i)\,|v_i \in H_1\left(Y\right),\ i = 1, 2, 3\}$$

$$H_{\text{per}}\left(Y\right) = \left\{\mathbf{v} \in H_1\left(Y\right)|\mathbf{v} \text{ takes equal values on opposite sides of Y}\right\}$$

This homogenization formulation has no limitations regarding volume fraction or the shape of the composite constituents, and is based upon assumptions of periodicity of the microstructure and the separation of the microstructure scale from the component scale through an asymptotic expansion.

Now consider a composite material under dynamic excitation (electrical or mechanical). If the operational wavelength is larger than the unit cell dimensions, it seems natural that homogenization equations can be applied. This situation is called a *static case*. If operational wavelength is smaller than the unit cell dimensions, then unit cell scale will affect the calculation of effective properties, that is, the effective properties will have a dispersive behavior as before. Essentially, what happens is that if the wavelength is smaller than unit cell dimensions, there will be wave reflections inside of the unit cell and this effect must be taken into account in the homogenization equations. This situation is called a *dynamic case* and homogenization equations must be developed again, originating the so-called homogenization equations for the dynamic case.

For a piezoelectric medium, the homogenization theory for piezoelectricity considering the static case (where the operational wavelength is much larger than the unit cell dimensions) was developed by Telega [63]. Galka *et al.* [64] present the homogenization equations and effective properties for thermopiezoelectricity considering the static case. For dynamic applications (wavelength is of the same size as, or smaller than, the unit cell dimensions), Turbé and Maugin [65] developed a homogenization formulation to obtain the dynamical effective properties of the piezoelectric medium. In the limit of the static (and low-frequency) case, they recovered the expressions derived by Telega [63]. Finally, Otero *et al.* [66] developed general homogenized equations and effective properties for (heterogeneous and periodic) piezoelectric medium by considering terms of infinite order in the homogenization asymptotic expansion.

The homogenization equations for piezoelectric medium considering the static case are [63]

$$\begin{aligned}
&\frac{1}{|Y|}\int_Y \left[c^E_{ijkl}(\mathbf{x},\mathbf{y})\left(\delta_{im}\delta_{jn} + \frac{\partial \chi_i^{(mn)}}{\partial y_j}\right) + e_{ikl}(\mathbf{x},\mathbf{y})\frac{\partial \psi^{(mn)}}{\partial y_i}\right] S_{kl}(\mathbf{v})\, dY = 0 \\
&\forall \mathbf{v} \in H_{\mathrm{per}}\left(Y, R^3\right) \\
&\frac{1}{|Y|}\int_Y \left[e_{ikl}(\mathbf{x},\mathbf{y})\left(\delta_{im}\delta_{jn} + \frac{\partial \chi_i^{(mn)}}{\partial y_j}\right) - \varepsilon^S_{ik}(\mathbf{x},\mathbf{y})\frac{\partial \psi^{(mn)}}{\partial y_i}\right] \frac{\partial \varphi}{\partial y_k} dY = 0 \\
&\forall \varphi \in H_{\mathrm{per}}(Y)
\end{aligned} \tag{8.48}$$

and

$$\begin{aligned}
&\frac{1}{|Y|}\int_Y \left[c^E_{klij}(\mathbf{x},\mathbf{y})\frac{\partial \Phi_k^{(m)}}{\partial y_l} + e_{kij}(\mathbf{x},\mathbf{y})\left(\delta_{mk} + \frac{\partial R^{(m)}}{\partial y_k}\right)\right] S_{ij}(\mathbf{v})\, dY = 0 \\
&\forall \mathbf{v} \in H_{\mathrm{per}}\left(Y, R^3\right) \\
&\frac{1}{|Y|}\int_Y \left[e_{kij}(\mathbf{x},\mathbf{y})\frac{\partial \Phi_i^{(m)}}{\partial y_j} - \varepsilon^S_{ik}(\mathbf{x},\mathbf{y})\left(\delta_{mi} + \frac{\partial R^{(m)}}{\partial y_i}\right)\right] \frac{\partial \varphi}{\partial y_k} dY = 0 \\
&\forall \varphi \in H_{\mathrm{per}}(Y)
\end{aligned} \tag{8.49}$$

These equations are equivalent to Eq. (8.47) for elastic medium. They are obtained by expanding piezocomposite displacement $\mathbf{u}$ and electrical potential ϕ in zeroth-order and first-order terms. The zeroth-order terms represent the "average" values of these quantities over the piezocomposite domain scale $(\mathbf{x})$, while first-order terms represent the variation of these quantities in the unit cell domain scale $(\mathbf{y})$, that is,

$$\begin{aligned}
\mathbf{u}^\varepsilon &= \mathbf{u}_0(\mathbf{x}) + \varepsilon \mathbf{u}_1(\mathbf{x},\mathbf{y}) \\
\phi^\varepsilon &= \phi_0(\mathbf{x}) + \varepsilon \phi_1(\mathbf{x},\mathbf{y})
\end{aligned} \tag{8.50}$$

The characteristic functions χ_i, ψ, Φ_i, R represent the displacement and electrical response of the unit cell to the applied load cases (Figure 8.17).

By using FEM formulation to solve Eqs (8.48) and (8.49), the following matrix system is obtained [67]:

$$\begin{bmatrix} \mathbf{K}_{\mathbf{uu}} & \mathbf{K}_{\mathbf{u}\phi} \\ \mathbf{K}_{\mathbf{u}\phi}^{t} & -\mathbf{K}_{\phi\phi} \end{bmatrix} \begin{Bmatrix} \hat{\chi}^{(mn)} \\ \hat{\psi}^{(mn)} \end{Bmatrix} = \begin{Bmatrix} \mathbf{F}^{(mn)} \\ \mathbf{Q}^{(mn)} \end{Bmatrix}$$

$$\begin{bmatrix} \mathbf{K}_{\mathbf{uu}} & \mathbf{K}_{\mathbf{u}\phi} \\ \mathbf{K}_{\mathbf{u}\phi}^{t} & -\mathbf{K}_{\phi\phi} \end{bmatrix} \begin{Bmatrix} \hat{\Phi}^{(mn)} \\ \hat{R}^{(mn)} \end{Bmatrix} = \begin{Bmatrix} \mathbf{F}^{(mn)} \\ \mathbf{Q}^{(mn)} \end{Bmatrix} \tag{8.51}$$

where $\hat{\chi}$, $\hat{\psi}$, $\hat{\Phi}$, $\hat{R}$ are the corresponding nodal values of the characteristics functions χ_i, ψ, Φ_i, R respectively, and

$$F_{iI}^{e(mn)} = -\int_{\Omega^e} c_{ijmn}^{E} \frac{\partial N_I}{\partial y_j} d\Omega^e \quad Q_I^{e(mn)} = -\int_{\Omega^e} e_{kmn} \frac{\partial N_I}{\partial y_k} d\Omega^e$$

$$F_{iI}^{e(m)} = -\int_{\Omega^e} e_{mij} \frac{\partial N_I}{\partial y_j} d\Omega^e \quad Q_I^{e(m)} = -\int_{\Omega^e} \varepsilon_{mj}^{S} \frac{\partial N_I}{\partial y_j} d\Omega^e \tag{8.52}$$

The other terms are defined in Section 8.4.2. By analyzing Eq. (8.52) we conclude that for the three-dimensional problem, there are nine load cases to be solved independently. Six of them come from Eq. (8.48), where the indices m and n can be 1, 2, or 3, resulting in the following mn combinations: 11, 22, 33, 12 or 21, 23 or 32, and 13 or 31. The remaining three load cases come from Eq. (8.49) where the index m can be 1, 2, and 3. For example, in the two-dimensional problem, there are five load cases to be solved independently (Figure 8.17). Three of them come from Eq. (8.48), where the indices m and n can be 1 or 3, resulting in the combinations 11, 33, and 13 or 31, for mn. The other two load cases come from Eq. (8.49) where the index m can be 1 or 3. All load cases must be solved by enforcing periodic boundary conditions in the unit cell for the displacements and electrical potentials. The displacements and electrical potential of some point of the cell must be prescribed to overcome the nonunique solution of the problem; otherwise, the problem will be ill posed [68]. The choice of the point of the prescribed values does not affect the homogenized coefficients since only derivatives of the characteristic functions are used for their computation. The numerical solution of matrix system (Eq. (8.52)) has already been discussed in Section 8.4.2.

The applied load cases for the bidimensional problem considering elastic material (Eq. (8.47)) and piezoelectric material (Eqs (8.48) and (8.49)) are described in Figures 8.16 and 8.17, respectively.

After solving for the characteristic displacements and electrical potentials, the effective properties are computed by using Eq. (8.53):

$$c_{pqrs}^{H}(\mathbf{x}) = \frac{1}{|Y|}\int_Y \left[c_{pqrs}^{E}(\mathbf{x},\mathbf{y}) + c_{pqkl}^{E}(\mathbf{x},\mathbf{y}) \frac{\partial \chi_k^{(m)}}{\partial y_l} + e_{kpq}(\mathbf{x},\mathbf{y}) \frac{\partial \psi^{(rs)}}{\partial y_k} \right] dY$$

$$e_{prs}^{H}(\mathbf{x}) = \frac{1}{|Y|}\int_Y \left[e_{prs}(\mathbf{x},\mathbf{y}) + e_{pij}(\mathbf{x},\mathbf{y}) \frac{\partial \chi_i^{(rs)}}{\partial y_j} - \varepsilon_{pk}^{S}(\mathbf{x},\mathbf{y}) \frac{\partial \psi^{(rs)}}{\partial y_k} \right] dY$$

$$\varepsilon_{pq}^{H}(\mathbf{x}) = \frac{1}{|Y|}\int_Y \left[\varepsilon_{pq}^{S}(\mathbf{x},\mathbf{y}) + \varepsilon_{pj}^{S}(\mathbf{x},\mathbf{y}) \frac{\partial R^{(q)}}{\partial y_j} - e_{pij}(\mathbf{x},\mathbf{y}) \frac{\partial \Phi_i^{(q)}}{\partial y_j} \right] dY \tag{8.53}$$

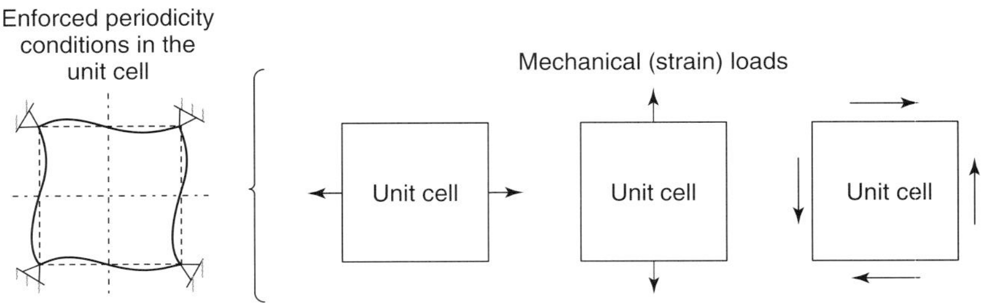

Figure 8.16 Load cases for homogenization of elastic materials.

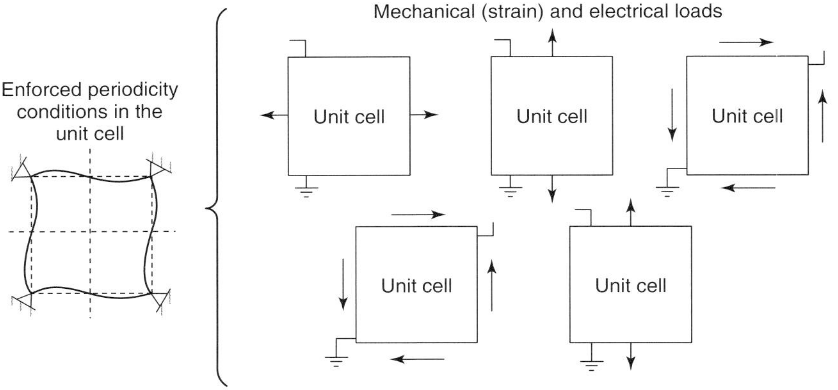

Figure 8.17 Load cases for homogenization of piezoelectric materials.

The concept of the continuum distribution of design variable based on the CAMD method discussed in Section 8.5.2 is also considered here.

8.6.3
Examples

Of the currently available configurations, the 2–2 piezocomposite has been the focus of most studies, which consists of alternating layers of piezoceramic PZT and polymer as shown in Figure 8.18. To illustrate the FGM concept for piezoelectricity, the results of calculated effective properties for 2–2 piezocomposite are presented.

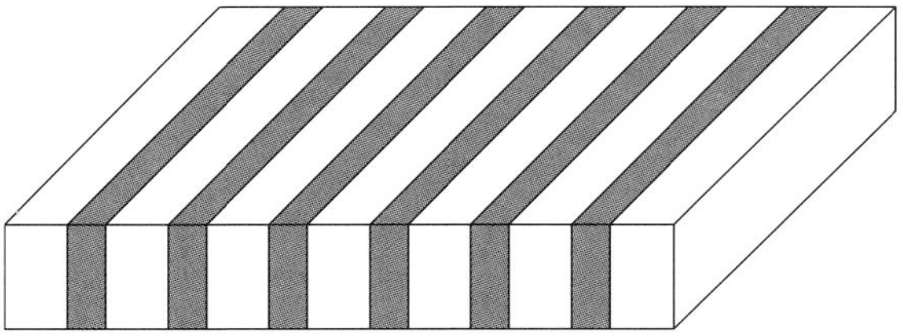

Figure 8.18 2–2 piezocomposite.

Table 8.1 Material properties.

Properties		PZT-5A	Epoxy polymer
Dielectric properties (F m^{-1})	ε_0	8.85×10^{-12}	8.85×10^{-12}
	ε_{11}^S	$916 \times \varepsilon_0$	$3.6 \times \varepsilon_0$
	ε_{33}^S	$830 \times \varepsilon_0$	$3.6 \times \varepsilon_0$
Piezoelectric properties (C m^{-2})	e_{31}	−5.4	0.0
	e_{33}	15.8	0.0
	e_{15}	12.3	0.0
Elastic properties (N m^{-2})	C_{11}^E	12.1×10^{10}	9.34×10^9
	C_{12}^E	7.54×10^{10}	9.34×10^9
	C_{13}^E	7.52×10^{10}	9.34×10^9
	C_{33}^E	11.1×10^{10}	9.34×10^9
	C_{44}^E	2.11×10^{10}	9.34×10^9
	C_{66}^E	2.28×10^{10}	9.34×10^9
Density (kg m^{-3})		7500	1340

Here, "2–2" designates the connectivity of the piezocomposite material; however, many other connectivities, such as 3–1 and 1–1, are also possible [59]. The example considers a bidimensional model (plane strain) of a 2–2 piezocomposite made of PZT-5A/ Epoxy (Table 8.1). The "volume fraction" (vol%) was set to 20% of PZT-5A, located in a vertical line in the middle, and 80% of epoxy polymer, distributed in the rest of the unit cell. Three cases are considered: a non-FGM unit cell and two FGM unit cells, with linear and sinusoidal gradations. These three cases are compared with the full unit cell of PZT-5A, in order to quantify the effects of the FGM on the performance characteristics.

Figures 8.19–8.21 show the graphics of the PZT-5A distribution in the x direction, where 1 refers to pure PZT-5A and 0 refers to pure epoxy polymer. The images in the (b) show the distributed material in the unit cell, where the PZT-5A is

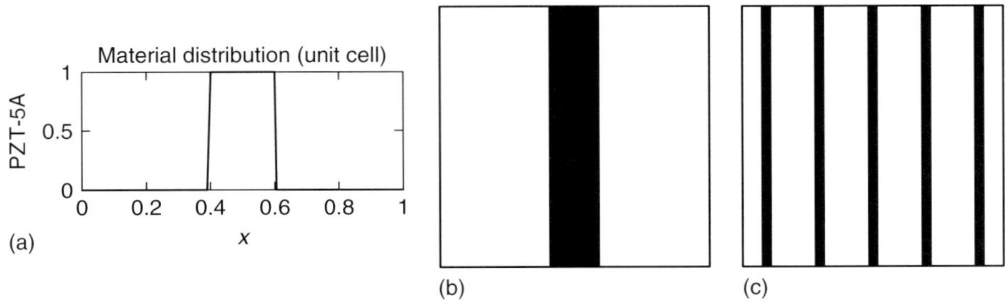

Figure 8.19 Non-FGM 2–2 piezocomposite. (a) Material distribution graphic of PZT-5A; (b) unit cell; and (c) periodic array.

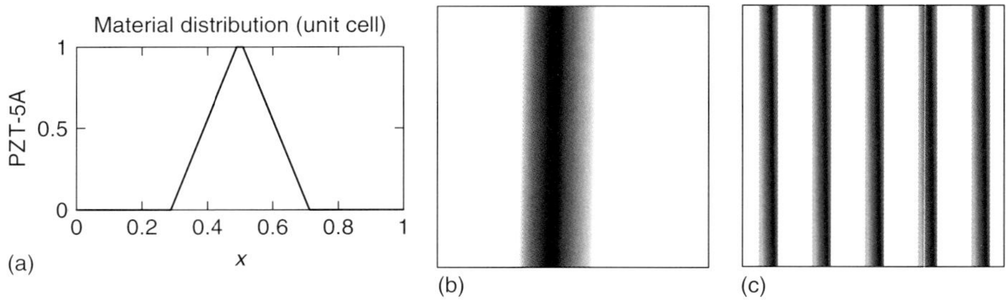

Figure 8.20 Linear FGM 2–2 piezocomposite. (a) Material distribution graphic of PZT-5A; (b) unit cell; and (c) periodic array.

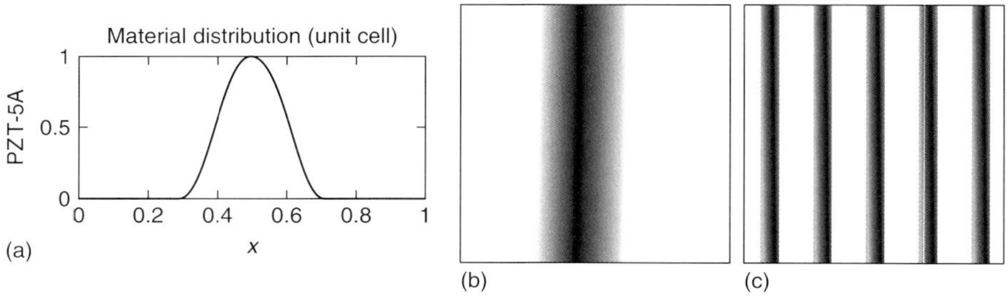

Figure 8.21 Sinusoidal FGM 2–2 piezocomposite. (a) Material distribution graphic of PZT-5A; (b) unit cell; and (c) periodic array.

represented as black and epoxy as white. The images (c) of these figures represent the periodic array of the unit cells.

Table 8.2 presents the performance characteristics for each type of unit cell. From Table 8.2, it is noticed that each performance characteristic is maximized by using different topologies of the unit cell. Considering the four topologies analyzed here, the topology that maximizes d_h and k_t is the traditional non-FGM 2-2 piezocomposite (Figure 8.19). However, $d_h g_h$ is maximized by using the linear or sinusoidal 2-2 piezocomposite (Figures 8.20 and 8.21, respectively). This fact indicates that it is possible to achieve higher performance characteristics by using 20% of PZT-5A in the unit cell distributed in a functionally graded way. The advantages, in addition to better performance of the piezocomposite, are the weight reduction and cost savings in the final material, as the epoxy polymer is lighter and chapter than the PZT-5A ceramic. the exception is k_h, which is maximized by using the pure PZT-5A material. However, the linear and sinusoidal 2-2 piezocomposite present a near value of k_h (0.177 and 0.122, respectively) and a lighter density (3496 kg/m^3) than the pure PZT-5A ceramic material (7500 kg/m^3) because the density of epoxy polymer is 56 times lighter then PZT-5A (see Table 8.1). Therefore, there is a trade off in choosing the topology of the unit cell among performance, weight and, consequently, cost in the final application. This trade off is also noticed in the performance characteristics d_h and k_t.

Table 8.2 Performance characteristics of the unit cells.

Material distribution	Density (kg/m^3)	Low-frequency applications			High-frequency applications
		d_h(pC/N)	$d_h g_h$(pm^2/N)	k_h	k_t
100% PZT-5A (full)	7500	68.196	0.222	0.145	0.361
Figure 8.19	3496	92.536	1.038	0.092	0.477
Figure 8.20	3496	73.804	1.700	0.117	0.422
Figure 8.21	3496	80.864	1.584	0.112	0.443

From these results, it is possible to conclude that FGM concept can be applied to obtain materials with the same or even higher properties of regular materials with weight reduction. Also the choice of the function applied for gradation has an important role in the design of the unit cells. The design of FGM unit cells is not trivial and requires optimization tools to avoid trial-and-error approaches.

8.7 Conclusion

The results give an idea about the potential of applying the FGM concept to design smart materials, both in micro- and macroscales. It is observed that piezoelectric transducers, designed according to FGM concept, have improved their performance in relation to nongraded ones; for instance, in ultrasonic applications, the FGM concept allows designing piezoelectric transducer with small time waveform or large bandwidth, which is desirable for obtaining high imaging resolution for medical and nondestructive testing applications. Likewise, it is observed that when the unit cell of 2–2 piezocomposite is designed on the basis of the FGM concept, high values of d_h, $d_h g_h$, and k_h are achieved, while the density value is significantly reduced in relation to the nongraded unit cell, which is desirable in applications to hydrophones. Additionally, from the examples, it is clear that both in micro- and macroscales the gradation function defines the piezoelectric behavior and, hence, optimization techniques must be used for designing graded piezoelectric structures. Specifically, the TOM arises as a general and powerful approach to find the optimal gradation function for achieving user-defined goals. In conclusion, the practical use of the proposed approach (to design piezoelectric structures considering gradation in micro- and macroscales) can broaden the range of application in the field of smart structures.

Acknowledgments

Wilfredo Montealegre Rubio and Sandro Luis Vatanabe thank FAPESP (São Paulo State Foundation Research Agency) for supporting them in their graduate studies through the fellowship No. 05/01762-5 and No. 08/57086-6, respectively. Emílio Carlos Nelli Silva is thankful for the financial support received from both CNPq (National Council for Research and Development, Brazil, No. 303689/2009-9). Gláucio H. Paulino acknowledges FAPESP for providing him the visiting scientist award at the University of São Paulo (USP) through project number 2008/5070-0.

References

1. Miyamoto, Y., Kaysser, W.A., Rabin, B.H., Kawasaki, A., and Ford, R.G. (1999) *Functionally Graded Materials: Design, Processing and Applications*, Kluwer Academic Publishers, London.
2. Suresh, S. and Mortensen, A. (1988) *Fundamentals of Functionally Graded Materials*, IOM Communications, London.
3. Paulino, G.H., Jin, Z.-H., and Dodds, R.H. Jr. (2003) Failure of functionally graded materials. *Comprehensive Structural Integrity*, **2**, 607–644.
4. Yamada, K., Sakamura, J.-I., and Nakamura, K. (1998) Broadband ultrasound transducers using effectively graded piezoelectric materials. IEEE Ultrasonics Symposium, pp. 1085–1089.
5. Samadhiya, R. and Mukherjee, A. (2006) Functionally graded piezoceramic ultrasonic transducers. *Smart Materials and Structures*, **15** (6), 627–631.
6. Chakraborty, A., Gopalakrishnan, S., and Kausel, E. (2005) Wave propagation analysis in inhomogeneous piezo-composite layer by the thin layer method. *International Journal for Numerical Methods in Engineering*, **64** (5), 567–598.
7. Guo, H., Cannata, J., Zhou, Q., and Shung, K. (2005) Design and fabrication of broadband graded ultrasonic transducers with rectangular kerfs. *IEEE Transaction on Ultrasonics, Ferroelectrics, and frequency Control*, **52** (11), 2096–2102.
8. Ichinose, N., Miyamoto, N., and Takahashi, S. (2004) Ultrasonic transducers with functionally graded piezoelectric ceramics. *Journal of the European Ceramic Society*, **24**, 1681–1685.
9. Torquato, S. (2002) *Random Heterogeneous Materials – Microstructure and Macroscopic Properties*, Springer, New York, ISBN-10:0387951679.
10. Ikeda, T. (1996) *Fundamentals of Piezoelectricity*, Oxford University Press, Oxford.
11. Nye, J.F. (1993) *Physical Properties of Crystals: Their Representation by Tensors and Matrices*, Clarendon Press, Oxford.
12. Kino, G. (2000) *Acoustic Waves: Devices, Imaging, and Analog Signal Processing*, Corrected Edition, Prentice-Hall, Inc., New Jersey.
13. ANSI/IEEE (1996) Publication and proposed revision of ANSI/IEEE standard 176-1987 "ANSI/IEEE Standard on Piezoelectricity". *IEEE Transactions on Ultrasonics, Ferroelectrics, and Frequency Control*, **43** (5), 717–772.
14. Naillon, M., Coursant, R.H., and Besnier, F. (1983) Analysis of piezoelectric structures by a finite element method. *Acta Electronica*, **25** (4), 341–362.
15. Kieback, B., Neubrand, A., and Riedel, H. (2003) Processing techniques for functionally graded materials. *Materials Science and Engineering: A*, **362**, 81–105.
16. Yin, H.M., Sun, L.Z., and Paulino, G.H. (2004) Micromechanics-based elastic model for functionally graded materials with particle interactions. *Acta Marterialia*, **52** (12), 3535–3543.
17. Koizumi, M. (1997) FGM activities in Japan. *Composites Part B: Engineering*, **28** (1), 1–4.
18. Birman, V. and Byrd, W.L. (2007) Modeling and analysis of functionally graded materials and structures. *Applied Mechanical Review*, **60** (5), 195–216.
19. Qiu, J.H., Tani, J., Ueno, T., Morita, T., Takahashi, H., and Du, H.J. (2003) Fabrication and high durability of functionally graded piezoelectric bending actuators. *Smart Materials and Structures*, **12** (1), 115–121.

20. Huang, J. and Rapoff, A.J. (2003) Optimization design of plates with holes by mimicking bones through non-axisymmetric functionally graded material. *Journal of Materials: Design and Applications*, **217** (1), 23–27.
21. Hedia, H.S. (2005) Comparison of one-dimensional and two-dimensional functionally graded materials for the backing shell of the cemented acetabular cup. *Journal of Biomedical Materials Research Part B: Applied Biomaterials*, **74B** (2), 732–739.
22. Huang, M., Wang, R., Thompson, V., Rekow, D., and Soboyejo, W. (2007) Bioinspired design of dental multilayers. *Journal of Material Science: Materials in Medicine*, **18** (1), 57–64.
23. Silva, E.C.N., Walters, M.C., and Paulino, G.H. (2006) Modeling bamboo as a functionally graded material: Lessons for the analysis of affordable materials. *Journal of Materials Science*, **41**, 6991–7004.
24. Lee, W.Y., Stinton, D.P., Berndt, C.C., Erdogan, F., Lee, Y.D., and Mutasim, Z. (1996) Concept of functionally graded materials for advanced thermal barrier coating applications. *Journal of the American Ceramic Society*, **79** (12), 3003–3012.
25. Pompe, W., Worch, H., Epple, M., Friess, W., Gelinsky, M., Greil, P., Hempel, U., Scharnweber, D., and Schulte, K. (2003) Functionally graded materials for biomedical applications. *Materials Science and Engineering: A*, **362** (1–2), 40–60.
26. Almajid, A., Taya, M., and Hudnut, S. (2001) Analysis of out-of-plane displacement and stress field in a piezocomposite plate with functionally graded microstructure. *International Journal of Solids and Structures*, **38** (19), 3377–3391.
27. Shi, Z.F. and Chen, Y. (2004) Functionally graded piezoelectric cantilever beam under load. *Archive of Applied Mechanics*, **74**, 237–247.
28. Taya, M., Almajid, A.A., Dunn, M., and Takahashi, H. (2003) Design of bimorph piezocomposite actuators with functionally graded microstructure. *Sensors and Actuators A-Physical*, **107** (3), 248–260.
29. Zhu, X.H. and Meng, Z.Y. (1995) Operational principle, fabrication and displacement characteristics of a functionally gradient piezoelectric ceramic actuator. *Sensors and Actuators A-Physical*, **48** (3), 169–176.
30. Rubio, W.M., Buiochi, F., Adamowski, J.C., and Silva, E.C.N. (2009) Modeling of functionally graded piezoelectric ultrasonic transducers. *Ultrasonics*, **49** (4-5), 484–494.
31. Wang, B.L. and Noda, N. (2001) Design of a smart functionally graded thermopiezoelectric composite structure. *Smart Materials and Structures*, **10** (2), 189–193.
32. Krautkramer, J. and Krautkramer, H. (1977) *Ultrasonic Testing of Materials*, 2nd edn, Springer-Verlag, New York.
33. Tiersten, H.F. (1967) Hamilton's principle for linear piezoelectric media. *Proceedings of the IEEE*, **55** (8), 1523–1524.
34. Zienkiewicz, O.C. and Taylor, R.L. (1991) *The Finite Element Method*, 4th edn, McGraw-Hill, New York.
35. Zuiker, J. and Dvorak, G. (1994) The effective properties of functionally graded composites-I. Extension of Mori-Tanaka method to linearly varying fields. *Composite Engineering*, **4** (1), 19–35.
36. Kim, J.H. and Paulino, G.H. (2002) Isoparametric graded finite elements for nonhomogeneous isotropic and orthotropic materials. *Journal of Applied Mechanics: Transactions of the ASME*, **69** (4), 502–514.
37. Silva, E.C.N., Carbonari, R.C., and Paulino, G.H. (2007) On graded elements for multiphysics applications. *Smart Materials and Structures*, **16**, 2408–2428.
38. Mills, D. and Smith, S.W. (2002) Multi-layered PZT/polymer composites to increase signal-to-noise ratio and resolution for medical ultrasound transducers part II: thick film technology. *IEEE Transactions on Ultrasonics, Ferroelectrics and Frequency Control*, **49** (7), 1005–1014.
39. Lamberti, N., Caliano, G., Iula, A., and Pappalardo, M. (1997) A new approach for the design of ultrasono-therapy

transducers. *IEEE Transactions on Ultrasonics, Ferroelectrics and Frequency Control*, **44** (1), 77–84.

40. Ludwig, G.D. (1950) The velocity of sound through tissues and the acoustic impedance of tissues. *The Journal of the Acoustical Society of America*, **22** (6), 862–866.
41. Rubio, W.M., Silva, E.C.N., and Paulino, G.H. (2009) Toward optimal design of piezoelectric transducers based on multifunctional and smoothly graded hybrid material systems. *Journal of Intelligent Material Systems and Structures*, **20** (14), 1725–1746.
42. Carbonari, R.C., Silva, E.C.N., and Paulino, G.H. (2009) Multi-actuated functionally graded piezoelectric micro-tools design: A multiphysics topology optimization approach. *International Journal for Numerical Methods in Engineering*, **77** (3), 301–336.
43. Carbonari, R.C., Silva, E.C.N., and Paulino, G.H. (2007) Topology optimization design of functionally graded bimorph-type piezoelectric actuators. *Smart Materials and Structures*, **16** (6), 2607–2620.
44. Bendsøe, M.P. and Sigmund, O. (2003) *Topology Optimization: Theory, Methods and Applications*, Springer, Berlin.
45. Rubio, W.M., Buiochi, F., Adamowski, J.C., and Silva, E.C.N. (2010) Topology optimized design of functionally graded piezoelectric ultrasonic transducers. *Physics Procedia*, **3** (1), 891–896.
46. Guo, N., Cawley, P., and Hitchings, D. (1992) The finite element analysis of the vibration characteristics of piezoelectric discs. *Journal of Sound and Vibration*, **159** (1), 115–138.
47. Kim, T.S. and Kim, Y.Y. (2000) MAC-based mode-tracking in structural topology optimization. *Computers and Structures*, **74** (3), 375–383.
48. Matsui, K. and Terada, K. (2004) Continuous approximation of material distribution for topology optimization. *International Journal for Numerical Methods in Engineering*, **59** (14), 1925–1944.
49. Guest, J.M., Prevost, J.H., and Belytschko, T. (2004) Achieving minimum length scale in topology optimization using nodal design variables and projection functions. *International Journal for Numerical Methods in Engineering*, **61** (2), 238–254.
50. Smith, W.A. (1989) The role of piezocomposites in ultrasonic transducers. Proceedings of IEEE Ultrasonic Symposium, pp. 755–765.
51. Smith, W.A. (1990) The application of 1–3 piezocomposites in acoustic transducers. Proceedings of 7th International Symposium on Applications of Ferroelectrics, pp. 145–152.
52. Safari, A. (1994) Development of piezoelectric composites for transducers. *Journal de Physique III*, **4** (7), 1129–1149.
53. Sanchez-Palencia, E. (1980). *Non-Homogeneous Media and Vibration Theory*, Lectures Notes in Physics, vol. 127, Springer, Berlin.
54. Guedes, J.M. and Kikuchi, N. (1990) Preprocessing and postprocessing for materials based on the homogenization method with adaptive finite element methods. *Computer Methods in Applied Mechanics and Engineering*, **83**, 143–198.
55. Cherkaev, A. and Kohn, R. (1997) *Topics in the Mathematical Modeling of Composite Materials*, Birkhauser, Boston, ISBN-10: 3764336625.
56. Allaire, G. (2002) *Shape Optimization by the Homogenization Method*, Applied Mathematical Sciences, vol. 146, Springer, New York, ISBN-10: 0387952985.
57. Guedes, J.M. and Kikuchi, N. (1990) Preprocessing and postprocessing for materials based on the homogenization method with adaptive finite-element methods. *Computer Methods in Applied Mechanics and Engineering*, **83** (2), 143–198.
58. Smith, W.A. (1993) Modeling 1-3 composite piezoelectrics: hydrostatic response. *IEEE Transactions on Ultrasonics, Ferroelectrics, and Frequency Contrl.*, **40** (1), 41–49.
59. Newnham, R.E., Skinner, D.P., and Cross, L.E. (1978) Connectivity and piezoelectric-pyroelectric composites. *Materials Research Bulletin*, **13** (5), 525–536.
60. Smith, W.A. and Auld, B.A. (1991) Modeling 1-3 composite piezoelectrics:

thickness-mode oscillations. *IEEE Transactions on Ultrasonics, Ferroelectrics, and Frequency Control*, **38** (1), 40–47.

61. Avellaneda, M. and Swart, P.J. (1994) The role of matrix porosity and poisson's ratio in the design of high-sensitivity piezocomposite transducers. *Adaptive Structures and Composites Materials: Analysis and Application – ASME Aerospace Division AD*, **45**, 59–66.
62. Sanchez-Palencia, E. and Zaoui, A. (1985) *Homogenization Techniques for Composite Media*, Lecture Notes in Physics, Springer-Verlag.
63. Telega, J.J. (1990) Piezoelectricity and homogenization: application to biomechanics. *Continuum Models and Discrete Systems*, **2**, 220–230.
64. Galka, A., Telega, J.J., and Wojnar, R. (1992) Homogenization and thermopiezoelectricity. *Mechanics Research Communications*, **19** (4), 315–324.
65. Turbé, N. and Maugin, G.A. (1991) On the linear piezoelectricity of composite materials. *Mathematical Methods in the Applied Sciences*, **14** (6), 403–412.
66. Otero, J.A., Catillero, J.B., and Ramos, R.R. (1997) Homogenization of heterogeneous piezoelectric medium. *Mechanics Research Communications*, **24** (1), 75–84.
67. Silva, E.C.N., Fonseca, J.S.O., and Kikuchi, N. (1997) Optimal design of piezoelectric microstructures. *Computational Mechanics*, **19** (5), 3987–3410.
68. Silva, E.C.N., Fonseca, J.S.O., and Kikuchi, N. (1998) Optimal design of periodic piezocomposites. *Computer Methods in Applied Mechanics and Engineering*, **159**, 49–77.

9
Variational Foundations of Large Strain Multiscale Solid Constitutive Models: Kinematical Formulation

Eduardo A. de Souza Neto and Raúl A. Feijóo

9.1
Introduction

The modeling of the constitutive behavior of solids by means of *multiscale* theories has been a subject of growing interest over the past decade. Various aspects of this approach to the description and prediction of the constitutive response of solids under a wide range of conditions are discussed in recent publications by Hori and Nemat-Nasser [1], Michel *et al.* [2], Nemat-Nasser [3], Ibrahimbegović and Marković [4], Miehe *et al.* [5], Miehe [6, 7], Miehe and Koch [8], Clayton and McDowell [9], Reese [10], Terada and Kikuchi [11], Terada *et al.* [12], Kouznetsova *et al.* [13], Matsui *et al.* [14], Michel and Suquet [15], Yang and Becker [16], Bilger *et al.* [17], Gotkepe and Miehe [18], Ibrahimbegović *et al.* [19], Li and E [20], Okumura *et al.* [21], Nakamachi *et al.* [22], Speirs *at al.* [23], and Giusti *et al.* [24, 25]. Pioneering contributions to the analysis of solid behavior taking into account information at two length scales are provided by Hill [26], Mandel [27], Gurson [28], Bensoussan *et al.* [29], Germain *et al.* [30], and Sanchez-Palencia [31], among others. Since then, the increasing demand for more accurate constitutive models capable of describing more complex inelastic behavior has led to the continuous development of theories relying on multiscale concepts. This demand has been partly driven by the more recent application of computer simulation methods – mainly based on finite-element procedures – outside the usual range of traditional engineering disciplines including, among others, the analysis of metallic alloys undergoing extreme straining as well as nonconventional materials such as natural composites and living biological tissues. The typical microstructure of a biological material, for instance, can be extremely elaborate, resulting in a macroscopic constitutive response of difficult representation by means of conventional phenomenological theories (see, for example, Holzapfel *et al.* [32] for an assessment of hyperelastic models of arterial wall behavior). The problem becomes far more pronounced when inelastic phenomena play a significant role in the overall material response. Mechanical dissipation in such cases arises from (possibly intricate) irreversible interactions among the microstructural constituents. Often, the modeling of such phenomena by means of purely macroscopic (phenomenological) theories results in

Advanced Computational Materials Modeling: From Classical to Multi-Scale Techniques.
Edited by Miguel Vaz Júnior, Eduardo A. de Souza Neto, and Pablo A. Munoz-Rojas

ISBN: 978-3-527-32479-8

substantial discrepancies between the predicted and observed constitutive response, which may significantly worsen with the complexity of the strain history. This is a well-known fact, for instance, in the modeling of the Bauschinger effect by means of classical kinematic hardening plasticity theories. The development and use of multiscale theories in such cases appear to be a very promising alternative to circumvent the major weaknesses of the classical phenomenological approach and provide more realistic descriptions of the constitutive response.

This chapter focuses on the family of multiscale constitutive theories of solids based on the volume averaging of the microscopic strain and stress fields over a *representative volume element* (RVE) [2, 3, 8, 12, 33]. Such models have been enjoying increasing popularity in recent years, markedly within the computational mechanics community, given their suitability for implementation in finite-element-based computer simulation systems. However, multiscale theories of this type are often presented in a rather ad hoc manner, making it particularly difficult to distinguish the essential assumptions from their consequences to the resulting theory. Our aim here is to fill this gap by providing a complete (kinematical) variational foundation for this family of theories. The formulation presented provides a clearly structured axiomatic framework where the entire theory can be rigorously derived from five basic statements: (i) the deformation gradient averaging relation; (ii) the requirement that the actual functional set of admissible displacement fluctuations of the RVE be a subspace of minimally constrained space of displacement fluctuations compatible with the deformation gradient averaging assumption; (iii) the principle of virtual work (equilibrium) for the RVE; (iv) the stress averaging relation; and (v) the Hill–Mandel principle of macrohomogeneity [26, 27] that establishes the energy consistency between the macro- and microscales. Within this framework, any class of constitutive models is completely defined by the choice of the functional space of virtual displacements of the RVE (which embodies the assumed kinematical constraints imposed upon the RVE). In this context, the variational statement of the Hill–Mandel principle of macrohomogeneity [26, 27] plays a crucial role. It allows the external load system of the RVE (body force and surface traction fields) to be viewed as mere reactions to the prescribed RVE kinematical constraints. Four well-known classes of multiscale models are cast within this framework: (i) the classical Taylor or homogeneous microscopic strain model, commonly known as the rule of mixtures; (ii) the linear boundary displacement model; (iii) the periodic boundary fluctuation displacements model; and (iv) the minimum kinematical constraint model, also referred to as the *uniform boundary traction model*. General canonical formulae of homogenized constitutive tangent operators are also presented. These are first derived in the (time- and space-) continuum setting and later specialized to the time-discrete (and space-continuum) case. The latter formulae are generally applicable to problems where the microscopic constitutive response is described by internal variable-based theories and the corresponding problem of evolution is approximated by some incremental numerical integration scheme in time – such as return-mapping algorithms for classical plasticity [34, 35] – regardless of the spatial discretization method to be adopted. Fully discrete (in time and space) counterparts, such as those derived by

Miehe *et al.* [36] in the finite-element context, can be obtained by simply specializing the time-discrete/space-continuum formulae according to the underlying spatial discretization method.

The chapter is organized as follows. The essential axioms that form the basis of the proposed variational framework are stated in Section 9.2, the variational setting summarized in this section is described in detail in Ref. [42]. In Section 9.3, the general family of multiscale constitutive models based on the axioms of Section 9.2 is defined. In Section 9.4, well-known classes of multiscale theories are cast within the proposed variational framework. The conditions of equivalence between stress averaging in the deformed and reference configurations of the RVE are briefly discussed in Section 9.5. In Section 9.6, general canonical forms of constitutive tangent operators for the multiscale models are derived in detail. These are specialized in Section 9.7 for the time-discrete case. Section 9.8 outlines the specialization of the proposed framework to the case of linearized kinematics. Section 9.9 presents some concluding remarks.

9.2 Large Strain Multiscale Constitutive Theory: Axiomatic Structure

In the classical, purely mechanical constitutive theory, the axioms of *constitutive determinism* and *local action* [37] establish that the stress tensor at any point of the continuum is uniquely determined by the history of the deformation gradient tensor at that point. This implies that there exists a tensor-valued functional $\mathfrak{P}$ such that, at any instant t, the first Piola–Kirchhoff stress tensor, $\mathbf{P}$, is given by

$$\mathbf{P}(t) = \mathfrak{P}(\mathbf{F}^t) \tag{9.1}$$

where $\mathbf{F}^t$ denotes the history of the deformation gradient,

$$\mathbf{F} \equiv \mathbf{I} + \nabla \mathbf{u} \tag{9.2}$$

at the material point of interest up to instant t, with ∇ the material (or reference) gradient operator, $\mathbf{u}$ the displacement field, and $\mathbf{I}$ the second-order identity tensor.

Widely used specializations of Eq. (9.1) accounting for inelastic behavior include, among other theories, hypoelastic-based and multiplicative hyperelastic-based finite strain elastoplasticity and elastoviscoplasticity [38, 35]. In such cases, the history of $\mathbf{F}$ is parameterized by a set of internal variables whose evolution in time is governed by ordinary differential equations. Such models are typically *phenomenological* in that their internal variables are usually associated with *macroscopic* phenomena observable by means of relatively simple experiments (see Ref. [39] for a detailed discussion on this topic).

The stress-constitutive functionals for the family of the so-called multiscale constitutive models that are the main subject of this chapter are also specializations of Eq. (9.1). For the present models, however, the description in terms of a system of ordinary differential equations typical of conventional phenomenological theories is replaced by the assumption that the deformation gradient and the stress tensor at an arbitrary material point $\mathbf{x}$ of the continuum are, respectively,

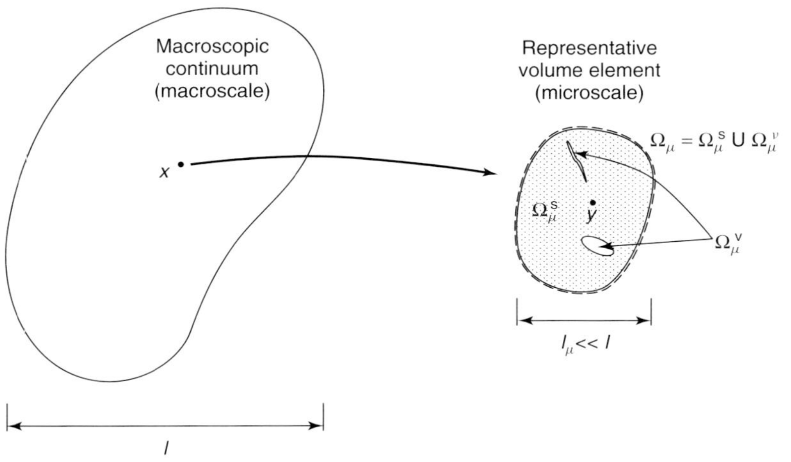

Figure 9.1 Multiscale constitutive models. Macrocontinuum and the RVE.

the volume average of the deformation gradient and stress fields over a local RVE or *microscopic cell* (Figure 9.1). The characteristic length, l_μ, of the RVE is assumed much smaller than the characteristic length, l, of the macroscopic continuum. It should be emphasized that the actual size of what is referred to here as the *microscopic* scale – the RVE – depends fundamentally on the nature of the material to be modeled and on the level of detail to be included in the model. This size is not necessarily microscopic in the conventional sense. For example, a polycrystalline metal can be modeled within this framework having a microscale of the size of a few grains (order of microns), whereas a model for concrete may be assigned a microscale of the size of a sufficiently representative sample of the cement/aggregate composite (order of centimeters).

The domain Ω_μ (with boundary $\partial\Omega_\mu$) of the reference configuration of the RVE is assumed to consist in general of a solid part, Ω_μ^s, and a void part, Ω_μ^v:

$$\Omega_\mu = \Omega_\mu^s \cup \Omega_\mu^v \tag{9.3}$$

What we refer to as the *void* part here may consist of cracks and pores, which can, in general, be subjected to self-contact or may be filled with a pressurized fluid. For simplicity, in what follows, we consider only RVEs whose void part does not intersect the RVE boundary, that is, it is assumed that

$$\partial\Omega_\mu \cap \overline{\Omega}_\mu^v = \emptyset \tag{9.4}$$

where $\overline{\Omega}_\mu^v$ denotes the closure of the set Ω_μ^v.[1)]

1) The extension of the material presented here to the more general case where we may have $\partial\Omega_\mu \cap \overline{\Omega}_\mu^v \neq \emptyset$ is conceptually straightforward. However, the introduction of this generalization makes the derivation of the theory unnecessarily cumbersome.

Constitutive models of this type have been used, for instance, by Miehe *et al.* [36], Terada *et al.* [12], and Matsui *et al.* [14], in the analysis of generally inelastic solids undergoing finite strains. The underlying theory is a geometrically nonlinear extension of the infinitesimal strain theory originating in the work of Sanchez-Palencia [40] and Germain *et al.* [30] and used in the computational context, among others, by Suquet [33] and Michel *et al.* [2]. Our key contribution here is the establishment of a clearly structured axiomatic variational framework for this family of constitutive models. The essential axioms are as follows: (i) the volume averaging relation linking the macroscopic and microscopic deformation gradients; (ii) the requirement that the actual functional set of admissible displacement fluctuations of the RVE be a subspace of the minimally constrained space of displacement fluctuations compatible with the strain averaging assumption; (iii) the equilibrium of the RVE, here stated by means of the principle of virtual work; (iv) the volume averaging relation linking the macroscopic and microscopic stresses; and (v) the Hill–Mandel principle of macrohomogeneity, which establishes the energy consistency between the macro- and microscales. As we shall see, this family of multiscale constitutive theories is entirely derived from these five fundamental statements, which are conveniently summarized in Box 9.1. The basic axioms together with their main consequences to the resulting theory are discussed in detail in the following.

Box 9.1: Finite Strain Multiscale Constitutive Models: Basic Axioms

(i) Deformation gradient averaging relation:

$$\mathbf{F} \equiv \frac{1}{V_\mu} \int_{\Omega_\mu} \mathbf{F}_\mu \, \mathrm{d}V$$

(ii) The actual set $\tilde{\mathscr{K}}_\mu$ of kinematically admissible displacement fluctuation fields of the RVE is a subspace of its minimally constrained counterpart,

$$\tilde{\mathscr{K}}_\mu^* \equiv \left\{ \mathbf{v}, \text{sufficiently regular} \middle| \int_{\partial\Omega_\mu} \mathbf{v} \otimes \mathbf{n} \, \mathrm{d}A = \mathbf{0} \right\}$$

(iii) Equilibrium of the RVE:

$$\int_{\Omega_\mu} \mathbf{P}_\mu(\mathbf{y}, t) : \nabla \boldsymbol{\eta} \, \mathrm{d}V - \int_{\Omega_\mu} \mathbf{b}(\mathbf{y}, t) \cdot \boldsymbol{\eta} \, \mathrm{d}V - \int_{\partial\Omega_\mu} \mathbf{t}^e(\mathbf{y}, t) \cdot \boldsymbol{\eta} \, \mathrm{d}A = 0 \; \forall \boldsymbol{\eta} \in \mathscr{V}_\mu$$

where $\mathscr{V}_\mu$ is the space of virtual displacements of the RVE.

(iv) First Piola–Kirchoff stress averaging relation:

$$\mathbf{P} \equiv \frac{1}{V_\mu} \int_{\Omega_\mu} \mathbf{P}_\mu \, \mathrm{d}V$$

(v) Hill–Mandel principle of macrohomogeneity:

$$\mathbf{P} : \dot{\mathbf{F}} = \frac{1}{V_\mu} \int_{\Omega_\mu} \mathbf{P}_\mu : \dot{\mathbf{F}}_\mu \, \mathrm{d}V$$

for any kinematically admissible velocity field of the RVE.

9.2.1
Deformation Gradient Averaging and RVE Kinematics

The first basic axiom – the starting point of the theory – establishes that the deformation gradient tensor $\mathbf{F}$ at a point $\mathbf{x}$ of the macrocontinuum is the *volume average* of the *microscopic deformation gradient field*, $\mathbf{F}_\mu$, over the RVE associated with $\mathbf{x}$:

$$\mathbf{F}(\mathbf{x}, t) = \frac{1}{V_\mu} \int_{\Omega_\mu} \mathbf{F}_\mu(\mathbf{y}, t) \, \mathrm{d}V \tag{9.5}$$

where V_μ is the volume of the RVE in its reference configuration, $\mathbf{y}$ denotes the reference coordinate of points within the RVE and

$$\mathbf{F}_\mu \equiv \mathbf{I} + \nabla \mathbf{u}_\mu \tag{9.6}$$

with $\mathbf{u}_\mu$ being the displacement field of the RVE. The tensor $\mathbf{F}$ is referred to as the *macroscopic* or *homogenized deformation gradient*. It should be noted here that, in the presence of voids in the microcell (Figure 9.1), expression (9.5) for the homogenized deformation gradient is valid only in a generalized sense since the microscopic displacement and deformation gradient fields are not defined over the void subdomain $\Omega_\mu^{\mathrm{v}} \subset \Omega_\mu$.

9.2.1.1 Consequence: Minimum RVE Kinematical Constraints

The main consequence of the basic axiom (9.5) to the kinematical variational formulation of this family of models is a constraint on the possible displacement fields of the RVE. For a given macroscopic deformation gradient $\mathbf{F}$, this axiom implies that only microscopic displacement fields that satisfy Eqs (9.5), (9.6) can be admissible. This gives rise to the definition of the *minimally constrained set of kinematically admissible microscopic displacements*, $\mathscr{K}_\mu^*$, as[2)]

$$\mathscr{K}_\mu^* \equiv \left\{ \mathbf{v}, \text{sufficiently regular} \middle| \int_{\Omega_\mu} \nabla \mathbf{v} \, \mathrm{d}V = V_\mu (\mathbf{F} - \mathbf{I}) \right\} \tag{9.7}$$

so that a necessary condition for a field $\mathbf{u}_\mu$ to be *kinematically admissible* is

$$\mathbf{u}_\mu \in \mathscr{K}_\mu^* \tag{9.8}$$

2) Here, and in what follows, we avoid the precise definition of regularity of functions and refer to fields as *sufficiently regular* meaning that all operations in which they are involved in the present context make sense.

An equivalent expression for the above kinematical constraint, exclusively in terms of RVE boundary displacements, can be obtainedby considering the general tensor relation[41]

$$\int_{\Omega} \mathbf{S}\,(\nabla \mathbf{v})^{\mathrm{T}}\,\mathrm{d}V = \int_{\partial\Omega} (\mathbf{S}\mathbf{n}) \otimes \mathbf{v}\,\mathrm{d}A - \int_{\Omega} (\operatorname{div}\mathbf{S}) \otimes \mathbf{v}\,\mathrm{d}V \tag{9.9}$$

valid for sufficiently regular tensor fields $\mathbf{S}$, vector fields $\mathbf{v}$, and domains Ω, where $\operatorname{div}(\cdot)$ denotes the divergence of $(\cdot)$ and $\mathbf{n}$ is the outward unit normal to the boundary $\partial\Omega$ of Ω. By specializing the above formula and identifying the generic tensor $\mathbf{S}$ with $\mathbf{I}$, the vector field $\mathbf{v}$ with $\mathbf{u}_\mu$, and the domain Ω with Ω_μ, we obtain in the absence of voids ($\Omega_\mu^{\mathrm{v}} = \emptyset$),

$$\mathbf{F} = \mathbf{I} + \frac{1}{V_\mu} \int_{\partial\Omega_\mu} \mathbf{u}_\mu \otimes \mathbf{n}\,\mathrm{d}A \tag{9.10}$$

so that constraint (9.10) is equivalent to Eqs (9.5), (9.6) and we can then define $\mathscr{K}_\mu^*$ alternatively to Eq. (9.7), exclusively in terms of RVE boundary displacements, as

$$\mathscr{K}_\mu^* \equiv \left\{ \mathbf{v}, \text{ sufficiently regular} \mid \int_{\partial\Omega_\mu} \mathbf{v} \otimes \mathbf{n}\,\mathrm{d}A = V_\mu\,(\mathbf{F} - \mathbf{I}) \right\} \tag{9.11}$$

Definition (9.11) is equivalent to Eq. (9.7) also in the presence of voids [42].

9.2.1.2 Minimum Constraint on Displacement Fluctuations

Without loss of generality, any microscopic displacement field, $\mathbf{u}_\mu$, may be decomposed as a sum

$$\mathbf{u}_\mu(\mathbf{y}, t) = \mathbf{u}(\mathbf{x}, t) + [\mathbf{F}(\mathbf{x}, t) - \mathbf{I}]\,\mathbf{y} + \pi_\mu(\mathbf{y}, t) \tag{9.12}$$

of a uniform (constant within the RVE) displacement $\mathbf{u}$ – the displacement of the corresponding point $\mathbf{x}$ of the macrocontinuum – a *linear displacement* (i.e., a displacement that varies linearly in $\mathbf{y}$), $(\mathbf{F} - \mathbf{I})\mathbf{y}$, and a *displacement fluctuation*, $\tilde{\mathbf{u}}_\mu$. Note that in stating Eq. (9.12), we have *defined* the fluctuation field as

$$\tilde{\mathbf{u}}_\mu(\mathbf{y}, t) \equiv \mathbf{u}_\mu(\mathbf{y}, t) - \{\mathbf{u}(\mathbf{x}, t) + [\mathbf{F}(\mathbf{x}, t) - \mathbf{I}]\,\mathbf{y} \tag{9.13}$$

Accordingly, the microscopic strain field (9.6) is decomposed into the sum

$$\mathbf{F}_\mu(\mathbf{y}, t) = \mathbf{F}(\mathbf{x}, t) + \tilde{\mathbf{H}}_\mu(\mathbf{y}, t) \tag{9.14}$$

of a *homogeneous* (constant in $\mathbf{y}$) deformation gradient coinciding with the macroscopic deformation gradient, $\mathbf{F}$, and a *displacement gradient fluctuation* field,

$$\tilde{\mathbf{H}}_\mu \equiv \nabla \tilde{\mathbf{u}}_\mu \tag{9.15}$$

that generally varies in $\mathbf{y}$. A straightforward manipulation taking Eq. (9.12) into account shows that the minimum constraint Eqs (9.8), (9.11) is equivalent to stating that any kinematically admissible displacement *fluctuation*, $\tilde{\mathbf{u}}_\mu$, satisfies

$$\tilde{\mathbf{u}}_\mu \in \tilde{\mathscr{K}}_\mu^* \tag{9.16}$$

where

$$\tilde{\mathscr{K}}_\mu^* \equiv \left\{ \mathbf{v}, \text{sufficiently regular} \middle| \int_{\partial\Omega_\mu} \mathbf{v} \otimes \mathbf{n}\, \mathrm{d}A = \mathbf{0} \right\} \tag{9.17}$$

is the *minimally constrained vector space of kinematically admissible displacement fluctuations* of the RVE. From the above definition, it follows that the minimally constrained set of kinematically admissible microscopic displacements, $\mathscr{K}_\mu^*$, can be equivalently defined as

$$\mathscr{K}_\mu^* = \left\{ \mathbf{u}_\mu = \mathbf{u} + (\mathbf{F} - \mathbf{I})\mathbf{y} + \tilde{\mathbf{u}}_\mu \middle| \tilde{\mathbf{u}}_\mu \in \tilde{\mathscr{K}}_\mu^* \right\} \tag{9.18}$$

Then, for a given macroscopic displacement, $\mathbf{u}$, and deformation gradient, $\mathbf{F}$, the set $\mathscr{K}_\mu^*$ is a *translation* [43] of space $\tilde{\mathscr{K}}_\mu^*$.

9.2.2
Actual Constraints: Spaces of RVE Velocities and Virtual Displacements

In establishing the kinematical theory of multiscale constitutive models, we may, in general, impose further constraints upon the RVE kinematics. Such constraints will define the *actual* set $\mathscr{K}_\mu$ of kinematically admissible displacements of the RVE, which, according to Eq. (9.8), must satisfy

$$\mathscr{K}_\mu \subset \mathscr{K}_\mu^* \tag{9.19}$$

As we shall see later, different constraints imposed upon a given RVE (i.e., different definitions of $\mathscr{K}_\mu$) will lead, in general, to different classes of macroscopic constitutive models.

At this point, we introduce another basic assumption of the theory: we require that any further constraints to be imposed on the kinematics of the RVE be such that actual set of kinematically admissible displacement fluctuations,

$$\tilde{\mathscr{K}}_\mu \subset \tilde{\mathscr{K}}_\mu^* \tag{9.20}$$

is a *subspace* of $\tilde{\mathscr{K}}_\mu^*$. In this case, the actual set $\mathscr{K}_\mu$ of kinematically admissible microscopic displacements is a *translation* of space $\tilde{\mathscr{K}}_\mu$ and is given by

$$\mathscr{K}_\mu = \left\{ \mathbf{u}_\mu = \mathbf{u}(\mathbf{x}) + (\mathbf{F} - \mathbf{I})\,\mathbf{y} + \tilde{\mathbf{u}}_\mu \middle| \tilde{\mathbf{u}}_\mu \in \tilde{\mathscr{K}}_\mu \right\} \subset \mathscr{K}_\mu^* \tag{9.21}$$

The set $\mathscr{K}_\mu$, together with the associated *space of virtual kinematically admissible displacements of the RVE*, denoted by $\mathscr{V}_\mu$, plays a fundamental role in the variational characterization of the equilibrium of the RVE. In the present context, $\mathscr{V}_\mu$ can be defined as

$$\mathscr{V}_\mu \equiv \left\{ \boldsymbol{\eta} = \mathbf{v}_1 - \mathbf{v}_2 | \mathbf{v}_1, \mathbf{v}_2 \in \mathscr{K}_\mu \right\} \tag{9.22}$$

In view of Eq. (9.21) and the fact that $\tilde{\mathscr{K}}_\mu$ is itself a vector space, it follows trivially from Eq. (9.22) that

$$\mathscr{V}_\mu = \tilde{\mathscr{K}}_\mu \tag{9.23}$$

Further, the same arguments applied to the rate form

$$\dot{\mathbf{u}}_\mu = \dot{\mathbf{u}} + \dot{\mathbf{F}}\,\mathbf{y} + \dot{\tilde{\mathbf{u}}}_\mu \tag{9.24}$$

of the additive split (9.12), establish that any *kinematically admissible fluctuation velocity*, $\dot{\tilde{\mathbf{u}}}_\mu$, satisfies

$$\dot{\tilde{\mathbf{u}}}_\mu \in \mathscr{V}_\mu \tag{9.25}$$

In summary, we have established in the above that, as a consequence of the assumption that $\tilde{\mathscr{K}}_\mu$ is a subspace of $\tilde{\mathscr{K}}_\mu^*$, the functional spaces of virtual displacements, kinematically admissible displacement fluctuations and fluctuation velocities of the RVE coincide.

9.2.3 Equilibrium of the RVE

The next fundamental axiom of the theory states that the RVE must be in equilibrium at each instant t of its deformation history. Let $\mathbf{P}_\mu = \mathbf{P}_\mu(\mathbf{y}, t)$ denote the first Piola–Kirchhoff stress field of the RVE. We shall refer to $\mathbf{P}_\mu$ as the *microscopic Piola–Kirchhoff stress.* With the RVE subjected in general to a *reference body force* field $\mathbf{b} = \mathbf{b}(\mathbf{y}, t)$ and a *reference external traction* field $\mathbf{t}^e = \mathbf{t}^e(\mathbf{y}, t)$ exerted upon the RVE across its external boundary $\partial\Omega_\mu$, the *principle of virtual work* establishes that the RVE is in equilibrium if and only if

$$\int_{\Omega_\mu} \mathbf{P}_\mu(\mathbf{y}, t) : \nabla\boldsymbol{\eta}\,\mathrm{d}V - \int_{\Omega_\mu} \mathbf{b}(\mathbf{y}, t)\cdot\boldsymbol{\eta}\,\mathrm{d}V - \int_{\partial\Omega_\mu} \mathbf{t}^e(\mathbf{y}, t)\cdot\boldsymbol{\eta}\,\mathrm{d}A = 0 \forall\boldsymbol{\eta} \in \mathscr{V}_\mu \tag{9.26}$$

holds at each t, where $\mathscr{V}_\mu$ is the (yet to be defined) space of virtual displacements of the RVE, which must comply with the general constraints discussed in Section 9.2.2.

Remark 9.1. *At this point, we recall that what we refer to as voids are not necessarily empty and may, in general, exert an influence upon the mechanical state of the RVE. A typical example appears in the modeling of saturated porous media and multiphase materials in general, where the presence of a pressurized fluid phase within the voids (pores in this case) has an important contribution to the overall mechanical state of the microcell.*

In view of the above remark, it is convenient to rewrite Eq. (9.26), making the distinct contributions from the solid and void parts explicit. Then, a straightforward manipulation taking into account that the void portion of Ω_μ is also in equilibrium gives

$$\int_{\Omega_\mu^\mathrm{v}} \mathbf{P}_\mu(\mathbf{y}, t) : \nabla\boldsymbol{\eta}\,\mathrm{d}V - \int_{\Omega_\mu^\mathrm{v}} \mathbf{b}(\mathbf{y}, t)\cdot\boldsymbol{\eta}\,\mathrm{d}V - \int_{\partial\Omega_\mu^\mathrm{v}} \mathbf{t}^\mathrm{v}(\mathbf{y}, t)\cdot\boldsymbol{\eta}\,\mathrm{d}A = 0\ \forall\boldsymbol{\eta} \in \mathscr{V}_\mu \tag{9.27}$$

gives the following convenient expression for the equilibrium of the solid part of the RVE:

$$\int_{\Omega_\mu^s} \mathbf{P}_\mu(\mathbf{y},t) : \nabla\boldsymbol{\eta}\, \mathrm{d}V - \int_{\Omega_\mu^s} \mathbf{b}(\mathbf{y},t)\cdot\boldsymbol{\eta}\, \mathrm{d}V - \int_{\partial\Omega_\mu} \mathbf{t}^e(\mathbf{y},t)\cdot\boldsymbol{\eta}\, \mathrm{d}A \\ - \int_{\partial\Omega_\mu^v} \mathbf{t}^v(\mathbf{y},t)\cdot\boldsymbol{\eta}\, \mathrm{d}A = 0 \quad \forall\boldsymbol{\eta}\in\mathscr{V}_\mu \tag{9.28}$$

where the *internal traction* field, $\mathbf{t}^v = \mathbf{t}^v(\mathbf{y},t)$, has been defined as the reference traction exerted upon the *solid* part of Ω_μ across the solid–void interface, $\partial\Omega_\mu^v$.

Remark 9.2. *Obviously, for porous materials consisting of solid phases only (empty pores), the internal tractions* $\mathbf{t}^v$ *vanish, whereas in the presence of a pressurized fluid phase in the voids (as referred to in Remark 9.1) tractions are usually prescribed as a uniform (possibly time-dependent) pressure distribution over the solid–void interface. Another important source of nonzero internal tractions could be the frictional contact between the opposing sides of collapsing voids or closing microcracks. In this case, the traction* $\mathbf{t}^v$ *is a given functional of the displacement history on the solid–void interface.*

9.2.3.1 Strong Form of Equilibrium

For sufficiently regular data fields, the variational equations of equilibrium Eqs (9.27) and (9.28), respectively, for the void and solid parts of the RVE, can be equivalently written in differential form as

$$\begin{cases} \operatorname{div}\mathbf{P}_\mu(\mathbf{y},t) = \mathbf{b}(\mathbf{y},t) & \forall\,\mathbf{y}\in\Omega_\mu^s \\ \operatorname{div}\mathbf{P}_\mu(\mathbf{y},t) = \mathbf{b}(\mathbf{y},t) & \forall\,\mathbf{y}\in\Omega_\mu^v \\ \mathbf{P}_\mu(\mathbf{y},t)\,\mathbf{n} = \mathbf{t}^e(\mathbf{y},t) & \forall\,\mathbf{y}\in\partial\Omega_\mu \\ [\![\mathbf{P}_\mu(\mathbf{y},t)\,\mathbf{n}]\!] = \mathbf{0} & \forall\,\mathbf{y}\in\partial\Omega_\mu^v \end{cases} \tag{9.29}$$

where $\mathbf{n}$ is the unit normal vector to the reference solid domain boundary and $[\![\mathbf{P}_\mu(\mathbf{y},t)\,\mathbf{n}]\!]$ denotes the *jump* of vector field $\mathbf{P}_\mu\,\mathbf{n}$ across the solid–void interface, $\partial\Omega_\mu^v$.

9.2.3.2 Solid–Void/Pore Interaction

In the present work, we restrict the analysis to situations where the *body forces acting on* Ω_μ^v *can be neglected*, that is, we assume

$$\mathbf{b}(\mathbf{y},t) = \mathbf{0} \quad \forall\,\mathbf{y}\in\Omega_\mu^v \tag{9.30}$$

In this case, only the surface traction field $\mathbf{t}^v$ over $\partial\Omega_\mu^v$ need be accounted for in the possible contribution of the void/pore to the overall mechanical state of the RVE.

Remark 9.3. *This assumption can be adopted, for example, in problems where the void/pore contains a fluid phase mathematically described as a problem of an intrinsically different nature from that of the solid phase. In such cases, the influence of* *the fluid phase upon the solid matrix of the RVE will, under the above assumption, be accounted for solely through the surface tractions resulting from the solution of the fluid problem.*

9.2.4
Stress Averaging Relation

Another axiom defining the family of multiscale constitutive theories discussed in this chapter is the *stress averaging relation*. Here, we postulate that analogously to Eq. (9.5), the first Piola–Kirchhoff stress tensor, $\mathbf{P}$, at a point $\mathbf{x}$ of the macrocontinuum is the volume average of the *microscopic first Piola–Kirchhoff stress* field, $\mathbf{P}_\mu$, which acts over the RVE associated with $\mathbf{x}$:[3)]

$$\mathbf{P}(\mathbf{x}, t) \equiv \frac{1}{V_\mu} \int_{\Omega_\mu} \mathbf{P}_\mu(\mathbf{y}, t)\, \mathrm{d}V = \frac{1}{V_\mu} \int_{\Omega_\mu^{\mathrm{s}}} \mathbf{P}_\mu(\mathbf{y}, t)\, \mathrm{d}V + \frac{1}{V_\mu} \int_{\Omega_\mu^{\mathrm{v}}} \mathbf{P}_\mu(\mathbf{y}, t)\, \mathrm{d}V \tag{9.31}$$

In what follows, $\mathbf{P}$ is referred to as the *macroscopic* or *homogenized first Piola–Kirchhoff stress.*

9.2.4.1 Macroscopic Stress in Terms of RVE Boundary Tractions and Body Forces

The macroscopic stress defined by Eq. (9.31) can be alternatively expressed in terms of RVE boundary traction and body force averages. To derive the alternative expression, we must first consider the tensor relation (9.9), which, specialized with the replacement of $\mathbf{S}$ with $\mathbf{P}_\mu$, $\mathbf{v}$ with $\mathbf{y}$, and Ω with Ω_μ, and combined with identity $\nabla\mathbf{y} = \mathbf{I}$, gives

$$\begin{aligned}\int_{\Omega_\mu} \mathbf{P}_\mu\, \mathrm{d}V &= \int_{\Omega_\mu^{\mathrm{s}}} \mathbf{P}_\mu\, \mathrm{d}V + \int_{\Omega_\mu^{\mathrm{v}}} \mathbf{P}_\mu\, \mathrm{d}V \\ &= \int_{\partial\Omega_\mu} (\mathbf{P}_\mu \mathbf{n}) \otimes \mathbf{y}\, \mathrm{d}A - \int_{\Omega_\mu^{\mathrm{s}}} (\operatorname{div} \mathbf{P}_\mu) \otimes \mathbf{y}\, \mathrm{d}V \\ &\quad - \int_{\Omega_\mu^{\mathrm{v}}} (\operatorname{div} \mathbf{P}_\mu) \otimes \mathbf{y}\, \mathrm{d}V + \int_{\partial\Omega_\mu^{\mathrm{v}}} [\![\mathbf{P}_\mu \mathbf{n}]\!] \otimes \mathbf{y}\, \mathrm{d}A\end{aligned} \tag{9.32}$$

Then, with the introduction of the strong form Eq. (9.29) of equilibrium equation in the above and considering the assumption (9.30), we obtain the following expression for the homogenized stress, exclusively in terms of RVE boundary tractions and body forces:

$$\mathbf{P}(\mathbf{x}, t) = \frac{1}{V_\mu} \left[\int_{\partial\Omega_\mu} \mathbf{t}^e(\mathbf{y}, t) \otimes \mathbf{y}\, \mathrm{d}A - \int_{\Omega_\mu^{\mathrm{s}}} \mathbf{b}(\mathbf{y}, t) \otimes \mathbf{y}\, \mathrm{d}V \right] \tag{9.33}$$

3) The choice of stress averaging relation is not unique in the finite strain context. We could, for instance, have postulated that macroscopic *Cauchy* stress tensor instead is the volume average of its microscopic counterpart over the deformed configuration of the RVE. This issue is briefly discussed in Section 9.5.

9.2.5
The Hill–Mandel Principle of Macrohomogeneity

The final axiom needed to establish the axiomatic variational framework is the *Hill–Mandel principle of macrohomogeneity* [26, 27], which states that *the macroscopic stress power equals the volume average of the microscopic stress power over the RVE*. In the large strain setting, it requires that at any state of the RVE characterized by an equilibrium stress field $\mathbf{P}_\mu$, the identity

$$\mathbf{P} : \dot{\mathbf{F}} = \frac{1}{V_\mu} \int_{\Omega_\mu} \mathbf{P}_\mu : \dot{\mathbf{F}}_\mu \, dV \tag{9.34}$$

holds for any kinematically admissible microscopic deformation gradient rate field, $\dot{\mathbf{F}}_\mu$. Note that a microscopic deformation gradient rate is said to be kinematically admissible if

$$\dot{\mathbf{F}}_\mu = \nabla \dot{\mathbf{u}}_\mu = \dot{\mathbf{F}} + \nabla \dot{\tilde{\mathbf{u}}}_\mu; \quad \dot{\tilde{\mathbf{u}}}_\mu \in \mathscr{V}_\mu \tag{9.35}$$

where $\mathscr{V}_\mu$ is the space of kinematically admissible RVE velocities, which has been shown in Section 9.2.2 to coincide with the (yet to be defined) spaces of kinematically admissible displacement fluctuations and virtual displacements of the RVE.

The following proposition is fundamental in the present context.

Proposition 9.1. **(variational statement of the Hill–Mandelprinciple)** *The Hill–Mandel principle of macrohomogeneity holds if and only if the virtual work of the external surface traction and body force field of the RVE vanish. That is, the Hill–Mandel principle is equivalent to the following variational equations:*

$$\int_{\partial\Omega_\mu} \mathbf{t}^e \cdot \boldsymbol{\eta} \, dA = 0; \quad \int_{\Omega_\mu^s} \mathbf{b} \cdot \boldsymbol{\eta} \, dV = 0 \ \ \forall \boldsymbol{\eta} \in \mathscr{V}_\mu \tag{9.36}$$

Proof *By introducing Eq. (9.35) on the right-hand side of Eq. (9.34), we obtain*

$$\begin{aligned} \frac{1}{V_\mu} \int_{\Omega_\mu} \mathbf{P}_\mu : \dot{\mathbf{F}}_\mu \, dV &= \frac{1}{V_\mu} \int_{\Omega_\mu} \mathbf{P}_\mu : (\dot{\mathbf{F}} + \nabla \dot{\tilde{\mathbf{u}}}_\mu) dV \\ &= \mathbf{P} : \dot{\mathbf{F}} + \frac{1}{V_\mu} \int_{\Omega_\mu} \mathbf{P}_\mu : \nabla \dot{\tilde{\mathbf{u}}}_\mu dV \end{aligned} \tag{9.37}$$

Hence, identity (9.34) holds if and only if

$$\int_{\Omega_\mu} \mathbf{P}_\mu : \nabla \dot{\tilde{\mathbf{u}}}_\mu \, dV = 0 \, \forall \, \dot{\tilde{\mathbf{u}}}_\mu \in \mathscr{V}_\mu \tag{9.38}$$

Integration by parts of the left-hand side of Eq. (9.38), admitting the presence of voids ($\Omega_\mu^v \neq 0$), gives

$$\begin{aligned} \int_{\Omega_\mu} \mathbf{P}_\mu : \nabla \dot{\tilde{\mathbf{u}}}_\mu \, dV &= \int_{\partial\Omega_\mu} \mathbf{P}_\mu \, \mathbf{n} \cdot \dot{\tilde{\mathbf{u}}}_\mu \, dA - \int_{\Omega_\mu^s} (\operatorname{div} \mathbf{P}_\mu) \cdot \dot{\tilde{\mathbf{u}}}_\mu \, dV \\ &\quad - \int_{\Omega_\mu^v} (\operatorname{div} \mathbf{P}_\mu) \cdot \dot{\tilde{\mathbf{u}}}_\mu \, dV + \int_{\partial\Omega_\mu^v} [\![\mathbf{P}_\mu \mathbf{n}]\!] \cdot \dot{\tilde{\mathbf{u}}}_\mu \, dA \end{aligned} \tag{9.39}$$

or, in view of the strong equilibrium equations (9.29) together with assumption (9.30),

$$\int_{\Omega_\mu} \mathbf{P}_\mu : \nabla\dot{\tilde{\mathbf{u}}}_\mu \, \mathrm{d}V = \int_{\partial\Omega_\mu} \mathbf{t}^e \cdot \dot{\tilde{\mathbf{u}}}_\mu \, \mathrm{d}A - \int_{\Omega_\mu^s} \mathbf{b} \cdot \dot{\tilde{\mathbf{u}}}_\mu \, \mathrm{d}V \tag{9.40}$$

Thus, the Hill–Mandel principle is equivalent to the following variational equation:

$$\int_{\partial\Omega_\mu} \mathbf{t}^e \cdot \dot{\tilde{\mathbf{u}}}_\mu \, \mathrm{d}A - \int_{\Omega_\mu^s} \mathbf{b} \cdot \dot{\tilde{\mathbf{u}}}_\mu \, \mathrm{d}V = 0 \quad \forall \dot{\tilde{\mathbf{u}}}_\mu \in \mathscr{V}_\mu \tag{9.41}$$

Further, since $\mathscr{V}_\mu$ is a vector space (recall Section, 9.2.2), the above variational equation holds if and only if each of its integrals vanish individually. Hence, Eq. (9.41) is equivalent to Eq. (9.36), and Proposition 9.1 is proved.

Remark 9.4. (RVE reactive load systems) *Equation (9.36) states that the Hill–Mandel principle is equivalent to requiring that the external surface traction $\mathbf{t}^e$ and body force field $\mathbf{b}$ of the RVE be purely reactive; that is, they are a reaction to the kinematical constraints (embodied in the choice of $\mathscr{V}_\mu$) imposed upon the RVE and cannot be prescribed independently. Note that $\mathbf{t}^e$ and $\mathbf{b}$ belong to the functional space orthogonal to $\mathscr{V}_\mu$. Thus, once $\mathscr{V}_\mu$ is chosen (i.e., the kinematical constraints to be imposed upon the RVE are defined), the space to which $\mathbf{t}^e$ and $\mathbf{b}$ belong is automatically specified. This point is often overlooked in the literature where assumptions such as vanishing body force [44] and antiperiodic or uniform external tractions are frequently introduced in addition to kinematical constraints, without a clear causal relation between the choice of RVE kinematical constraints and the resulting reactions.*

9.3 The Multiscale Model Definition

To complete the definition of the present family of multiscale constitutive models, it is essential that the constitutive response of the RVE material be characterized. Here, attention is focused on multiscale models for which the microscopic stress $\mathbf{P}_\mu$ over the solid domain is related to the history of $\mathbf{F}_\mu$ by means of a generic continuum tensor-valued constitutive functional of the type (9.1), which includes internal variable-based phenomenological inelastic theories as particular cases. In other words, at an arbitrary point $\mathbf{y}$ of Ω_μ^s, we have

$$\mathbf{P}_\mu(\mathbf{y}, t) = \mathfrak{P}_y(\mathbf{F}_\mu^t(\mathbf{y})) \tag{9.42}$$

where, again, the superscript t denotes the history up to instant t. We remark that the material response is assumed generally nonuniform throughout the RVE, that is, the constitutive model describing the material response may vary from point to point of the microcell. The subscript y is used in Eq. (9.42) to emphasize that $\mathfrak{P}_y$ is the tensor-valued constitutive functional *at point* $\mathbf{y}$ of Ω_μ^s.

9.3.1
The Microscopic Equilibrium Problem

The variational equilibrium equation (9.28), in conjunction with the microscopic constitutive hypothesis (9.42) and the assumption that the solid–void/pore mechanical interaction is accounted for only through the traction $\mathbf{t}^{\mathrm{v}}$, leads naturally to a *microscopic equilibrium problem*, which, in view of statement (9.36) of the Hill–Mandel principle, consists in finding, for a given history $\mathbf{F} = \mathbf{F}(\mathbf{x}, t)$ of the deformation gradient at a point $\mathbf{x}$ of the macrocontinuum, a kinematically admissible microscopic displacement field $\mathbf{u}_\mu \in \mathscr{K}_\mu$ such that

$$\int_{\Omega_\mu^{\mathrm{s}}} \mathfrak{P}_y\{[\mathbf{I} + \nabla \mathbf{u}_\mu(\mathbf{y}, t)]^t\} : \nabla^{\mathrm{s}} \boldsymbol{\eta} \, \mathrm{d}V - \int_{\partial \Omega_\mu^{\mathrm{v}}} \mathbf{t}^{\mathrm{v}}(\mathbf{y}, t) \cdot \boldsymbol{\eta} \, \mathrm{d}A = 0 \, \forall \boldsymbol{\eta} \in \mathscr{V}_\mu \tag{9.43}$$

Equivalently, in view of definition (9.21) and identity (9.23), the equilibrium problem can be stated in terms of the displacement fluctuation field as follows: Find $\tilde{\mathbf{u}}_\mu \in \mathscr{V}_\mu$ such that, for each t,

$$\begin{aligned} G(\mathbf{F}, \tilde{\mathbf{u}}_\mu, \boldsymbol{\eta}) \equiv & \int_{\Omega_\mu^{\mathrm{s}}} \mathfrak{P}_y\{[\mathbf{F}(\mathbf{x}, t) + \nabla \tilde{\mathbf{u}}_\mu(\mathbf{y}, t)]^t\} : \nabla^{\mathrm{s}} \boldsymbol{\eta} \, \mathrm{d}V \\ & - \int_{\partial \Omega_\mu^{\mathrm{v}}} \mathbf{t}^{\mathrm{v}}(\mathbf{y}, t) \cdot \boldsymbol{\eta} \, \mathrm{d}A = 0 \, \forall \boldsymbol{\eta} \in \mathscr{V}_\mu \end{aligned} \tag{9.44}$$

where G is the virtual work functional.

9.3.2
The Multiscale Model: Well-Posed Equilibrium Problem

As it stands, for a given microscopic domain $\Omega_\mu = \Omega_\mu^{\mathrm{s}} \cup \Omega_\mu^{\mathrm{v}}$, microscopic response functionals $\mathfrak{P}_y$ for all points $\mathbf{y}$ of the RVE and field $\mathbf{t}^{\mathrm{v}}$, but with unspecified space $\mathscr{V}_\mu$, the equilibrium problem (9.44) is obviously ill-posed. The formulation of any multiscale constitutive model within the present framework is completed with the choice of an appropriate space $\mathscr{V}_\mu$, that is, with the choice of kinematical constraints to be imposed on the RVE. The essential requirements in defining the space $\mathscr{V}_\mu$ are

- According to Eqs (9.20) and (9.23),

$$\mathscr{V}_\mu \subset \tilde{\mathscr{K}}_\mu^* \tag{9.45}$$

- Space $\mathscr{V}_\mu$ must make the microscopic equilibrium problem well-posed.

Once an appropriate $\mathscr{V}_\mu$ is defined, the resulting multiscale model will deliver the macroscopic stress as a functional of the history of the macroscopic strain as follows:

1) Given the history of the macroscopic strain tensor $\mathbf{F}$ at the generic point $\mathbf{x}$ of the macrocontinuum, solve the microscopic equilibrium problem, defined by variational equation (9.44), for the microscopic displacement fluctuation $\tilde{\mathbf{u}}_\mu$.

2) With the solution $\tilde{\mathbf{u}}_\mu$ at hand, the macroscopic stress can be promptly obtained by homogenizing the resulting distribution of microscopic stress

$$\boldsymbol{\sigma}_\mu(\mathbf{y}, t) = \mathfrak{P}_y\{[\mathbf{F}(\mathbf{x}, t) + \nabla\tilde{\mathbf{u}}_\mu(\mathbf{y}, t)]^t\} \tag{9.46}$$

according to Eq. (9.31).[4)]

Remark 9.5. *Operations (1) and (2) above effectively define a macroscopic constitutive functional of the type (9.1). The general definition of a multiscale constitutive model within the present variational framework is summarized in Box 9.2.*

Box 9.2: General Definition of a Homogenization-Based Large Strain Multiscale Constitutive Model

(i) Define the domain $\Omega_\mu = \Omega_\mu^s \cup \Omega_\mu^v$, the microscopic constitutive functionals $\mathfrak{P}_y$ for all points $\mathbf{y} \in \Omega_\mu^s$ and the field $\mathbf{t}^v$ over $\partial\Omega_\mu^v$.

(ii) Choose a suitable space $\mathscr{V}_\mu$ of kinematically admissible RVE displacement fluctuations such that

$$\mathscr{V}_\mu \subset \tilde{\mathscr{K}}_\mu^*$$

where

$$\tilde{\mathscr{K}}_\mu^* \equiv \left\{\mathbf{v}, \text{sufficiently regular} \mid \int_{\partial\Omega_\mu} \mathbf{v} \otimes \mathbf{n}\, dA = \mathbf{0}\right\}$$

(iii) Resulting multiscale stress-constitutive functional

1) Microscopic equilibrium problem: *Given the history of the macroscopic deformation gradient,* $\mathbf{F}$, *find the field* $\tilde{\mathbf{u}}_\mu \in \mathscr{V}_\mu$ *such that, for each t,*

$$\int_{\Omega_\mu^s} \mathfrak{P}_y\{[\mathbf{F}(\mathbf{x}, t) + \nabla\tilde{\mathbf{u}}_\mu(\mathbf{y}, t)]^t\} : \nabla\boldsymbol{\eta}\, dV - \int_{\partial\Omega_\mu^v} \mathbf{t}^v(\mathbf{y}, t) \cdot \boldsymbol{\eta}\, dA = 0\, \forall \boldsymbol{\eta} \in \mathscr{V}_\mu''$$

2) Evaluate the macroscopic stress tensor

$$\mathbf{P} \equiv \frac{1}{V_\mu}\int_{\Omega_\mu} \mathbf{P}_\mu\, dV = \frac{1}{V_\mu}\int_{\Omega_\mu^s} \mathfrak{P}_y[(\mathbf{F} + \nabla\tilde{\mathbf{u}}_\mu)^t]\, dV$$

4) Alternatively, in this step, one may determine the reactive forces $\mathbf{t}^e$ and $\mathbf{b}$ by freeing the space of virtual displacements after solving Eq. (9.44) and then obtaining the macroscopic stress from expression (9.33) exclusively in terms of external forces. This approach is particularly attractive in the computational context, leading to a substantial reduction in the number of operations involved in the computation of the macroscopic stress.

9.4 Specific Classes of Multiscale Models: The Choice of $\mathcal{V}_\mu$

In this section, we cast four well-known classes of multiscale constitutive models within the variational framework presented above. Namely, we derive

1) The *Taylor* or homogeneous microcell strain model. This model is also commonly referred to as the *rule of mixtures*;
2) The linear (or affine) RVE boundary displacement model;
3) Periodic RVE boundary displacement fluctuations model; and
4) The minimum kinematical constraint, or uniform RVE boundary traction, model.

The RVE domain, microscopic constitutive functional, and possible internal traction field $\mathbf{t}^{\mathrm{v}}$ are left as arbitrary. Their characterization will depend on the particular microstructure in question. It is important to emphasize that *the different classes of multiscale models differ from one another solely in the definition of the space* $\mathscr{V}_\mu \subset \tilde{\mathscr{K}}_\mu^*$. The corresponding definitions are summarized in Box 9.3 and a more detailed derivation of each class is described in the following.

Box 9.3: Common $\mathscr{V}_\mu$ Definitions

a. Taylor model

$$\mathscr{V}_\mu = {}^{\mathrm{Taylor}}\mathscr{V}_\mu \equiv \{\mathbf{0}\}$$

b. Linear boundary displacements model

$$\mathscr{V}_\mu = {}^{\mathrm{lin}}\mathscr{V}_\mu \equiv \{\tilde{\mathbf{u}}_\mu \in \tilde{\mathscr{K}}_\mu^* \,|\, \tilde{\mathbf{u}}_\mu(\mathbf{y}, t) = \mathbf{0}\,\forall\,\mathbf{y} \in \partial\Omega_\mu\}$$

c. Periodic boundary displacements fluctuations model

$$\mathscr{V}_\mu = {}^{\mathrm{per}}\mathscr{V}_\mu \equiv \left\{\tilde{\mathbf{u}}_\mu \in \tilde{\mathscr{K}}_\mu^* \,|\, \tilde{\mathbf{u}}_\mu(\mathbf{y}^+, t) = \tilde{\mathbf{u}}_\mu(\mathbf{y}^-, t)\forall \text{ pairs } \{\mathbf{y}^+, \mathbf{y}^-\}\right\}$$

d. Minimum kinematical constraint (or uniform boundary traction)

$$\mathscr{V}_\mu = {}^{\mathrm{uni}}\mathscr{V}_\mu \equiv \tilde{\mathscr{K}}_\mu^*$$

9.4.1 Taylor Model

This is the simplest two-scale model of solid. It is obtained by choosing

$$\mathscr{V}_\mu = {}^{\mathrm{Taylor}}\mathscr{V}_\mu \equiv \{\mathbf{0}\} \tag{9.47}$$

That is, the kinematical constraint on the RVE is

$$\tilde{\mathbf{u}}_\mu = \mathbf{0} \quad \forall\,\mathbf{y} \in \Omega_\mu^{\mathrm{s}} \tag{9.48}$$

This choice implies that the associated total microscopic displacement field is linear in $\mathbf{y}$:

$$\mathbf{u}_\mu(\mathbf{y}, t) = \mathbf{u}(\mathbf{x}, t) + [\mathbf{F}(\mathbf{x}, t) - \mathbf{I}]\,\mathbf{y} \quad \forall\, \mathbf{y} \in \Omega_\mu^{\mathrm{s}} \tag{9.49}$$

and the microcell deformation gradient is *homogeneous*,

$$\mathbf{F}_\mu(\mathbf{y}, t) = \mathbf{F}(\mathbf{x}, t) \tag{9.50}$$

and coincides with the macroscopic deformation gradient at the corresponding material point $\mathbf{x}$ of the macrocontinuum. The hypothesis of homogeneous strain over a generally heterogeneous microstructure is often referred to as the *Taylor assumption* and can be traced back to the work of Taylor [45] in the context of experimental analysis of the plastic deformation of polycrystalline metals.

The RVE equilibrium problem (9.44) in this case has the trivial solution (9.48) and the microscopic stress is obtained directly from the microcell constitutive equation with prescribed deformation gradient, $\mathbf{F}$:

$$\mathbf{P}_\mu(\mathbf{y}, t) = \mathfrak{P}_y(\mathbf{F}^t) \tag{9.51}$$

Here, the body force $\mathbf{b}$, external surface traction $\mathbf{t}^e$, and the internal surface traction $\mathbf{t}^{\mathrm{v}}$ are all reactions to the full kinematical constraint imposed by the choice ${}^{\mathrm{Taylor}}\mathscr{V}_\mu$. They belong to the spaces of all sufficiently regular fields over the corresponding domain.

Remark 9.6. *An important limitation of the Taylor approximation is the fact that it does not consider the (possibly intricate) mechanical interaction among the different solid phases or between the solid phases and voids. It is a well-known fact that the strain field in the surroundings of voids and near the interface between different solid phases can be quite complex and may have a crucial impact upon the macroscopic mechanical response. When using a Taylor model, one must be aware of this fact. Another important shortcoming of this class of models is that the traction* $\mathbf{t}^{\mathrm{v}}$ *on* $\partial\Omega_\mu^{\mathrm{v}}$ *produces no virtual work and is, therefore, a mere reaction to the Taylor kinematical constraint (which includes fully constrained displacements on* $\partial\Omega_\mu^{\mathrm{v}}$*). Thus,* $\mathbf{t}^{\mathrm{v}}$ *cannot be prescribed independently in this case, making Taylor-type models unsuitable, for example, to describe the solid–fluid interaction in saturated porous media or crack closure effects in cracked media – alluded to in Remark 9.2.*

9.4.1.1 The Taylor-Based Constitutive Functional: the Rule of Mixtures

The macroscopic stress-constitutive functional for the two-scale model under the Taylor assumption is obtained from Eqs (9.31) and (9.51) as

$$\mathbf{P}(t) = {}^{\mathrm{Taylor}}\mathfrak{P}(\mathbf{F}^t) = \frac{1}{V_\mu}\int_{\Omega_\mu^{\mathrm{S}}} \mathfrak{P}_y(\mathbf{F}^t)\,\mathrm{d}V \tag{9.52}$$

An instance of the above functional of particular practical interest arises when the solid part of the microcell is made of a single material with constitutive response functional $\mathfrak{P}$ independent of $\mathbf{y}$. In this case, the macroscopic stress predicted by the Taylor-based model is simply

$$\mathbf{P}(t) = {}^{\text{Taylor}}\mathfrak{P}(\mathbf{F}^t) = \frac{\mathfrak{P}(\mathbf{F}^t)}{V_\mu} \int_{\Omega_\mu^{\mathrm{s}}} \mathrm{d}V = v^{\mathrm{s}}\, \mathbf{P}_\mu \tag{9.53}$$

where

$$v^{\mathrm{s}} \equiv \frac{V_\mu^{\mathrm{s}}}{V_\mu} \tag{9.54}$$

is the *solid volume fraction,* where V_μ^{s} denotes the solid volume of the RVE, and $\mathbf{P}_\mu = \mathfrak{P}(\mathbf{F}^t)$ is the stress resulting from the imposed strain history on the solid material. Trivially, in the absence of voids ($v^{\mathrm{s}} = 1$) we have $\mathbf{P} = \mathbf{P}_\mu = \mathfrak{P}(\mathbf{F}^t)$, that is, the macroscopic and microscopic constitutive functionals coincide. Another interesting particular specialization of Eq. (9.52) occurs when the solid part of the microcell is made of a number of distinct materials. To see this, let us assume that the microcell solid domain is divided into k nonoverlapping subdomains $\Omega_\mu^1, \ldots, \Omega_\mu^k$, with corresponding volumes $V_\mu^1, \ldots, V_\mu^k$. We then have

$$\Omega_\mu^{\mathrm{s}} = \bigcup_{i=1}^{k} \Omega_\mu^i, \qquad V_\mu^{\mathrm{s}} = \sum_{i=1}^{k} V_\mu^i \tag{9.55}$$

Each subdomain i is assumed to be made of a material modeled by a constitutive functional $\mathfrak{P}^i$. By introducing the partitioning (9.55) with the corresponding $\mathfrak{P}^i$'s into Eq. (9.52), we then promptly obtain the following macroscopic stress-constitutive functional:

$$\mathbf{P}\,(t) = {}^{\text{Taylor}}\mathfrak{P}(\mathbf{F}^t) = \sum_{i=1}^{k} v_i\, \mathfrak{P}^i(\mathbf{F}^t) = \sum_{i=1}^{k} v_i\, \mathbf{P}_\mu^i \tag{9.56}$$

where

$$v_i \equiv \frac{V_\mu^i}{V_\mu} \tag{9.57}$$

is the *volume fraction* and $\mathbf{P}_\mu^i$ is the (uniform) microscopic first Piola–Kirchhoff stress of phase i. That is, the macroscopic stress in this case is simply the weighted average of the stresses acting at the different solid phases. This rule is commonly known as the *rule of mixtures* and is often used as a first constitutive approximation.

Remark 9.7. (Symmetry properties of Taylor-type models) *The following result is trivial. Let all phases of the RVE share the same symmetry group, $\mathscr{S}$, that is, for $i = 1, \ldots, k$,*

$$\mathfrak{P}^i(\mathbf{F}^t) = \mathfrak{P}^i([\mathbf{F}\,\mathbf{Q}]^t) \tag{9.58}$$

for all time-independent orthogonal tensors $\mathbf{Q} \in \mathscr{S}$. Then, the resulting homogenized Taylor constitutive functional also has $\mathscr{S}$ as its symmetry group, that is,

$${}^{\text{Taylor}}\mathfrak{P}(\mathbf{F}^t) = {}^{\text{Taylor}}\mathfrak{P}([\mathbf{F}\,\mathbf{Q}]^t) \tag{9.59}$$

for all time-independent $\mathbf{Q} \in \mathscr{S}$. The above is true regardless of the geometry of phase arrangement within the microcell. This is another important shortcoming of Taylor-based *models. Note, for example, that directional geometrical arrangements of isotropic phases with distinct mechanical responses are physically bound to produce anisotropic macroscopic response. In such cases, however, Taylor-based multiscale models will predict isotropic homogenized response.*

9.4.2
Linear RVE Boundary Displacement Model

This class of models is derived by assuming that the RVE boundary displacement fluctuations vanish, that is, the set $\mathscr{V}_\mu$ taking part in problem (9.44) is chosen as

$$\mathscr{V}_\mu = {}^{\text{lin}}\mathscr{V}_\mu \equiv \{\tilde{\mathbf{u}}_\mu \in \tilde{\mathscr{K}}_\mu^* | \tilde{\mathbf{u}}_\mu(\mathbf{y}, t) = \mathbf{0}\,\forall\, \mathbf{y} \in \partial\Omega_\mu\} \tag{9.60}$$

This choice renders the RVE boundary displacement *linear* in $\mathbf{y}$:

$$\mathbf{u}_\mu(\mathbf{y}, t) = [\mathbf{F}(\mathbf{x}, t) - \mathbf{I}]\,\mathbf{y} \text{ on } \partial\Omega_\mu \tag{9.61}$$

In this case, the external surface traction, $\mathbf{t}^e$, orthogonal to $\mathscr{V}_\mu$ belongs to the space of all sufficiently regular fields over $\partial\Omega_\mu$, whereas the only body force field orthogonal to $\mathscr{V}_\mu$ is simply

$$\mathbf{b}(\mathbf{y}, t) = \mathbf{0} \text{ in } \Omega_\mu \tag{9.62}$$

9.4.3
Periodic Boundary Displacement Fluctuations Model

This class of constitutive models is appropriate to describe the behavior of materials with periodic microstruture. In this case, it is possible to define an RVE whose periodic repetition generates the macrostructure [2]. For simplicity, we focus the description on two-dimensional problems. Here, we follow the notation adopted in Ref. [36]. Consider, for example, the square or hexagonal RVEs, as illustrated in Figure 9.2. In this case, each pair i of cell sides consists of the equally sized subsets

$$\Gamma_i^+ \text{ and } \Gamma_i^-$$

of $\partial\Omega_\mu$, with respective unit normals

$$\mathbf{n}_i^+ \text{ and } \mathbf{n}_i^-$$

such that

$$\mathbf{n}_i^- = -\mathbf{n}_i^+ \tag{9.63}$$

Note that a one-to-one correspondence exists between the points of Γ_i^+ and Γ_i^-, that is, each point $\mathbf{y}^+ \in \Gamma_i^+$ has a corresponding pair $\mathbf{y}^- \in \Gamma_i^-$.

The key kinematical assumption for this class of models is that the *displacement fluctuation is periodic on the boundary* of the RVE, that is, for each pair $\{\mathbf{y}^+, \mathbf{y}^-\}$ of boundary points, we have

$$\tilde{\mathbf{u}}_\mu(\mathbf{y}^+, t) = \tilde{\mathbf{u}}_\mu(\mathbf{y}^-, t) \tag{9.64}$$

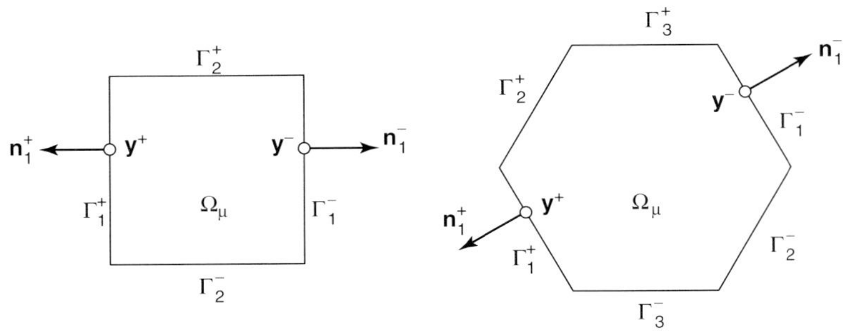

Figure 9.2 RVEs for periodic media. Square and hexagonal cells.

This assumption ensures the intercell displacement compatibility within the periodic medium. In addition, without loss of generality, to ensure the well-posedness of the equilibrium problem (9.44), we assume the origin of the RVE coordinate system to lie on $\partial\Omega^s_\mu$ and prescribe

$$\tilde{\mathbf{u}}_\mu(\mathbf{0}, t) = \mathbf{0} \tag{9.65}$$

Accordingly, the space $\mathscr{V}_\mu$ is defined as

$$\begin{array}{rcl} \mathscr{V}_\mu = {}^{\text{per}}\mathscr{V}_\mu & \equiv & \{\tilde{\mathbf{u}}_\mu \in \tilde{\mathscr{K}}^*_\mu | \tilde{\mathbf{u}}_\mu(\mathbf{0}, t) = \mathbf{0}, \\ \tilde{\mathbf{u}}_\mu(\mathbf{y}^+, t) & = & \tilde{\mathbf{u}}_\mu(\mathbf{y}^-, t)\ \forall\,\text{pairs}\{\mathbf{y}^+, \mathbf{y}^-\}\} \end{array} \tag{9.66}$$

To see that Eq. (9.45) holds for the above choice, note that under the present assumed geometrical partitioning of the RVE boundary, the constraint of set $\tilde{\mathscr{K}}^*_\mu$ defined in Eq. (9.17), can be equivalently written as

$$\sum_i \left(\int_{\Gamma_i^+} \tilde{\mathbf{u}}_\mu \otimes \mathbf{n}_i^+ \mathrm{d}A + \int_{\Gamma_i^-} \tilde{\mathbf{u}}_\mu \otimes \mathbf{n}_i^- \mathrm{d}A \right) = \mathbf{0} \tag{9.67}$$

Then, by simply replacing Eq. (9.63) together with Eq. (9.64) into the left-hand side of Eq. (9.67) we find that Eq. (9.67) is indeed satisfied.

The external surface traction, $\mathbf{t}^e$, is orthogonal to ${}^{\text{per}}\mathscr{V}_\mu$:

$$\int_{\partial\Omega_\mu} \mathbf{t}^e \cdot \boldsymbol{\eta}\, \mathrm{d}A = 0 \quad \forall \boldsymbol{\eta} \in {}^{\text{per}}\mathscr{V}_\mu \tag{9.68}$$

This implies that $\mathbf{t}^e$ is *antiperiodic* on $\partial\Omega_\mu$, that is,

$$\mathbf{t}^e(\mathbf{y}^+, t) = -\mathbf{t}^e(\mathbf{y}^-, t) \quad \forall\,\text{pairs}\{\mathbf{y}^+, \mathbf{y}^-\}\ \text{of}\ \partial\Omega_\mu \tag{9.69}$$

Finally, the body force field orthogonal to ${}^{\text{per}}\mathscr{V}_\mu$ here coincides with that of the linear boundary displacement model – the zero body force (9.62).

9.4.4 Minimum Kinematical Constraint: Uniform Boundary Traction

This class of models is obtained by assuming *minimum kinematical constraint* on the RVE. Again we impose Eq. (9.65) to ensure the well-posedness of the microscopic equilibrium problem, and then define the space of fluctuations as

$$\mathscr{V}_\mu = {}^{\text{uni}}\mathscr{V}_\mu \equiv \{\tilde{\mathbf{u}}_\mu \in \tilde{\mathscr{K}}_\mu^* | \tilde{\mathbf{u}}_\mu(\mathbf{0}, t) = \mathbf{0}\} \tag{9.70}$$

As for the linear and periodic boundary condition models, the above choice of $\mathscr{V}_\mu$ implies Eq. (9.62) – zero body force. The reactive external surface traction fields $\mathbf{t}^e$ compatible with the present model satisfy

$$\mathbf{t}^e(\mathbf{y}, t) = \mathbf{P}_\mu(\mathbf{y}, t)\mathbf{n}(\mathbf{y}) = \mathbf{P}\,(\mathbf{x}, t)\mathbf{n}(\mathbf{y}) \quad \forall \mathbf{y} \in \partial\Omega_\mu \tag{9.71}$$

where $\mathbf{P}$ is the *macroscopic* first Piola–Kirchhoff stress at the corresponding point $\mathbf{x}$ of the macrocontinuum. That the above is true is shown in the Appendix. Here and in what follows, we refer to fields $\mathbf{t}^e$ satisfying Eq. (9.71) as *uniform boundary traction* fields and the present class of models is also referred to as *uniform boundary traction model*. Note that, within the present kinematically based variational framework, the traction boundary condition (9.71) is *not* imposed a priori. The so-called uniform traction condition is a mere consequence of the choice of kinematically admissible fluctuations space ${}^{\text{uni}}\mathscr{V}_\mu$.

Remark 9.8. *The use of different definitions of $\mathscr{V}_\mu$ for a given RVE produces, in general, different estimates of the corresponding macroscopic constitutive response. The general rule is as follows. First, note that upon inspection of definitions (9.47), (9.60), (9.66), and (9.70) we see that*

$$^{\text{Taylor}}\mathscr{V}_\mu \subset {}^{\text{lin}}\mathscr{V}_\mu \subset {}^{\text{per}}\mathscr{V}_\mu \subset {}^{\text{uni}}\mathscr{V}_\mu \tag{9.72}$$

that is, the Taylor model gives the stiffest (most kinematically constrained) solution to the microscopic equilibrium problem, followed in order of decreasing stiffness, by the linear boundary displacement, the periodic displacement fluctuation, and the uniform boundary traction model. The uniform traction model produces the most compliant (least kinematically constrained) solution.

9.5
Models with Stress Averaging in the Deformed RVE Configuration

For the sake of completeness, it is worth remarking that an alternative family of large strain multiscale constitutive models can be obtained by postulating that the macroscopic *Cauchy* stress,

$$\boldsymbol{\sigma} \equiv (\det \mathbf{F}\,)^{-1}\mathbf{P}\,\mathbf{F}^T \tag{9.73}$$

is the volume average of the microscopic Cauchy stress field,

$$\boldsymbol{\sigma}_\mu \equiv (\det \mathbf{F}_\mu\,)^{-1}\mathbf{P}_\mu\mathbf{F}_\mu^T \tag{9.74}$$

over the *deformed* configuration of the RVE, that is, in this case, axiom (9.31) is replaced by

$$\boldsymbol{\sigma}(\mathbf{x}, t) \equiv \frac{1}{v_\mu}\int_{\varphi(\Omega_\mu)} \boldsymbol{\sigma}_\mu(\mathbf{y}, t)\mathrm{d}v \tag{9.75}$$

where $\varphi(\Omega_\mu)$ and v_μ denote, respectively, the deformed configuration and the volume of the deformed configuration of the RVE.

A crucial observation here is that, because of the nonlinearity of the relation between the Cauchy and first Piola–Kirchhoff stress tensors, the *material averaging* of the stress (9.31) is, in general, *not mechanically equivalent* to the *spatial averaging* (9.75), that is,

$$\begin{aligned}
\boldsymbol{\sigma} &\equiv \frac{1}{v_\mu}\int_{\varphi(\Omega_\mu)} \boldsymbol{\sigma}_\mu \, \mathrm{d}v = \frac{1}{v_\mu}\int_{\varphi(\Omega_\mu)} (\det \mathbf{F}_\mu)^{-1}\mathbf{P}_\mu \mathbf{F}_\mu^{\mathrm{T}} \, \mathrm{d}v \\
&\neq (\det \mathbf{F})^{-1}\mathbf{P}\,\mathbf{F}^{\mathrm{T}} \\
&\equiv \det\left(\frac{1}{V_\mu}\int_{\Omega_\mu} \mathbf{F}_\mu^{\mathrm{T}} \, \mathrm{d}V\right)^{-1}\left(\frac{1}{V_\mu}\int_{\Omega_\mu} \mathbf{P}_\mu \, \mathrm{d}V\right)\left(\frac{1}{V_\mu}\int_{\Omega_\mu} \mathbf{F}_\mu^{\mathrm{T}} \, \mathrm{d}V\right)
\end{aligned} \tag{9.76}$$

This issue has been discussed in detail by Nemat-Nasser [3]. The above inequality implies that, for a given RVE, macroscopic constitutive models resulting from the spatial averaging of the stress will, in general, differ from the analogous models based on the material averaging of stress. However, it is demonstrated by the authors [46] that under certain RVE kinematical constraints of practical interest, both formulations are equivalent. The equivalence holds, in particular, for the Taylor, linear boundary fluctuations and periodic boundary fluctuations models. In such cases, it is immaterial whether the stress averaging is defined on the material or on spatial configuration. For the uniform boundary tractions model, though, equivalence does not hold and the models resulting from material and spatial stress averaging do not coincide in general.

9.6 Problem Linearization: The Constitutive Tangent Operator

The linearization of nonlinear problems plays an important role in theoretical and computational continuum solid mechanics [47]. In the theoretical context, linearization can be essential in the determination of crucial properties, such as the stability of solutions for instance. In the computational setting, linearization becomes particularly important in the solution of approximate (discretised) nonlinear problems – typically undertaken by iterative numerical methods relying on the sequential solution of linearized problems. In particular, the widely used Newton–Raphson iterative algorithm, whose key advantage is its quadratic rate of asymptotic convergence, requires the exact linearization of the problem at each iteration [35, 48, 49]. Our main concern here is the derivation of an exact canonical form of constitutive tangent operator for multiscale constitutive models of the present type. We remark that the tangent operator will be derived in the continuum setting, that is, before any temporal or spatial discretization is introduced. The specific format taken by the tangent operators under different discretization schemes can be determined by simply introducing the relevant numerical approximations into the continuum canonical expressions. This is illustrated in Section 9.7, where the canonical formulae derived here are specialized to the time-discrete (but space-continuum) case.

First, we review the notion of tangent constitutive operator. To this end, consider the perturbation of $\mathbf{F}$,

$$\mathbf{F}_\epsilon(t) \equiv \mathbf{F}(t) + \epsilon\, \Delta\mathbf{F}(t) \tag{9.77}$$

by a scalar factor ϵ in the direction of a generic deformation gradient function $\Delta\mathbf{F}$ over the considered time history. The value of the generic tensor-valued constitutive functional (9.1) at the strain history $\mathbf{F}^t_\epsilon$ can be expressed as

$$\mathfrak{P}(\mathbf{F}^t_\epsilon) = \mathfrak{P}(\mathbf{F}^t) + \epsilon\, \mathscr{D}\mathfrak{P}(\mathbf{F}^t)[\Delta\mathbf{F}^t] + o(\epsilon) \tag{9.78}$$

where

$$\mathscr{D}\mathfrak{P}(\mathbf{F}^t)[\Delta\mathbf{F}^t] \equiv \left.\frac{\mathrm{d}}{\mathrm{d}\epsilon}\right|_{\epsilon=0} \mathfrak{P}(\mathbf{F}^t_\epsilon) \tag{9.79}$$

denotes the *directional derivative*, of the functional $\mathfrak{P}$ at $\mathbf{F}^t$ in the direction of $\Delta\mathbf{F}^t$ and $o(\cdot)$ denotes a tensor that vanishes faster than $(\cdot)$, that is, for any scalar a,

$$\lim_{a\to 0} \frac{o(a)}{a} = \mathbf{0} \tag{9.80}$$

The first two terms on the right-hand side of Eq. (9.78) define the linearization of the functional $\mathfrak{P}$ about the history $\mathbf{F}^t$. If the representation (9.78) holds for any $\Delta\mathbf{F}^t$, then the functional $\mathfrak{P}$ is said to be *differentiable* at $\mathbf{F}^t$ and the operator $\mathscr{D}\mathfrak{F}(\mathbf{F}^t)$ defined by Eq. (9.79) is the gradient – or *tangent constitutive operator* – of $\mathfrak{P}$ at $\mathbf{F}^t$. The operator $\mathscr{D}\mathfrak{P}(\mathbf{F}^t)$ is a tensor-valued *linear* functional that maps deformation gradient histories into stresses.

9.6.1
Homogenized Constitutive Functional

For multiscale constitutive models based on the volume averaging of the microscopic first Piola–Kirchhoff stress, the tensor-valued constitutive functional that delivers the macroscopic stress $\mathbf{P}$ – the *homogenized constitutive functional*, denoted by ${}^{\mathrm{hom}}\mathfrak{P}(\mathbf{F}^t)$ – is defined in generic form as

$${}^{\mathrm{hom}}\mathfrak{P}(\mathbf{F}^t) \equiv \frac{1}{V_\mu}\int_{\Omega^s_\mu} \mathfrak{P}_y[(\mathbf{F} + \nabla\tilde{\mathbf{u}}_\mu)^t]\,\mathrm{d}V \tag{9.81}$$

where the history of deformation gradient fluctuation $\nabla\tilde{\mathbf{u}}_\mu$ is itself a functional of the history of the macroscopic deformation gradient, $\mathbf{F}$, and is determined through the solution of the microscopic equilibrium problem (9.44). Note that for the simplest case – the Taylor model – for which $\tilde{\mathbf{u}}_\mu = \mathbf{0}$, the homogenized functional (9.81) reduces to Eq. (9.52). The (generally nonlinear) relationship between the histories of $\mathbf{F}$ and $\nabla\tilde{\mathbf{u}}_\mu$ defined through Eq. (9.44) is represented here by a (generally nonlinear) functional $\mathfrak{E}$:

$$(\nabla\tilde{\mathbf{u}}_\mu)^t = \mathfrak{E}(\mathbf{F}^t) \tag{9.82}$$

so that Eq. (9.81) can be more precisely expressed as

$$^{\text{hom}}\mathfrak{P}(\mathbf{F}^t) \equiv \frac{1}{V_\mu}\int_{\Omega_\mu^s} \mathfrak{P}_y[(\mathbf{F}^t + \mathfrak{E}(\mathbf{F}^t)]\,\mathrm{d}V \tag{9.83}$$

9.6.2
The Homogenized Tangent Constitutive Operator

The *homogenized tangent constitutive operator* is the tangent operator associated with functional $^{\text{hom}}\mathfrak{P}$. It is obtained by applying the directional derivative formula (9.79) to Eq. (9.83). This gives[5)]

$$\begin{aligned}
\mathscr{D}^{\text{hom}}\mathfrak{P}(\mathbf{F}^t)[\Delta\mathbf{F}^t] &\equiv \left.\frac{\mathrm{d}}{\mathrm{d}\epsilon}\right|_{\epsilon=0} {}^{\text{hom}}\mathfrak{P}(\mathbf{F}^t_\epsilon)\\
&= \frac{1}{V_\mu}\int_{\Omega_\mu^s} \mathscr{D}\,\mathfrak{P}_y(\mathbf{F}^t_\mu)\,\{\Delta\mathbf{F}^t + \mathscr{D}\,\mathfrak{E}(\mathbf{F}^t)[\Delta\mathbf{F}^t]\}\,\mathrm{d}V\\
&= \left\{\frac{1}{V_\mu}\int_{\Omega_\mu^s} \mathscr{D}\,\mathfrak{P}_y(\mathbf{F}^t_\mu)\mathrm{d}V\right\}[\Delta\mathbf{F}^t]\\
&\quad + \left\{\frac{1}{V_\mu}\int_{\Omega_\mu^s} \mathscr{D}\,\mathfrak{P}_y(\mathbf{F}^t_\mu)\mathscr{D}\,\mathfrak{E}(\mathbf{F}^t)\mathrm{d}V\right\}[\Delta\mathbf{F}^t]
\end{aligned} \tag{9.84}$$

where $\mathscr{D}\,\mathfrak{P}_y(\mathbf{F}^t_\mu)\,[\Delta\mathbf{F}^t]$ denotes the directional derivative of $\mathfrak{P}_y$ at the generic point $\mathbf{y}$ of Ω_μ^s evaluated at the microscopic deformation gradient history $\mathbf{F}^t_\mu(\mathbf{y})$ and $\mathscr{D}\,\mathfrak{E}(\mathbf{F}^t)\,[\Delta\mathbf{F}^t]$ is the directional derivative of the functional defined in Eq. (9.82) evaluated at $\mathbf{F}^t$.

To obtain a more convenient expression for the tangent operator, we proceed to explore the directional derivative of $\mathfrak{E}$. Then, in addition to the perturbed strain (9.77), consider the microscopic displacement fluctuation function

$$\tilde{\mathbf{u}}_\epsilon(\mathbf{y},t) \equiv \tilde{\mathbf{u}}_\mu(\mathbf{y},t) + \epsilon\,\Delta\tilde{\mathbf{u}}_\mu(\mathbf{y},t) \tag{9.85}$$

perturbed by the scalar factor ϵ in the direction of a generic field $\Delta\tilde{\mathbf{u}}_\mu \in \mathscr{V}_\mu$. Perturbation of both sides of Eq. (9.82), in conjunction with the directional derivative concept of Eqs (9.78) and (9.79), gives

$$(\nabla\tilde{\mathbf{u}}_\mu)^t + \epsilon(\nabla\Delta\tilde{\mathbf{u}}_\mu)^t = \mathfrak{E}(\mathbf{F}^t) + \epsilon\,\mathscr{D}\,\mathfrak{E}(\mathbf{F}^t)[\Delta\mathbf{F}^t] + o(\epsilon) \tag{9.86}$$

from which the following tangential relationship emerges:

$$(\nabla\Delta\tilde{\mathbf{u}}_\mu)^t = \mathscr{D}\,\mathfrak{E}(\mathbf{F}^t)[\Delta\mathbf{F}^t] \tag{9.87}$$

5) The linear operator represented as the integral of $\mathscr{D}\,\mathfrak{P}_y(\mathbf{F}^t)$ on the last right-hand side of Eq. (9.84) is *defined* such that

$$\left\{\frac{1}{V_\mu}\int_{\Omega_\mu^s}\mathscr{D}\,\mathfrak{P}_y(\mathbf{F}^t)\mathrm{d}V\right\}[\Delta\mathbf{F}^t] = \frac{1}{V_\mu}\int_{\Omega_\mu^s}\mathscr{D}\,\mathfrak{P}_y(\mathbf{F}^t)\,[\Delta\mathbf{F}^t]\mathrm{d}V$$

for any deformation gradient history $[\Delta\mathbf{F}^t]$. The linear operator represented as the integral of $\mathscr{D}\,\mathfrak{P}_y(\mathbf{F}^t_\mu)\mathscr{D}\,\mathfrak{E}(\mathbf{F}^t)$ has an analogous definition.

That is, $\mathscr{D}\,\mathfrak{E}$ maps increments of macroscopic deformation gradient histories linearly into increments of microscopic displacement fluctuation histories. Now, note that any *admissible pair* $\{\mathbf{F}_\epsilon, \tilde{\mathbf{u}}_\epsilon\}$ – that is, combination of macroscopic deformation gradient and displacement fluctuation functions that solves the equilibrium problem defined in Eq. (9.44) – satisfies

$$G(\mathbf{F}_\epsilon, \tilde{\mathbf{u}}_\epsilon, \boldsymbol{\eta}) = 0 \, \forall\, \boldsymbol{\eta} \in \mathscr{V}_\mu \tag{9.88}$$

Then, the *linearisation* of Eq. (9.88) at an admissible pair $\{\mathbf{F}, \tilde{\mathbf{u}}_\mu\}$ (at $\epsilon = 0$) is given by the variational equation

$$\begin{aligned} \mathscr{L}G(\mathbf{F}_\epsilon, \tilde{\mathbf{u}}_\epsilon, \boldsymbol{\eta}) &\equiv G(\mathbf{F}, \tilde{\mathbf{u}}_\mu, \boldsymbol{\eta}) + \left.\frac{\partial}{\partial \epsilon}\right|_{\epsilon=0} G(\mathbf{F}_\epsilon, \tilde{\mathbf{u}}_\epsilon, \boldsymbol{\eta}) \\ &= \left.\frac{\partial}{\partial \epsilon}\right|_{\epsilon=0} G(\mathbf{F}_\epsilon, \tilde{\mathbf{u}}_\epsilon, \boldsymbol{\eta}) = 0 \, \forall\, \boldsymbol{\eta} \in \mathscr{V}_\mu \end{aligned} \tag{9.89}$$

where $\mathscr{L}G$ denotes the linearized virtual work functional. By applying standard rules of differentiation to definition (9.44) of G, we see that the above relation defines the problem of finding, for a given function $\Delta\mathbf{F}$, the field $\Delta\tilde{\mathbf{u}}_\mu \in \mathscr{V}_\mu$ that solves the *linear* variational equation

$$\int_{\Omega^s_\mu} \mathscr{D}\,\mathfrak{P}_y(\mathbf{F}^t_\mu)[(\nabla \Delta\tilde{\mathbf{u}}_\mu)^t] : \nabla \boldsymbol{\eta} \,\mathrm{d}V = -\int_{\Omega^s_\mu} \mathscr{D}\,\mathfrak{P}_y(\mathbf{F}^t_\mu)[\Delta\mathbf{F}^t] : \nabla \boldsymbol{\eta} \,\mathrm{d}V \forall \boldsymbol{\eta} \in \mathscr{V}_\mu \tag{9.90}$$

This equation defines the linear operator $\mathscr{D}\,\mathfrak{E}(\mathbf{F}^t)$ of Eq. (9.87), that is, the linear mapping $\mathscr{D}\,\mathfrak{E}$ represents the solution of the problem defined by Eq. (9.90).

Remark 9.9. *Clearly, $\mathscr{D}\,\mathfrak{E}$ depends in general on the choice of space $\mathscr{V}_\mu$. For example, for the Taylor model ($\mathscr{V}_\mu = \{\mathbf{0}\} \Rightarrow \Delta\tilde{\mathbf{u}}_\mu = \mathbf{0}$) the tangent $\mathscr{D}\,\mathfrak{E}$ maps any macroscopic deformation gradient history increment into the zero tensor field. In this case, the general tangent operator (9.84) reduces to*

$$\mathscr{D}^{\mathrm{hom}}\,\mathfrak{P}(\mathbf{F}^t) = \mathscr{D}^{\mathrm{Taylor}}\mathfrak{P}(\mathbf{F}^t) \equiv \frac{1}{V_\mu}\int_{\Omega^s_\mu} \mathscr{D}\,\mathfrak{P}_y(\mathbf{F}^t)\mathrm{d}V \tag{9.91}$$

that is, the Taylor model tangent operator is the volume average of the microscopic constitutive tangent. Obviously, the same result can be arrived at by simply linearizing the Taylor homogenized stress-constitutive functional (9.52).

Finally, by substituting Eq. (9.91) into Eq. (9.84), we obtain the general canonical form of the homogenized tangent constitutive operator

$$\mathscr{D}^{\mathrm{hom}}\mathfrak{P}(\mathbf{F}^t) = \mathscr{D}^{\mathrm{Taylor}}\mathfrak{P}(\mathbf{F}^t_\mu) + \mathfrak{L}(\mathbf{F}^t_\mu) \tag{9.92}$$

where $\mathfrak{L}$ is the linear operator

$$\mathfrak{L}(\mathbf{F}^t_\mu) \equiv \frac{1}{V_\mu}\int_{\Omega^s_\mu} \mathscr{D}\,\mathfrak{P}_y(\mathbf{F}^t_\mu)\mathscr{D}\,\mathfrak{E}(\mathbf{F}^t_\mu)\mathrm{d}V \tag{9.93}$$

Remark 9.10. *Equation (9.90) has an appealing interpretation. Its right-hand side can be seen as the virtual work of the reactive body force field,* $\mathbf{b}_T$, *defined by*

$$\int_{\Omega_\mu} \mathbf{b}_T \cdot \boldsymbol{\eta}\, dV = \int_{\Omega_\mu} \mathscr{D}\,\mathfrak{P}_y(\mathbf{F}^t_\mu)[\Delta\mathbf{F}^t] : \nabla\boldsymbol{\eta}\, dV\, \forall\boldsymbol{\eta} \in \mathscr{V}_\mu \tag{9.94}$$

that would result if the RVE (with linearized constitutive equation) were subjected to a Taylor-type increment of microscopic deformation gradient field (i.e., a uniform deformation gradient increment equal to the macroscopic deformation gradient increment). The corresponding increment $\Delta\tilde{\mathbf{u}}_\mu$ *of microscopic displacement fluctuation on the left-hand side of Eq. (9.90) is the field of* $\mathscr{V}_\mu$ *that would generate stresses (through the linearized microscopic constitutive functional) in equilibrium with this reactive body force.*

Remark 9.11. (the structure of $\mathscr{D}^{\text{hom}}\mathfrak{P}$) *Expression (9.92) for the homogenized tangent constitutive operator reveals an interesting structure. The tangent* $\mathscr{D}^{\text{hom}}\mathfrak{P}$ *consists of the sum of two linear operators: (i) a contribution* $\mathscr{D}^{\text{Taylor}}\mathfrak{P}$, *which measures how the macroscopic stress varies under the Taylor assumption at the state defined by the microscopic deformation gradient history* $\mathbf{F}^t_\mu$; *and (ii) a contribution* $\mathfrak{L}(\mathbf{F}^t_\mu)$ *that measures the variation of microscopic stress that balances the body force reaction (refer to Remark 10) needed to maintain the Taylor constraint over an RVE whose displacement fluctuations are constrained to be in* $\mathscr{V}_\mu$. *The contribution* $\mathfrak{L}(\mathbf{F}^t_\mu)$ *is obviously generally dependent upon the choice of space* $\mathscr{V}_\mu$.

9.7 Time-Discrete Multiscale Models

In this section, we focus our attention on time-discrete approximations to constitutive functionals derived from a multiscale description of the material response. Time-discrete constitutive functions are particular instances of the generic tensor-valued functional $\mathfrak{P}$ of Eq. (9.1) where, for a given suitably chosen set of variables α_n defining the state at a time t_n and a given time increment Δt, the only event in the history of $\mathbf{F}$ required to determine the stress at time $t_{n+1} = t_n + \Delta t$ is the instantaneous value $\mathbf{F}_{n+1}$ of the deformation gradient at t_{n+1}, that is,

$$\mathbf{P}\,(t_{n+1}) = \mathfrak{P}(\mathbf{F}^{t_{n+1}}) \equiv \hat{\mathbf{P}}(\mathbf{F}_{n+1}, \Delta t; \alpha_n) \tag{9.95}$$

The function $\hat{\mathbf{P}}$ (for fixed Δt and α_n) is analogous to a (generally nonlinear) *elastic* constitutive function and is commonly referred to as the *algorithmic* or *incremental* constinutive function for the stress tensor. Time-discrete functions of the type (9.95) arise almost invariably in the finite-element solution of initial boundary value problems characterized by dissipative constitutive models, such as elastoplastic and viscoplastic. The time-discrete approximate constitutive law is obtained by introducing some form of time discretization in the original time-continuum constitutive equations [35].

The material presented in what follows is a particular case of what has already been seen in the preceding sections, obtained by simply specializing the relevant history constitutive functionals for the stress tensor as incremental constitutive functions of the type (9.95).

9.7.1 The Incremental Equilibrium Problem

By replacing the history functional $\mathfrak{P}_y$ with its incremental counterpart $\hat{\mathbf{P}}_\mu$ [6], the general (time-continuum) microscopic equilibrium problem (9.44) is transformed into a sequence of incremental (or time-discrete) equilibrium problems. Each step of this sequence is defined as follows. Knowing the displacement field and the set of variables $\boldsymbol{\alpha}_n$ for all points of the RVE at t_n and given the macroscopic deformation gradient $\mathbf{F}_{n+1}$, the time increment Δt and, if applicable, the reference internal surface traction $\mathbf{t}^{\mathrm{v}}_{n+1}$, the incremental equilibrium problem at a generic time step $n+1$ consists in determining a microscopic displacement fluctuation field $\tilde{\mathbf{u}}_\mu|_{n+1} \in \mathscr{V}_\mu$ such that

$$\hat{G}(\mathbf{F}_{n+1}, \tilde{\mathbf{u}}_\mu|_{n+1}, \boldsymbol{\eta}) \equiv \int_{\Omega^{\mathrm{s}}_\mu} \hat{\mathbf{P}}_\mu(\mathbf{F}_{n+1} + \nabla\tilde{\mathbf{u}}_\mu|_{n+1}, \Delta t; \boldsymbol{\alpha}_n) : \nabla\boldsymbol{\eta}\, \mathrm{d}V - \int_{\partial\Omega^{\mathrm{v}}_\mu} \mathbf{t}^{\mathrm{v}}_{n+1} \cdot \boldsymbol{\eta}\, \mathrm{d}A = 0\, \forall \boldsymbol{\eta} \in \mathscr{V}_\mu \tag{9.96}$$

where $\hat{G}$ is the *incremental* virtual work functional of step $n+1$.

9.7.2 The Homogenized Incremental Constitutive Function

Let us start by considering the simplest multiscale constitutive model – the Taylor model. In this case, we have

$$\mathscr{V}_\mu = \{\mathbf{0}\} \Longrightarrow \mathbf{u}_\mu|_{n+1} = \mathbf{u}(\mathbf{x}, t_{n+1}) + (\mathbf{F}_{n+1} - \mathbf{I})\,\mathbf{y} \Longleftrightarrow \tilde{\mathbf{u}}_\mu|_{n+1} = \mathbf{0} \tag{9.97}$$

and the homogenized constitutive function ${}^{\mathrm{Taylor}}\mathfrak{P}$ of Eq. (9.52) specializes as

$${}^{\mathrm{Taylor}}\hat{\mathbf{P}}(\mathbf{F}_{n+1}, \Delta t; \overline{\boldsymbol{\alpha}}_n) \equiv \frac{1}{V_\mu} \int_{\Omega^{\mathrm{s}}_\mu} \hat{\mathbf{P}}_\mu(\mathbf{F}_{n+1}, \Delta t; \boldsymbol{\alpha}_n) \mathrm{d}V \tag{9.98}$$

where $\overline{\boldsymbol{\alpha}}_n$ denotes the field of variables $\boldsymbol{\alpha}_n$ over the *entire domain* Ω^{s}_μ, that is, at each point $\mathbf{y} \in \Omega^{\mathrm{s}}_\mu$,

$$\boldsymbol{\alpha}_n = \overline{\boldsymbol{\alpha}}_n(\mathbf{y}) \tag{9.99}$$

Note that, since the constitutive model describing the microcell response may vary from point to point, $\boldsymbol{\alpha}$ may represent different sets of variables at distinct points of Ω^{s}_μ. For a microcell made of a number k of distinct solid phases, we have,

6) As the functional $\mathfrak{P}_y$, the incremental constitutive function in general varies from point to point of the RVE.

analogously to Eq. (9.56), the homogenized function

$$^{\text{Taylor}}\hat{\mathbf{P}}(\mathbf{F}_{n+1}, \Delta t; \overline{\boldsymbol{\alpha}}_n) = \sum_{i=1}^{k} v_i \, \hat{\mathbf{P}}^i_\mu (\mathbf{F}_{n+1}, \Delta t; \boldsymbol{\alpha}^i_n) \tag{9.100}$$

with

$$\overline{\boldsymbol{\alpha}}_n \equiv \{\boldsymbol{\alpha}^1_n, \ldots, \boldsymbol{\alpha}^k_n\} \tag{9.101}$$

In the general case, the homogenized incremental constitutive function for the stress is defined implicitly through the incremental microscopic equilibrium equation (9.96). The stress $\mathbf{P}_{n+1}$ is obtained by first solving Eq. (9.96) and then, with $\tilde{\mathbf{u}}_\mu|_{n+1}$ at hand, computing

$$\mathbf{P}_{n+1} = \frac{1}{V_\mu} \int_{\Omega^s_\mu} \hat{\mathbf{P}}_\mu (\mathbf{F}_{n+1} + \nabla \tilde{\mathbf{u}}_\mu \Big|_{n+1}, \Delta t; \boldsymbol{\alpha}_n) \mathrm{d}V \tag{9.102}$$

that is, the incremental macroscopic stress constitutive function is defined as

$$^{\text{hom}}\hat{\mathbf{P}}(\mathbf{F}_{n+1}, \Delta t; \overline{\boldsymbol{\alpha}}_n) \equiv \frac{1}{V_\mu} \int_{\Omega^s_\mu} \hat{\mathbf{P}}_\mu (\mathbf{F}_{n+1} + \nabla \tilde{\mathbf{u}}_\mu \Big|_{n+1}, \Delta t; \boldsymbol{\alpha}_n) \mathrm{d}V \tag{9.103}$$

where $\tilde{\mathbf{u}}_\mu|_{n+1}$ is itself a function *solely* of $\mathbf{F}_{n+1}$ (for given Δt and $\boldsymbol{\alpha}_n$) defined as the solution of Eq. (9.96). Note that, under the Taylor assumption (9.97), definition (9.103) recovers the Taylor model incremental constitutive function (9.98).

9.7.3 Time-Discrete Homogenized Constitutive Tangent

The time-discrete homogenized constitutive tangent operator is the incremental specialization of the linearized constitutive functional $\mathscr{D}^{\text{hom}}\mathfrak{P}$ discussed in Section 9.5. In this case, consider the redefined perturbed macroscopic deformation gradient

$$\mathbf{F}_\epsilon = \mathbf{F}_{n+1} + \epsilon \, \Delta\mathbf{F} \tag{9.104}$$

Here, $\mathbf{F}_\epsilon$ and $\Delta\mathbf{F}$ denote tensors and not tensor *functions* over the time domain as in Eq. (9.77). The incremental tangent, denoted by $^{\text{hom}}\mathsf{A}$ – a specialization of the linear functional $\mathscr{D}^{\text{hom}}\mathfrak{P}$ – is a *fourth-order tensor* that expresses the tangential relationship between the macroscopic first Piola–Kirchhoff stress tensor and the macroscopic deformation gradient at t_{n+1}, consistently with the homogenised incremental constitutive function (9.103). In other words, for any macroscopic deformation gradient direction $\Delta\mathbf{F}$, we have

$$^{\text{hom}}\hat{\mathbf{P}}(\mathbf{F}_\epsilon, \Delta t; \overline{\boldsymbol{\alpha}}_n) = {}^{\text{hom}}\hat{\mathbf{P}}(\mathbf{F}_{n+1}, \Delta t; \overline{\boldsymbol{\alpha}}_n) + \epsilon \, {}^{\text{hom}}\mathsf{A} : \Delta\mathbf{F} + o(\epsilon) \tag{9.105}$$

where $^{\text{hom}}\mathsf{A} : \Delta\mathbf{F}$ is the directional derivative of the incremental constitutive function $^{\text{hom}}\hat{\mathbf{P}}$ in the direction $\Delta\mathbf{F}$:

$$^{\text{hom}}\mathsf{A} : \Delta\mathbf{F} \equiv \frac{\mathrm{d}}{\mathrm{d}\epsilon}\Big|_{\epsilon=0} {}^{\text{hom}}\hat{\mathbf{P}}(\mathbf{F}_{n+1} + \epsilon \, \Delta\mathbf{F}, \Delta t; \overline{\boldsymbol{\alpha}}_n) \tag{9.106}$$

The operator ${}^{\text{hom}}\mathbf{A}$ is simply

$$ {}^{\text{hom}}\mathbf{A} \equiv \left.\frac{\partial}{\partial \mathbf{F}}\right|_{\mathbf{F}_{n+1}} {}^{\text{hom}}\hat{\mathbf{P}}(\mathbf{F}, \Delta t; \overline{\boldsymbol{\alpha}}_n) \tag{9.107} $$

9.7.3.1 Taylor Model

For the incremental version of the Taylor model, we obtain by specializing Eq. (9.91) to the time-discrete case (or by differentiating Eq. (9.98)), the following homogenized tangent operator:

$$ {}^{\text{hom}}\mathbf{A} = {}^{\text{Taylor}}\mathbf{A} \equiv \frac{1}{V_\mu}\int_{\Omega_\mu^s} \mathbf{A}_\mu \, \mathrm{d}V \tag{9.108} $$

where

$$ \mathbf{A}_\mu = \left.\frac{\partial}{\partial \mathbf{F}_\mu}\right|_{\mathbf{F}_\mu|_{n+1}} \hat{\mathbf{P}}_\mu(\mathbf{F}_\mu, \Delta t; \alpha b_n) \tag{9.109} $$

is the tangent operator consistent with the microscopic incremental constitutive law; that is, for the Taylor model, the incremental homogenized tangent tensor is the volume average of the microscopic incremental constitutive tangent tensor.

9.7.3.2 The General Case

To obtain a canonical expression for the general case, let us first consider the perturbed microscopic displacement fluctuation

$$ \tilde{\mathbf{u}}_\epsilon(\mathbf{y}) \equiv \tilde{\mathbf{u}}_\mu|_{n+1}(\mathbf{y}) + \epsilon\, \Delta\tilde{\mathbf{u}}_\mu(\mathbf{y}) \tag{9.110} $$

Note that, different from Eq. (9.85), the vectors $\tilde{\mathbf{u}}_\epsilon$ and $\Delta\tilde{\mathbf{u}}_\mu$ here are fields over Ω_μ^s only (and not over the time history).

The tangential relation between $\Delta\mathbf{F}$ and $\Delta\tilde{\mathbf{u}}_\mu$ is obtained from the linearization of the incremental equilibrium problem defined by Eq. (9.96), which results in the following specialization of the problem associated with Eq. (9.90): Given $\Delta\mathbf{F}$, find the field $\Delta\tilde{\mathbf{u}}_\mu \in \mathscr{V}_\mu$ that solves the linear variational equation

$$ \int_{\Omega_\mu^s} \nabla\boldsymbol{\eta} : \mathbf{A}_\mu : \nabla\Delta\tilde{\mathbf{u}}_\mu \mathrm{d}V = -\left[\int_{\Omega_\mu^s} \nabla\boldsymbol{\eta} : \mathbf{A}_\mu \mathrm{d}V\right] : \Delta\mathbf{F}\, \forall \boldsymbol{\eta} \in \mathscr{V}_\mu \tag{9.111} $$

A more compact expression for the above time-discrete tangent relation can be obtained by adopting the decomposition used by Michel *et al.* [2] in the infinitesimal strain context as follows. Let us write $\Delta\mathbf{F}$ in Cartesian component form:

$$ \Delta\mathbf{F} = \Delta F_{ij}\, \mathbf{e}_i \otimes \mathbf{e}_j \tag{9.112} $$

where $\{\mathbf{e}_i\}$ is an orthonormal basis of the three-dimensional Euclidean space and the scalars ΔF_{ij} are the corresponding Cartesian components of $\Delta\mathbf{F}$. In addition, let the vector field $\Delta\tilde{\mathbf{u}}_{ij} \in \mathscr{V}_\mu$ be the solution of

$$ \int_{\Omega_\mu^s} \nabla\boldsymbol{\eta} : \mathbf{A}_\mu : \nabla\Delta\tilde{\mathbf{u}}_{ij}\, \mathrm{d}V = -\left[\int_{\Omega_\mu^s} \nabla\boldsymbol{\eta} : \mathbf{A}_\mu\, \mathrm{d}V\right] : \mathbf{e}_i \otimes \mathbf{e}_j\, \forall \boldsymbol{\eta} \in \mathscr{V}_\mu \tag{9.113} $$

The vector fields $\Delta\tilde{\mathbf{u}}_{ij}$ are referred to as *tangential displacement fluctuations*. Since Eq. (9.111) is linear, $\Delta\tilde{\mathbf{u}}_\mu$ can be constructed by superposition of tangential solutions $\Delta\tilde{\mathbf{u}}_{ij}$ and can be expressed by the following linear combination:

$$\Delta\tilde{\mathbf{u}}_\mu = \Delta F_{ij}\, \Delta\tilde{\mathbf{u}}_{ij} \tag{9.114}$$

To obtain a final expression for the homogenized operator ${}^{\mathrm{hom}}\mathsf{A}$, we specialize Eq. (9.84) to the time-discrete setting, which gives

$$\begin{aligned} {}^{\mathrm{hom}}\mathsf{A} : \Delta\mathbf{F} &= \left[\frac{1}{V_\mu}\int_{\Omega_\mu^{\mathrm{s}}} \mathsf{A}_\mu \,\mathrm{d}V\right] : \Delta\mathbf{F} + \frac{1}{V_\mu}\int_{\Omega_\mu^{\mathrm{s}}} \mathsf{A}_\mu : \nabla\Delta\tilde{\mathbf{u}}_\mu \,\mathrm{d}V \\ &= {}^{\mathrm{Taylor}}\mathsf{A} : \Delta\mathbf{F} + \left[\frac{1}{V_\mu}\int_{\Omega_\mu^{\mathrm{s}}} \mathsf{A}_\mu : \nabla\Delta\tilde{\mathbf{u}}_{ij} \,\mathrm{d}V\right] \Delta F_{ij} \end{aligned} \tag{9.115}$$

where we have used Eq. (9.114) and expression (9.108) for the Taylor model incremental tangent operator. To further simplify the expression, note that in Cartesian component form, the second-order tensor $\mathsf{A}_\mu : \nabla\Delta\tilde{\mathbf{u}}_{ij}$ can be expressed as

$$\mathsf{A}_\mu : \nabla\Delta\tilde{\mathbf{u}}_{ij} = (\mathsf{A}_\mu)_{klpq}(\nabla\Delta\tilde{\mathbf{u}}_{ij})_{pq}\mathbf{e}_k \otimes \mathbf{e}_l \tag{9.116}$$

so that the second summand on the last right-hand side of Eq. (9.115) can be written as

$$\left[\frac{1}{V_\mu}\int_{\Omega_\mu^{\mathrm{s}}} \mathsf{A}_\mu : \nabla\Delta\tilde{\mathbf{u}}_{ij} \,\mathrm{d}V\right] \Delta F_{ij} = \tilde{\mathsf{A}} : \Delta\mathbf{F} \tag{9.117}$$

where $\tilde{\mathsf{A}}$ is the time-discrete version of the linear operator $\mathfrak{L}$ of Eq. (9.93):

$$\tilde{\mathsf{A}} \equiv \left[\frac{1}{V_\mu}\int_{\Omega_\mu^{\mathrm{s}}} (\mathsf{A}_\mu)_{ijpq}\,(\nabla\Delta\tilde{\mathbf{u}}_{kl})_{pq} \,\mathrm{d}V\right] \mathbf{e}_i \otimes \mathbf{e}_j \otimes \mathbf{e}_k \otimes \mathbf{e}_l \tag{9.118}$$

Finally, with the above at hand, we arrive at the following compact canonical formula for the general homogenized incremental constitutive tangent operator, analogous to Eq. (9.92),

$${}^{\mathrm{hom}}\mathsf{A} = {}^{\mathrm{Taylor}}\mathsf{A} + \tilde{\mathsf{A}} \tag{9.119}$$

Only the contribution $\tilde{\mathsf{A}}$ depends on the choice of space $\mathscr{V}_\mu$ (note that the solutions $\Delta\tilde{\mathbf{u}}_{ij}$ of Eq. (9.113) taking part in Eq. (9.118) depend on this choice in general). Obviously, under the Taylor assumption (9.97), $\Delta\tilde{\mathbf{u}}_{ij} = \mathbf{0}$, so that $\tilde{\mathsf{A}}$ vanishes and Eq. (9.119) recovers the Taylor tangent (9.108). For convenience, the essential expressions defining ${}^{\mathrm{hom}}\mathsf{A}$ are summarized in Box 9.4. We remark that expression (9.119), which generalizes the infinitesimal homogenized elasticity tensor derived by Michel *et al.* [2], has been recently presented by Larsson and Runesson [44], derived on the basis of a primal and a dual approach by direct linearization of the time-discrete equations. Here, this result is attained by the specialization of the canonical continuum form Eq. (9.91).

Box 9.4: The Incremental Homogenized Constitutive Tangent Operator

(i) Taylor contribution

$$^{\text{Taylor}}\mathsf{A} = \frac{1}{V_\mu}\int_{\Omega^s_\mu} \mathsf{A}_\mu \,\mathrm{d}V$$

(ii) Tangential displacement fluctuations: For $i, j = 1, 2, 3$, find $\Delta\tilde{\mathbf{u}}_{ij} \in \mathscr{V}_\mu$ such that

$$\int_{\Omega^s_\mu} \nabla\boldsymbol{\eta} : \mathsf{A}_\mu : \nabla\Delta\tilde{\mathbf{u}}_{ij}\,\mathrm{d}V = -\left[\int_{\Omega^s_\mu} \nabla\boldsymbol{\eta} : \mathsf{A}_\mu\,\mathrm{d}V\right] : \mathbf{e}_i \otimes \mathbf{e}_j \forall\, \boldsymbol{\eta} \in \mathscr{V}_\mu$$

(iii) Tangential fluctuation contribution

$$\tilde{\mathsf{A}} \equiv \left[\frac{1}{V_\mu}\int_{\Omega^s_\mu} (\mathsf{A}_\mu)_{ijpq}\,(\nabla\Delta\tilde{\mathbf{u}}_{kl})_{pq}\,\mathrm{d}V\right] \mathbf{e}_i \otimes \mathbf{e}_j \otimes \mathbf{e}_k \otimes \mathbf{e}_l$$

(iv) The incremental homogenized tangent operator

$$^{\text{hom}}\mathsf{A} = ^{\text{Taylor}}\mathsf{A} + \tilde{\mathsf{A}}$$

9.8 The Infinitesimal Strain Theory

An entirely analogous axiomatic variational structure can be established for infinitesimal strain multiscale constitutive models by simply constraining the above finite strain framework to the case of linearized kinematics. For completeness, the corresponding basic axioms are listed in Box 9.4, where $\boldsymbol{\varepsilon}$ and $\boldsymbol{\varepsilon}_\mu$ denote, respectively, the macroscopic and microscopic infinitesimal strain tensors; $\boldsymbol{\sigma}$ and $\boldsymbol{\sigma}_\mu$ are the macroscopic and microscopic stress tensors; and ∇^s and $\otimes_s$ denote, respectively, the symmetric gradient operator and the symmetric tensor product.

The general model definition follows that of Box 9.2, with the definition of $\tilde{\mathscr{K}}^*_\mu$, the equilibrium equation and the stress averaging relation replaced by those of Box 9.5. The corresponding functional spaces for the Taylor, linear boundary displacements, and periodic boundary fluctuations have the same definitions as those of Box 9.3, whereas the minimally constrained space of kinematically admissible fluctuations follows the definition of item (ii) of Box 9.5. Note that the displacement fluctuation field in the present context of infinitesimal strains is defined, analogously to Eq. (9.13), as

$$\tilde{\mathbf{u}}_\mu(\mathbf{y}, t) \equiv \mathbf{u}_\mu(\mathbf{y}, t) - [\mathbf{u}(\mathbf{x}, t) + \boldsymbol{\varepsilon}\,\mathbf{y}] \tag{9.120}$$

where $\boldsymbol{\varepsilon}\mathbf{y}$ represents a displacement field that causes uniform straining of the RVE, coinciding with the macroscopic strain. The uniform boundary traction,

$$\mathbf{t}(\mathbf{y}) = \boldsymbol{\sigma}_\mu(\mathbf{y})\,\mathbf{n}(\mathbf{y}) = \boldsymbol{\sigma}\,\mathbf{n}(\mathbf{y}) \quad \forall \mathbf{y} \in \partial\Omega_\mu \tag{9.121}$$

resulting from the minimal constraint in this case can be proved analogously to the proof given in the Appendix for the large strain model. A step-by-step proof specific for the small strain case is given in Ref. [42].

Box 9.5: Infinitesimal Strain Multiscale Constitutive Models: Basic Axioms

(i) Infinitesimal strain averaging relation:

$$\varepsilon \equiv \frac{1}{V_\mu} \int_{\Omega_\mu} \varepsilon_\mu \, \mathrm{d}V$$

(ii) The actual set $\tilde{\mathscr{K}}_\mu$ of kinematically admissible displacement fluctuation fields of the RVE is a subspace of its minimally constrained counterpart,

$$\tilde{\mathscr{K}}_\mu^* \equiv \left\{ \mathbf{v}, \text{sufficiently regular} \Big| \int_{\partial\Omega_\mu} \mathbf{v} \otimes_s \mathbf{n} \, \mathrm{d}A = \mathbf{0} \right\}$$

(iii) Equilibrium of the RVE:

$$\int_{\Omega_\mu} \boldsymbol{\sigma}_\mu(\mathbf{y}, t) : \nabla^s \boldsymbol{\eta} \, \mathrm{d}V - \int_{\Omega_\mu} \mathbf{b}(\mathbf{y}, t) \cdot \boldsymbol{\eta} \, \mathrm{d}V - \int_{\partial\Omega_\mu} \mathbf{t}^e(\mathbf{y}, t) \cdot \boldsymbol{\eta} \, \mathrm{d}A = 0 \, \forall \boldsymbol{\eta} \in \mathscr{V}_\mu$$

where $\mathscr{V}_\mu$ is the space of virtual displacements of the RVE.

(iv) Stress averaging relation:

$$\boldsymbol{\sigma} \equiv \frac{1}{V_\mu} \int_{\Omega_\mu} \boldsymbol{\sigma}_\mu \, \mathrm{d}V$$

(v) Hill–Mandel principle of macrohomogeneity:

$$\boldsymbol{\sigma} : \dot{\varepsilon} = \frac{1}{V_\mu} \int_{\Omega_\mu} \boldsymbol{\sigma}_\mu : \dot{\varepsilon}_\mu \, \mathrm{d}V$$

for any kinematically admissible velocity field of the RVE.

Finally, the incremental homogenized constitutive tangent operator of the infinitesimal theory is given by completely analogous expressions to those of Box 9.4, with ∇ replaced with ∇^s in items (ii) and (iii) and $\otimes$ replaced with $\otimes_s$ in item (ii).

9.9 Concluding Remarks

The formulation of a wide family of small and large strain generally inelastic multiscale constitutive models has been reviewed and cast within an axiomatic kinematical variational framework. The axiomatic structure is defined by (i) the strain averaging relation; (ii) a simple constraint upon the possible choices of sets of kinematically admissible displacement fluctuation fields of the RVE; (iii) the equilibrium of the RVE; (iv) the stress averaging relation; and (v) the Hill–Mandel principle of macrohomogeneity. Four well-known classes of multiscale theories have been presented within this framework: the Taylor model; the linear boundary displacement fluctuations model; the periodic boundary fluctuations model; and the minimally constrained, or uniform boundary traction, model. Canonical formulae for constitutive tangent operators have been derived first in the fully (space- and time-) continuum setting and then specialized to the space-continuum/time-discrete case. Fully discrete tangent operators can be obtained by simply specializing the derived time-discrete formulae further according to the chosen spatial discretisation method. The possible mechanical equivalence between large strain models based on the reference volume averaging of the first Piola–Kirchhoff stress and the spatial averaging of the Cauchy stress has been briefly discussed. Finally, we remark that a completely analogous statical variational formulation of the present family of models – dual to the kinematical formulation presented here – is straightforward and will be the subject of a forthcoming publication (see Ref. [50] for a dual formulation under infinitesimal strains).

Appendix

Here we show that Eq. (9.71) is indeed true. That is, we demonstrate that if $\mathbf{t}^e$ lies in the space orthogonal to ${}^{\mathrm{uni}}\mathscr{V}_\mu$ of Eq. (9.70), then Eq. (9.71) holds – the reference external RVE boundary tractions are uniform.

We start by noting that, in general, as the RVE is in equilibrium, we have

$$\mathbf{t}^e(\mathbf{y}, t) = \mathbf{P}_\mu(\mathbf{y}, t)\, \mathbf{n}(\mathbf{y}) \tag{A.9.1}$$

Further, if $\mathbf{t}^e$ is orthogonal to ${}^{\mathrm{uni}}\mathscr{V}_\mu$, we have

$$\int_{\partial\Omega_\mu} \mathbf{t}^e \cdot \boldsymbol{\eta}\, \mathrm{d}A = 0\, \forall \boldsymbol{\eta} \in {}^{\mathrm{uni}}\mathscr{V}_\mu \tag{A.9.2}$$

or, equivalently, in view of the identity (A.9.1),

$$\int_{\partial\Omega_\mu} \mathbf{P}_\mu : [\boldsymbol{\eta} \otimes \mathbf{n}]\, \mathrm{d}A = 0\, \forall \boldsymbol{\eta} \in {}^{\mathrm{uni}}\mathscr{V}_\mu \tag{A.9.3}$$

In the present demonstration, we make use of the fact that an arbitrary tensor field $\mathbf{P}_\mu$ over Ω_μ can be split as a sum

$$\mathbf{P}_\mu(\mathbf{y}, t) = \Sigma(t) + \tilde{\Sigma}(\mathbf{y}, t) \tag{A.9.4}$$

of a tensor Σ constant in $\mathbf{y}$ and a tensor $\tilde{\Sigma}$ such that, in the two-dimensional case,[7)]

$$\int_{\partial\Omega_\mu} [\tilde{\Sigma}_n(\mathbf{n}\otimes\mathbf{n}) + \tilde{\Sigma}_s(\mathbf{s}\otimes\mathbf{n})]\,\mathrm{d}A = 0 \tag{A.9.5}$$

with $\tilde{\Sigma}_n$ and $\tilde{\Sigma}_s$ denoting, respectively, the normal and tangential components of the vector field $\tilde{\Sigma}\mathbf{n}$ in the local oriented orthonormal basis $\{\mathbf{n}(\mathbf{y}), \mathbf{s}(\mathbf{y})\}$ consisting of the unit normal and tangential vectors to $\partial\Omega_\mu$:

$$\tilde{\Sigma}_n \equiv \tilde{\mathbf{\Sigma}} : (\mathbf{n}\otimes\mathbf{n}), \quad \tilde{\Sigma}_s \equiv \tilde{\mathbf{\Sigma}} : (\mathbf{s}\otimes\mathbf{n}) \tag{A.9.6}$$

In the split Eqs (A.9.4–A.9.5),

$$\mathbf{\Sigma}(t) = \mathsf{R}^{-1} : \Psi, \quad \tilde{\Sigma}(\mathbf{y},t) = \mathbf{P}_\mu(\mathbf{y},t) - \Sigma(t) \tag{A.9.7}$$

where Ψ is the second-order tensor defined by

$$\Psi \equiv \int_{\partial\Omega_\mu} \mathbf{P}_\mu : (\mathbf{n}\otimes\mathbf{n}\otimes\mathbf{n}\otimes\mathbf{n} + \mathbf{s}\otimes\mathbf{n}\otimes\mathbf{s}\otimes\mathbf{n})\,\mathrm{d}A \tag{A.9.8}$$

and R is the fourth-order tensor

$$\mathsf{R} \equiv \int_{\partial\Omega_\mu} (\mathbf{n}\otimes\mathbf{n}\otimes\mathbf{n}\otimes\mathbf{n} + \mathbf{s}\otimes\mathbf{n}\otimes\mathbf{s}\otimes\mathbf{n})\,\mathrm{d}A \tag{A.9.9}$$

that depends exclusively on the geometry of the RVE boundary and is invertible for closed boundaries $\partial\Omega_\mu$[8)] (which is always the case for the family of multiscale constitutive models discussed in this chapter). Note that Eq. (A.9.4) follows trivially from Eq. (A.9.7) and the inversibility of R, and that Eq. (A.9.5) holds can be established by means of straightforward manipulations after the substitution of Eqs (A.9.6)–(A.9.9) on the left-hand side of Eq. (A.9.5). Hence, the additive split (A.9.4, A.9.5) is indeed true.

Now, let us consider the following vector field over $\partial\Omega_\mu$

$$\eta^*(\mathbf{y},t) \equiv \tilde{\Sigma}_n(\mathbf{y},t)\,\mathbf{n}(\mathbf{y}) + \tilde{\Sigma}_s(\mathbf{y},t)\,\mathbf{s}(\mathbf{y}) - \mathbf{c}(t) \tag{A.9.10}$$

where $\mathbf{c}(t)$ is the (constant in $\mathbf{y}$) vector

$$\mathbf{c}(t) = \tilde{\Sigma}_n(\mathbf{0},t)\,\mathbf{n}(\mathbf{0}) + \tilde{\Sigma}_s(\mathbf{0},t)\,\mathbf{s}(\mathbf{0}) \tag{A.9.11}$$

7) For simplicity, the demonstration here focuses on the two-dimensional case. The three-dimensional case is completely analogous.

8) For example, for a square-shaped RVE, it can be easily established that

$$\mathsf{R} = a\mathsf{I}$$

where the scalar a equals half of the perimeter of the RVE and I is the fourth-order identity tensor with cartesian components

$$\mathsf{I}_{ijkl} = \delta_{ik}\delta_{jl}$$

From the above definition, trivially,

$$\boldsymbol{\eta}^*(\mathbf{0}, t) = \mathbf{0} \tag{A.9.12}$$

In addition,

$$\int_{\partial\Omega_\mu} \boldsymbol{\eta}^* \otimes \mathbf{n}\, dA = \int_{\partial\Omega_\mu} [\tilde{\Sigma}_n(\mathbf{n} \otimes \mathbf{n}) + \tilde{\Sigma}_s(\mathbf{s} \otimes \mathbf{n})]\, dA - \mathbf{c} \otimes \int_{\partial\Omega_\mu} \mathbf{n}\, dA \tag{A.9.13}$$

which, in view of Eq. (A.9.5) and the fact that $\int_{\partial\Omega_\mu} \mathbf{n}\, dA = \mathbf{0}$ for closed domains $\partial\Omega_\mu$, is equivalent to

$$\int_{\partial\Omega_\mu} \boldsymbol{\eta}^* \otimes \mathbf{n}\, dA = \mathbf{0} \tag{A.9.14}$$

Hence, according to definition (9.69), we have that

$$\boldsymbol{\eta}^* \in {}^{\text{uni}}\mathscr{V}_\mu \tag{A.9.15}$$

so that the orthogonality condition (A.9.3) requires, in particular, that

$$\int_{\partial\Omega_\mu} \mathbf{P}_\mu : (\boldsymbol{\eta}^* \otimes \mathbf{n}) dA = 0 \tag{A.9.16}$$

By expanding Eq. (A.9.16), taking Eqs (A.9.4) and (A.9.10) into account, we obtain

$$\begin{aligned}\int_{\partial\Omega_\mu} \mathbf{P}_\mu : (\boldsymbol{\eta}^* \otimes \mathbf{n})\, dA = \boldsymbol{\sigma} : \int_{\partial\Omega_\mu} (\tilde{\Sigma}_n \mathbf{n} \otimes \mathbf{n} + \tilde{\Sigma}_s \mathbf{s} \otimes \mathbf{n})\, dA + \int_{\partial\Omega_\mu} \tilde{\Sigma}_n^2\, dA \\ + \int_{\partial\Omega_\mu} \tilde{\Sigma}_s^2\, dA + \mathbf{c} \cdot \int_{\partial\Omega_\mu} \mathbf{P}_\mu \mathbf{n}\, dA = 0\end{aligned} \tag{A.9.17}$$

In view of Eq. (A.9.5) and the fact that equilibrium with $\mathbf{b} = \mathbf{0}$ (the body force field for the present class of models) requires that $\int_{\partial\Omega_\mu} \mathbf{P}_\mu \mathbf{n} dA = 0$, the above identity implies that

$$\tilde{\Sigma}_n(\mathbf{y}, t) = \tilde{\Sigma}_s(\mathbf{y}, t) = 0\ \forall\, \mathbf{y} \in \partial\Omega_\mu \tag{A.9.18}$$

which together with Eq. (A.9.4) gives

$$\mathbf{P}_\mu(\mathbf{y}, t)\, \mathbf{n}(\mathbf{y}) = \boldsymbol{\Sigma}(t)\, \mathbf{n}(\mathbf{y})\ \forall\, \mathbf{y} \in \partial\Omega_\mu \tag{A.9.19}$$

The substitution of the above into Eq. (A.9.1) establishes that, for the present class of constitutive models, the reference external surface traction $\mathbf{t}^e$ is related to the constant (in $\mathbf{y}$) tensor Σ according to

$$\mathbf{t}^e = \boldsymbol{\Sigma}\, \mathbf{n} \tag{A.9.20}$$

Finally, by combining Eqs (A.9.20), (9.33), and the fact that $\mathbf{b} = \mathbf{0}$ for the present class of models, we obtain

$$\mathbf{P} = \Sigma \left(\frac{1}{V_\mu} \int_{\partial\Omega_\mu} \mathbf{n} \otimes \mathbf{y}\, dA \right) = \Sigma\, \mathbf{I} = \Sigma \tag{A.9.21}$$

This completes the demonstration.

Acknowledgments

The present research was supported by the Brazilian funding agencies CNPq (Grant nos. 381.924/2004-1, 478502/2006-0, 305525/2006-9, 550780/2007-6 and 573710/2008-2) and FAPERJ (Grant nos. E-26/100.605/2007 and E-26/112.023/2008). This support is gratefully acknowledged.

References

1. Hori, M. and Nemat-Nasser, S. (1999) On two micromechanics theories for determining micro-macro relations in heterogeneous solids. *Mechanics of Materials*, **31**, 667–682.
2. Michel, J.C., Moulinec, H., and Suquet, P. (1999) Effective properties of composite materials with periodic microstructure: a computational approach. *Computer Methods in Applied Mechnics and Engineering*, **172**, 109–143.
3. Nemat-Nasser, S. (1999) Averaging theorems in finite deformation plasticity. *Mechanics of Materials*, **31**, 493–523.
4. Ibrahimbegović, A. and Markovič, D. (2003) Strong coupling methods in multi-phase and multi-scale modeling of inelastic behavior of heterogeneous structures. *Computer Methods in Applied Mechnics and Engineering*, **192**, 3089–3107.
5. Miehe, C., Shotte, J., and Lambrecht, M. (2002) Homogenization of inelastic solid materials at finite strains based on incremental minimization principles. Application to the texture analysis of polycrystals. *Journal of the Mechanics and Physics of Solids*, **50**(10), 2123–2167.
6. Miehe, C. (2003) Computational micro-to-macro transitions for discretized micro-structures of heterogeneous materials at finite strains based on the minimization of averaged incremental energy. *Computer Methids in Applied Mechanics and Engineering*, **192**, 559–591.
7. Miehe, C. (2002) Strain-driven homogenization of inelastic microstructures and composites based on an incremental variational formulation. *International Journal of Numerical Methods in Engineering*, **55**, 1285–1322.
8. Miehe, C. and Koch, A. (2002) Computational micro-to-macro transitions of discretized microstructures undergoing small strains. *Archive of Applied Mechanics*, **72**, 300–317.
9. Clayton, J.D. and McDowell, D.L. (2003) A multiscale multiplicative decomposition for elastoplasticity of polycrystals. *International Journal of Plasticity*, **19**, 1401–1444.
10. Reese, S. (2003) Meso-macro modelling of fibre-reinforced rubber-like composites exhibiting large elastoplastic deformations. *International Journal of Solids and Structures*, **40**, 951–980.
11. Terada, K. and Kikuchi, N. (2001) A class of general algorithms for multi-scale analysis of heterogeneous media. *Computer Methods in Applied Mechanics and Engineering*, **190**, 5427–5464.
12. Terada, K., Saiki, I., Matsui, K., and Yamakawa, Y. (2003) Two-scale kinematics and linearization for simultaneous two-scale analysis of periodic heterogeneous solids at finite strains. *Computer Methods in Applied Mechanics and Engineering*, **192**, 3531–3563.
13. Kouznetsova, V.G., Geers, M.G.D., and Brekelmans, W.A.M. (2004) Multi-scale second order computational homogenization of multi-phase materials: a nested finite element solution strategy. *Computer Methods in Applied Mechnics and Engineering*, **193**, 5525–5550.
14. Matsui, K., Terada, K., and Yuge, K. (2004) Two-scale finite element analysis of heterogeneous solids with periodic microstructure. *Computers and Structures*, **82**, 593–606.
15. Michel, J.C. and Suquet, P. (2004) Computational analysis of nonlinear composite structures using the

nonuniform transformation field analysis. *Computer Methods in Aplied Mechnics and Engineering*, **193**, 5477–5502.
16. Yang, Q.-S. and Becker, W. (2004) Effective stiffness and microscopic deformation of an orthotropic plate containing arbitrary holes. *Computers and Structures*, **82**, 2301–2307.
17. Bilger, N., Auslender, F., Bornert, M., Michel, J.-C., Moulinec, H., Suquet, P., and Zaoui, A. (2005) Effect of a nonuniform distribution of voids on the plastic response of voided materials: a computational and statistical analysis. *International Journal of Solids and Structures*, **42**, 517–538.
18. Goktepe, S. and Miehe, C. (2005) A micro-macro approach to rubber-like materials. Part III: the micro-sphere model of anisotropic mullins-type damage. *Journal of the Mechanics and Physics of Solids*, **53**, 2259–2283.
19. Ibrahimbegović, A., Grešovnik, I., Markovič, D., Melnyk, S., and Rodič, T. (2005) Shape optimization of two-phase inelastic material with microstructure. *Engineering Computations*, **22**(5/6), 605–645.
20. Li, X. and Wienan, E. (2005) Multiscale modeling of the dynamics of solids at finite temperature. *Journal of the Mechanics and Physics of Solids*, **53**, 1650–1685.
21. Okumura, D., Higashi, Y., Sumida, K., and Ohno, N. (2007) A homogenization theory of strain gradient single crystal plasticity and its finite element discretization. *International Journal of Plasticity*, **23**(7), 1148–1166.
22. Nakamachi, E., Tam, N.N., and Morimoto, H. (2007) Multi-scale finite element analyses of sheet metals by using SEM-EBSD measured crystallographic RVE models. *International Journal of Plasticity*, **23**(3), 450–489.
23. Speirs, D.C.D., de Souza Neto, E.A., and Perić, D. (2008) An approach to the mechanical constitutive modelling of arterial tissue based on homogenization and optimization. *Journal of Biomechanics*, **41**, 2673–2680.
24. Giusti, S.M., Novotny, A.A., de Souza Neto, E.A., and Feijóo, R.A. (2008) Sensitivity of the macroscopic elasticity tensor to topological micrsotructural changes. *Journal of the Mechanics and Physics of Solids*, **57**, 555–570.
25. Giusti, S.M., Novotny, A.A., and de Souza Neto, E.A. (2010) Sensitivity of the macroscopic response of elastic microstructures to the insertion of holes. *Proceedings of the Royal Society A*, **466**(2118), 1703–1723.
26. Hill, R. (1965) A self-consistent mechanics of composite materials. *Journal of the Mechanics and Physics of Solids*, **13**(4), 213–222.
27. Mandel, J. (1971) *Plasticité Classique et Viscoplasticité*, CISM Lecture Notes, Udine, Italy, Springer-Verlag.
28. Gurson, A.L. (1977) Continuum theory of ductile rupture by void nucleation and growth – Part I: yield criteria and flow rule for porous media. *Journal of Engineering Materials and Technology*, **99**, 2–15.
29. Bensoussan, A., Lions, J.L., and Papanicolau, G. (1978) *Asymptotic Analysis of Periodic Structures*, North-Holland.
30. Germain, P., Nguyen, Q.S., and Suquet, P. (1983) Continuum thermodynamics. *ASME Journal of Applied Mechanics*, **50**, 1010–1020.
31. Sanchez-Palencia, E. and Zaoui, A. (eds) (1987) *Homogenization Techniques for Composite Media*, Lecture Notes in Physics, vol. 272, Springer-Verlag, Berlin.
32. Holzapfel, G., Gasser, T.C., and Ogden, R.W. (2000) A new constitutive framework for arterial wall mechanics and a comparative study of material models. *Journal Of Elasticity*, **61**, 1–48.
33. Suquet, P.M. (1987) Elements of homogenization for inelastic solid mechanics, in *Homogenization Techniques for Composite Media* (eds E. Sanchez-Palencia and A. Zaoui), Lecture Notes in Physics, vol. 272, Springer-Verlag, Berlin, pp. 194–278.
34. Simo, J.C. and Hughes, T.J.R. (1998) *Computational Inelasticity*, Springer-Verlag, New York.
35. de Souza Neto, E.A., Perić, D., and Owen, D.R.J. (2008) *Computational*

Methods for Plasticity: Theory and Application, Wiley, Chichester.

36. Miehe, C., Schotte, J., and Schröder, J. (1999) Computational micro-macro transitions and overall moduli in the analysis of polycrystals at large strains. *Computational Materials Science*, **16**, 372–382.
37. Truesdell, C. (1969) *Rational Thermodynamics*, McGraw-Hill, New York.
38. Lubliner, J. (1990) *Plasticity Theory*, Macmillan, New York.
39. Lemaitre, J. and Chaboche, J.L. (1990) *Mechanics of Solid Materials*, Cambridge University Press.
40. Sanchez-Palencia, E. (1980) *Non-homogeneous Media and Vibration, Theory Lecture Notes in Physics*, Springer, Berlin.
41. Gurtin, M.E. (1981) *An Introduction to Continuum Mechanics*, Academic Press.
42. de Souza Neto, E.A. and Feijóo, R.A. (2006) Variational foundations of multi-scale constitutive models of solid: small and large strain kinematical formulation, LNCC Research & Development Report, No. 16/2006, National Laboratory for Scientific Computing, Petrópolis, Brazil.
43. Oden, J.T. (1979) *Applied Functional Analysis*, Prentice-Hall, Englewood Cliffs, New Jersey.
44. Larsson, F. and Runesson, K. (2007) RVE computations with error control and adaptivity: the power of duality. *Computational Mechanics*, **39**, 647–661.
45. Taylor, G.I. (1938) Plastic strains in metals. *Journal Institute of Metals*, **62**, 307–324.
46. de Souza Neto, E.A. and Feijóo, R.A. (2008) On the equivalence between spatial and material volume averaging of stress in large strain multi-scale solid constitutive models. *Mechanics of Materials*, **40**(10), 803–811.
47. Hughes, T.J.R. and Pister, K. (1978) Consistent linearization in mechanics of solids and structures. *Computers and Structures*, **8**, 391–397.
48. Simo, J.C. and Taylor, R.L. (1985) Consistent tangent operators for rate-independent elastoplasticity. *Computer Methods in Applied Mechanics and Engineering*, **48**, 101–118.
49. Crisfield, M.A. (1997) *Non-linear Finite Element Analysis of Solids and Structures*, vol. 2, Advanced Topics, Wiley, Chichester.
50. Blanco, P.J. (2008) Kinematical incompatibility, domain immersion and constitutive modelling: Connection to the modelling of the cardiovascular system, PhD Thesis, LNCC, Petrópolis, Brazil (in portuguese).
51. Haug, E.J. Céa, J. (1981 *Proceedings: Optimization of Distributed Parameters Structures*, EUA, Iowa.

10
A Homogenization-Based Prediction Method of Macroscopic Yield Strength of Polycrystalline Metals Subjected to Cold-Working

Kenjiro Terada, Ikumu Watanabe, Masayoshi Akiyama, Shigemitsu Kimura, and Kouichi Kuroda

10.1
Introduction

The plastic forming of polycrystalline metals, which are typical history-dependent materials, is inevitably accompanied by a change in strength characteristics. The evolution of the yield strength during cold-working, which is a plastic-forming operation, is one of the most important characteristics in practice, since it eventually determines the strength or durability of the final metal products. In the spirit of Computer-Aided Engineering (CAE), it has been expected that the yield strength after cold-working can be predicted by metal forming (numerical) simulations, which, in most cases, crucially rely on phenomenological constitutive models in the classical theory of plasticity.

It is known that the yield strength of polycrystalline metals during and after cold-working often reveals anisotropy and its underlying mechanism is twofold. One is due to the residual stress inside the material, and the other is the change in the flow stress of the slip surfaces and the crystallographic orientations of the grains. The former causes the so-called Bauschinger effect [1, 2] and is thought to be due to various types of microscale heterogeneities as is demonstrated by microscale analyses [3]. On the other hand, the latter mechanism in effect becomes influential when the deformation due to plastic forming is relatively large and the resulting texture development brings the directional dependence of the deformation as well as the strength characteristics [4]. To accurately simulate the arbitrary cold-working of polycrystalline metals and at the same time predict the post-forming strength, these two features must be properly reflected in the constitutive models. In this context, numerous attempts have been made since the finite-element method (FEM) was applied to metal-forming simulations. Although they are too many to discuss comprehensively, we briefly review some representative developments below.

The Bauschinger effect has been incorporated into the constitutive equations by the introduction of kinematic hardening models, which are basically represented by the movement of the yield surface, that is, the center of the yield surface moves

Advanced Computational Materials Modeling: From Classical to Multi-Scale Techniques.
Edited by Miguel Vaz Júnior, Eduardo A. de Souza Neto, and Pablo A. Munoz-Rojas

ISBN: 978-3-527-32479-8

according to so-called back stress without changing the size of the surface. A considerable number of studies have been made on the modeling of kinematic hardening. Among them, the model proposed by Armstrong and Frederick [5] is one of the most popular ones, while the multisurface models by Mroz [6] and Iwan [7] are also worthy of mention. The development of similar models is still an active area of research [8–10].

To introduce the anisotropic strength characteristics due to texture developments to the constitutive model, Hill [11] proposed a quadratic yield function with parameters controlling its noncircular shape as a generalization of the von-Mises yield condition in J_2 flow theory. Gotoh and Ishise [12] extended the idea and represented the yield condition by a quartic function they found to be more reasonable considering their experimental observations. Various improved models have been proposed, and some of them have been designed specifically as numerical simulations for sheet metal forming [13–16]. However, these models are for the initial state of the base material and their anisotropic features do not come from their deformation histories during plastic forming. Therefore, they cannot represent the evolution of the yield strength during and after the cold-working. Although a number of articles have been devoted to the proposal of more realistic models that can represent the change of the shapes of anisotropic yield functions [17–20], they are still under development, but issues remain regarding their accuracy and the methods of parameter identification.

Thus, the phenomenological constitutive models in the classical theory of plasticity have not been able to provide sufficient accuracy to predict the yield stress of polycrystalline metals after plastic forming. In view of this fact, there have been studies based on the mechanism underlying the scale of crystal grains, that is, the macroscopically observed material behavior of polycrystalline metals can be thought to be the average of various microscale phenomena, including the change in flow stress of the crystal grains due to the motion and accumulation of dislocations, the residual stress inside or between the grains, and the nonuniform distribution of crystallographic orientations. Studies from this aspect have been made following the pioneering work done by Taylor [21] who made some assumptions on the slippage and hardening characteristics of single crystals. The so-called Taylor model is an analytical expression of the averaged response of polycrystalline metals, in which the above-mentioned microscale phenomena of crystal grains are taken into account, under the assumption that all the crystal grains are subjected to constant and uniform strain in an aggregate. Although a modification of the Taylor model was made by Bishop and Hill [22, 23] to investigate the macroscopic yield surface, the Bauschinger effect, or equivalently, the effect of residual stress usually caused by nonuniform stress and strain, cannot be represented by the model because of the above-mentioned assumption. On the other hand, a class of self-consistent models [24–26] derived by another analytical averaging method based on Eshelby's theory of equivalent inclusions [27] enables us to consider the interaction between neighboring grains so that the Bauschinger effect can be reproduced [28]. Although these models can be used for the macroscopic constitutive responses in finite element (FE) simulations of the plastic forming of metal products, the reliability of the quantitative evaluations of macroscopic mechanical behavior are still open to question; see Refs. [29–32] for attempts with the Taylor model within the framework of crystal plasticity, and Refs. [33, 34] and [35] for the compact review and the incorporation of the texture component into the FEM.

In recent years, there has been a renewal of interest in the characterization of the macroscopic mechanical behavior of polycrystalline metals by the FE simulations of crystal aggregates, that is, by preparing an FE model of a representative volume element (RVE) composed of crystal grains and employing the crystal plasticity model, its equilibrium problem for microscopic stress under certain boundary conditions can be solved [36–39]. Although the macroscopic and microscopic responses predicted by this approach are much more realistic and reasonable than those by the phenomenological constitutive models in the classical theory of plasticity, it does not provide the functional form of the macroscopic constitutive law and, therefore, does not enable us to perform metal forming analyses for macrostructures. As a result, the evaluation of the macroscopic yield strength during and after plastic forming is also not possible.

The methodology that can overcome the discrepancy of the RVE approach is brought about by the mathematical theory of homogenization [40–43], since the theory provides the so-called two-scale boundary value problem (BVP) for arbitrary plastic forming, in which the microscale BVP is equivalent to that of the RVE approach with a periodic boundary condition. In this two-scale analysis method based on the homogenization theory, the macroscopic material behavior is implicitly evaluated at each integration point of the macroscale FE model by solving the equilibrium problem of the corresponding RVE, or equivalently, the microscale BVP, without having the explicit form of the macroscopic constitutive equation. This kind of solution scheme used to solve the two-scale BVP is referred to as the *micro–macro* (or *global–local*) *coupling scheme* and is typified in Refs. [44–46] for general heterogeneous media and followed by many authors [47–50]; see in particular Nakamachi *et al.* [51] for their application to polycrystalline metals with a dynamic explicit FE code.

Although the micro–macro coupling scheme is promising in the sense that various types of macroscopic material behavior can be captured without knowing the explicit functional forms of the material models, the method, by its nature, requires a significant amount of computational cost. Therefore, the decoupling of micro- and macroscale BVPs is indispensable and this has been realized by Watanabe and Terada [52, 53] with a view to approximately solving the two-scale BVP encountered in practice. The solution method, called the *micro–macro decoupling scheme*, consists of numerical material tests (NMTs) on a periodic microstructure, namely, a unit cell, for the parameter identification of an assumed approximate macroscopic constitutive model, and the decoupled macro- and microscale analyses. It is expected that the approximate solution with the decoupling scheme can be applied to the evaluation of the macroscopic yield strength after plastic forming.

This chapter presents a homogenization-based evaluation method of the macroscopic yield strength of polycrystalline metals after cold-working. The procedure of the method is as follows: first, the approximate macroscopic constitutive model and its material parameters are assumed and then the decoupled macroscale analysis is performed to simulate the macroscopic forming process. Next, the macroscopic deformation histories are applied to unit cells at certain macroscopic points of interest to obtain the "numerical specimens" that have experienced the cold-working process, and finally, the NMT is conducted on them to evaluate the macroscopic postforming yield strengths.

In Section 10.2, after the two-scale BVP is formulated in a general context, both the micro–macro coupling and decoupling schemes are introduced for two-scale analyses. Then we propose the method to evaluate the macroscopic yield strength, which hinges on the solution of the microscale BVP, or equivalently, the numerical specimen subjected to macroscopic deformation histories during plastic forming. Section 10.3 is devoted to the preparation of the numerical specimen for NMTs, which is a unit cell of a polycrystalline aggregate composed of several crystal grains, and Section 10.4 provides the parameter identification for the assumed approximate macroscopic constitutive model. In Section 10.5, we validate the proposed method by taking the three-step forming process as an example of macroscopic plastic forming in this study. The decoupling scheme is applied to solve the corresponding two-scale BVP and, in turn, the NMTs are carried out on the "numerical specimens" obtained at the last step of the two-scale analysis to evaluate the macroscopic yield strength. Since the method relies on the solution of the two-scale BVP, approximated by applying the micro–macro decoupling scheme, the method is validated by comparing the results with those obtained by the equivalent two-scale analysis in combination with the coupling scheme. Section 10.6 presents a numerical example of the pilger mill rolling process, which imposes very complex deformation histories to polycrystalline aggregates to demonstrate the capability and promise of the proposed method for practical applications. This is followed by the conclusion in Section 10.7.

10.2 Two-Scale Modeling and Analysis Based on Homogenization Theory

In this section, after providing the two-scale BVP, which can be derived for a general heterogeneous medium with periodic microstructures, we introduce two separate solution methods for solving it. One is the micro–macro coupling solution scheme, which requires us to solve the microscale BVP whenever macroscopic stress is evaluated. The other is the micro–macro decoupling scheme, in which the micro- and macroscopic BVPs are solved separately and the macroscopic constitutive equation is assumed before the macroscopic analysis.

10.2.1
Two-Scale Boundary Value Problem

We introduce the two-scale BVP that governs the coupled mechanical behavior of a general heterogeneous medium. The heterogeneous medium is assumed to be composed of periodic microstructures, each of which is identified with an RVE associated with its overall or macroscopic mechanical behavior and is commonly referred to as a *unit cell* within the framework of mathematical homogenization theory. The spatial size of the unit cell is characterized by parameter ϵ, which is assumed to be very fine compared with the overall structure [46].

Let $\mathcal{B}_0^\epsilon \subset \mathcal{R}^{n_{\dim}}$ ($n_{\dim} = 1, 2,$ or 3) be the reference configuration of a continuum body, where $\mathcal{R}^{n_{\dim}}$ is the $n_{\dim}$-dimensional real space. We identify in the body a particle labeled by its position vector $\mathbf{X} \in \mathcal{B}_0^\epsilon$ relative to the standard basis in $\mathcal{R}^{n_{\dim}}$. On the other hand, a point $\mathbf{x}$ in the current configuration $\mathcal{B}^\epsilon$ of the body is obtained via the mapping $\varphi : \mathcal{B}_0^\epsilon \to \mathcal{B}^\epsilon \subset \mathcal{R}^{n_{\dim}}$, defined as $\mathbf{x} = \varphi(\mathbf{X})$ for all $\mathbf{X} \in \mathcal{B}_0^\epsilon$. Here and in the following formulation, the parameter ϵ is used to indicate the dependency of variables or notations on the heterogeneity.

As is conventional in mathematical homogenization [40, 41], we decompose the domain into two; $\mathcal{B}_0$ with invisible heterogeneities and $\epsilon\mathcal{R}^{n_{\dim}}$, which is assumed to represent an assembly of microstructures, that is, the actual domain $\mathcal{B}_0^\epsilon$ can be regarded as a product space $\mathcal{B}_0 \times \epsilon\mathcal{R}^{n_{\dim}}$. To measure the spatial changes in the domains, $\mathcal{B}_0$ and $\mathcal{R}^{n_{\dim}}$, we introduce two separate scales: a macroscale $\mathbf{X} \in \mathcal{B}_0$ for the former and a microscale $\mathbf{Y} \in \mathcal{Y}_0$ for the latter, in which $\mathcal{Y}_0$ is identified with the physical domain for microstructures. These are related to each other by $\mathbf{Y} = \mathbf{X}/\epsilon$. Then, the domain of this problem can formally be represented as $\mathcal{B}_0^\epsilon = \mathcal{B}_0 \times \mathcal{Y}_0$.

Field variables can be represented in terms of the two scales; for example, the displacement $\mathbf{u}(\mathbf{X}, \mathbf{Y})$ and the nominal stress $\mathbf{P}(\mathbf{X}, \mathbf{Y})$ are measured by the two scales. In addition, the microstructures are assumed to be periodically arranged with a period $\epsilon\mathcal{Y}_0$, and the bounded domain $\mathcal{Y}_0$ can be defined as an RVE and thus called a *unit cell*. Then, the field variables such as displacement fields could be periodic in $\mathbf{Y}$, namely, $\mathcal{Y}_0$-periodic. After applying the results in the two-scale convergence theory [43], we can separate the description of the micro- and macroscopic mechanical behaviors by means of field variables with their own spatial scales.

Macroscopic deformation is given as a function of the initial position $\mathbf{X} \in \mathcal{B}_0$ of the body, which accommodates a microstructure where the variables are measured by the initial position $\mathbf{Y} \in \mathcal{Y}_0$. In addition, the macroscopic variables are commonly obtained as the volume average of the corresponding microscopic variables. For instance, the macroscopic nominal stress is defined as

$$\tilde{\mathbf{P}}(\mathbf{X}) := \frac{1}{|\mathcal{Y}_0|} \int_{\mathcal{Y}_0} \mathbf{P}^0(\mathbf{X}, \mathbf{Y}) \mathrm{d}Y \tag{10.1}$$

where $|\mathcal{Y}_0|$ is the volume of the RVE in the initial configuration. On the other hand, the microscopic displacement field is defined as

$$\mathbf{w}(\mathbf{X}, \mathbf{Y}) = \nabla_X \mathbf{u}^0(\mathbf{X}) \cdot \mathbf{Y} + \mathbf{u}^1(\mathbf{X}, \mathbf{Y}) \tag{10.2}$$

Then, the microscopic displacement gradient $\nabla_Y \mathbf{w}(\mathbf{X},\mathbf{Y}) = \mathbf{H}^0(\mathbf{X},\mathbf{Y})$ is given by

$$\mathbf{H}^0(\mathbf{X},\mathbf{Y}) = \nabla_X \mathbf{u}^0(\mathbf{X}) + \nabla_Y \mathbf{u}^1(\mathbf{X},\mathbf{Y}) \tag{10.3}$$

from which the microscopic deformation gradient is identified with

$$\mathbf{F}^0 = \nabla_Y(\mathbf{w}(\mathbf{X},\mathbf{Y}) + \mathbf{Y}) = \mathbf{H}^0 + \mathbf{1} = \tilde{\mathbf{H}} + \hat{\mathbf{H}} + \mathbf{1} \tag{10.4}$$

where $\tilde{\mathbf{H}}(\mathbf{X}) := \nabla_X \mathbf{u}^0(\mathbf{X})$, $\hat{\mathbf{H}}(\mathbf{X},\mathbf{Y}) := \nabla_Y \mathbf{u}^1(\mathbf{X},\mathbf{Y})$ and $\mathbf{1}$ is the second-order identity tensor. Here, ∇_X and ∇_Y are the gradient operators with respect to macro- and microscales, respectively. Because of the periodicity of $\mathbf{u}^1(\mathbf{X},\mathbf{Y})$ with respect to $\mathbf{Y}$, namely $\mathcal{Y}_0$-periodic, the following averaging relationship holds:

$$\tilde{\mathbf{H}}(\mathbf{X}) := \frac{1}{|\mathcal{Y}_0|}\int_{\mathcal{Y}_0} \mathbf{H}^0(\mathbf{X},\mathbf{Y})\mathrm{d}Y \tag{10.5}$$

which associates the macroscopic displacement gradient at $\mathbf{X}$ with the microscopic displacement gradient over the whole unit cell domain $\mathcal{Y}_0$. For an understanding of the idea of separation of micro- and macroscopic kinematic representations in detail, see Ref. [46].

The microscopic motion associated with the displacement $\mathbf{w}(\mathbf{X},\mathbf{Y})$ should be understood as a one-to-one mapping $\psi : \mathcal{Y}_0 \to \mathcal{Y} \subset \mathcal{R}^{n_{\mathrm{dim}}}$ where $\mathcal{Y}$ is the current configuration and is defined as $\psi(\mathbf{Y}) = \mathbf{Y} + \mathbf{w}(\mathbf{X},\mathbf{Y})$, for all $\mathbf{Y} \in \mathcal{Y}_0$, with the local condition $J_Y(\mathbf{Y}) := \det(\nabla_Y \psi(\mathbf{Y})) = \det(\mathbf{F}^0(\mathbf{X},\mathbf{Y})) > 0$. Accordingly, the spatial version of the microscopic BVP is given for the RVE as

$$\int_{\mathcal{Y}} \tau^0(\mathbf{x},\mathbf{y}) : \nabla_y \eta^1 \frac{\mathrm{d}y}{J_Y} = 0, \forall \eta^1 \in W^{1,p}_{\mathrm{per}}(\mathcal{Y}) \tag{10.6}$$

along with a selected constitutive equation that relates the microscopic deformation to the microscopic Kirchhoff stress $\tau^0(\mathbf{x},\mathbf{y})$. Here, $\mathrm{d}y$ denotes the differential volume of the current configuration of the RVE, namely, $\mathrm{d}y = J_Y \mathrm{d}Y$, and $W^{1,p}$ is the Sobolev space of $\mathcal{Y}_0$-periodic functions. It is to be noted that, in this general setting, any constitutive law is usable for constituents in the unit cell, though the crystal plasticity model [34] is chosen for the specific purpose in this study.

On the other hand, the macroscopic BVP is governed by the following variational formulation of spatial description:

$$\int_{\mathcal{B}} \tilde{\tau} : \nabla_x \boldsymbol{\eta}^0 \frac{\mathrm{d}x}{\tilde{J}} - g_{\mathrm{ext}}(\boldsymbol{\eta}^0) = 0 \quad \forall \boldsymbol{\eta}^0 \in \mathcal{V}_{\mathcal{B}} \tag{10.7}$$

along with

$$\tilde{\tau}(\mathbf{x}) = \frac{1}{|\mathcal{Y}|}\int_{\mathcal{Y}} \tau^0(\mathbf{X},\mathbf{Y})\frac{\mathrm{d}y}{J_Y} \tag{10.8}$$

which is equivalent to Eq. (10.1). Here, $|\mathcal{Y}|$ is the volume of the RVE in the current configuration, and $\mathcal{V}_{\mathcal{B}}$ is defined as $\mathcal{V}_{\mathcal{B}} = \left\{\mathbf{v} : \mathcal{B} \mapsto \mathcal{R}^{n_{\mathrm{dim}}} \,\middle|\, v_i \in W^{1,p}, \mathbf{v} = \mathbf{0} \text{ on } \Gamma_u\right\}$. In addition, $\tilde{J} := \det[\tilde{\mathbf{F}}]$ is the macroscopic volumetric change, whereas $\mathrm{d}y = J_Y \mathrm{d}Y$ is the microscopic counterpart. The set of these two BVPs constitutes the so-called two-scale BVP, in which each BVP requires the solution of the other.

Since the theory of homogenization does not have the function of deriving nonlinear constitutive equations for the macroscopic BVP and just provides the averaging relationships (10.1) (or equivalently Eq. (10.8)) and (10.5), we have to solve the two-scale BVP either without the macroscopic constitutive equation or by simply assuming its explicit functional form that suitably can approximate the macroscopic material behavior characterized with the corresponding unit cell. The corresponding approaches are, respectively, referred to as the *micro–macro coupling* and *decoupling schemes*, each of which is explained below.

10.2.2 Micro–Macro Coupling and Decoupling Schemes for the Two-Scale BVP

In the two-scale BVP, the macroscopic constitutive equation is an implicit function of the solutions of the microscale BVP and, thus, the microscale BVP indirectly represents the macroscopic material response, that is, it is not until the microscale equilibrated stress is determined that the macroscopic stress can be calculated in view of Eq. (10.1). Therefore, if the two-scale coupling analysis is performed by the FEM, the microscale BVP must be associated with an integration point located in a macroscale FE model and solved for the microscale equilibrated stress to evaluate the macroscale stress by the averaging relation (10.1), which must satisfy the macroscale BVP at the same time. In particular, when an implicit and incremental solution method with a Newton–Raphson-type iterative procedure is employed to solve the two-scale BVP, the microscale BVP is to be solved in every iteration to attain the macroscale equilibrium state at every loading step. Needless to say, the microscale BVP is also nonlinear and therefore requires the iterative method. This type of solution scheme to solve the two-scale BVP is referred to as the *micro–macro* (or *global–local*) *coupling scheme* and is typified in Refs. [44–46].

The micro–macro coupling scheme is promising in the sense that various types of macroscopic material behavior can be captured without knowing the explicit functional forms of material models if the unit cell is eligible for an RVE. However, the method, by nature, requires a significant amount of computational cost. In fact, the model size of the macroscale BVP raises the number of microscale BVPs to the second power, since each macroscale integration point is associated with its own microscale BVP. Although some parallel algorithms can reduce the cost to some extent [54], we are bound to say that the coupling scheme is all but useless in most practical applications. Therefore, the decoupling of micro- and macroscale BVPs is indispensable for applying the two-scale approach based on homogenization to various problems encountered in practice [52, 53].

In this context, let us recall that the homogenization procedure for linear problems [55], which consists of microscale numerical analyses performed on the unit cell to identify the homogenized coefficients. For example, the homogenization analysis for 3D linear elasticity problems requires us to carry out six separate microscopic analyses, in which a unit macroscopic strain tensor with only one nonzero component is used, and provides us with the components of the homogenized elasticity tensor, which are identified according to the volume average of the self-equilibrated microscopic stress tensors. Therefore, even for nonlinear problems, if a relevant macroscopic constitutive equation is prepared so as to approximate the actual macroscopic material behavior characterized by the microstructure, homogenization is just a set of microscale analyses that are followed by parameter identification. Each microscale analysis for this purpose can be referred to as a *numerical material test* (*NMT*) and the numerical model of the unit cell can be called a *numerical specimen*. This idea enables us to decouple the micro- and macroscale BVPs; that is, each problem can be solved separately and in series. The corresponding solution method is called the *micro–macro decoupled analysis scheme* in this study [52, 53] and is summarized as follows. Note that the homogenization for linear problems can be considered to be a special case of the decoupled analyses.

The concrete procedure of the decoupling scheme is described as follows:

1) An appropriate constitutive model relevant for the macroscopic material behavior under consideration is assumed. We call this the *approximate macroscopic constitutive model*.
2) A series of NMTs is conducted on a unit cell model (FE mesh), which is regarded as a *numerical specimen*, to obtain the homogenized or macroscopic material behavior. Note that the loading patterns to be considered here depend on the selected constitutive model.
3) Material parameters of the assumed constitutive model are identified by means of the "empirical" data obtained from the NMTs and an appropriate curve fitting scheme.
4) FE analyses are carried out to solve the macroscale BVP using the assumed constitutive model with identified material parameters. Let us call each of them the *decoupled macroscale analysis*.
5) If necessary, after extracting the time series of macroscopic deformation history from the macroscopic analysis result and applying it as a series of boundary conditions, the localization analyses are performed to evaluate what has actually been happening inside the unit cell during the macroscopic deformation process. We call this process the *decoupled microscale analysis*, which corresponds to the localization process in the theory of homogenization.

The precondition of decoupling is that we are able to pick up a constitutive model to properly characterize the macroscopic material behavior. However, the assumed approximate macroscopic constitutive model does not always properly represent macroscopic material behavior, and indeed, the decoupling scheme is just an approximate scheme. Thus, the coupling scheme should be used rather than the decoupling one if the highest level of accuracy is desired irrespective of computational cost.

10.2.3
Method of Evaluating Macroscopic Yield Strength after Cold-Working

We propose here a method of evaluating the macroscopic yield strengths of polycrystalline metals subjected to arbitrary patterns of cold-working. The method is simply to use the results of the decoupled microscale analysis in Step (5), or equivalently, of the localization analysis. To be more specific, the NMTs are again performed on the numerical specimen obtained after the final process of the micro–macro decoupling scheme. Since the numerical specimen is generated as if it has been embedded in the macroscopic structure during the macroscopic plastic forming simulated in Step (4), the expectation is that it possesses as much information about the macroscopic deformation histories as that in actual specimens.

Note, however, that the approximate macroscopic constitutive model may not be reliable, as mentioned before. Nonetheless, the decoupling scheme must be capable of evaluating the macroscopic yield strength of polycrystalline metals after cold-working with reasonable quantitative accuracy, since most of the metal forming processes are defined by prescribing the macroscopic deformation and dominated by plastic deformation rather than elastic deformation. That is, although there is no guarantee that the macroscopic stress response evaluated in the decoupled macroscale analysis in Step (4) is the same as that obtained by the corresponding coupling scheme, it is expected that the macroscopic deformation during cold-working should be almost the same, and should also be reliable enough for the decoupled microscale analysis in Step (5). Since the numerical specimen thus obtained has been subjected to the reliable deformation history of plastic forming, it can be considered a good approximation of that obtained by the coupled analysis, and therefore, the macroscopic yield strength can be evaluated after arbitrary plastic-forming processes.

It is also worth remarking that if we can find a constitutive law that accurately predicts the macroscopic stress as well as the deformation of arbitrary cold-working, the macroscopic yield strength can also be predicted from the evolution of the macroscopic yield function, meaning that neither the two-scale analysis nor the proposed method of evolution of the macroscopic yield strength makes sense. The motivation of this study comes from the common recognition that there are as yet no phenomenological constitutive models that can predict not only the stress and deformation histories but also the evolution of yield surface for arbitrary patterns of cold-working. In this respect, the most important requirement for the strength evaluation based on the micro–macro decoupling scheme is the reliability of the microscale numerical analyses, which strongly relies on the capability of the constitutive model employed for the material in the unit cell. In this study, we assume that the so-called crystal plasticity model [34] is reliable enough to represent the mechanical behavior of single crystals in a polycrystalline metal, when the periodic microstructure, namely, the unit cell, can be defined as an aggregate of crystal grains.

10.3 Numerical Specimens: Unit Cell Models with Crystal Plasticity

In this study, we prepare finite element models of two types of numerical specimens of polycrystalline metals, as shown in Figure 10.1. The preparation includes the

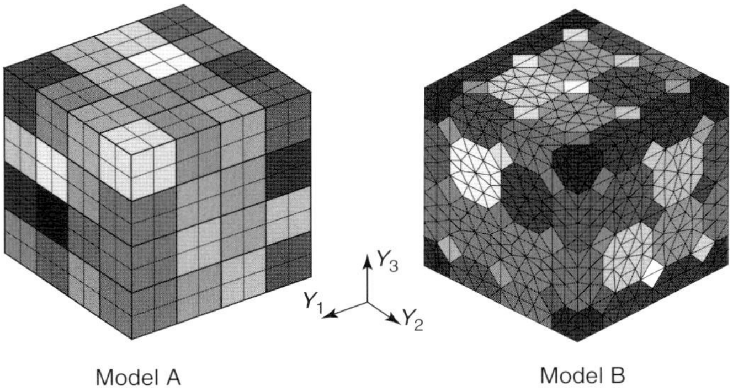

Figure 10.1 Finite element models of polycrystalline aggregate. Model A: Eight-node hexahedral elements: 512 Nodes: 729. Model B: Four-node tetrahedral elements: 5127 Nodes: 1303

setting of plasticity models for single crystals within the framework of crystal plasticity and their material parameters [34].

Model A is the unit cell's FE model of a polycrystalline aggregate composed of 64 crystal grains, and its crystal grains have the same geometrical feature (cubic). Each grain is assumed to be a face-centered cubic (FCC) and is modeled by eight eight-node hexahedral solid elements. On the other hand, Model B is composed of 54 FCC grains, each of which has a truncated-octahedron shape, and its FE mesh is generated with four-node tetrahedral elements. It is also assumed that both models satisfy the geometrical periodicity condition and have randomly set crystallographic orientations.

In the following description of the models, we confine ourselves to a single material point in an RVE so that only a single spatial scale $\mathbf{Y}$ is used for field variables. Omitting the distinction between the micro- and macroscopic variables introduced in Section 10.2.1, we denote the microscopic total deformation gradient by $\mathbf{F}$ instead of $\mathbf{F}^0$. We also assume the multiplicative split of $\mathbf{F}$ into an elastic $\mathbf{F}^e$ and a plastic part $\mathbf{F}^p$ with $\det\mathbf{F}^p = 1$ such that $\mathbf{F} = \mathbf{F}^e\mathbf{F}^p$, in the kinematic description of the models in single crystal plasticity.

Let $\mathbf{s}_0^{(\alpha)}$ ($\alpha = 1, \ldots, n_{\text{slip}}$) be the vector representing the slip direction of the αth slip system in the reference configuration of a single crystal and be defined on the slip plane with an outward unit normal vector given by $\mathbf{m}_0^{(\alpha)}$. Here, n_{slip} is the number of slip systems at a single material point in a single grain. The two vectors for the αth slip system are, respectively, related to $\mathbf{m}_t^{(\alpha)}$ and $\mathbf{s}_t^{(\alpha)}$ in the current configuration via the relationships $\mathbf{s}_t^{(\alpha)} = \mathbf{F}^e\mathbf{s}_0^{(\alpha)}, \mathbf{m}_t^{(\alpha)} = \mathbf{F}^{e-T}\mathbf{m}_0^{(\alpha)}$.

The rate-independent model, typified in [56], is employed for Model A, and the following yield function is used for the αth slip system:

$$\phi^{(\alpha)} = \left|\mathbf{s}^{(\alpha)} \cdot \left(\tau \mathbf{m}^{(\alpha)}\right)\right| - \tau_Y - \sum_{\beta=1}^{n_{\text{slip}}} h_{\alpha\beta}\xi^{(\beta)} \leq 0 \quad \forall\alpha \in \left\{1, \cdots, n_{\text{slip}}\right\} \tag{10.9}$$

where τ is the Kirchhoff stress, τ_Y is the initial critical resolved shear stress (CRSS), n_{slip} is the number of slip systems ($n_{\text{slip}} = 12$ for FCC), and $\xi^{(\alpha)}$ is the accumulated slip of slip system α. Here, $h_{\alpha\beta}$ is the hardening modulus and is defined as the following nonlinear equation:

$$h_{\alpha\beta} = \left[(h_0 - h_{\text{sat}}) \exp\left(-p \sum_{\lambda=1}^{n_{\text{slip}}} \xi^{(\lambda)} \right) + h_{\text{sat}} \right] \left[\delta_{\alpha\beta} + q(1 - \delta_{\alpha\beta}) \right] \qquad (10.10)$$

where $\delta_{\alpha\beta}$ is the Kronecker's delta symbol and p is the sensitivity parameter for the exponential function. In addition, h_0, h_{sat}, and q are the moduli that characterize the initial hardening, asymptotically saturating hardening, and latent hardening, respectively. Here, the latent hardening, which represents the coupling effect with other slip systems, induces anisotropy in the deformation and strength characteristics [57], and its modulus is known to have the value $p \in [1, 1.4]$ [58].

Since an FCC crystal reveals anisotropy in elastic deformation, we employ the St. Venant–Kirchhoff model for the elastic constitutive model $\hat{\mathbf{S}} = \hat{\mathbb{C}}^{\text{e}} : \mathbf{E}^{\text{e}}$. Here, the second Piola–Kirchhoff stress in the intermediate configuration $\hat{\mathbf{S}}$ and the elastic Green–Lagrange strain tensor $\mathbf{E}^{\text{e}}$ are, respectively, defined as $\hat{\mathbf{S}} := \mathbf{F}^{\text{e}-1} \tau \mathbf{F}^{\text{e}-\text{T}}$ and $\mathbf{E}^{\text{e}} := \frac{1}{2}(\mathbf{F}^{\text{eT}}\mathbf{F}^{\text{e}} - \mathbf{1})$, and $\hat{\mathbb{C}}^{\text{e}}$ is the forth-order elasticity tensor in the intermediate configuration with constant components. The material parameters chosen for the single crystal grains for Model A (Figure 10.1a) are given in Table 10.1.

On the other hand, we utilize the rate-dependent model introduced in [33] for Model B. The model is suitable especially when the deformation is so large and process is so complex, as in some practical situations, that the rate-independent models suffer from the lack of uniqueness in the return mapping problem. In this model, the slip rate $\dot{\gamma}^{(\alpha)}$ of αth slip system is given by

$$\dot{\gamma}^{(\alpha)} = \dot{\gamma}_0 \frac{\tau^{(\alpha)}/g^{(\alpha)}}{\left| \tau^{(\alpha)}/g^{(\alpha)} \right|^{1-n}} \qquad (10.11)$$

where $\dot{\gamma}_0$ is the reference strain rate, $\tau^{(\alpha)}$ is the resolved shear stress, and n is the material rate sensitivity parameter. Here, $g^{(\alpha)}$ is a function of the sum of the

Table 10.1 Material constants for crystal grains in Model A.

Elastic modulus	$\hat{\mathbb{C}}^{\text{e}}_{1111}$(GPa)	230
Elastic modulus	$\hat{\mathbb{C}}^{\text{e}}_{1122}$(GPa)	130
Elastic modulus	$\hat{\mathbb{C}}^{\text{e}}_{1212}$(GPa)	100
Initial CRSS	$\tau_Y^{(\alpha)}$(GPa)	0.12
Initial hardening modulus	h_0(GPa)	0.4
Saturation hardening modulus	h_{sat}(GPa)	0.05
Sensitivity parameter	p	1.0
Ratio of latent hardening	q	1.1

Table 10.2 Material constants for crystal grains in Model B.

Elastic modulu	$\hat{\mathbb{C}}^{e}_{1111}$(GPa)	127.51
Elastic modulu	$\hat{\mathbb{C}}^{e}_{1122}$(GPa)	111.35
Elastic modulu	$\hat{\mathbb{C}}^{e}_{1212}$(GPa)	85.31
Initial hardening modulu	H_0(MPa)	220.0
Saturated stress	τ_s(MPa)	340
Initial CRSS	τ_0 (MPa)	93.0
Material sensitivity parameter	n	30
Ratio of latent hardening	q	1.4

accumulated slip defined by

$$\gamma = \sum_{\alpha=1}^{n_{\text{slip}}} \int_0^t \left|\dot{\gamma}^{(\alpha)}\right| \mathrm{d}t \tag{10.12}$$

so that its rate is evaluated as

$$\dot{g}^{(\alpha)} = \sum_{\alpha=1}^{n_{\text{slip}}} h_{\alpha\beta}\dot{\gamma}^{(\beta)} \tag{10.13}$$

In this expression, $h_{\alpha\beta}$ is the hardening modulus such that

$$h_{\alpha\beta} = qH(\gamma) + (1-q)H(\gamma)\delta_{\alpha\beta} \tag{10.14}$$

together with $H(\gamma) = H_0 \sinh^2 \left|\gamma/(\tau_s - \tau_0)\right|$, where H_0 is is the initial hardening modulus, τ_s is the saturated strength, and $\tau_0 := g^{(\alpha)}(0)$ is the initial CRSS. The material parameters used for this constitutive model is given in Table 10.2.

Since each NMT is just an FE analysis for the unit cell, subjected to the periodic boundary condition, it is performed by arbitrary FEM software. In this study, ABAQUS [59] with the user-material subroutine is utilized for the rate-dependent crystal plasticity model explained above to conduct NMTs for Model B for evaluating initial- and postforming strengths, whereas our own code is used for the rate-independent one for Model A.

10.4
Approximate Macroscopic Constitutive Models

Approximate macroscopic constitutive models are introduced and NMTs are conducted to identify their material parameters.

10.4.1
Definition of Macroscopic Yield Strength

Let us first define the macroscopic yield strength from the data "measured" in NMTs. Although all the components of the macroscopic stress and strain are contained in each data set, we obtain the single stress–strain curve for each NMT that can be identified with the relationship between the macroscopic equivalent stress and strain defined, respectively, as

$$\sigma^* := \sqrt{\frac{3}{2}\mathrm{dev}\,[\tilde{\sigma}] : \mathrm{dev}\,[\tilde{\sigma}]} \tag{10.15}$$

$$\varepsilon^* := \sqrt{\frac{2}{3}\mathrm{dev}\,[\tilde{\varepsilon} - \tilde{\varepsilon}_0] : \mathrm{dev}\,[\tilde{\varepsilon} - \tilde{\varepsilon}_0]} \tag{10.16}$$

where $\tilde{\sigma}$ is the macroscopic Cauchy stress, $\tilde{\varepsilon}_0$ is the macroscopic strain when the loading is applied, and $\tilde{\varepsilon}$ is the macroscopic logarithmic strain (in the current configuration) defined as

$$\tilde{\varepsilon} := \frac{1}{2}\ln\left[\tilde{\mathbf{F}}\tilde{\mathbf{F}}^{\mathrm{T}}\right] \tag{10.17}$$

Here, $\tilde{\mathbf{F}} = \tilde{\mathbf{H}} + \mathbf{1}$ is the macroscopic deformation gradient.

In this study, with the stress–strain curve thus obtained, we define the macroscopic yield strength as the stress value, σ^*, that satisfies the following relationship:

$$W^{\mathrm{p}} - \int \sigma^* \mathrm{d}\varepsilon^{\mathrm{p}*} = 0 \tag{10.18}$$

where W^{p} is the value of the plastic work to be prescribed and $\varepsilon^{\mathrm{p}*}$ is the macroscopic equivalent plastic strain. The integration of the second term in above equation is calculated from the macroscopic equivalent stress–plastic strain curve transformed with the following equation:

$$\varepsilon^{\mathrm{p}*} := \varepsilon^* - \frac{\sigma^*}{E^*} \tag{10.19}$$

where E^* is the elastic proportional constant of the macroscopic equivalent stress–strain curve.

10.4.2
Macroscopic Yield Strength at the Initial State

Before choosing the approximate macroscopic constitutive model, we evaluate here the macroscopic yield strength at the initial state by conducting NMTs on the numerical specimen generated above with various proportional loading paths.

For the NMTs, to evaluate the macroscopic yield strength at the initial state at which $\tilde{\varepsilon}_0 = \mathbf{0}$, 12 paths of proportional loading are set within the space of the macroscopic deviatoric stress, as depicted in Figure 10.2(a). Then, after carrying out

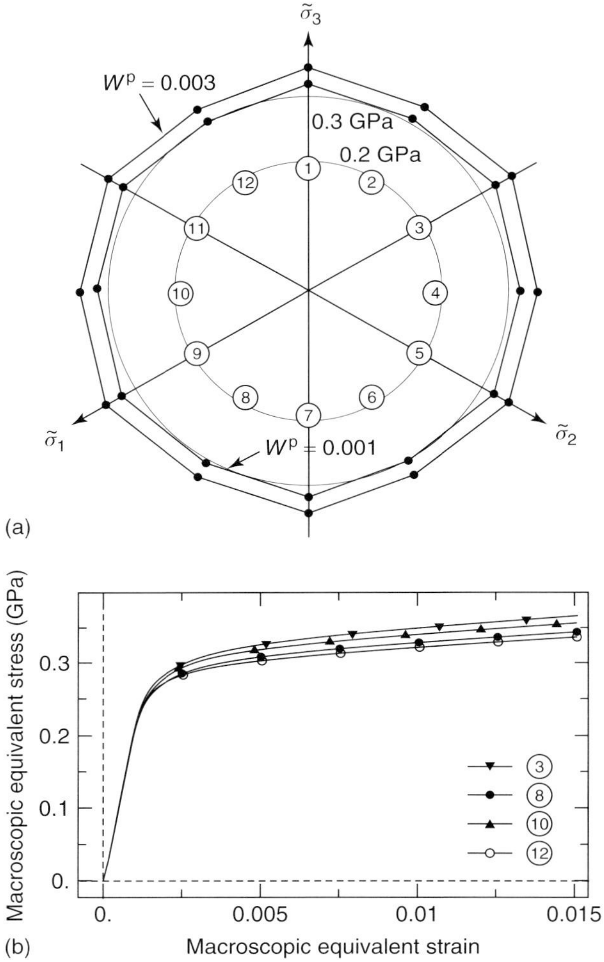

Figure 10.2 Yield strengths in initial state: (a) initial yield surface and (b) macroscopic stress–strain curves.

the NMTs, we obtained 12 sets of macroscopic stress–strain curves. Figure 10.2(b) shows only four of them, each of which exhibits relatively large deviation from the others. As can be seen, the macroscopic responses slightly deviate from isotropy. This is probably due to the fact that a relatively small number of crystal grains are used in the unit cell model, which is expected to be an RVE. However, the mismatch is not so large that it must not be influential on the evaluation of the macroscopic yield strength.

After choosing $W^p = 0.001$ and 0.003 in Eq. (10.18), we evaluated the macroscopic yield strength from the macroscopic stress–strain curves and obtained the yield surface, as depicted in Figure 10.2(a). The surface looks like a circle in the deviatoric stress plane so that the macroscopic initial yield condition could be identified with that of von-Mises, which is classical and standard for many kinds of metals. As can be seen from the comparison between the surfaces, the macroscopic hardening characteristics are also isotropic. Thus, the FE model of the polycrystalline aggregate, shown in Figure 10.1(a), provides the isotropy in the macroscopic yield strength at the initial state and can be a relevant numerical specimen for the NMTs in this study. The same conclusion also applies to the results for Model B, though it is not presented here.

10.4.3
Approximate Macroscopic Constitutive Model

In view of the evaluation above, the standard phenomenological constitutive model can be chosen as an approximate macroscopic constitutive model for the decoupled macroscale analysis, that is, we employ the classical J_2 flow theory with the combined (isotropic and kinematic) hardening model. More specifically, the yield function employed is of the form

$$\tilde{\phi} := \sqrt{\frac{3}{2}\text{dev}[\tilde{\tau} - c\tilde{\alpha}] : \text{dev}[\tilde{\tau} - c\tilde{\alpha}]} - \tilde{\tau}_Y - \tilde{q} \tag{10.20}$$

where variables with "$\sim$" are the macroscopic ones, $\tilde{\alpha}$ is the tensor-valued internal variable representing plastic deformation for the Prager's linear kinematic hardening model, and c is its coefficient. For isotropic hardening, the following nonlinear function [60] is employed:

$$\tilde{q} = H\tilde{\xi} + (\tilde{\tau}_\infty - \tilde{\tau}_Y)\left(1 - \exp\left[-\delta\tilde{\xi}\right]\right) \tag{10.21}$$

where $\tilde{\xi}$ is the scalar-valued internal variable and H is its coefficient. In addition, $\tilde{\tau}_Y$ is the initial yield stress, $\tilde{\tau}_\infty$ is the asymptotically saturating stress, and δ is the sensitivity parameter for the exponential function of the isotropic hardening curve.

For the evaluation of the stress by the elastic deformation, we employ the isotropic St. Venant–Kirchhoff elasticity model $\tilde{\mathbf{S}} = \tilde{\mathbb{C}}^e : \tilde{\mathbf{E}}$, where $\tilde{\mathbb{C}}^e$ is the fourth-order elasticity tensor of the form $\tilde{\mathbb{C}}^e = \left(\kappa - \frac{2}{3}\mu\right)\mathbf{1} \otimes \mathbf{1} + 2\mu\mathbb{I}$, where $\mathbb{I}$ is the fourth-order symmetric identify tensor. Here, μ and κ are the elastic shear and bulk moduli, respectively, and are related to Young's modulus and Poisson's ratio by the relations $\mu = \frac{E}{2(1+\nu)}$ and $\kappa = \frac{E}{3(1-2\nu)}$.

10.4.4
Parameter Identification for Approximate Macroscopic Constitutive Model

To identify the material parameters of the above-defined approximate macroscopic constitutive equation, we conducted three NMTs on the unit cell models, as shown in Figure 10.1, by applying macroscopically uniaxial, cyclic loadings in

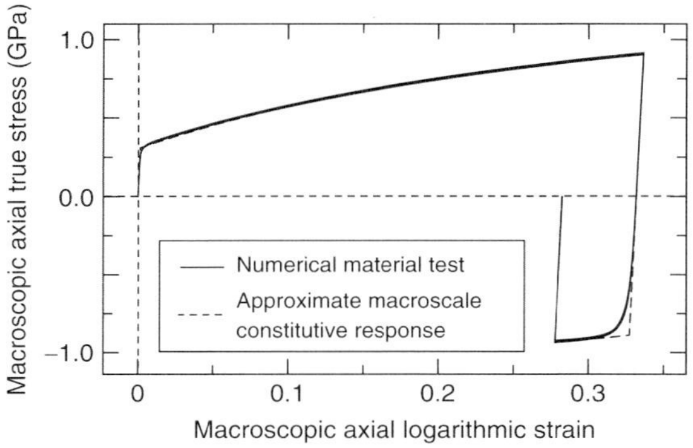

Figure 10.3 Macroscopic stress–strain curves for Model A.

Table 10.3 Material constants for the approximate macroscopic constitutive model.

Young modulus	E(GPa)	197
Poisson ratio	ν	0.3
Yield stress	$\tilde{\tau}_Y$(GPa)	0.31
Isotropic hardening modulus	H(GPa)	0.4
Saturation yield stress	$\tilde{\tau}_{sat}$(GPa)	0.76
Sensitivity	δ	4
Kinematic hardening modulu	c(GPa)	0.02

three directions (X_1, X_2, X_3) that are specifically set to conform with the microscale coordinate axes (Y_1, Y_2, Y_3).

The NMTs are conducted on Model A to obtain the macroscopic stress–strain curves, as shown by the solid lines in Figure 10.3. As can be seen, the three lines are almost identical, again implying that the macroscopic material behavior is isotropic. Although it may be possible to perform the homogenization analysis for linear elasticity to determine the elastic constants E and ν, we use the median of the Hashin–Shtrikman's lower and upper bounds [61] in this study. In addition, the material parameters in Eqs (10.20) and (10.21) are determined, as indicated in Table 10.3, by trial and error instead of using a certain optimization scheme. The macroscopic stress–strain curve obtained with the identified parameters is shown in Figure 10.3 along with the curves by the NMTs. As can be seen, the macroscopic constitutive response is reasonable in the sense that the two curves conform to some extent. Note, however, that this conformity does not necessarily imply the reliability of the evaluation of the internal variables for plastic deformation, and, in turn, the macroscopic yield strength, as was mentioned in Section 10.2.2. The NMTs on the numerical specimen of Model B also provide isotropy and the stress–strain curves only in one direction, as shown in Figure 10.4. Young's modulus, Poisson's ratio, and the initial yield stress identified are set at 193 GPa, 0.3, and 275 MPa, respectively.

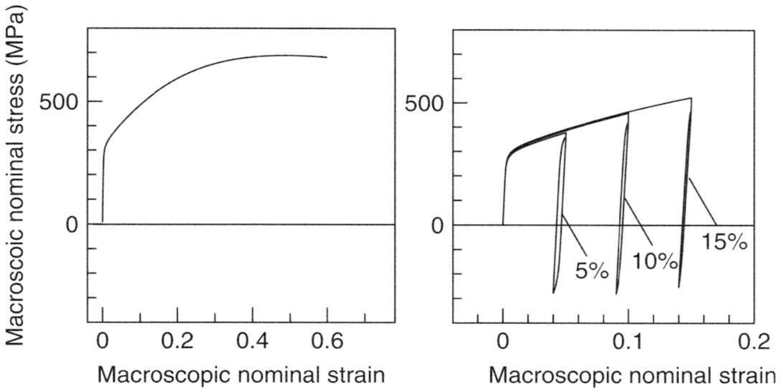

Figure 10.4 Macroscopic stress–strain curves for Model B.

10.5 Macroscopic Yield Strength after Three-Step Plastic Forming

In this section, using Model A as a numerical specimen, we validate the proposed method of evaluation of macroscopic yield strength of polycrystalline metals subjected to three-step plastic forming. After setting up the three-step forming process, we solve the corresponding two-scale boundary value problem with both the micro–macro coupling and decoupling schemes to have two separate numerical specimens subjected to the macroscopic plastic forming under the same loading condition. Then, NMTs are conducted on them to compare the predicted macroscopic yield strengths.

10.5.1 Forming Condition

To bring computational costs down with the micro–macro coupling scheme, we consider a single finite element, shown in Figure 10.5(a), as a macroscopic structure subjected to three-step plastic forming. Each point indicated by × in the figure is an integration point, at which the macroscopic stress is evaluated. In the following discussions, the macro- and microscopic responses associated with Points A and B are used to compare the results with the coupling and decoupling schemes.

The macroscopic deformation pattern of the assumed three-step plastic forming process is schematized in Figure 10.5(b) and consists of the following three steps:

Step 1: The top surface of the macroscopic structure is moved up in the X_3-direction and at the same time shrunk in the X_2-direction, while the bottom surface is expanded in the X_2-direction. The amount of expansion and shrinkage is given in the figure.

Step 2: While the displacements at all the eight nodes in the X_2-direction are fixed at zero, the top surface is moved down in the X_3-direction.

Step 3: The reaction forces at all the nodes are released so that the macroscopic stress becomes almost zero at every integration point.

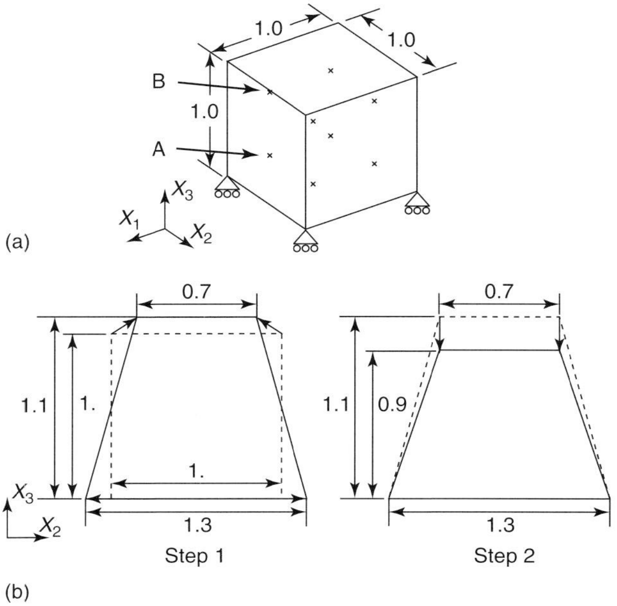

Figure 10.5 Macroscopic plastic forming problem. Three-step forming: (a) macroscopic finite element and (b) forming condition.

Note that during the course of this three-step plastic forming process, the stress-free condition is given on the opposing surfaces perpendicular to the X_1-axis. In addition, the dotted and solid lines in Figure 10.5(b) indicate the configurations at the beginning and the end of each step, respectively.

10.5.2 Two-Scale Analyses with Micro–Macro Coupling and Decoupling Schemes

First, the two-scale BVP corresponding to the three-step forming process defined above is solved with the micro–macro coupling scheme. The numerical specimen here is Model A, as prepared in Section 10.3. Figure 10.6(a) shows the distribution of the microscopic von-Mises equivalent stress with the deformed configurations of the unit cells associated with Points A and B in Figure 10.5 at the last stage of the three-step forming. Even though the macroscopic stress is almost zero at the end of Step III, the absolute value of the microscale equivalent stress is moderately high inside the crystal grains. This high stress value at microscale must be influential on the macroscopic yield strength, but is difficult to be incorporated into the macroscopic constitutive model in general.

Next, the micro–macro decoupling scheme is applied to solve the same two-scale BVP, that is, using the approximate macroscopic constitutive model with the material parameters in Table 10.3 (given in Section 10.4.4), we carry out the decoupled

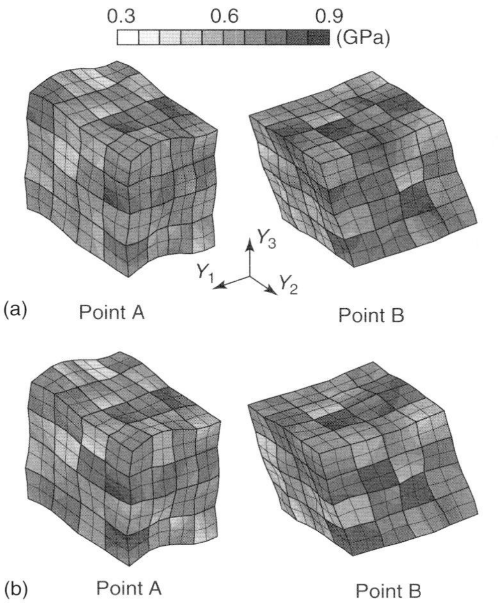

Figure 10.6 Microscopic von-Mises equivalent stress in unit cells after three-step forming: (a) coupled analysis and (b) decoupled analysis.

macroscale analysis, extract the data of the macroscopic deformation histories at Points A and B in Figure 10.5, and then apply these to the unit cell model in Figure 10.1(a) as the loading conditions to perform the decoupled microscale analyses (Step (v)). The last step provides the numerical specimens expected to be equivalent to those obtained by the coupled analysis and to possess sufficient information about macroscopic plastic forming. The stress distributions with deformed configurations of the unit cells at the last stage of the three-step forming are shown in Figure 10.6(b). It can be seen from the figure that the results of the decoupled analysis are in good agreement with those of the coupled analysis shown in Figure 10.6(a). On close comparison, however, there is a slight difference in the absolute value of microscopic stress. This is probably due to the doubtful accuracy of the decoupled macroscale analysis, especially when the forming process is controlled by the macroscopic stress rather than the macroscopic deformation. Note that the displacement constraints are removed in the unloading process of Step III; this means that the macroscopic stress-free condition is imposed on the unit cell. Nonetheless, it is expected that the unit cells thus obtained with the decoupled analysis possess enough information about the macroscopic deformation histories during the plastic forming and that the effect of this difference on the evaluation of the macroscopic yield strength is minor. This section is dedicated to examining this estimate.

The macroscopic loading paths at Points A and B obtained by the coupled and decoupled analyses are shown in Figure 10.7, which has three separate lines.

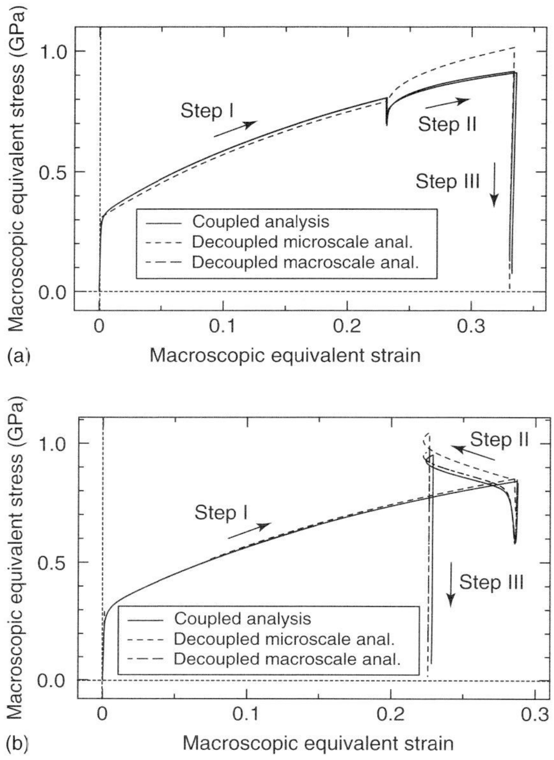

Figure 10.7 Macroscopic stress–strain curves during three-step forming: (a) Point A and (b) Point B.

In the figure, the path with the decoupled macroscale analysis is the response of the approximate macroscopic constitutive model, while that of the decoupled microscale analysis is the response with the volume average of the microscopic stress and strain evaluated in Step (v). As can be seen from the figure, on comparison with the macroscopic response predicted in the decoupled macroscale analysis, the path predicted by the decoupled microscale analysis shows fairly good agreement with that of the coupled analysis, which can be regarded as reference in this study. The deviation of the path of the decoupled macroscale analysis from the other two results is noticeable after the loading condition is changed in Step II.

10.5.3
Evaluation of Macroscopic Yield Strength after Three-Step Plastic Forming

In this section, the macroscopic yield strength after the three-step plastic forming is evaluated by conducting NMTs on the numerical specimens, namely, the unit cell models, on which the corresponding macroscopic deformation histories are imposed. The NMT patterns are the same as in Section 10.4.2. That is, 12 patterns of

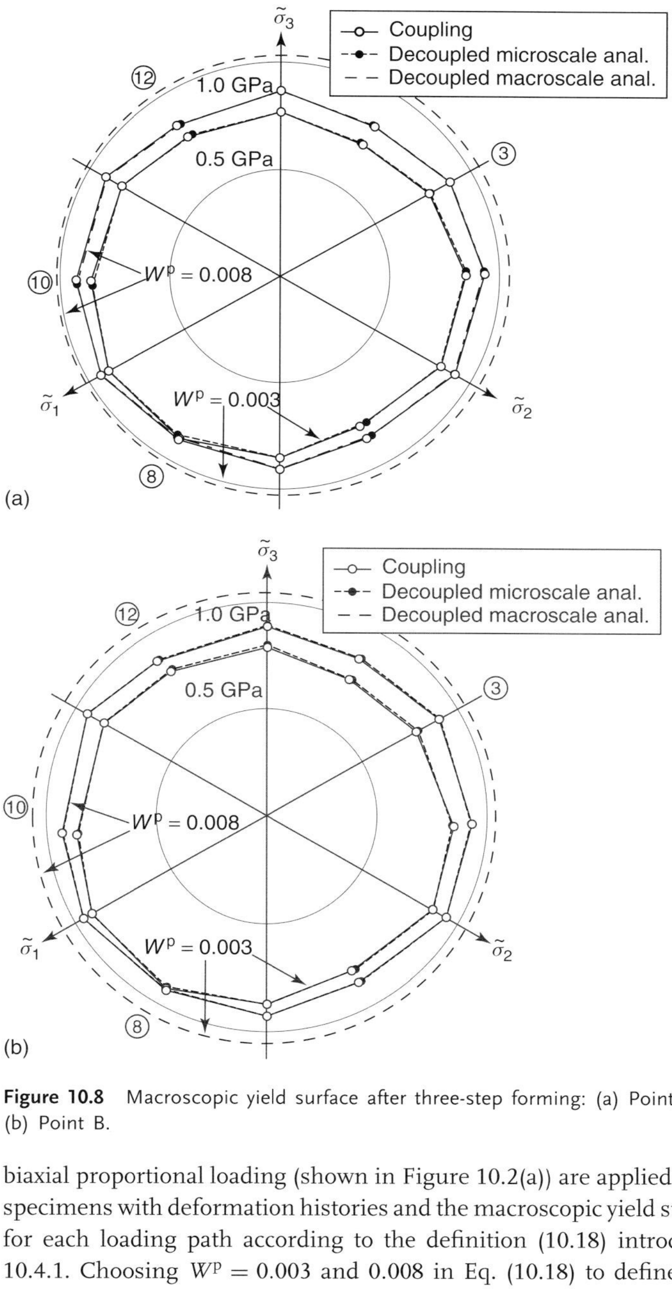

Figure 10.8 Macroscopic yield surface after three-step forming: (a) Point A and (b) Point B.

biaxial proportional loading (shown in Figure 10.2(a)) are applied to the numerical specimens with deformation histories and the macroscopic yield stress is calculated for each loading path according to the definition (10.18) introduced in Section 10.4.1. Choosing $W^p = 0.003$ and 0.008 in Eq. (10.18) to define yield strengths, and setting ε_0 at the total strain after the last step (the elastic unloading step in Step III) of the three-step forming process, we obtained the surfaces formed by the macroscopic yield strengths, as shown in Figure 10.8. As can be seen in the figure, the surfaces evaluated from the decoupled analysis are almost identical with those of the coupled analysis. This means that the numerical specimens prepared in the previous section by the decoupled microscale analysis possess almost the same information about the macroscopic deformation histories as those of the coupled analysis. In addition, as anticipated in the previous section, it is safe to conclude that the effect of the difference between the microscopic stress values in Figure 10.6(a) and (b) is fairly minor.

It is worthwhile to compare the yield surfaces obtained from the two-scale analysis with the one evaluated by the approximate macroscopic constitutive law, which is depicted in the same figure. For the latter case, no difference between the results with $W^{\mathrm{p}} = 0.003$ and 0.008 is observed, so only one surface is shown in the figure. Since we have assumed the classical J_2 flow theory with the combined (isotropic and kinematic) hardening model as the approximate constitutive law, it is natural that the obtained macroscopic yield surface is circular in the deviatoric stress plane. However, the diameter of the surface is entirely different from those obtained by the two-scale analyses, even though the difference in the three macroscopic loading paths in Figure 10.7 is not so significant, and the values approach almost zero at the last stage of plastic forming. In this regard, we have to point out that it is difficult to predict the anisotropy of the macroscopic yield strength of polycrystalline metals subjected to multistep plastic forming using the classical constitutive model. In other words, since the deformation histories are not properly reflected in the yield functions in classical plasticity, it cannot represent the dependence of the macroscopic yield strength on the directions and degrees of plastic forming processes.

Although the proposed method has been validated, it is interesting to characterize the macroscopic yield strength and further investigate the microscale mechanism associated with the anisotropy. Figure 10.9 shows four typical stress–strain curves obtained in the above NMTs on the unit cell associated with Point A. The numbers

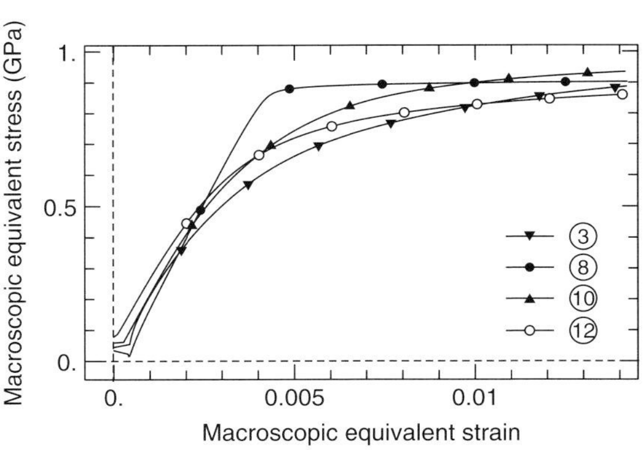

Figure 10.9 Macroscopic stress–strain curves obtained by the proposed method for evaluation of macroscopic post-forming strength. Results of numerical material tests on unit cell subjected to three-step forming at Point A.

3, 8, 10, and 12 in this figure correspond to the directions defined in Figure 10.2; see also Figure 10.8. As can be seen from the figure, the response in the loading direction 8 differs greatly from the other three. This is probably due to the direction of the macroscopic stress state just before Step III, namely, at the end of Step II, which lies along 8. That is, since the macroscopic stress has been released along the direction 8 during the macroscopic unloading process in Step III, no microscale plastic deformation tends to occur during the reloading in the same direction for strength evaluation. Therefore, the macroscopic yield strength is relatively high in comparison with the others. On the other hand, the reloading in the other three directions, which stimulates the unit cell under macroscopically stress-free condition, must be unstable even with slight excitation due to the microscopic residual stress stored in the crystal grains. Thus, lower strength is in effect expected during reloading in directions 3, 10, and 12.

It is also worth mentioning that the shapes of the yield surfaces for Points A and B depicted in Figure 10.8 are almost the same, even though the macroscopic deformation histories and loading paths are different, as illustrated in Figure 10.7. However, we have confirmed that the macroscopic stress state at Point B just before macroscopic unloading (Step III) lies along the same direction as that of Point A, that is, 8. Therefore, it seems reasonably safe to conclude that more recent deformation and stress states are more influential on the directional dependence of the macroscopic yield strength after plastic forming. This tendency can be understood as the effect of the fading memory of the material.

10.6
Application for Pilger Rolling of Steel Pipe

In this section, another numerical example is presented to demonstrate the applicability of the proposed method to practical problems. Assuming Model B is the numerical specimen of FCC steel, we take the pilger rolling of a steel pipe as the macroscopic problem and evaluate the macroscopic yield strength that reflects its macroscopic deformation history.

10.6.1
Forming Condition

A pilger mill rolling is one of the rolling processes used in actual cold-working of steel pipes, and its operation is schematized in Figure 10.10. The pilger mill rolling process employed in this particular example consists of the following steps:

Step 1: The roll is rotated about the transverse axis so that the pipe is translated and at the same time rolled.
Step 2: The mandrel is rotated so that the pipe is rotated by 63° about its central axis.
Step 3: The pipe is carried forward by certain distance (in this study, 10 mm).

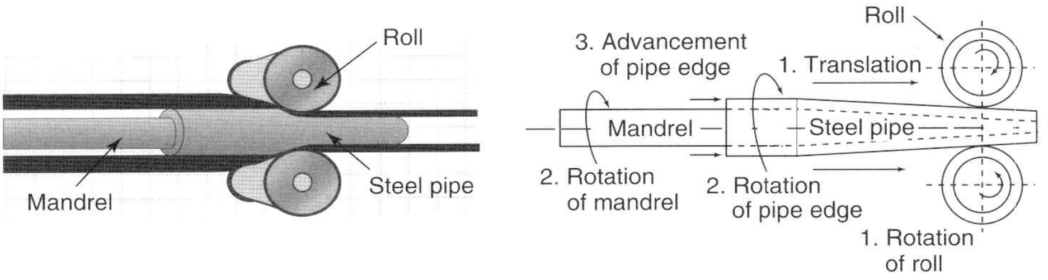

Figure 10.10 Pilger mill and pilger rolling process.

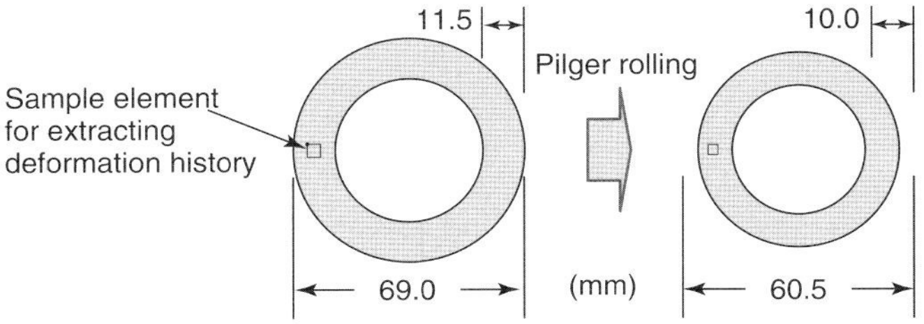

Figure 10.11 Forming condition by pilger rolling process.

By this rolling process, the eight 360° rotations are applied to the pipe so that its thickness and diameter are reduced, as is schematized in Figure 10.11. More specifically, the rolling process is conditioned so that the initial 69.0 mm outer diameter of the pipe becomes 60.5 mm, whereas the radial thickness of 11.5 mm is reduced to 10.0 mm.

The advantage of the proposed method utilizing the micro–macro decoupling scheme to solve the two-scale BVP is that any commercial software can be used to solve macroscopic forming problems. In this particular example, we utilize ELFEN [62], which is one of the general-purpose FEM codes, to carry out the corresponding decoupled macroscale analysis. The FE model with rolling tools are shown in Figure 10.12. The material parameters set in Section 10.4.4 are used for the assumed approximate macroscopic constitutive model and the friction coefficient is set at 0.05.

It took 15 s in pseudotime to complete the FE metal forming analysis. The trajectory of the node of the element shown in Figure 10.11 is depicted in Figure 10.13, which characterizes the pilger rolling process explained above. The corresponding histories of the position vector at this node and the macroscopic strain tensor at the integration point closest to this node are presented in Figures 10.14 and 10.15, respectively. It can be imagined from these results that the forming process must have imposed a very complex macroscopic deformation history on the microstructure of the steel pipe.

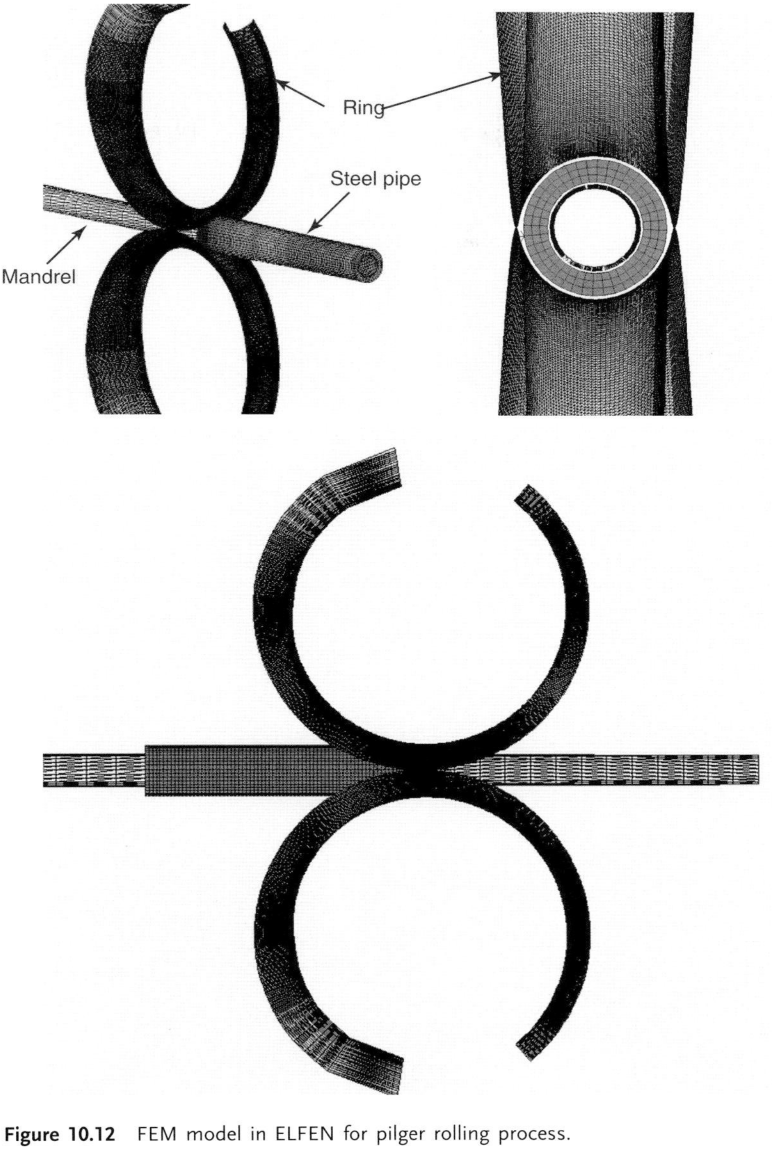

Figure 10.12 FEM model in ELFEN for pilger rolling process.

10.6.2 Decoupled Microscale Analysis

After the macroscopic analysis of the Pilger mill rolling of the steel pipe, we proceed to conduct a decoupled microscale analysis with ABAQUS by imposing the macroscopic deformation history obtained above for the pilger rolling process on the unit cell together with the periodic boundary condition.

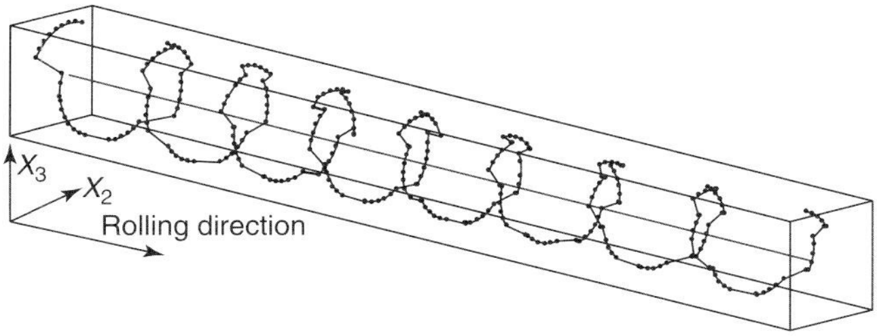

Figure 10.13 Trajectory of macroscopic material point in pilger rolling.

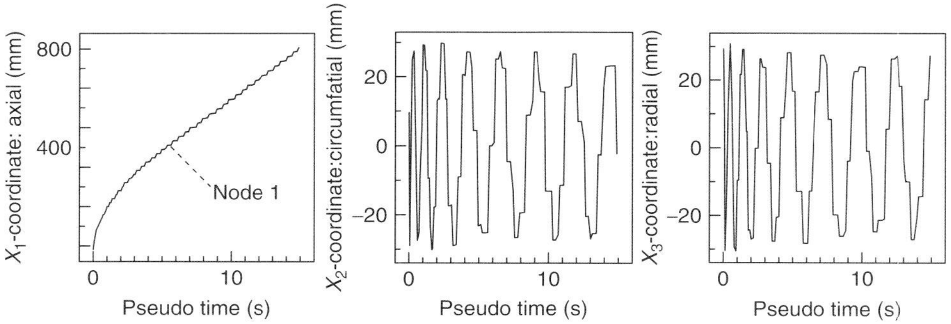

Figure 10.14 History of position of macroscopic material point.

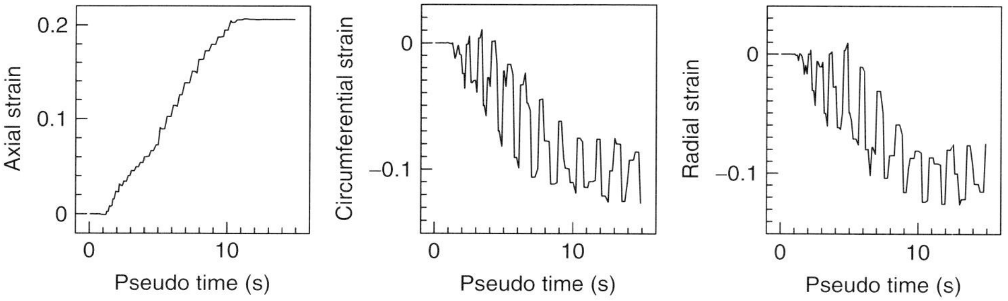

Figure 10.15 History of macroscopic strain.

Figure 10.16 shows the distributions of the microscopic stress and the accumulated plastic strain during the pilger rolling process. It should be noted that the unit cell model is gradually rolled so that its height is reduced by (11.5−10.0)/11.5 and, at the same time, is subjected to rigid-body rotation due to the macroscopic rotation of the pipe made in Step 2 of the pilger rolling process explained above. This means that the Schmid factor for each slip system is continuously changed during the rolling, since the macroscopic loading and unloading directions are changed by the rotation.

Figure 10.17 shows the macroscopic true stress–strain curves calculated as the volume-averaged responses of the microscopic stress and strain components

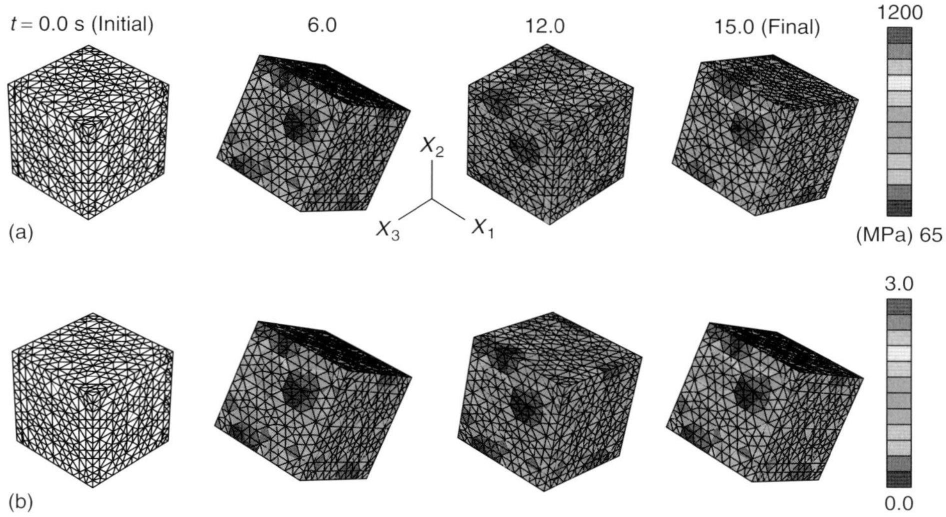

Figure 10.16 Microscopic stress and plastic strain during pilger rolling: (a) microscopic von-Mises equivalent stress and (b) microscopic accumulated plastic strain.

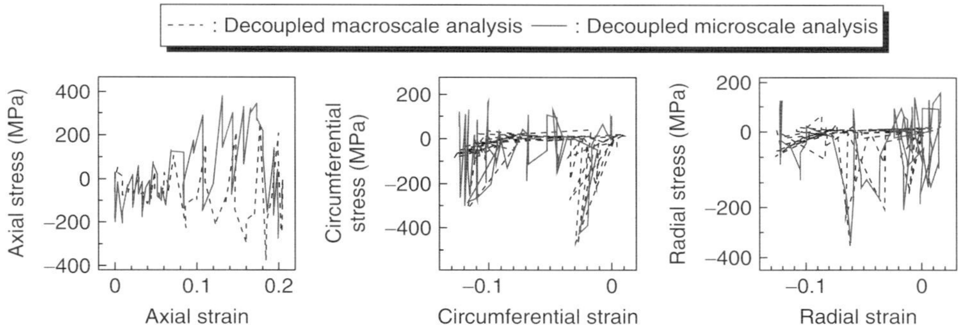

Figure 10.17 Microscopic stress–strain curves during pilger rolling.

(solid lines) together with those obtained by the decoupled macroscale analysis with the approximate macroscopic constitutive model (dotted lines). In addition, the equivalent information can be the stress path in the deviatoric stress plane as shown in Figure 10.18. As can be seen from these figures, the macroscopic stress path obtained by the decoupled *microscale* analysis is totally different from that of the decoupled *macroscale* analysis with the approximate macroscopic constitutive model. Although experimental verification is difficult, the former should be considered more reliable, according to the results of the verification study discussed in Section 10.4.

In addition, Figure 10.19 shows the {111} polefigure obtained from the unit cell after the final step of the forming. Here, the Y_1 coincides with X_1 or the

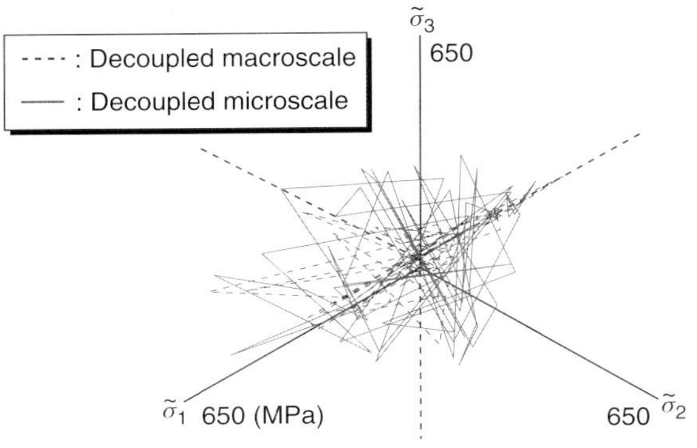

Figure 10.18 Microscopic stress paths during pilger rolling.

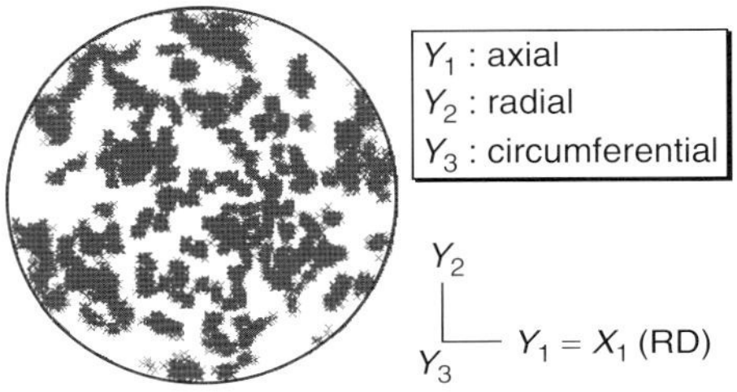

Figure 10.19 The {111} polefigure.

axial direction that is the rolling direction (RD), and Y_2 and Y_3-directions are along the radial and circumferential directions of the pipe, respectively. As a consequence of the reasonably low rolling ratio in this particular example, the texture development is not prominent. Thus, the effect of the rolling texture on the macroscopic postforming strength is likely to be minor.

10.6.3
Evaluation of Macroscopic Yield Strength after Pilger Rolling Process

To investigate the anisotropy in the macroscopic yield strength of the steel after the pilger mill cold-rolling process, we conduct the NMT on the numerical specimen, namely, the unit cell model that has been obtained in the decoupled microscale analysis above. Prior to the investigation, it is to be noted that the unit cell located inside the steel pipe reveals the macroscopic residual stress. This fact can be clearly observed from the stress–strain curve shown in Figure 10.20, in which Point P corresponds to the final step of the rolling. Assuming that a specimen is cut from the pipe, as is the case on actual experiments, we simulate the equivalent cutting-out process by applying macroscopically stress-free (unloading) condition to the numerical specimen. In the corresponding microscale analyses, the constraints in the axial, circumferential, and radial directions are removed in order. The figure

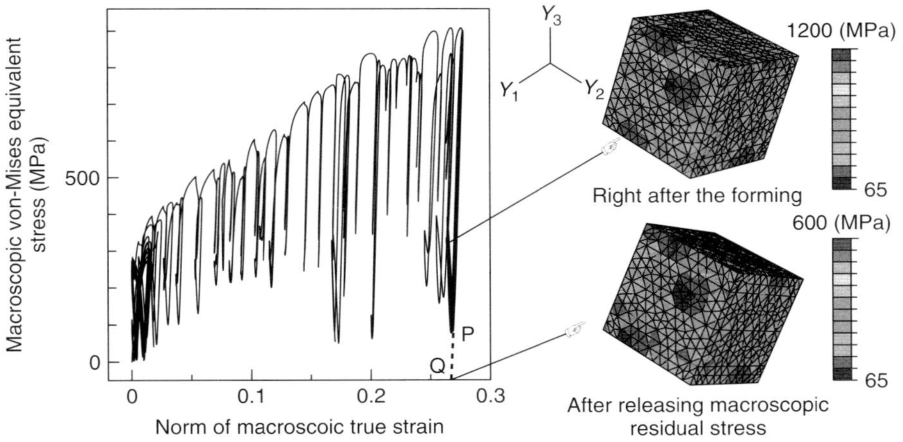

Figure 10.20 Macroscopic stress–strain curve during pilger rolling and after cutout process.

also provides the stress distributions in the numerical specimens at States P and Q. It is noted that even though the macroscopic residual stress is zero at State Q, its microscopic residual stress is still very high. The unloaded state (State Q) is taken into consideration in the investigation below, as is the one just after the forming (State P).

Uniaxial uniform tensile loadings are applied to the two numerical specimens, which correspond to States P and Q, in the Y_1, Y_2, and Y_3-directions separately. Figure 10.21(a) and (b) shows the corresponding macroscopic stress–strain curves, from which the anisotropy in the strengths can be observed. For quantitative investigation, we evaluate the 0.2% offset bearing stress in each direction, which can be an approximation to the macroscopic yield strength. The values of 0.2% offset bearing stress are put down with the curves in Figure 10.21 (a) and (b). As can be seen from the figures, the yield strength for each of the States P and Q in the Y_1-direction, or equivalently in the RD, is much higher than that in the circumferential direction. This tendency is in agreement with actual experimental knowledge. In addition, the yield strength in the radial direction, which can rarely be evaluated in experiments, takes a median value between the others.

It is, however, noted that the numerical specimen in State P is in nature embedded in the steel pipe and the macroscopic residual stress can be neither recognized nor measured. Therefore, the macroscopic bearing stress should be the value subtracted from that evaluated above by the macroscopic residual stress in each direction. From Figure 10.21(a), this consideration leads to the opposite tendency for the bearing strengths in the axial and circumferential directions, and thus the result illustrates an important aspect of the strength evaluation of polycrystalline metals after cold-working. In other words, it is suggested that a

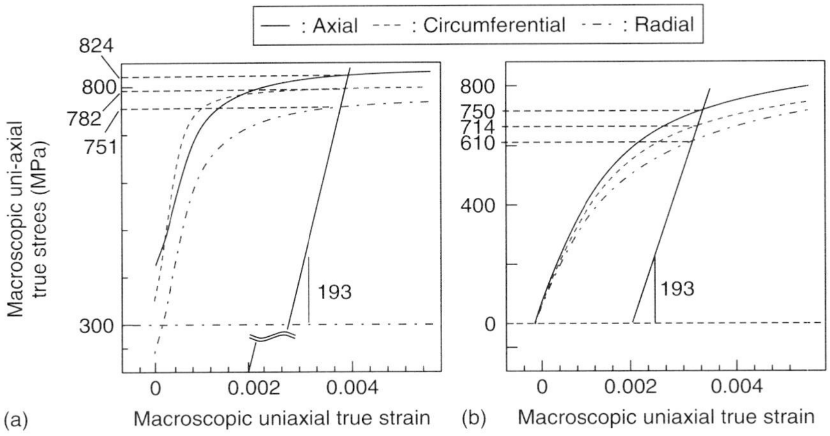

Figure 10.21 Results of NMTs on numerical specimens after pilger rolling: (a) for State P with macroscopic residual stress and (b) for State Q without macroscopic residual stress.

metal product after cold-working must have different value of bearing stress than those of actual experiments on specimens without macroscopic residual stress.

10.7
Conclusion

In this chapter, a homogenization-based prediction method of the macroscopic yield strength of polycrystalline metals after cold-working has been proposed. The method crucially relies on the solution of the two-scale BVP, which can be derived by the mathematical theory of homogenization. To solve the two-scale BVP, we employed the micro–macro decoupling scheme [52], by which micro- and macroscale BVPs are separately solved in two-scale FE analyses.

In the validation study, assuming a three-step plastic forming, we solved the corresponding two-scale BVP with the micro–macro coupling and decoupling schemes and compared the microscale responses and the macroscopic yield strengths predicted by the proposed method. It was demonstrated that the decoupling scheme enables us to obtain the approximate solution of the two-scale BVP, which conforms with that of the coupling scheme. As a result, the microscale responses obtained by the decoupled microscale analysis were almost identical with those obtained by the coupled analysis. This implies that the reliability of the numerical specimen prepared in the localization process is sufficient for evaluating the macroscopic yield strength after the three-step plastic forming. In fact, the yield surfaces predicted by conducting a series of NMTs on the prepared numerical specimens are almost identical for the coupled and decoupled analyses. In addition, to demonstrate the capability and promise for practical applications, the proposed method was successfully applied to the pilger rolling process, a process that imposes very complex deformation histories on the microstructures. The results suggested that the bearing stress in an actual metal product subjected to cold-working should be different from that measured in experiments that use cutout specimens.

The reliability of the proposed method hinges on the accuracy of the prediction of the macroscopic deformation history made by the decoupled macroscopic analysis, while the stress response evaluated by the approximate macroscopic constitutive model is unnecessary. Of course, a more reliable constitutive model is preferred, especially for forming processes controlled mainly by the macroscopic stress rather than macroscopic deformation. However, when a reliable constitutive model is found for polycrystalline metals, both the coupling and decoupling schemes for the two-scale analysis themselves are useless for the evaluation of the postforming strength of polycrystalline metals. In this sense, the two-scale analysis method with the decoupling scheme, accompanied by NMTs on numerical specimens of crystalline aggregates, can be considered an ideal "experimental device" to develop relevant constitutive models for strength evaluation in plastic-forming problems.

References

1. Bauschinger, J. (1881) Ueber die veranderung der elasticitatagrenze und dea elasticitatamoduls verschiadener metalle. *Zivilingur*, **27**, 289–347.
2. Sowerby, R., Uko, D.K., and Tomita, Y. (1979) A review of certain aspect of the Bauschinger effect in metals. *Materials Science and Engineering*, **41**, 43–58.
3. Terada, K., Matsui, K., Akiyama, M., and Kuboki, T. (2004) Numerical re-examination of the micro-scale mechanism of the Bauschinger effect in carbon steels. *Computational Materials Science*, **31**, 67–83.
4. Bunge, H.-J. (1982) *Texture Analysis in Materials Science*, Butterworths, London.
5. Armstrong, P.J. and Frederick, C.O. (1966) A mathematical representation of the multiaxial Bauschinger effect. CEGB Report RD/B/N731, Berkeley Nuclear Laboratories, Berkeley.
6. Mroz, Z. (1967) On the description of anisotropic workhardening. *Journal of the Mechanics and Physics of Solids*, **15**, 163–175.
7. Iwan, W.D. (1967) On a class of models for the yielding behavior of continuous and composite systems. *Journal of Applied Mechanics*, **34**, 612–617. ASME.
8. Chaboche, J.L. (1989) Constitutive equations for cyclic plasticity and cyclic viscoplasticity. *International Journal of Plasticity*, **5**, 247–302.
9. Uemori, T., Okada, T., and Yoshida, F. (1998) Simulation of springback in V-bending process by elasto-plastic finite element method with consideration of Bauschinger effect. *Metals and Materials*, **4**, 311–314.
10. Yoshida, F. and Uemori, T. (2002) A model of large-strain cyclic plasticity describing the Bauschinger effect and workhardening stagnation. *International Journal of Plasticity*, **18**, 661–686.
11. Hill, R. (1948) A theory of the yielding and plastic flow of anisotropic metals. *Proceedings of the Royal Society. London*, **A193**, 281–297.
12. Gotoh, M. and Ishise, F. (1978) A finite element analysis of rigid-plastic deformation of the flange in a deep-drawing. *International Journal of Mechanical Science*, **20**, 423–435.
13. Logan, R.W. and Hosford, W.F. (1980) Upper–bound anisotropic yield locus calculations assuming $< 111 >$–pencil glide. *International Journal of Mechanical Science*, **22**, 419–430.

14. Barlat, F. and Lian, J. (1989) Plastic behaviour and stretchability of sheet metals. I. A yield function for orthotropic sheets under plane stress conditions. *International Journal of Plasticity*, **5**, 51–66.
15. Hill, R. (1990) Constitutive modelling of orthotropic plasticity in sheet metals. *Journal of the Mechanics and Physics of Solids*, **38**, 405–417.
16. Hill, R. (1993) A user-friendly theory of orthotropic plasticity in sheet metals. *International Journal of Mechanical Science*, **35**, 19–25.
17. Baltov, A. and Sawczuk, A. (1965) A rule of anisotropic hardening. *Acta Mechnica*, **1**, 81–92.
18. Wu, H.-C., Hong, H.-K. and Lu, J.-K. (1995) An endochronic theory accounted for deformation induced anisotropy. *International Journal of Plasticity*, **11**, 145–162.
19. Ishikawa, H. (1997) Subsequent yield surface probed from its current center. *International Journal of Plasticity*, **13**, 81–92.
20. Chiang, D.-Y., Su, K.-H., and Liao, C.-H. (2002) A study on subsequent yield surface based on the distributed-element model. *International Journal of Plasticity*, **18**, 51–70.
21. Taylor, G.I. (1938) Plastic strain in metals. *Journal of the Institute of Metals*, **62**, 307–324.
22. Bishop, J.F.W. and Hill, R. (1951) A theory of the plastic distortion of a polycrystalline aggregate under combined stresses. *Philosophical Magazine*, **42**, 414–427.
23. Bishop, J.F.W. and Hill, R. (1951) A theoretical derivation of the plastic properties of a polycrystalline face-centred metal. *Philosophical Magazine*, **42**, 1298–1307.
24. Kroner, E. (1961) Zur plastischen verformung des vielkristalls. *Acta Metallurgica*, **9**, 155–161.
25. Budiansky, B. and Wu, T.T. (1962) Theoretical prediction of plastic strains of polycrystal. *Proceedings of the 4th Congress on Applied Mechanics*, **2**, 1175–1185.
26. Hill, R. (1965) Continuum micro-mechanics of elasto-plastic polycrystals. *Journal of the Mechanics and Physics of Solids*, **13**, 89–101.
27. Eshelby, J.D. (1957) The determination of the elastic field of an ellipsoidal inclusion and related problems. *Proceedings of the Royal Society. London*, **A214**, 376–396.
28. Iwakuma, T. and Nemat-Nasser, S. (1983) Finite elastic-plastic deformation of polycrystalline metals. *Proceedings of the Royal Society. London*, **A394**, 87–119.
29. Mathur, K.K. and Dawson, P.R. (1989) On modeling the development of crystallographic texture in bulk forming processes. *International Journal of Plasticity*, **5**, 67–94.
30. Kalidindi, S.R., Bronkhorst, C.A., and Anand, L. (1992) Crystallographic texture evolution in bulk deformation processing of FCC metals. *Journal of the Mechanics and Physics of Solids*, **40**, 537–569.
31. Kalidindi, S.R. and Schoenfeld, S.E. (2000) On the prediction of yield surfaces by the crystal plasticity models for fcc polycrystals. *Materials Science and Engineering*, **A293**, 120–129.
32. Kuroda, M. and Tvergaard, V. (2001) Shear band development predicted by a non-normality theory of plasticity and comparison to crystal plasticity predictions. *International Journal of Solids and Structures*, **38**, 8945–8960.
33. Peirce, D., Asaro, R.J. and Needleman, A. (1982) An analysis of nonuniform and localized deformation in ductile single crystals. *Acta Metallurgica*, **30**, 1087–1119.
34. Asaro, R.J. (1983) Crystal plasticity. *Journal of Applied Mechanics*, **50**, 921–934. ASME.
35. Raabe, D., Klose, P., Engl, B. *et al.* (2002) Concepts for integrating plastic anisotropy into metal forming simulations. *Advanced Engineering Materials*, **4**, 169–180.
36. Takahashi, H., Motohashi, H., Tokuda, M., and Abe, T. (1994) Elastic–plastic finite element polycrystal model. *International Journal of Plasticity*, **10**, 63–80.

37. Becker, R. and Richmond, O. (1994) Incorporation of microstructural geometry in material modeling. *Modelling and Simulation in Materials Science and Engineering*, **2**, 439–454.
38. Quilici, S. and Cailletaud, G. (1999) FE simulation of macro-, meso- and micro-scales in polycrystalline plasticity. *Computational Materials Science*, **16**, 383–390.
39. Diard, O., Leclercq, S., Rousselier, G., and Cailletaud, G. (2005) Evaluation of finite element based analysis of 3D multicrystalline aggregates plasticity Application to crystal plasticity model identification and the study of stress and strain fields near grain boundaries. *International Journal of Plasticity*, **21**, 691–722.
40. Benssousan, A., Lions, J.-L., and Papanicoulau, G. (1978) *Asymptotic Analysis for Periodic Structures*, North-Holland.
41. Sanchez-Palencia, E. (1980) *Non-homogeneous Media and Vibration Theory*, Springer-Verlag, Berlin.
42. Suquet, P.M. (1987) Elements of homogenization theory for inelastic solid mechanics, in *Homogenization Techniques for Composite Media* (eds E. Sanchez-Palencia and A. Zaoui), Springer-Verlag, Berlin, pp. 193–278.
43. Allaire, G. (1992) Homogenization and two-scale convergence. *Journal of Mathematical Analysis*, **23**, 1482–1518. SIAM.
44. Terada, K. and Kikuchi, N. (1995) Nonlinear homogenization method for practical applications, in *Computational Methods in Micromechanics*, vol. AMD-212/MD-62 (eds S. Ghosh and M. Ostoja-Starzewski), AMSE, New York, pp. 1–16.
45. Terada, K. and Kikuchi, N. (2001) A class of general algorithms for multi-scale analyses of heterogeneous media. *Computer Methods in Applied Mechanics and Engineering*, **190**, 5427–5464.
46. Terada, K., Saiki, I., Matsui, K., and Yamakawa, Y. (2003) Two-scale kinematics and linearization for simultaneous two-scale analysis of periodic heterogeneous solids at finite strain. *Computer Methods in Applied Mechanics and Engineering*, **192**, 3531–3563.
47. Smit, R.J.M., Brekelmans, W.A.M., and Meijer, H.E.H. (1998) Prediction of the mechanical behavior of nonlinear heterogeneous systems by multi-level finite element modeling. *Computer Methods in Applied Mechanics and Engineering*, **155**, 181–192.
48. Wieckowski, Z. (2000) Dual finite element methods in homogenization for elastic-plastic fibrous composite material. *International Journal of Plasticity*, **16**, 199–221.
49. Zheng, S.F., Ding, K., Denda, M., and Weng, G.J. (2000) A dual homogenization and finite-element study on the in-plane local and global behavior of a nonlinear coated fiber composite. *Computer Methods in Applied Mechanics and Engineering*, **183**, 141–155.
50. Feyel, F. and Chaboche, J.-L. (2000) FE^2 multiscale approach for modelling the elastoviscoplastic behaviour of long fibre SiC/Ti composite materials. *Computer Methods in Applied Mechanics and Engineering*, **183**, 309–330.
51. Nakamachi, E., Tam, N.N., and Morimoto, H. (2007) Multi scale finite element analyses of sheet metals by using SEM-EBSD measured crystallographic RVE models. *International Journal of Plasticity*, **23**, 450–489.
52. Watanabe, I. and Terada, K. (2005) Decoupled micro–macro analysis method for two-scale BVPs in nonlinear homogenization theory. *Journal of Applied Mechanics*, **8**, 277–285. JSCE.
53. Watanabe, I. and Terada, K. (2010) A method of predicting macroscopic yield strength of polycrystalline metals subjected to plastic forming by micro–macro decoupling scheme. *International Journal of Mechanical Sciences*, **52**(2), 343–355.
54. Matsui, K., Terada, K., and Yuge, K. (2004) Two-scale finite element analysis of heterogeneous solids with periodic microstructures. *Computers and Structures*, **82**, 593–606.
55. Guedes, J.M. and Kikuchi, N. (1990) Preprocessing and postprocessing for materials based on the homogenization method with adaptive finite element

methods. *Computer Methods in Applied Mechanics and Engineering*, **83**, 143–198.

56. Terada, K. and Watanabe, I. (2007) Computational aspects of tangent moduli tensors in rate-independent crystal elastoplasticity. *Computational Mechanics*, **40**, 497–511.
57. Watanabe, I., Terada, K., and Akiyama, M. (2005) Two-scale analysis for deformation-induced anisotropy of polycrystalline metals. *Computational Materials Science*, **32**, 240–250.
58. Kocks, U.F. (1970) The relation between polycrystal deformation and single crystal deformation. *Metallurgical Transactions*, **1**, 1121–1143.
59. ABAQUS (2007) ABAQUS User's Manual, SIMULIA, Dassault Systémes, Providence, RI, USA.
60. Voce, E. (1955) A practical strain hardening function. *Metallurgia*, **51**, 219–226.
61. Hashin, Z. and Shtrikman, S. (1962) A variational approach to the theory of the elastic behaviour of polycrystals. *Journal of the Mechanics and Physics of Solids*, **10**, 343–352.
62. ELFEN, Rockfield Software Limited, Technium, King's Road, Prince of Wales Dock, Swansea, SA1 8PH, Wales, U.K.

Index

Advanced Computational Materials Modeling: From Classical to Multi-Scale Techniques.
Edited by Miguel Vaz Júnior, Eduardo A. de Souza Neto, and Pablo A. Munoz-Rojas

ISBN: 978-3-527-32479-8

u

v